ALGEBRAIC THEORY OF AUTOMATA NETWORKS

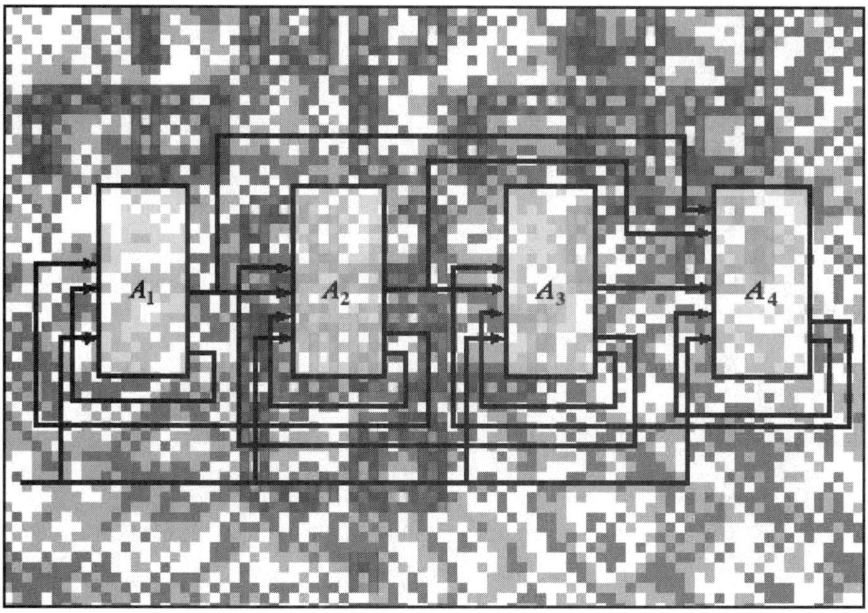

SIAM Monographs on Discrete Mathematics and Applications

The series includes advanced monographs reporting on the most recent theoretical, computational, or applied developments in the field; introductory volumes aimed at mathematicians and other mathematically motivated readers interested in understanding certain areas of pure or applied combinatorics; and graduate textbooks. The volumes are devoted to various areas of discrete mathematics and its applications.

Mathematicians, computer scientists, operations researchers, computationally oriented natural and social scientists, engineers, medical researchers, and other practitioners will find the volumes of interest.

Series Volumes

Dömösi, P., and Nehaniv, C. L., *Algebraic Theory of Automata Networks: An Introduction*
Murota, K., *Discrete Convex Analysis*
Toth, P. and Vigo, D., *The Vehicle Routing Problem*
Anthony, M., *Discrete Mathematics of Neural Networks: Selected Topics*
Creignou, N., Khanna, S., and Sudan, M., *Complexity Classifications of Boolean Constraint Satisfaction Problems*
Hubert, L., Arabie, P., and Meulman, J., *Combinatorial Data Analysis: Optimization by Dynamic Programming*
Peleg, D., *Distributed Computing: A Locality-Sensitive Approach*
Wegener, I., *Branching Programs and Binary Decision Diagrams: Theory and Applications*
Brandstädt, A., Le, V. B., and Spinrad, J. P., *Graph Classes: A Survey*
McKee, T. A. and McMorris, F. R., *Topics in Intersection Graph Theory*
Grilli di Cortona, P., Manzi, C., Pennisi, A., Ricca, F., and Simeone, B., *Evaluation and Optimization of Electoral Systems*

ALGEBRAIC THEORY OF AUTOMATA NETWORKS

AN INTRODUCTION

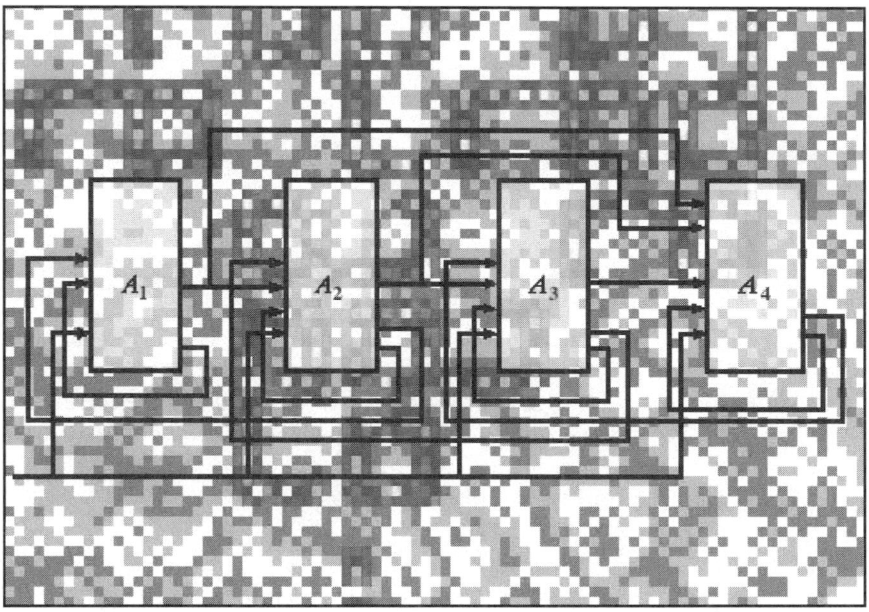

PÁL DÖMÖSI
University of Debrecen
Debrecen, Hungary

CHRYSTOPHER L. NEHANIV
University of Hertfordshire
Hatfield, United Kingdom

siam

Society for Industrial and Applied Mathematics
Philadelphia

10 9 8 7 6 5 4 3 2 1

Library of Congress Cataloging-in-Publication Data

Dömösi, Pál.
Algebraic theory of automata networks : an introduction / Pál Dömösi, Chrystopher L. Nehaniv.
p. cm. -- (SIAM monographs on discrete mathematics and applications)
Includes bibliographical references and index.
ISBN 0-89871-569-5
1. Computer networks. 2. Machine theory. 3. Semigroups. 4. Algebra, Abstract. I. Nehaniv, Chrystopher L., 1963– II. Title. III. Series.

QA276.D653 2005
004.6--dc22 2004057838

Cover shows an α_2-product of automata against the background of an asynchronous cellular automata network.

To our teachers,
Ferenc Gécseg and John L. Rhodes

Contents

Preface and Overview

"Networks are everywhere. The brain is a network of nerve cells connected by axons, and cells themselves are networks of molecules connected by biochemical reactions. Societies, too, are networks of people linked by friendships, familial relationships and professional ties. On a larger scale, food webs and ecosystems can be represented as networks of species. And networks pervade technology: the Internet, power grids and transportation systems are but a few examples. Even the language we are using to convey these thoughts to you is a network, made up of words connected by syntactic relationships.

"Yet despite the importance and pervasiveness of networks, scientists have had little understanding of their structure and properties. How do the interactions of several malfunctioning nodes in a complex genetic network result in cancer? How does diffusion occur so rapidly in certain social and communications systems, leading to epidemics of diseases and computer viruses? How do some networks continue to function even after the vast majority of their nodes have failed?"

—*Albert-László Barabási and Eric Bonabeau,* Scientific American*, May 2003*

An automata network is a collection of automata connected together according to a directed graph $\mathcal{D}$. The vertices of $\mathcal{D}$ are considered as automata and the edges indicate the existence of communication links. Thus $\mathcal{D}$ has no parallel edges. Each automaton can change its state at discrete time steps as a local transition function of the states and a global input, and synchronous action of the local state transitions defines a global transition on the entire network. We investigate automata networks as algebraic structures and develop their theory in line with other algebraic theories, such as those of semigroups, groups, rings, and fields.

In this monograph we restrict ourselves almost exclusively to finite automata networks (with notable exceptions in the study of asynchronous networks) for two reasons. This introductory monograph is devoted to the most fundamental cases. These occur when the network is finite: there are only finitely many component automata in the network (i.e., the interconnection digraph $\mathcal{D}$ is finite) and all component automata are also finite, having only finitely many states. On the other hand, finiteness is a natural constraint arising for real-world networks, including those in computational and technical applications.

Algebraic interpretations arise from consideration of the semigroup of transformations induced on the set of states by all possible finite sequences of inputs, but they also enter the subject in other ways when we study division relations of automata and the completeness of networks.

We also investigate automata networks as "products of automata," i.e., as compositions of automata obtained by cascading without feedback or with feedback of various restricted types, or, most generally, with the feedback dependencies controlled by an arbitrary directed graph. We survey and extend the fundamental results in regard to automata networks, including the main decomposition theorems of Letichevsky, of Krohn and Rhodes, and others. These theorems also indicate what basic components are necessary and sufficient for the synthesis of particular finite computational structures using particular types of networking, including limitations on feedback and number of local connections in the (directed graph of) communication links.

We deal with classes of finite automata that are complete in one or several of four different senses. In particular, we consider completeness with respect to homomorphic representation, isomorphic representation, homomorphic simulation, and isomorphic simulation. In all four types of completeness, it is understood that not necessarily is the class itself complete but rather that its closure under a given type of product is complete.

In other words, we investigate complete classes of automata, i.e., classes of automata whose closure under various notions of products allow the simulation of ordinary automata in various senses. The questions arise naturally when one tries to understand how to decompose or synthesize complicated automata as combinations of simpler ones. An issue of central importance is to understand the behavior and computational power of automata networks having given restricted types of communication links.

This monograph is an effort in this direction. We characterize the structure of the communication links of those whose finite automata networks (as several product types of automata) which have as simple a structure as possible, and representational power preserves the representational power of the most general networks or of the (general) cascade networks. Real-world examples of automata networks include computer networks, electrical networks, transport networks, the Internet, and genetic regulatory networks. Many of these can modify their internal structure in the course of their functioning. Our book covers the foundations of what is currently known about automata networks, leading the reader to the forefront of research in many areas. It also lays a foundation upon which a rigorous theory of dynamic automata networks (that change their topology, member components, and functioning) may one day be constructed. We also pay some attention to the case in which network can cyclically modify its inner structure during its work as well as to asynchronous automata networks.

Section 1.1 introduces standard notation that we use throughout. Basic algebraic concepts of semigroup, monoid, and group, as well as some useful results concerning these, are collected in Section 1.2. Section 1.3 introduces their actions on sets (transformation semigroups and permutation groups), which are intimately related to automata, and the fundamental concepts of division (which allows us to compare computational power), direct and wreath products, and their connections to the decomposition of transformation semigroups and permutation groups. These results are used repeatedly in the book, e.g., in the study of feed-forward (or cascade) products of finite automata.

Directed graphs (digraphs) and their power to support arbitrary (penultimate) permutational computations are studied and characterized in terms of simple structural graph-theoretic properties in Section 2.1. Section 2.2 introduces automata, associated concepts, and automata mappings. Relationships between automata and semigroup-theoretic simulation and division are studied and described in Section 2.3. Section 2.4 introduces various types of automata products used in the construction of automata networks over directed graphs and having restrictions on the number of connections and types of feedback that may exist between component automata. The Gluškov and Letichevsky criteria are introduced, and important decomposition theorems characterizing isomorphic and homomorphic completeness, respectively, for classes of automata under the general, unrestricted product are presented.

Chapter 3 introduces Krohn–Rhodes theory, which is concerned with the feedback-free (cascade) decomposition and synthesis of finite automata by homomorphic simulation using cascades of simple permutation groups and identity-reset automata. The fundamental theorem is proved via a new proof of the deep holonomy decomposition theorem, which yields relatively efficient cascade decompositions (Section 3.1). The remainder of the chapter treats related results and characterizes homomorphically complete classes under the quasi-direct products, cascade products, and α_1-products of automata.

Chapter 4 studies classes and networks of automata that fail to satisfy the Letichevsky criterion. The study of these classes is naturally divided into those classes that satisfy the weaker semi-Letichevsky criterion and those that fail even this criteria. For both kinds of class of automata without Letichevsky's criterion, constructive proofs show how very restricted types of networking can already realize as much computational power as unrestricted types of networking. Examples and counterexamples reveal the sharpness of many of the results.

Realizing computation with automata and classes of automata that satisfy Letichevsky's criterion is studied more deeply in Chapter 5. Results include profound strengthenings of the Letichevsky decomposition theorem, such as the remarkable Ésik–Horváth characterization theorem showing that α_2-products already yield all finite automata that can be homomorphically represented by networks of finite automata from a given class $\mathcal{K}$ of finite automata. We give a proof of the Letichevsky decomposition theorem at the end of Chapter 5.

Chapter 6 further strengthens the Letichevsky decomposition by a (sharp) theorem showing that primitive products—whose interconnection digraphs satisfy a strict graph-theoretic outerplanarity condition severely limiting the local connectivity and guaranteeing nice embeddability properties—already suffice to achieve the full computational power of arbitrary networks constructed from automata classes satisfying Letichevsky's criterion. Temporal products of automata are also studied in this chapter (Section 6.4). They are simple models of automata networks that can (cyclically) change the structure of their communication links in the course of computation. We show, contrary to what might be expected from their simple structure, that they have a very strong representational power.

Chapter 7 develops the algebraic theory of state-homogeneous automata networks, i.e., networks whose constituent automata all have the same state set at each node on a given directed graph where edges correspond to permissible intercommunication links. Such automata networks are natural generalizations of cellular automata, and their consideration is useful to the design of computer networks. In particular, we consider finite state-homogeneous automata networks algebraically and characterize by a simple graph-theoretic condition those size n network topologies that are complete with respect to simulation via projection, that is, those capable of simulation (via projection) of every size n network. Network completeness (for simulation by projection) for such networks is characterized in terms of graph-theoretic properties of their interconnectivity graphs (Section 7.2), and those with a minimal numbers of edges (communication links) are characterized in Sections 7.3 and 7.4. Section 7.5 shows how arbitrary automata networks (including possibly infinite ones over locally finite digraphs) can be emulated by asynchronously updated ones (over essentially the same underlying undirected graph) derived by a simple construction that keeps only an extra copy of the most recent previous local state and a new local cyclical counter at each node. As a consequence, many results on automata networks (including, e.g., cellular automata) have nontrivial and automatic generalizations to the asynchronous realm.

Other connections, results, and open problems related to the covered topics are included. We give a self-contained treatment of all results, except for citations of a very small number of well-known theorems. Bibliographic remarks can be found at the end of each chapter, and further pointers to the literature and an index are given at the end of the book.

In this volume we overview some (theoretical) basic properties of automata networks (including products of automata) and we do not pay direct attention to the applications. We plan to cover applications and more advanced results in a later volume. The monograph gives an abstract theoretical background to computational network synthesis and design. It is devoted to computer scientists, electrical engineers, communication engineers, system scientists, and anyone for whom the concepts and capabilities of networked processes is important. It is also useful to researchers and postgraduate students working in the structure theory of automata, universal algebra, or semigroup theory, since automata networks are strongly related to these areas.

Acknowledgments

The work of the first author was supported by grants from Dirección General de Universidades, Secretaría de Estado de Educación y Universidades, Ministerio de Educación, Cultura y Deporte (SAB2001-0081), España, Xerox Foundation UAC grant (1478-2004), U.S.A., and the Hungarian National Foundation for Scientific Research (OTKA T030140). The work of the second author was supported by the Algorithms Research Group at the University of Hertfordshire. The work was also supported by grants from the University of Aizu ("Algebra & Computation" and "Automata Networks" projects (R-10-1, R-10-4)), the "Automata & Formal Languages" project of the Hungarian Academy of Sciences, the Japanese Society for Promotion of Science (No. 15), the Hungarian National Foundation for Scientific Research (OTKA T030140), and the "Formal Systems" joint Hungarian–German project supported by the Hungarian Ministry of Education and the German National Science Foundation (D39/2000).

We are grateful to Ferenc Gécseg, Balázs Imreh, Masami Ito, Manfred Kudlek, Carlos Martin-Vide, Satoshi Okawa, and John L. Rhodes for their help and support, as well as to Attila Egri-Nagy, László Kovács, and Johanna Hunt for help with the preparation of the manuscript.

We thank Peter Hammer, Alexa Epstein, Louis Primus, and the staff at SIAM for their work on bringing this book to press, as well as the referees for their valuable comments, which helped improve the manuscript.

Pál Dömösi
Debrecen, Hungary

Chrystopher L. Nehaniv
Hatfield, Hertfordshire, UK
March 2004

Chapter 1

Preliminaries

In this chapter we overview some basic concepts and results that are important later in the monograph. Our approach in this book is algebraic, so we overview all the "pure" algebraic concepts and results that are necessary to understand our explanations. To introduce the reader to the structure of proofs in the further chapters, we also provide elementary proofs.

1.1 Basic Notation and Notions

First we need to fix some standard terminology. We start with a discussion of some set-theoretic notation. The set S consisting of all the elements that have the *property* P is written as $S = \{s \mid s \text{ has property } P\}$.[1] If s is an element of S, we write $s \in S$. The opposite case is expressed by $s \notin S$. And if $s \in S$ implies that $s \in T$, then S is a *subset* of T and we write $S \subseteq T$. $S \setminus T = \{s \mid s \in S \text{ and } s \notin T\}$ is the *set-theoretic difference* of S and T. Two sets S and T are *equal,* i.e., $S = T$, if $S \subseteq T$ and $T \subseteq S$. Moreover, S is a *proper subset* of T, denoted $S \subsetneq T$, if $S \subseteq T$ and $S \neq T$. The set containing no elements, the *empty* or *void* set, is denoted by $\emptyset$. The *intersection* of S and T is the set consisting of all the elements in both S and T and we write $S \cap T = \{s \mid s \in S \text{ and } s \in T\}$. The *union* of S and T is the set consisting of the elements in either S or T. In symbols, $S \cup T = \{s \mid s \in S \text{ or } s \in T\}$. The set operations extend naturally to families of sets $\{S_i \mid i \in I\}$, where I is referred to as an *index set*:[2]

$$\bigcup_{i \in I} S_i = \{s \mid s \in S_i \text{ for some } i \in I\},$$

$$\bigcap_{i \in I} S_i = \{s \mid s \in S_i \text{ for all } i \in I\}.$$

[1]This way of specifying sets suffices for this monograph and will not lead us into any foundational difficulties. To avoid misunderstanding, sometimes we also use the form $S = \{s : s \text{ has property } P\}$ instead of $S = \{s \mid s \text{ has property } P\}$.

[2]An index set may be empty, but for intersection, it is required that the index set I be nonempty. In this monograph we consider only nonempty index sets.

Two sets are *disjoint* if $S \cap T = \emptyset$, and the family of sets $\{S_i \mid i \in I\}$ is *disjoint* if the sets are *pairwise disjoint*: $S_i \cap S_j \neq \emptyset$ implies $i = j$ for all $i, j \in I$. The cardinality of a set S is denoted by $|S|$. S is called *finite* if it has finitely many elements. Thus $|S|$ denotes the number of elements for a finite set S. In particular, if $|S| = 1$, then S is called a *singleton*.

Let S and T be sets. A (well-defined) *function* f of S into T, written $f : S \to T$, assigns to every element $s \in S$ one and only one element $t \in T$, written $f(s) = t$. Then t is the *image* of s, and s is an *inverse image* or *preimage* of t under f. S is called the *source* and T the *target* of f. We put $f^{-1}(t) = \{s \mid f(s) = t, s \in S\}$ for every $t \in T$. We will also use the notation $f(S') = \{f(s) \mid s \in S'\}$ and $f^{-1}(T') = \bigcup_{t \in T'} f^{-1}(t)$ for any $S' \subseteq S$, $T' \subseteq T$. The function f is sometimes called a *map* or *mapping* from S to T. The set $f(S) = \{f(s) \mid s \in S\}$ is called the *image* of $f : S \to T$. The *rank* of f is the cardinality of its image. If $f(S) = T$, then f is an *onto* or *surjective* function. If f is surjective, we may also write $f : S \twoheadrightarrow T$. The function f is *one-to-one* or *injective* if for every $s_1, s_2 \in S$, $s_1 \neq s_2$ implies that $f(s_1) \neq f(s_2)$. If f is injective, we sometimes write $f : S \hookrightarrow T$. If f is surjective and injective, then it is called *bijective*. A bijective map from any set S to S is a *permutation* and is said to *permute* the elements of S. If $|f(S)| = 1$, then f is a *constant function*, or *constant* for short.

Let $f : A \to B$, $g : C \to D$ be functions with $C \subseteq A$ and $g(c) = f(c)$ for each $c \in C$. Then we say that f is an *extension* of g (to A) and that g is a restriction of f (to C), and sometimes we write $g = f|_C$. The *(right) composite* or *(right) product* fg of functions $f : S \to T$, $g : T \to U$ is the function $h : S \to U$ with $h(s) = g(f(s))$ for all $s \in S$.

The *cartesian product* of the finite sequence of sets $S_1, \ldots, S_n$ is the set $S_1 \times \cdots \times S_n = \{(s_1, \ldots, s_n) \mid s_1 \in S_1, \ldots, s_n \in S_n\}$. It is also defined for a not necessarily finite family $\{S_i \mid i \in I\}$ of sets as the set of all functions $\varphi : I \to \bigcup_{i \in I} S_i$ such that, for every $i \in I$, $\varphi(i)$ is in S_i. For this concept we use the notation $\prod_{i \in I} S_i$. (For finite index set I, it is more convenient to think of the elements of a cartesian product as a set of n-tuples as defined above.)

A *relation* between a set S and a set T is a subset ρ of $S \times T$. For $(s, t) \in \rho$ we write $s \; \rho \; t$. Thus $\rho = \{(s, t) \mid s \; \rho \; t\}$.

A relation ρ between S and S itself is called simply a relation on S. It is called *reflexive* if for all $s \in S$, $s \; \rho \; s$; *symmetric* if for every $s, t \in S$, $s \; \rho \; t$ implies $t \; \rho \; s$; *antisymmetric* if for every $s, t \in S$, $s \; \rho \; t$ and $t \; \rho \; s$ implies $s = t$; and *transitive* if $s \; \rho \; t$ and $t \; \rho \; u$ implies $s \; \rho \; u$ for every $s, t, u \in S$. A relation ρ on S is an *equivalence relation* on S if ρ is reflexive, symmetric, and transitive. If ρ is an equivalence relation on S, then for every $s \in S$, the set $s/\rho = \{t \mid s \; \rho \; t\}$ is the *equivalence class* of s under ρ. This notation is extended to an arbitrary subset S' of S by $S'/\rho = \{s'/\rho \mid s' \in S'\}$. A *partition* π on S is a collection of disjoint subsets of S whose set union is S. Then, in symbols, $\pi = \{S_i \mid i \in I\}$ such that $S_i \cap S_j = \emptyset$ for $i \neq j, i, j \in I$, and $\bigcup_{i \in I} S_i = S$. Sometimes we refer to the elements of π as *blocks*. For every $s \in S$, $\pi(s)$ will denote the block containing the element s. It is clear that if ρ is an equivalence relation on S, then S/ρ is a partition of S, and that every partition of S can be given in this way.

A reflexive, antisymmetric, and transitive relation ρ on a set S is a *partial ordering* on S. A *preorder* is a reflexive and transitive relation on S. For a preorder, identifying s with t in S whenever both $s \; \rho \; t$ and $t \; \rho \; s$ hold results in a partial ordering. A partial ordering ρ on a set S is called a *linear ordering* (or *total ordering*) or, in short, an *ordering* if $s \; \rho \; t$ or $t \; \rho \; s$ for every pair $s, t \in S$. If an arbitrary set S is supplied with a partial ordering, then we speak of a *partially ordered set*. Similarly, if an arbitrary set S is supplied with an ordering, then S is called an *ordered set*. Given a partially ordered set S with partial order $\leq$, $s \in S$ is called a *minimal element* of S (with respect to the partial ordering $\leq$) if $s' \leq s''$ implies $s'' \neq s$ for every distinct $s', s'' \in S$. It is easy to see that every finite nonempty partially ordered set S has a minimal element; moreover, this minimal element is unambiguously determined if S is an ordered set.

Let $f : X_1 \times \cdots \times X_n \to X$ be a mapping having n variables for some positive integer n; moreover, let $t \in \{1, \ldots, n\}$. It is said that f *really depends* on its tth variable if there exist distinct $x_t, x_t' \in X_t$ such that for some $x_1 \in X_1, \ldots, x_{t-1} \in X_{t-1}, x_{t+1} \in X_{t+1}, \ldots, x_n \in X_n$, $f(x_1, \ldots, x_{t-1}, x_t, x_{t+1}, \ldots, x_n) \neq f(x_1, \ldots, x_{t-1}, x_t', x_{t+1}, \ldots, x_n)$. If f does not have this property, then we also say that f is *really independent* of its tth variable. Moreover, if there is no danger of confusion, then sometimes we omit the attribute "really."

For a given nonempty set, X and positive integer n denote by X^n the nth cartesian power of X. Given a k-element subset H of $\{1, \ldots, n\}$, $H = \{i_1, \ldots, i_k\}$ $(i_1 < \cdots < i_k)$, the *H-projection* of X^n is the mapping $pr_H : X^n \to X^k$ defined by $pr_H(x_1, \ldots, x_n) = (x_{i_1}, \ldots, x_{i_k})$, where $(x_1, \ldots, x_n) \in X^n$. If it is well defined, the function $pr_H(F) : X^k \to X^k$ with $pr_H(F)(pr_H(x_1, \ldots, x_n)) = pr_H(F(x_1, \ldots, x_n))$, for $(x_1, \ldots, x_n) \in X^n$, is called the *$H$-projection* of $F : X^n \to X^n$. If $H = \{h\}$ for some $h \in \{1, \ldots, n\}$, i.e., H is a singleton, then sometimes we use the expression *h-projection* (of a vector or function) in the same sense as the concept "H-projection" (and in this case we sometimes use the notation pr_h instead of $pr_{\{h\}}$). Moreover, for an arbitrary $i \in \{1, \ldots, n\}$, we define the ith *component* of $F : X^n \to X^n$ as the function $cp_i(F) : X^n \to X$ with $cp_i(F)(x_1, \ldots, x_n) = pr_i(F(x_1, \ldots, x_n))$ for $(x_1, \ldots, x_n) \in X^n$.

For any pair $F_i : X^n \to X^n$, $i = 1, 2$, one denotes by $F_1 \circ F_2 : X^n \to X^n$ the function $F_1 \circ F_2(x_1, \ldots, x_n) = F_2(F_1(x_1, \ldots, x_n))$ for $(x_1, \ldots, x_n) \in X^n$. (This is just the (right) product of functions defined above.)

A *word* (over X) is a finite sequence of elements of some nonempty set X. We call the set X an *alphabet,* the elements of X *letters.* If $u = x_1 \cdots x_k$ $(x_1, \ldots, x_k \in X)$ and $v = x_{k+1} \cdots x_n$ $(x_{k+1}, \ldots, x_n \in X)$ are words over an alphabet X, then their *catenation* $uv = x_1 \cdots x_k x_{k+1} \cdots x_n$ is also a word over X. In particular, for every word u over X, $u\lambda = \lambda u = u$, where λ denotes the empty word having no letters. The set of all words over X is denoted X^*. If $p = qr$ for some $q, r \in X^*$, then q is said to be a *prefix* and r a *suffix* of p. If u, v, w are words, then v is a *subword* of uvw. For every $x_1, \ldots, x_k \in X$, the *reverse* of $p = x_1 \ldots x_k$ is defined by $\bar{p} = x_k \ldots x_1$. Thus the reverse of any letter is the letter itself. Similarly, the reverse of the empty word is the empty word itself. The reverse of a word is also called its *mirror image.* In addition, for any word p, we set $p^0 = \lambda$ and $p^n = p^{n-1}p$ $(n > 0)$. Thus p^k $(k \geq 0)$ is the kth *power* of p. The *length* $|w|$ of a word w is the number of letters in w, where each letter is counted as many times as it occurs. Thus the length of the empty word λ is zero by definition.

If there is no danger of confusion, we shall sometimes denote an n-tuple $(a_1, \ldots, a_n)$ with each $a_i \in X$ by the word $a_1 \ldots a_n$. Therefore, for any alphabet X and nonnegative integer n, X^n also denotes the set of n-length words over the alphabet X. $X^0 = \{\lambda\}$, and $X^* = \bigcup_{n \geq 0} X^n$. By the *free monoid X^* generated by X* we mean the set of all words (including the *empty word* λ) having catenation as multiplication. We set $X^+ = X^* \setminus \{\lambda\}$, where the subsemigroup X^+ of X^* is said to be the *free semigroup generated by X*. (See below for definitions of monoid and semigroup.)

Throughout this monograph, for integers k, n ($n \geq 2$), $k \pmod n$ denotes the least positive integer k' such that n divides $k - k'$. (In particular, $0 \pmod n = n$.)

Finally, given an alphabet A, let us consider a word $a_1 \dots a_n \in A^+$ and an integer $k \in \{1, \dots, n\}$. We will denote by $c(a_1 \dots a_n, k)$ the kth *cyclic permutation* of $a_1 \dots a_n$. To be precise, let

$$c(a_1 \dots a_n, k) = \begin{cases} a_{k+1} \dots a_n a_1 \dots a_k & \text{if } k < n, \\ a_1 \dots a_n & \text{if } k = n. \end{cases}$$

1.2 Semigroups, Monoids, and Groups

A nonempty set S with a mapping μ from the cartesian product $S \times S$ into S forms a *semigroup*[3] if the binary operation μ satisfies the *associative law*: $\mu(s_1, \mu(s_2, s_3)) = \mu(\mu(s_1, s_2), s_3)$ for all $s_1, s_2, s_3 \in S$. The mapping μ is called *multiplication* in S, and $\mu(s_1, s_2)$ is the *product* of s_1 and s_2 (in that order). If S is finite and it has n elements, then we also say that S is a *finite semigroup of order* n. S is called *commutative* or *abelian* if $\mu(s_1, s_2) = \mu(s_2, s_1)$ holds for every $s_1, s_2 \in S$. Otherwise we say that S is noncommutative or nonabelian. One writes instead of $s_1 = \mu(s_2, s_3)$ simply $s_1 = s_2 s_3$ ($s_1, s_2, s_3 \in S$). Associativity guarantees that products written without parentheses have well-defined values in S. S' is a *subsemigroup* of S if $S' \subseteq S$, and, moreover, for every $s_1, s_2 \in S'$, $s_1 s_2 \in S'$. For every $s \in S$ and subset $H \subseteq S$ we put $sH = \{sh \mid h \in H\}$, $Hs = \{hs \mid h \in H\}$. Furthermore, for any pair $H_1, H_2 \subseteq S$, we write $H_1 H_2 = \{h_1 h_2 \mid h_1 \in H_1, h_2 \in H_2\}$. In particular, if $H_1 = H_2$, then we put $H^2 = HH$ and, in general, let $H^1 = H$, $H^k = H^{k-1}H$ for every positive integer $k > 1$. Associativity of S guarantees that $H_1(H_2 H_3) = (H_1 H_2) H_3$ for every choice of subsets $H_1, H_2, H_3 \subseteq S$, so $H_1 H_2 H_3$, written without any parentheses, is a well-defined subset of S. Therefore the set of all subsets of S is itself a semigroup under this multiplication. The *subsemigroup of S generated by a subset* $H \subseteq S$ is $\langle H \rangle = \bigcup_{n \geq 1} H^n$. We also say that $H \subseteq S$ *generates* a subset $H' \subseteq S$ if $H' \subseteq \langle H \rangle$. In the case of singleton $H = \{h\}$ or finite $H = \{h_1, \dots, h_n\}$, we write $\langle h \rangle$ for $\langle \{h\} \rangle$ and may write $\langle h_1, \dots, h_n \rangle$ for $\langle \{h_1, \dots, h_n\} \rangle$, respectively. In addition, if $H \subseteq S$, $s \in S$, and $s \in \langle H \rangle$, then sometimes it is said too that H generates s. H is a *generating system* for S if $\langle H \rangle = S$. A generating system H is *minimal* if for every $h \in H$, $H \setminus \{h\}$ is not a generating system. A minimal generating system is also called a *basis*. In addition, a semigroup S is *finitely generated* if it has a generating system with finitely many elements. If H is a finite generating system for S, then for an appropriate $B \subseteq H$, B is a basis of S.

Let S and T be semigroups having a mapping $\psi : S \to T$ such that $\psi(s_1 s_2) = \psi(s_1)\psi(s_2)$ for all $s_1, s_2 \in S$. Then we say that ψ is a *homomorphism* from S to T. If ψ is surjective we also say that S *can be mapped homomorphically onto* T and that T is a *homomorphic image* of S. If in addition ψ is bijective, then we say that S is *isomorphic* to T (or S and T are isomorphic) and ψ is called an *isomorphism*. If the semigroup S_1 is a homomorphic (isomorphic) image of the semigroup S_2 and S_2 is a homomorphic (isomorphic) image of the semigroup S_3, then S_1 is homomorphic (isomorphic) image of S_3. An *automorphism* is an isomorphism whose source and target coincide. A map of semigroups $\varphi : S \to T$ is an *antihomomorphism* if $\psi(s_1 s_2) = \psi(s_2)\psi(s_1)$ for all $s_1, s_2 \in S$.

[3]Empty semigroups may also be allowed. (See, for example, Eilenberg [1974, 1976].) In this monograph we consider only nonempty semigroups.

A bijective antihomomorphism is called an *anti-isomorphism*. It is easily verified that if T' is a subsemigroup of T and $\psi : S \to T$ is a homomorphism of semigroups, then $S' = \psi^{-1}(T')$ is a subsemigroup of S. A semigroup S *divides* a semigroup T if T has a subsemigroup T' such that S is a homomorphic image of T'. It is easy to verify that division is also a transitive relation. That is, for any semigroups T, S, and U, if T divides S and S divides U, then T divides U.

If $n \geq 1$ and $s \in S$, then the nth *power of* s, denoted s^n, is defined inductively, by $s^1 = s$ and $s^{n+1} = s^n s$. An element $e \in S$ is an *idempotent* if $e^2 = e$. If $\psi : S \to T$ is a homomorphism, then $\psi(e) = \psi(e^2) = \psi(e)\psi(e)$ so $\psi(e)$ is also an idempotent. Moreover, if $f^2 = f$ is an idempotent of T, then its preimage $\varphi^{-1}(f)$ is of course a subsemigroup of S.

Proposition 1.1. *In a finite semigroup, every element s has a unique idempotent among its powers.*

Proof. The list $s, s^2, s^3, \ldots$ must be finite. Hence $s^m = s^{m+a}$ for some $m \geq 1, a > 0$. Letting m be the least such, take the least $a > 0$ for which this equation holds for m. Then $s^m = s^{m+a}$ implies, inductively, that $s^m = s^{m+ra}$ for all $r \geq 0$, whence also $s^t = s^{t+ra}$ for all $t \geq m$. The distinct powers of s are thus $s, \ldots, s^m, \ldots, s^{m+a-1}$. Let $k > 0$ be the least integer such that $ka \geq m$. Taking $n = ka$, $(s^n)^2 = s^{n+ka} = s^n$ is an idempotent power of s. To see uniqueness, if $(s^t)^2 = s^t$ for some $t > 0$, then $t \geq m$ (whence $ta \geq m$, and so by leastness of k, $t \geq k$), and so $s^t = (s^t)^a = s^{ta} = s^{ka+(t-k)a} = s^{ka} = s^n$. □

An element $r \in S$ is called a *right-zero element* of S if $sr = r$ for all $s \in S$. Symmetrically, $\ell \in S$ is a *left-zero element* if $\ell s = \ell$ for all $s \in S$. In addition, $o \in S$ is the *zero element* if $os = so = o, s \in S$. If the semigroup has both a left-zero element ℓ and a right-zero element r, then it has an unambiguously determined zero element $o = \ell r = \ell = r$. It follows that the zero element of a semigroup S is uniquely determined if it exists. A semigroup is a *monoid* if it has an element 1 for which $s1 = 1s = s, s \in S$. Just as for a zero element, the element 1 of a semigroup is uniquely determined if it exists and is called the *identity element* of S. (Similarly, left-identities and right-identities are defined analogously to left- and right-zeros.) Clearly, left- and right-zeros and identities are all idempotents. If $\psi : S \to T$ is a homomorphism, and S and T are both monoids with respective identity elements 1_S and 1_T, it need *not* be the case that $\psi(1_S) = 1_T$. The monoid with two right-zero elements or *flip-flop monoid* is $\mathbf{F} = \{e, \ell, r\}$ with $ee = e, e\ell = \ell e = \ell\ell = r\ell = \ell, er = re = rr = \ell r = r$. Then, for example, the constant function from $\mathbf{F}$ to $\mathbf{F}$ taking every element of $\mathbf{F}$ to ℓ is a homomorphism but does not take the identity element e to itself. A homomorphism between monoids that does take the identity to the identity is called a *monoid homomorphism*.

A monoid is a *group* if for every $s \in S$ there is an $s^{-1} \in S$ such that $ss^{-1} = s^{-1}s = 1$. Inverses in a group are unique, since if $t, r \in S$ were both inverses to s, then $t = t1 = t(sr) = (ts)r = 1r = r$. Then for every $s \in S$ there is exactly one s^{-1} satisfying the above equalities, and s^{-1} is called the *inverse* of s. Obviously then, s is the inverse of s^{-1}, so the operation on any group G, given by $g \mapsto g^{-1}$ $(g \in G)$ is a permutation of the elements of G. If X is any subset of G, then X^{-1} denotes the set $\{x^{-1} \mid x \in X\}$. If g_0 is any element of G, it is easy to check that *conjugation by* g_0, defined by $g \mapsto g_0 g g_0^{-1}$, is an isomorphism from G onto itself. A monoid or group is said to be *trivial* if it consists of only its identity element. However, if G is a nontrivial group, then it cannot have any left-zero or right-zero (or zero) element. In fact, the identity element of a group is its unique idempotent (since $1 = e^{-1}e = e^{-1}e^2 = e$ for any $e^2 = e \in G$). Therefore any homomorphism from a monoid to a group is necessarily a monoid homomorphism. If n is the least nonnegative integer such that $g^n = 1$, then we say g is an element of finite *order* n, and we write $o(g) = n$. If no such n exists, then g is of infinite order, $o(g) = \infty$. It is easy to see that $o(g)$ is the cardinality in the subsemigroup generated by g. Obviously, $\langle g^{-1}, g\rangle$ is a group, the smallest one containing g, and, moreover, $\langle g^{-1}, g\rangle = \langle g\rangle$ if and only if $o(g)$ is finite.

A *congruence relation* ρ on a semigroup S is an equivalence relation such that $s_1 \,\rho\, s_2$ and $s_1' \,\rho\, s_2'$ implies $s_1s_1' \,\rho\, s_2s_2'$. If $s_1\rho s_2$ for some $s_1, s_2 \in S$, then sometimes we write $s_1 \equiv s_2 (\text{mod}\, \rho)$ or, in short, $s_1 \equiv s_2$. (Then we say that s_1 *is congruent to* s_2 *modulo* ρ or, in short, s_1 *is congruent to* s_2.) Let S and T be semigroups having a homomorphism $\psi : S \to T$. Then ψ determines a congruence relation ρ with $s \equiv s'(\text{mod}\, q)$ if and only if $\psi(s) = \psi(s')(s, s' \in S)$. A partition S/ρ of the semigroup S is called *compatible* if ρ is a congruence relation. In this case, multiplication in S *induces* a semigroup structure on S/ρ: Letting $[s]$ denote the ρ equivalence class s/ρ of $s \in S$, compatibility means that the multiplication $[s_1][s_2] = [s_1s_2]$ is well defined for all $s_1, s_2 \in S$ and is associative. It follows that $s \mapsto [s]$ is a homomorphism from S onto the *quotient semigroup* S/ρ. Moreover, if S is a monoid or group, then so is S/ρ.

Considering semigroups (monoids, groups) as algebraic structures, we can speak about a subsemigroup (submonoid, subgroup) of a given semigroup (monoid, group). In speaking of a submonoid of a monoid, it is required that the identity element of the submonoid coincide with that of the monoid. (For subgroups of groups, this condition is obtained automatically.) Thus a subset H of a group G is a *subgroup* of G if H is a subsemigroup of G, H contains the identity element 1, and $g^{-1} \in H$ whenever $g \in H$. If H is subsemigroup of a finite group G, then H is necessarily a group: for any $h \in H$, h^n is idempotent for some $n > 1$, since G is finite. Thus H contains $h^n = 1$, the unique idempotent of G, and $h^n = h^{n-1}h = 1 = hh^{n-1}$, so H contains 1 and $h^{-1} = h^{n-1}$ for each $h \in H$. If G is not finite, a subsemigroup need not be closed under taking inverses, i.e., it might contain an element h but not h^{-1}, and so may fail to be a subgroup. This is the case, for example, with the natural numbers considered as a subsemigroup of the group of integers under the operation of addition. We say G is *generated as a group* by a nonempty set $X \subseteq G$ if $X \cup X^{-1}$ generates G. (Of course, if G is finite, then X generates G as a group if and only if $X \cup X^{-1}$ generates G.) This is equivalent to saying that the smallest subgroup of G containing X is G itself. Then we put $\langle X\rangle_{\text{group}} = G$. Then, for example, $\langle g\rangle_{\text{group}} = \langle g^{-1}, g\rangle$ holds for any g in any group G. And, in general, in any group G, the subgroup generated as a group by a subset X is $\langle X\rangle_{\text{group}} = \langle X \cup X^{-1}\rangle$ for any $X \subseteq G$. By the above discussion, if G is finite, or more generally when every element of G has finite order, then $\langle X\rangle_{\text{group}} = \langle X\rangle$. It is common to speak of $\langle X\rangle_{\text{group}}$ as "the group generated by X," and, if no confusion can result, to write $\langle X\rangle$ for $\langle X\rangle_{\text{group}}$.

Let G be a group and H a subgroup of G. For every $g \in G$, we define $Hg = \{hg \mid h \in H\}$. This Hg is called the *right coset* of H by g. The *left coset* of H by g is defined symmetrically: $gH = \{gh \mid h \in H\}$. Both $\{Hg \mid g \in G\}$ and $\{gH \mid g \in G\}$ are partitions of G with the same cardinality. We say that $g \in G$ *normalizes* H if gH and Hg coincide.

A subgroup H of G is *normal* if its right and left cosets coincide, i.e., $gH = Hg$ for every $g \in G$. (Equivalently, $gHg^{-1} = H$ for all $g \in G$.) Then $\{gH \mid g \in G\}$ $(= \{Hg \mid g \in G\})$

is called a *partition of G by H*. We write $H \lhd G$ if H is normal in G. If H is a normal subgroup, then the partition of G by H is compatible. Using normality and $H^2 = H$, we have

$$Hg_1Hg_2 = (g_1H)g_2H = g_1H(Hg_2) = (g_1H)g_2 = Hg_1g_2.$$

On the other hand, every compatible partition of G can be given by a normal subgroup. Then the set $G/H = \{gH \mid g \in G\}$ forms a *factor group* of G having $Hg_1Hg_2 = Hg_1g_2$, $Hg_1, Hg_2 \in G/H$, where H $(= 1H)$ is the identity element of G/H and $(Hg)^{-1} = Hg^{-1}$ for each $Hg \in G/H$. However, if $\psi : G \to T$ is a surjective homomorphism, then the *kernel* of ψ, $\ker\psi = K = \{g \in G : \psi(g) = \psi(1)\}$, is easily seen to be a normal subgroup, and, moreover, G/K is isomorphic to T via the map $gK \mapsto \psi(gK) = \psi(g)$. Thus G maps homomorphically onto G/H when H is normal and every homomorphic image of G is of this form.

Obviously, G and $\{1\}$ are normal subgroups of G. If N is a normal subgroup of G such that $\{1\} \subsetneq N \subsetneq G$, then it is called a *proper normal subgroup of* G, and then G/N is said to be a *proper factor group of* G. If G is nontrivial and has no proper normal subgroups, then it is called *simple*. If G is simple, then every homomorphic image of G is isomorphic to $\{1\}$ or G. A proper normal subgroup $N \lhd G$ is *maximal* if $N \subseteq H \lhd G$ implies $H = N$ or $H = G$. This holds if and only if G/N is simple.

In an abelian group, every subgroup is normal. A group is *cyclic* if it can be generated as a group by one of its elements. An elementary exercise shows that every cyclic group is necessarily abelian, but not conversely. If an abelian group G is simple, it contains an element $g \neq 1$, and since $\langle g\rangle$ is normal in G, we have that $\langle g\rangle = G$. Thus G is cyclic. If $o(g) = \infty$, G must be isomorphic to the group $\mathbb{Z}$ of integers under addition, but this is not simple since the even integers form a (normal) subgroup. Hence, $o(g) \neq \infty$, but then the order of g must be prime: for if $o(g) = nm$, for integers $n, m > 0$, then $\langle g\rangle = G$ has nm elements. Suppose $m \neq 1$; then by definition of order of g, $g^n \neq 1$ and so the normal subgroup $\langle g^n\rangle$ must be G. Now since $(g^n)^m = g^{nm} = 1$, the group $\langle g^n\rangle = \{g^n, \ldots, (g^n)^m\} = G$ has no more than m elements. So it can only be that $n = 1$, whence $o(g)$ is prime. Thus an abelian group is simple if and only if it is a finite cyclic group of prime order. It is immediate that for every simple (finite or infinite) group G, G is nonabelian if and only if G is not cyclic.

If any two semigroups S_1 and S_2 are considered, the ordered pairs (s_1, s_2) with s_1 from S_1 and s_2 from S_2 form a semigroup according to the rule that defines products of pairs componentwise, i.e., $(s_1', s_2')(s_1'', s_2'') = (s_1's_1'', s_2's_2'')$, $s_1', s_1'' \in S_1, s_2', s_2'' \in S_2$. This semigroup is called the *direct product of* S_1 and S_2 and is written $S_1 \times S_2$. It is easy to see that $S_1 \times S_2$ and $S_2 \times S_1$ are isomorphic. If e_2 is an idempotent in S_2, then the ordered pairs (s_1, e_2) with $s_1 \in S_1$ make up a subsemigroup of $S_1 \times S_2$ isomorphic to S_1. In particular, if S_1 has an identity e_1 and S_2 has an identity e_2, then (e_1, e_2) is an identity for $S_1 \times S_2$. Thus the direct product of monoids is obviously a monoid, and the component monoids are isomorphic to submonoids of the direct product. The analogous assertions hold for groups since inverses in the direct product of groups can be obtained by taking inverses in each component.

Next we prove the following theorem.

Theorem 1.2. *Let $G = \{g_1, \ldots, g_n\}$ be a (finite) order n group. Put $P_G = \{g_{P(1)} \cdots g_{P(n)} \mid$ P is a permutation over $\{1, \ldots, n\}\}$. If G is simple and noncommutative, then there exists a positive integer m with $P_G^m = G$.*

Proof. First, for every positive integer t and $g \in P_G$, we have $|P_G^{t+1}| \geq |gP_G^t| = |P_G^t|$. Since the group is finite, such growth must eventually finish, i.e., there exists a positive t_0 such that $t \geq t_0$ implies $|P_G^t| = |P_G^{t_0}|$.

Of course, since taking inverses permutes the elements of G, for every $g = g_{P(1)} \cdots g_{P(n)} \in P_G$, we have $g^{-1} = (g_{P(n)})^{-1} \cdots (g_{P(1)})^{-1} \in P_G$. Thus $e \in P_G^m$ for all m even, where e denotes the identity element of the group G. Let $m \geq t_0$ be such that $e \in P_G^m$.

Now $P_G^{2m} = P_G^m P_G^m$ and $P_G^m = eP_G^m \subseteq P_G^{2m}$. Since $2m, m \geq t_0$, P_G^m and P_G^{2m} have the same number of elements, so it follows that $P_G^m P_G^m = P_G^m$. Therefore, P_G^m is a subgroup of G.

Since conjugation permutes elements of G, if r is an arbitrary element of G, then $rg_{P(1)} \cdots g_{P(n)}r^{-1} = (rg_{P(1)}r^{-1}) \cdots (rg_{P(n)}r^{-1}) \in P_G$. Thus $rP_Gr^{-1} = P_G$, and inductively, for all $t \geq 1$, it follows that $rP_G^{t+1}r^{-1} = (rP_G^tr^{-1})(rP_Gr^{-1}) = P_G^tP_G = P_G^{t+1}$. In particular, $rP_G^mr^{-1} = P_G^m$. Therefore, every element of G normalizes P_G^m, and thus P_G^m is a normal subgroup in G.

Since G is noncommutative, there are $g_i, g_j \in G$ with $g_ig_j \neq g_jg_i$. Without loss of generality, suppose these elements are g_1 and g_2. Then $g_1g_2g_3 \cdots g_n \neq g_2g_1g_3 \cdots g_n$ are two distinct elements of P_G. Since $|P_G^m| \geq |P_G| \geq 2$, P_G^m is not the trivial subgroup. Therefore, by the simplicity of G, $P_G^m = G$ necessarily holds. □

Let G be a group. An element $g \in G$ is called a *commutator* if $g = aba^{-1}b^{-1}$ for some elements $a, b \in G$. The smallest subgroup that contains all commutators of G is called the *commutator subgroup* (or *derived subgroup*) of G and is denoted by G'. It is easy to check that G' is normal in G and that it is nontrivial if and only if G is noncommutative. In particular, $G = G'$ whenever G is simple and noncommutative. Thus we can also get our previous result as a direct consequence of the following well-known theorem.

Theorem 1.3 (Dénes–Hermann theorem). *Let $G = \{g_1, \ldots, g_n\}$ be a (finite) order n noncommutative group and denote G' its commutator subgroup. Put $P_G = \{g_{P(1)} \cdots g_{P(n)} \mid P$ is a permutation over $\{1, \ldots, n\}\}$. There exists a $g \in G$ with $P_G = G'g$. Thus $P_G = G$, whenever $G = G'$.* □

1.3 Transformation Semigroups, Division, and Wreath Products

Let A be a nonvoid set. A mapping $\varphi : A \to A$ is called a *transformation* of A. Recall that for every pair $s_i : A \to A$ of transformations we define the (right) *multiplication* s_1s_2 *of* s_1 *by* s_2 as the transformation $s : A \to A$ having $s(a) = s_2(s_1(a))$. The set S of all transformations of A form a semigroup under this (right) multiplication of mappings since function composition is associative. Then $T_S(A) = (A, S)$ is called the *(right) full transformation semigroup of* A. Sometimes we denote $s_i(a)$ as $a \cdot s_i$, and we have $(a \cdot s_1) \cdot s_2 = a \cdot (s_1s_2)$ for all $a \in A$, $s_1, s_2 \in S$. If H is a subsemigroup of S, then (A, H) is called a (right) *transformation semigroup on* A, and we say that H (and each element h of H) *acts* on (the right of) A. In particular, if $A = \{1, \ldots, n\}$ for some positive integer n, then $T_S(A)$ is the *(right) full transformation semigroup of degree n* and S is the *symmetric semigroup of degree n*. Note that if (A, H) is a transformation semigroup, then for all $s, s' \in H$, if for $a \in H$, $a \cdot s = a \cdot s'$

holds, then $s = s'$, since s and s' give the same transformation. If there exist $a, b \in A$ with $a \neq b$ and $s(a) = s(b) = a$, but with $s(u) = u$ for all $u \in A \setminus \{a, b\}$, then s is called an *elementary collapsing*.

Again, take a nonvoid set A. For finite A, if there is some $a \in A$ such that for each $a' \in A$ there is a positive integer n with $a' = \varphi^n(a)$, then φ is called a *cyclic permutation* of length $|A|$ on A. Clearly, φ is a cyclic permutation of length $|A|$ on A if and only if by an appropriate choice of indexing for $A = \{a_1, \ldots, a_n\}$, for example, $a_i = \varphi^i(a)$ $(1 \le i \le n)$, the permutation φ shifts all elements of A one position in the ordering given by the indices, with an element shifted off the end inserted back at the beginning. We may then write $\varphi(a_i) = a_{i+1\,(\mathrm{mod}\,n)}$.[4] (We remark that shifting each a_i $(1 \le i \le n)$ by any k $(1 \le k \le n-1)$ to $a_{i+k\,(\mathrm{mod}\,n)}$ defines a cyclic permutation of length n if and only if n and k are relatively prime.) Given a nonempty subset $B \subseteq A$, a transformation $\varphi : A \to A$ is called a *cyclic permutation of length* $|B|$ *on* A if $\varphi|_B$ is a cyclic permutation of length $|B|$ on B and, simultaneously, $\varphi(a) = a$ for every $a \in A \setminus B$. A cyclic permutation of length m is also called a *cycle* of length m. A cyclic permutation of length 2 is called a *transposition*. A transposition may be given using the notation $\tau_{a,b} : A \to A$ with

$$\tau_{a,b}(u) = \begin{cases} u & \text{if } u \in A \setminus \{a, b\}, \\ a & \text{if } u = b, \\ b & \text{if } u = a \end{cases}$$

for some $a, b \in A$ with $a \neq b$. The set P of all permutations of A forms a group under the (right) multiplication of mappings. $T_G(A) = (A, P)$ is called the *(right) full permutation group on* A. And if H is a subgroup of P, then (A, H) is a *permutation group on* A, and we say that H *acts* on (the right of) A *by permutations*. If $A = \{1, \ldots, n\}$ for some positive integer n, then $T_G(A)$ is the *(right) full permutation group of degree n* and P is the *symmetric group of degree n*. Permutation groups are also sometimes called *transformation groups*.

Given a permutation $p : \{1, \ldots, n\} \to \{1, \ldots, n\}$, the pair $p(i), p(j)$ is called an *inversion* if $(p(i) - p(j))(i - j) < 0$. A permutation is called *even* if the number of its inversions is even. Equivalently, p is even if it can be written as the product of an even number of transpositions. The set of all even permutations of $\{1, \ldots, n\}$ forms a group under the usual (right) multiplication of mappings, and this group is called the *alternating group of degree n*.

Theorem 1.4. *Given a positive integer $n > 1$, the alternating group A_n of degree n is the only maximal proper normal subgroup of the symmetric group of degree n. Moreover, A_n is simple if and only if $n \neq 4$.* □

We will use the following simple fact.

Proposition 1.5. *Let $n > 1$ and take the following three transformations of* $\{1, \ldots, n\}$:

- *the cyclic permutation* $\gamma_1(i) = i + 1(\mathrm{mod}\,n)$,
- *the transposition* $\gamma_2(1) = 2$, $\gamma_2(2) = 1$, *and* $\gamma_2(i) = i$ *if* $i > 2$,

[4]Recall that for arbitrary integers k, m $(m \ge 2)$, $k\,(\mathrm{mod}\,m)$ denotes the least positive integer k' such that m divides $k - k'$. In particular, $0\,(\mathrm{mod}\,m) = m\,(\mathrm{mod}\,m) = m$.

- *the elementary collapsing* $\gamma_3(1) = \gamma_3(2) = 1$ *and* $\gamma_3(i) = i$ *if* $i > 2$ *for all* i *(*$1 \leq i \leq n$*).*

Then $\{\gamma_1, \gamma_2\}$ *generates the full permutation group of degree* n*, and* $\{\gamma_1, \gamma_2, \gamma_3\}$ *generates the full transformation semigroup of degree* n*.*

Proof. Consider the following game. Let us have n distinct places and n distinct coins and let us place the n coins c_i, $i = 1, \ldots, n$, onto the places $1, \ldots, n$ so that at the start c_i is placed onto i:

Any transformation $t : \{1, \ldots, n\} \to \{1, \ldots, n\}$ is represented by moving the contents of position $t(i)$ to place i. More precisely, we have the following interpretation of transformations—*Interpretation* 1: *Simultaneously for all places i (*$1 \leq i \leq n$*), replace the contents of place i by a copy of the current contents of position* $t(i)$.

Obviously, the transformation t is completely determined by its effect on the coins in their initial configuration. Note that in the resulting configuration there may be no, one, or more than one copy of a given coin c_k depending on how many times t takes the value k. Also, in the resulting configuration, each place holds exactly one coin, and therefore this remains true if we apply any further transformations.

If $t' : \{1, \ldots, n\} \to \{1, \ldots, n\}$ is another transformation, let us observe what happens when we first carry out t and then t' under this interpretation. After t has been carried out, position i will contain $c_{t(i)}$, but, moreover, position $t'(i)$ will contain $c_{t(t'(i))}$. If we then carry out t', we must replace the contents of place i by a copy of the current contents of position $t'(i)$, i.e., by $c_{t(t'(i))}$. Thus, under this interpretation, t followed by t' results in putting the original contents of position $t(t'(i))$ into position i. Thus, *under this interpretation, t followed by t′ has the same effect as the transformation* $t' \circ t$.

To show that $\{\gamma_1, \gamma_2, \gamma_3\}$ generates all possible transformations, it suffices to show that for an arbitrary transformation t, from the initial configuration with each coin c_i on position i, we are able to obtain, by applying some finite sequence of moves representing the γ's, the configuration in which a copy of coin $c_{t(i)}$ is on position i for all $1 \leq i \leq n$.

First we prove that $\{\gamma_1, \gamma_2\}$ generates all permutations of $\{1, \ldots, n\}$. We use two types of moves: either exchange the coins in the first two places (this is the move corresponding to the transposition γ_2) or turn all the coins such that c_n is moved to 1, and for every $i \in \{1, \ldots, n-1\}$, c_i is moved to $(i+1)$ (this is the move corresponding to the cyclic permutation $\gamma_1^{-1} = \gamma_1^{n-1}$):

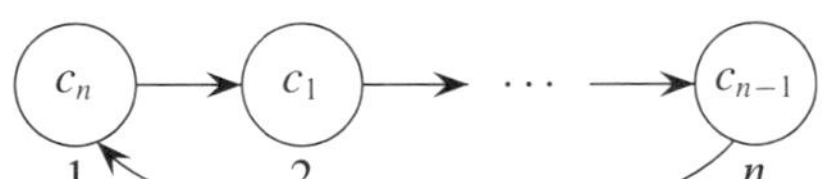

TURN 1 STEP

Thus, for every $i = 1, \dots, n-1$, we can exchange the coins in the places i and $(i+1)$ in the following way. First turn the coins in $(n+1-i)$ steps around the places $1, \dots, n$ so that c_i is moved to 1 and c_{i+1} is moved to 2:

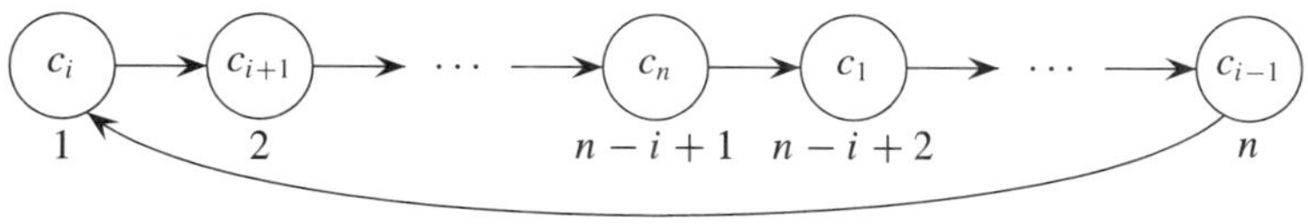

TURN $n+1-i$ TIMES

Then exchange c_i and c_{i+1} (using the transposition γ_2):

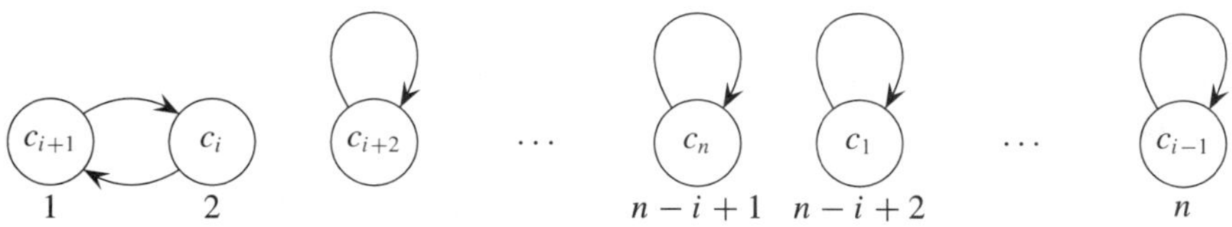

EXCHANGE CONTENTS OF PLACES 1 AND 2

Finally again turn all coins $i-1$ steps:

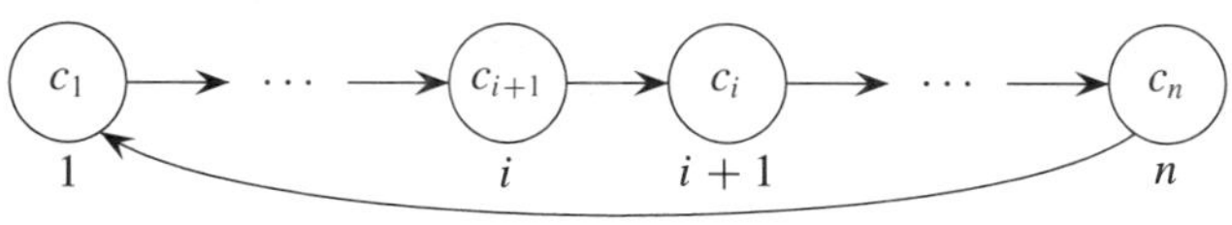

TURN $i-1$ TIMES

As a result, c_{i+1} has moved to i, c_i has moved to $(i+1)$, and all the others go back to their original places.

Now, let $p : \{1, \dots, n\} \to \{1, \dots, n\}$ be an arbitrary permutation and suppose again that first the n coins $c_1, \dots, c_n$ are placed onto the places $1, \dots, n$ so that c_i is placed onto i. If $p(1) = 1$, then leave c_1 on 1. Otherwise, successively exchange the coin $c_{p(1)}$ with its left neighbor $(p(1)-1)$ times so that it moves onto place 1. (The left neighbor of a coin on place $i+1$ is the coin on place i, for $1 \le i \le n-1$, and place i is to the left of place j if $i < j$.) Now if place 2 contains $c_{p(2)}$, do nothing; otherwise observe that $c_{p(2)}$ cannot be on place 1 (which now contains $c_{p(1)}$ since p is a permutation), and exchange $c_{p(2)}$ repeatedly with its left neighbor until it arrives at place 2. Then we repeat this procedure inductively with places $i = 3, 4, \dots, n-1$, in that order. At each stage, by induction, all places to the left of the position i that we are currently considering already contain the correct coins, and therefore, since p is a permutation, coin $c_{p(i)}$ now cannot be to the left of place i.

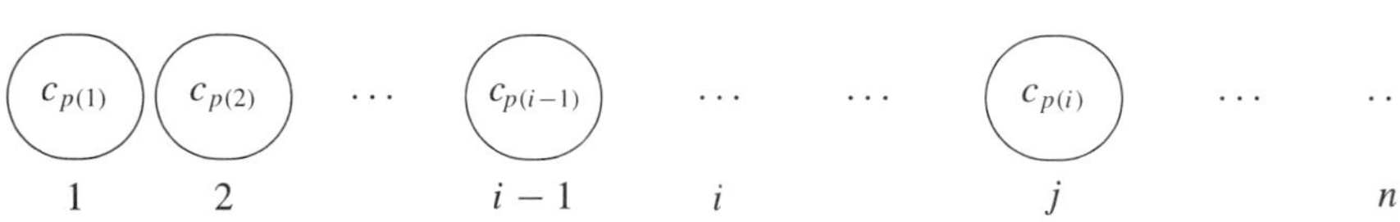

AT STAGE i IN THE INDUCTION COIN $c_{p(i)}$ MUST LIE ON A PLACE j WITH $i \le j \le n$

Therefore we can move $c_{p(i)}$ left to place i by zero or more exchanges with successive left neighbors as before without disturbing the coins already correctly positioned on places 1 to $i-1$. Finally, each place i will contain the coin $c_{p(i)}, i = 1, \ldots, n-1$, so, since p is permutation, place n must contain $c_{p(n)}$.

Now let $t : \{1, \ldots, n\} \to \{1, \ldots, n\}$ be an arbitrary transformation. If t is a permutation, then we are done. Otherwise, let $1 \leq k < n$ be the largest integer such that $t(i_1), \ldots, t(i_k)$ are pairwise distinct for some $i_1 < \cdots < i_k$ in $\{1, \ldots, n\}$. Then first we can move $c_{t(i_1)}$ to $i_1, \ldots, c_{t(i_k)}$ to i_k by any permutation on $\{1, \ldots, n\}$ that agrees with t on $\{i_1, \ldots, i_k\}$. This puts coin $c_{t(i_\ell)}$ on place i_ℓ for each $\ell \in \{1, \ldots, k\}$. Suppose that $t(j) = t(i_\ell)$ for some $j \notin \{i_1, \ldots, i_k\}, i_\ell \in \{i_1, \ldots, i_k\}$. We may assume $i_\ell, j \notin \{1, 2\}$, for otherwise the argument is similar with minor adjustments for the various possibilities. Then exchange the coins on places 2 and j; moreover, exchange the coin on 1 and the coin on i_ℓ. Now we remove the coin from place 2 and put a new copy of $c_{t(i_\ell)}$ onto 2 (the elementary collapsing γ_3). Then we exchange again the coins on 1 and i_ℓ, and similarly exchange again the coins on 2 and j. As a result, places j and i_ℓ both contain a copy of $c_{t(j)} = c_{t(i_\ell)}$ and all other positions are as they were. By repeating this procedure for the other elements not in $\{i_1, \ldots, i_k\}$, finally all places i will have the appropriate copies of the appropriate coins $c_{t(i)}$. The proof is complete. □

An interpretation of transformations other than the one used in the proposition above (which we refer to as interpretation 1) is possible. Namely, for interpretation 2, a transformation $t : \{1, \ldots, n\} \to \{1, \ldots, n\}$ is represented as follows—*Interpretation* 2: *Simultaneously, move all the current contents of each position i to position $t(i)$.* Note that t is determined by its effect on the above initial configuration—that is, by the resulting configuration with coin c_i in position $t(i)$. Under this interpretation, some positions may get no, one, or several coins, but each coin always has exactly one position (and is never duplicated or removed).

Observe that if p is a permutation; then under interpretation 1 applied to the initial configuration, place i contains $c_{p(i)}$, while under interpretation 2, place $p(j)$ contains c_j, and taking $j = p^{-1}(i)$, place $p(j) = i$ will contain $c_j = c_{p^{-1}(i)}$. Thus, p under interpretation 1 then has the same effect as p^{-1} has under interpretation 2, and vice versa.

Under interpretation 2, if t is followed by t', first place $t(i)$ gets the contents of position i, and then, if $j = t(i)$, position $t'(j) = t'(t(i))$ gets the current contents of position $j = t(i)$, i.e., gets the original contents of position i. Thus, *under interpretation 2, the effect of t followed by t' is the same as the effect of $t \circ t'$.*

Given a transformation t, let t^{I} denote its action on configurations of coins and positions under interpretation 1, and similarly let t^{II} denote its action on configurations under interpretation 2. Conversely, if T is a transformation of coins and positions arising from a transformation t under interpretation 1, then let $[T]_{\mathrm{I}} = t$, and similarly if T is a transformations of coins and positions arising from a transformation t under interpretation 2, then let $[T]_{\mathrm{II}} = t$.

As a consequence of these observations and the remarks in the discussion of interpretation 1 in the proposition, we have the following.

Fact 1.6. *Let $t, t_1, \ldots, t_m$ be transformations on $\{1, \ldots, n\}$ and p a permutation. Then*

$$[t^{\mathrm{I}}]_{\mathrm{I}} = t \quad \textit{and} \quad [t^{\mathrm{II}}]_{\mathrm{II}} = t,$$

$$[p^{\mathrm{I}}]_{\mathrm{II}} = p^{-1} \quad \textit{and} \quad [p^{\mathrm{II}}]_{\mathrm{I}} = p^{-1},$$
$$t_1^{\mathrm{I}} \circ \cdots \circ t_k^{\mathrm{I}} = (t_k \circ \cdots \circ t_1)^{\mathrm{I}},$$
$$t_1^{\mathrm{II}} \circ \cdots \circ t_k^{\mathrm{II}} = (t_1 \circ \cdots \circ t_k)^{\mathrm{II}}.$$

□

Corollary 1.7 (position-contents duality lemma).

(1) *Let* $p_1, \ldots, p_k$ *be permutations on* $\{1, 2, \ldots, n\}$. *Then*

$$[p_1^{\mathrm{I}} \circ \cdots \circ p_k^{\mathrm{I}}]_{\mathrm{II}} = p_1^{-1} \circ \cdots \circ p_k^{-1} \quad \textit{and} \quad [p_1^{\mathrm{II}} \circ \cdots \circ p_k^{\mathrm{II}}]_{\mathrm{I}} = p_k^{-1} \circ \cdots \circ p_1^{-1}$$

(2) *Let* $t_1, \ldots, t_k$ *be transformations on* $\{1, 2, \ldots, n\}$. *Then*

$$[t_1^{\mathrm{I}} \circ \cdots \circ t_k^{\mathrm{I}}]_{\mathrm{I}} = [t_k^{\mathrm{II}} \circ \cdots \circ t_1^{\mathrm{II}}]_{\mathrm{II}}.$$

Proof. For the first part, using the fact,

$$[p_1^{\mathrm{I}} \circ \cdots \circ p_k^{\mathrm{I}}]_{\mathrm{II}} = [(p_k \circ \cdots \circ p_1)^{\mathrm{I}}]_{\mathrm{II}} = (p_k \circ \cdots \circ p_1)^{-1} = p_1^{-1} \circ \cdots \circ p_k^{-1},$$
$$[p_1^{\mathrm{II}} \circ \cdots \circ p_k^{\mathrm{II}}]_{\mathrm{I}} = [(p_1 \circ \cdots \circ p_k)^{\mathrm{II}}]_{\mathrm{I}} = (p_1 \circ \cdots \circ p_k)^{-1} = p_k^{-1} \circ \cdots \circ p_1^{-1}.$$

The second part follows from the last two equations in the fact:

$$[t_1^{\mathrm{I}} \circ \cdots \circ t_k^{\mathrm{I}}]_{\mathrm{I}} = [(t_k \circ \cdots \circ t_1)^{\mathrm{I}}]_{\mathrm{I}} = t_k \circ \cdots \circ t_1$$
$$= [(t_k \circ \cdots \circ t_1)^{\mathrm{II}}]_{\mathrm{II}} = [t_k^{\mathrm{II}} \circ \cdots \circ t_1^{\mathrm{II}}]_{\mathrm{II}}.$$

□

Remark. *It is perhaps interesting to note that the second part of the position-contents duality lemma shows that switching between the two interpretations corresponds to reversing the direction of time, i.e., whether a sequence of transformations is carried out reading from left to right or, in the reverse order, reading from right to left.*

Remark. *In what follows, we will generally follow interpretation* 1 *of transformations as it appears to be more common in the automata-theoretic literature (although this is seldom made explicit). Should the reader encounter the other interpretation anywhere, it is possible to use Fact* 1.6 *and Corollary* 1.7 *to convert between the interpretations. Using the right product of functions and interpretation* 1 *means that (right) composition of functions corresponds to* left *composition of transformations of configurations.*

Let A and B be two (not necessarily disjoint) finite nonempty sets. Moreover, let H_A and H_B be two sets of permutations over A and B, respectively. For every $\varphi \in H_A \cup H_B$ let

$$\varphi|^{A \cup B}(x) = \begin{cases} \varphi(x) & \text{if } x \in A \text{ and } \varphi \in H_A \\ & \text{or } x \in B \text{ and } \varphi \in H_B, \\ x & \text{otherwise.} \end{cases}$$

Moreover, let $\langle H_A \cup H_B \rangle = \langle \{\varphi|^{A \cup B} : \varphi \in H_A \cup H_B\} \rangle$.
Next we prove the following lemma.

Lemma 1.8. *Let A and B be two finite sets with $A \cap B \neq \emptyset$, $A \setminus B \neq \emptyset$, and let $T_G(A)$ and $T_G(A \cup B)$ denote the full permutation groups of A and $A \cup B$, respectively. Then $\langle T_G(A) \cup \{c_B\}\rangle = T_G(A \cup B)$ holds for every cyclic permutation c_B of length $|B|$ on B.*

Proof. If $A = B$, there is nothing to prove. If not, let $A = \{a_1, \ldots, a_t\}$, $|B| = k$, and $a_1 \in B$. We shall inductively build a chain of sets

$$B = D_k \subsetneq D_{k+1} \subsetneq D_{k+2} \subsetneq \cdots \subsetneq D_n = A \cup B$$

such that $|D_m| = m$ and $c_{D_m} \in \langle T_G(A) \cup \{c_B\}\rangle$ is a cyclic permutation of length m for each $m = k, k+1, \ldots, n$, i.e.,

$$c_{D_m}(u) = \begin{cases} d_{i+1\,(\mathrm{mod}\, m)} & \text{if } u = d_i \in D_m,\ 1 \leq i \leq m, \\ u & \text{if } u \in (A \cup B) \setminus D_m \end{cases}$$

for an appropriate arrangement $d_1, \ldots, d_m$ of D_m depending on m.

For this, the base case $m = k$ is trivial since we may take $c_{D_k} = c_B$. Suppose then that D_m has been constructed with $|D_m| = m$ along with the length m cycle $c_{D_m} \in \langle T_G(A) \cup \{c_B\}\rangle$. If $D_m = A \cup B$, then the induction is complete; otherwise let $x = a_i$ denote the element in $A \setminus D_m$ with the least index. We note that $i > 1$ because $a_1 \in D_k \subseteq D_m$. Therefore $y = a_{i-1} \in A$ exists and lies in D_m. We may therefore write $y = d_j$ for some $1 \leq j \leq m$ in the ordering $d_1, \ldots, d_m$ of D_m. (Note that there exists such a pair D_m, D_{m+1} because $A \setminus B \neq \emptyset$ is assumed.) Now let $D_{m+1} = D_m \cup \{x\}$ and consider the transposition $\tau_{x,y}$ exchanging x and y. Since $x, y \in A$, we have $\tau_{x,y} \in T_G(A)$. Thus $\tau_{x,y}c_{D_m} \in \langle T_G(A) \cup \{c_B\}\rangle$ in view of the induction assumption $c_{D_m} \in \langle T_G(A) \cup \{c_B\}\rangle$. On the other hand,

$$(\tau_{x,y}c_{D_m})(u) = \begin{cases} u & \text{if } u \notin D_m \cup \{x\}, \\ d_{i+1\,(\mathrm{mod}\, m)} & \text{if } u = d_i \in D_m \text{ and } i \neq j, \\ x & \text{if } u = d_j (= y), \\ d_{j+1\,(\mathrm{mod}\, m)} & \text{if } u = x. \end{cases}$$

Hence $c_{D_{m+1}} = \tau_{x,y}c_{D_m}$ is a cyclic permutation of length $m+1$ on $A \cup B$. (The ordering of D_{m+1} is obtained by inserting x immediately after $y = d_j$ in the sequence $d_1, \ldots, d_m$.) Since $c_{D_{m+1}} = \tau_{x,y}c_{D_m} \in \langle T_G(A) \cup \{c_B\}\rangle$, the induction step is complete.

In the final step of our induction, the symbols x and y were adjacent in the cycle $c_{A\cup B}$. Thus, by Proposition 1.5, $\langle c_{A\cup B}, \tau_{x,y}\rangle = T_G(A \cup B)$. Since $c_{A\cup B}$ and $\tau_{x,y}$ both lie in $\langle T_G(A) \cup \{c_B\}\rangle$, the lemma is proved. □

Observe that the condition $A \setminus B \neq \emptyset$ in Lemma 1.8 is essential. For example, if $A = \{1, 3\}$, $B = \{1, 2, 3, 4\}$, then $T_G(A) \cup \{c_B\} \subsetneq T_G(A \cup B)$.

If (X, S) is a transformation semigroup, we denote by $(X, \overline{S})$ the transformation semigroup with transformations $\overline{S} = \{t \mid t \in S \text{ or } t \text{ is constant}\}$. Let 1_X denote the identity map on X. Also we denote by (X, S^λ) the transformation semigroup with $S^\lambda = \{t \mid t \in S \text{ or } t = 1_X\}$. If $X = \{1, 2\}$, then $(X, \overline{\{1_X\}})$ is called the *flip-flop*, and its transformations form a semigroup isomorphic to the flip-flop monoid **F** mentioned above.

The *wreath product* of transformation semigroups (X, S) and (Y, T), denoted $(Y, T) \wr (X, S)$, is the transformation semigroup $(Y \times X, W)$, where W is the set of all transformations w of $Y \times X$ satisfying for all $(y, x) \in Y \times X$

$$(y, x) \cdot w = (y \cdot f(x), x \cdot s)$$

for some fixed $f : X \to T$ and $s \in S$, both depending only on w. For each element w of W, f and s are uniquely determined. Thus we may identify W with the set of all (f, s) with $f : X \to T$ and $s \in S$. In W, the product of (f, s) followed by any $(g, s') \in W$ is given by function composition

$$(y, x) \cdot ((f, s)(g, s')) = (y \cdot f(x), x \cdot s)(g, s') = (y \cdot f(x)g(x \cdot s), x \cdot ss').$$

Thus the product is of the form (h, ss') with $h(x) = f(x)g(x \cdot s)$ for all $x \in X$. This shows W is closed under composition, so W is indeed a semigroup and $(Y \times X, W)$ is indeed a transformation semigroup. If S and T are monoids, then so is W since it has identity element $1_W = (i, 1_S)$, where 1_S is the identity element of S and $i : X \to T$ is the constant function $i(x) = 1_T$, the identity element of T, for all $x \in X$. If (X, S) and (Y, T) are permutation groups, then so is their wreath product. For $(f, s) \in W$ then let $f'(x) = (f(x \cdot s^{-1}))^{-1}$ for all $x \in X$. Then $(y, x) \cdot (f, s)(f', s^{-1}) = (y \cdot f(x)f'(x \cdot s), x \cdot ss^{-1}) = (y \cdot f(x)f((x \cdot s) \cdot s^{-1})^{-1}, x \cdot 1_S) = (y \cdot f(x)f(x \cdot (ss^{-1}))^{-1}, x) = (y \cdot f(x)f(x \cdot 1_S)^{-1}, x) = (y \cdot f(x)f(x)^{-1}, x) = (y \cdot 1_T, x) = (y, x) = (y, x) \cdot (i, 1_S)$. So $(f, s)(f', s^{-1}) = 1_W$. In particular, each (f, s) is seen to be injective. Similarly $(y, x) \cdot (f', s^{-1})(f, s) = (y \cdot f'(x)f(x \cdot s^{-1}), x \cdot s^{-1}s) = (y \cdot f(x \cdot s^{-1})^{-1}f(x \cdot s^{-1}), x \cdot 1_S) = (y \cdot 1_T, x \cdot 1_S) = (y, x)(i, 1_S)$. So $(f', s^{-1})(f, s) = 1_W$, whence (f, s) is surjective, hence a permutation of $Y \times X$, with inverse (f', s^{-1}) in W. This shows that W is a group and that $(Y \times X, W)$ is a permutation group.

Observe that $(Z, U) \wr ((Y, T) \wr (X, S)) = ((Z, U) \wr (Y, T)) \wr (X, S)$ since both have states $Z \times Y \times X$ and transformation semigroups consisting of exactly those transformations w of the form $(z, y, x) \cdot w = (z \cdot f_3(y, x), y \cdot f_2(x), x \cdot f_1)$ for some functions $f_3 : Y \times X \to U$, $f_2 : X \to T$, and $f_1 \in S$ uniquely determining w. We see that the wreath product is an associative operation on the class of transformation semigroups and also on the class of permutation groups. Similarly, the wreath product of $n > 0$ transformation semigroups (X_i, S_i), $i = 1, \dots, n$, is unambiguously defined. We have $(X_n, S_n) \wr \cdots \wr (X_1, S_1)$ with state set $X_n \times \cdots \times X_1$ and transformations consisting of all w of the form $(x_n, \dots, x_1) \cdot w = (x_n \cdot f_n(x_{n-1}, \dots, x_1), \dots, x_2 \cdot f_2(x_1), x_1 \cdot f_1)$, where $f_1 \in S_1$ and $f_i : X_{i-1} \times \cdots \times X_1 \to S_i$ for $2 \le i \le n$. We see that f_i determines the transformation in component i as a function of the components x_j of the state with $0 < j < i$. Thus transformations of this n-fold wreath product are in one-to-one correspondence with the n-tuples $(f_n, \dots, f_1)$.

We say (X, S) *embeds* in (X', S') if there exist $Y \subseteq X'$, $T \subseteq S'$, a bijective mapping $\psi_2 : X \to Y$, and an isomorphism of semigroups $\psi_1 : S \to T$ such that $\psi_2(x \cdot s) = \psi_2(x) \cdot \psi_1(s)$ for all $x \in X, s \in S$. It follows that (Y, T) is a transformation semigroup. We then write $(X, S) \hookrightarrow (X', S')$. In particular, since (Y, T) must be a transformation semigroup, for each $t_1, t_2 \in T$, $t_1 \neq t_2$ implies there exists a $y \in Y$ (not merely in X'!) such that $y \cdot t_1 \neq y \cdot t_2$. We say T acts *faithfully* on Y if this condition holds. Thus, a semigroup T of transformation does not act faithfully on a set Z if $t_1 \neq t_2$ but $t_1(z) = t_2(z)$ for all $z \in Z$.

The *direct product* of transformation semigroups is $(Y, T) \times (X, S) = (Y \times X, T \times S)$ with $(y, x) \cdot (t, s) = (y \cdot t, x \cdot s)$ for all $x \in X, y \in Y, s \in S, t \in T$.

Remark. *The direct product of transformation semigroups embeds in their wreath product.*

Proof. Let $\psi_2(y,x) = (y,x)$ and $\psi_1(t,s) = (c_t, s)$, where $c_t : X \to T$ is the constant function taking value t. Clearly these are bijective onto their images. Since for all $(y,x) \in Y \times X$, $t, t' \in T$, $s, s' \in S$, we have $(y,x)\cdot(c_t,s)(c_{t'},s') = (y \cdot c_t(x), x\cdot s)\cdot(c_{t'},s') = (y\cdot c_t(x)c_{t'}(x\cdot s), x\cdot ss') = (y\cdot tt', x\cdot ss') = (y,x)\cdot(c_{tt'}, ss')$, it follows that $\psi_1(t,s)\psi_1(t',s') = \psi_1(tt',ss')$, i.e., ψ_1 is a homomorphism. Finally $(y,x)\cdot(c_t,s) = (y\cdot c_t(x), x\cdot s) = (y\cdot t, x\cdot s)$, so $\psi_2(y,x)\cdot\psi_1(t,s) = \psi_2((y,x)\cdot(t,s))$, establishing the embedding $(Y,T)\times(X,S) \hookrightarrow (Y,T)\wr(X,S)$. □

We say that a transformation semigroup (X,S) *divides* (X',S') if for some subset $Y \subseteq X'$, subsemigroup T of S', with $Y\cdot T \subseteq Y$, there exist an onto function $\psi_2 : Y \twoheadrightarrow X$ and a surjective semigroup homomorphism $\psi_1 : T \twoheadrightarrow S$ satisfying $\psi_2(y\cdot t) = \psi_2(y)\cdot\psi_1(t)$ for all $y \in Y, t \in T$. We write $(X,S) < (X',S')$. Members of $\psi_2^{-1}(x)$ are called *lifts* of the state x $(x \in X)$, and members of $\psi_1^{-1}(s)$ are called *lifts* of transformation s $(s \in S)$. Division is more general than embedding: given an embedding, to construct a corresponding division one simply takes the (unique) lifts of states x and transformations s to be their respective images under the embedding.

A caveat: distinct transformations on a set X' might restrict to the *same* transformation of $Y \subseteq X'$. For example, if $Y = \{3\} \subsetneq \{1,2,3\} = X'$ and $\sigma \in T = \{\sigma, \sigma^2\}$ with σ interchanging 1 and 2 but leaving 3 fixed, then $\sigma \neq \sigma^2$ but σ and σ^2 act as the identity transformation on Y. Thus, in the definition of division, although each element $t \in T$ gives a well-defined action on Y, we cannot conclude that the pair (Y,T) is a transformation semigroup. However, considering the *restriction* $T|_Y$ *of* T *to* Y, *whose elements are the restrictions* of $t : X' \to X'$ to transformations of Y, we do have a transformation semigroup $(Y, T|_Y)$, and moreover, the $T|_Y$ is a homomorphic image of T under the congruence $t \equiv t'$ if and only if $y\cdot t = y\cdot t'$ for all $y \in Y$.

These concept are also defined for semigroups S and S' : $S < S'$ if S is a homomorphic image of a subsemigroup of S' as we have already defined. $S \hookrightarrow S'$ if S is isomorphic to a subsemigroup of S'.

Proposition 1.9. *Let (X,S) and (X',S') be transformation semigroups. (X,S) divides (X',S') if and only if there exist a nonempty set $Y \subseteq X'$ and functions $h : Y \to X$, $\varphi : S \to S'$ such that h is surjective, $y\cdot\varphi(s) \in Y$, and $h(y\cdot\varphi(s)) = h(y)\cdot s$ for all $y \in Y$ and $s \in S$. In addition, (X,S) embeds (X',S') if and only if we have the above properties such that h is a bijection and for every $s_1', s_2' \in \varphi(S)$, $s_1'|_Y = s_2'|_Y$ implies $s_1' = s_2'$.*[5]

Proof. First we assume $(X,S) < (X',S')$. Then there are a subset $Y \subseteq X'$, a subsemigroup T of S', with $Y\cdot T \subseteq Y$, an onto function $\psi_2 : Y \twoheadrightarrow X$, and a surjective semigroup homomorphism $\psi_1 : T \twoheadrightarrow S$ satisfying $\psi_2(y\cdot t) = \psi_2(y)\cdot\psi_1(t)$ for all $y \in Y, t \in T$. Let T' be a subset of the semigroup T such that for every $s \in S$ there exists exactly one $t \in T'$ with $\psi_1(t) = s$. Then we can construct a bijective mapping $\psi : S \to T'$ such that for every $s \in S$, $\psi(s) \in \psi_1^{-1}(s) \cap T'$. Put $h = \psi_2$ and define $\varphi : S \to S'$ by $\varphi(s) = \psi(s)$, $s \in S$. Then, of course, there exists a nonempty set $Y \subseteq X'$ and functions $h : Y \to X$, $\varphi : S \to S'$ such that h is surjective, and, moreover, $y\cdot\varphi(s) \in Y$ and $h(y\cdot\varphi(s)) = h(y)\cdot s$ for all $y \in Y$ and $s \in S$. In addition, if (X,S) embeds (X',S'),

[5] We need not assume the injectivity of φ because this fact will be a consequence of our conditions.

then ψ_2 should be bijective. However, we assumed $h = \psi_2$. On the other hand, ψ_1 is a semigroup isomorphism. Consequently, for every $s_1', s_2' \in \varphi(S)$, we have $s_1'|_Y = s_2'|_Y$ if and only if $\psi_2(y) \cdot \psi_1(s_1') = \psi_2(y) \cdot \psi_1(s_2')$, $y \in Y$. But then $\psi_1(s_1') = \psi_1(s_2')$, which leads to $s_1' = s_2'$. Therefore, $s_1'|_Y = s_2'|_Y$ implies $s_1' = s_2'$ for every $s_1', s_2' \in \varphi(S)$, as we stated.

Conversely, assume that (X, S) and (X', S') are transformation semigroups such that there exists a nonempty set $Y \subseteq X'$ and functions $h : Y \to X$, $\varphi : S \to S'$ such that h is surjective, and, moreover, $y \cdot \varphi(s) \in Y$ and $h(y \cdot \varphi(s)) = h(y) \cdot s$ for all $y \in Y$ and $s \in S$. First we show that φ is injective. Assume that, contrary of our assumptions, there are distinct $s_1, s_2 \in S$ such that $\varphi(s_1) = \varphi(s_2)$. Then $x \cdot s_1 \neq x \cdot s_2$ for some $x \in X$. Let $x' \in Y$ be arbitrary having $h(x') = x$. Then $h(x') \cdot s_1 \neq h(x') \cdot s_2$, which implies $h(x' \cdot \varphi(s_1)) \neq h(x' \cdot \varphi(s_2))$. But then $x' \cdot \varphi(s_1) \neq x' \cdot \varphi(s_2)$, i.e., $\varphi(s_1) \neq \varphi(s_2)$, which is a contradiction. Now we consider the subsemigroup T of S' generated by $\varphi(S)$. First we observe that $y \cdot t \in Y$ holds for every $y \in Y$ and $t \in T$ because T preserves this property of $\varphi(S)$. Indeed, if $y \cdot t_i \in Y$ holds for every $y \in Y$, $t_i \in \varphi(S)$, $i = 1, \ldots, n$, then $y \cdot t_1 \ldots t_n \in Y$ for every $y \in Y$ and $t_1, \ldots, t_n \in \varphi(S)$.

Let $y \in Y$, $t_1, t_2 \in T$ be arbitrary and put $y_1 = y \cdot t_1$, $y_2 = y_1 \cdot t_2$. Given a pair $s_1, s_2 \in S$, let $\varphi(s_1) = t_1$, $\varphi(s_2) = t_2$. Then $h(y \cdot t_1) = h(y) \cdot s_1$. Thus $h(y_1) = h(y) \cdot s_1$. Similarly, $h(y_1 \cdot t_2) = h(y_1) \cdot s_2$ and thus $h(y_2) = h(y_1) \cdot s_2$. Hence we get $h(y_2) = h(y) \cdot s_1 s_2$. Therefore, $h(y) \cdot s_1 s_2 = h(y \cdot t_1 t_2)$ holds for every $y \in Y$. On the basis of this observation, we can prove by an induction that $h(y \cdot \varphi(t_1) \cdots \varphi(t_n)) = h(y) \cdot t_1 \cdots t_n$, $y \in Y$, $t_1, \ldots, t_n \in T$. This means that $y \cdot \varphi(s_1) \ldots \varphi(s_u) = y \cdot \varphi(s_1') \ldots \varphi(s_v')$ implies $h(y) \cdot s_1 \ldots s_u = h(y) \cdot s_1' \ldots s_v'$ for every $y \in Y$, $s_1, \ldots, s_u, s_1', \ldots, s_v' \in S$. But then $\varphi(s_1) \ldots \varphi(s_u) = \varphi(s_1') \ldots \varphi(s_v')$ implies $s_1 \ldots s_u = s_1' \ldots s_v'$ for every $s_1, \ldots, s_u, s_1', \ldots, s_v' \in S$. Therefore, the surjective mapping $\psi_1 : T \to S$ is well defined having $\psi_1(t) = s_1 \cdots s_n$ whenever $\varphi(s_1) \cdots \varphi(s_n) = t$, $t \in T$, $s_1, \ldots, s_n \in S$. On the other hand, $\psi_1(t_1 t_2) = \psi_1(t_1)\psi_1(t_2)$, $t_1, t_2 \in T$ obviously holds. Thus ψ_1 is a homomorphism of T onto S. Put $\psi_2 = h$. Then we have that there are a subset $Y \subseteq X'$, a subsemigroup T of S', with $Y \cdot T \subseteq Y$, an onto function $\psi_2 : Y \twoheadrightarrow X$, and a surjective semigroup homomorphism $\psi_1 : T \twoheadrightarrow S$ satisfying $\psi_2(y \cdot t) = \psi_2(y) \cdot \psi_1(t)$ for all $y \in Y$, $t \in T$. Therefore, $(X, S) < (X', S')$.

Now we suppose that h is bijective and $s_1'|_Y = s_2'|_Y$ implies $s_1' = s_2'$ for every $s_1', s_2' \in \varphi(S)$. We now show that $\varphi(s_1)\varphi(s_2) = \varphi(s_1 s_2)$, $s_1, s_2 \in S$. Let $y_1 = y \cdot \varphi(s_1)$ and $y_2 = y_1 \cdot \varphi(s_2)$ for some $y \in Y$.

Then $h(y \cdot \varphi(s_1)) = h(y) \cdot s_1$. Thus $h(y_1) = h(y) \cdot s_1$. Similarly, $h(y_1 \cdot \varphi(s_2)) = h(y_1) \cdot s_2$ and thus $h(y_2) = h(y_1) \cdot s_2$. Hence we get $h(y_2) = h(y) \cdot s_1 s_2$. Therefore, $h(y) \cdot s_1 s_2 = h(y \cdot \varphi(s_1)\varphi(s_2))$ holds for every $y \in Y$. On the other hand, by our assumptions, $h(y) \cdot s_1 s_2 = h(y \cdot \varphi(s_1 s_2))$, $y \in Y$. Thus we obtain $h(y \cdot \varphi(s_1)\varphi(s_2)) = h(y \cdot \varphi(s_1 s_2))$. Hence we obtain $\varphi(s_1)\varphi(s_2)|_Y = \varphi(s_1 s_2)|_Y$ by the bijectivity of h. But then we assumed $\varphi(s_1)\varphi(s_2) = \varphi(s_1 s_2)$. Thus φ is a semigroup isomorphism of S onto $\varphi(S)$. Put $T = \varphi(S)$, $\psi_1 = \varphi^{-1}$, $\psi_2 = h$. Then we obtain that $Y \subseteq X'$, $T \subseteq S'$ is a subsemigroup of S', $\psi_2 : X \to Y$ is a bijective mapping, and $\psi_1 : S \to T$ is an isomorphism of semigroups, such that $\psi_2(x \cdot s) = \psi_2(x) \cdot \psi_1(s)$ for all $x \in X$, $s \in S$. The proof is complete. □

In practice, the following technique is useful when showing that (X, S) divides (X', S').

Proposition 1.10. *Let (X, S) and (X', S') be transformation semigroups. To show that (X, S) divides (X', S'), it suffices to choose one or more $\tilde{x} \in X'$ as lifts for each $x \in X$ and* *one or more $\tilde{s} \in S'$ as lifts for each $s \in S$, such that the following hold:*

(1) *Each member of X' (resp., S') is a lift of at most one element of X (resp., S).*

(2) *If $\tilde{x}$ is any lift of x and $\tilde{s}$ is any lift of s, then $\tilde{x} \cdot \tilde{s}$ is some lift of $x \cdot s$.*

Proof. Let Y be the set of lifts of elements of X and T be the subsemigroup of S' generated by lifts of members of S. Notice that (2) implies that $Y \cdot T \subseteq Y$. Let $\psi_2(\tilde{x}) = x$ and $\psi_1(\tilde{s}) = s$ for each $\tilde{x} \in X$ and each generator $\tilde{s}$ of T. These are well defined by (1). Now define $\psi_1(\tilde{s_n} \cdots \tilde{s_1}) = \psi_1(\tilde{s_n}) \cdots \psi_1(\tilde{s_1}) = s_n \cdots s_1$. Then $\psi_2 : Y \twoheadrightarrow X$, and $\psi_1 : T \twoheadrightarrow S$ is a homomorphism, provided that ψ_1 is well defined on all of T. To see that this extended ψ_1 is well defined, suppose $\tilde{s_n} \cdot \cdots \cdot \tilde{s_1} = \tilde{r_m} \cdots \tilde{r_1}$ for some $s_i, r_j \in S$ $(1 \leq i \leq n,\ 1 \leq j \leq m,\ m, n > 0)$. Then both these products determine the same transformation of X'. Now if $s_n \cdots s_1 \neq r_m \cdot \cdots \cdot r_1$, then since (X, S) is a transformation semigroup, there exists an $x \in X$ where these two transformations differ, i.e., $x \cdot s_n \cdots s_1 \neq x \cdot r_m \cdot \cdots \cdot r_1$. Consider any lift $\tilde{x}$ of x. By (2) applied inductively, $\tilde{x} \cdot \tilde{s_n} \cdots \tilde{s_1}$ is a lift of $x \cdot s_n \cdots s_1$, and $\tilde{x} \cdot \tilde{r_m} \cdots \tilde{r_1}$ is a lift of $x \cdot r_m \cdots r_1$. Hence, since no element of X' is a lift two distinct elements of X it follows that $\tilde{x} \cdot \tilde{s_n} \cdots \tilde{s_1} \neq \tilde{x} \cdot \tilde{r_m} \cdots \tilde{r_1}$, a contradiction. Therefore, $s_n \cdots s_1 = r_m \cdot \cdots \cdot r_1$, showing that ψ_1 is well-defined. Clearly, $\psi_2(\tilde{x} \cdot t) = \psi_2(\tilde{x}) \cdot \psi_1(t)$ for all $\tilde{x} \in Y$ and $t \in T$, establishing the division. □

Several times in this monograph we make use of the following.

Proposition 1.11. *Let G be a group and let (A, S) be a transformation semigroup such that G divides S, with S finite. Then G is a homomorphic image of some group $\widetilde{G}$ contained in S. Moreover, $\widetilde{G}$ acts by permutations on some $Z \subseteq A$ so that $(Z, \widetilde{G})$ is a permutation group.*

Proof. By definition of division $G \overset{\psi}{\twoheadleftarrow} S' \subseteq S$ for some subsemigroup S' of S and some surjective homomorphism ψ. We take S' to be minimal (under inclusion) among subsemigroups of S having this property. (S' exists since S is finite by hypothesis.) Let $e^2 = e$ be an idempotent in $\psi^{-1}(1)$ for $1 \in G$. (e exists since $\psi^{-1}(1) \subseteq S$ is a finite semigroup.) Then $eS'e \subseteq S'$ also has the property, since $\psi(s) = 1\psi(s)1 = \psi(e)\psi(s)\psi(e) = \psi(ese)$ for all $s \in S'$, so by minimality $S' = eS'e$, which is a monoid with identity e. Take $t \in eS'e$, and let e' be the unique idempotent power of t by Proposition 1.1. Then $e'S'e' \subseteq eS'e$ has the property as well, so $e'S'e' = eS'e$. But e' is an identity for $e'S'e'$ and e is an identity for $eS'e$, whence $e = e'$. Thus for each t, there exists $n > 1$ such that $t^n = e$, so $t^{n-1}t = tt^{n-1} = e$. This shows that $eS'e$ is a group.

Let $Z = A \cdot e = \{a \cdot e : a \in A\}$. Now e is idempotent, so it acts as the identity on Z. We have $Z \cdot eS'e \subseteq Z = Z \cdot e = Z \cdot eee \subseteq Z \cdot eS'e$, whence $Z = Z \cdot eS'e$. Moreover, each $t \in eS'e$ acts on Z as a permutation: taking $z \in Z$, we have $(z \cdot t^{n-1}) \cdot t = z \cdot e = z$, so t maps Z surjectively onto itself; and if $z \cdot t = z' \cdot t$ for $z, z' \in Z$, then $z = z \cdot e = z \cdot t^n = z' \cdot t^n = z'$, and so the action of t is also injective. Let $\widetilde{G} = eS'e$.

Finally, to show that $(Z, \widetilde{G})$ is a permutation group on Z, we must show that $\widetilde{G}$ embeds into the full permutation group on Z. We already know that the elements of $\widetilde{G}$ permute Z. It remains to show that distinct elements of $\widetilde{G}$ give rise to distinct permutations, i.e., for all $g_1, g_2 \in \widetilde{G}$, $g_1 \neq g_2$ implies that there exists a $z \in Z$ such that $z \cdot g_1 \neq z \cdot g_2$. Now suppose

$g_1 \neq g_2$. Since (A, S) is a transformation semigroup with $\widetilde{G} \subseteq S$, there exists $a \in A$ with $a \cdot g_1 \neq a \cdot g_2$. Letting $z = a \cdot e \in Z$, we have for $i = 1, 2$, $z \cdot g_i = (a \cdot e) \cdot g_i = a \cdot (eg_i) = a \cdot g_i$. Hence $z \cdot g_1$ and $z \cdot g_2$ are distinct. This completes the proof. □

We record three important observations that are clear from the proof of Proposition 1.11.

Corollary 1.12. *For every homomorphism φ of a finite semigroup S onto a group G, there exists a subgroup (i.e., a subsemigroup that is a group) $\widetilde{G}$ of S such that $\varphi(\widetilde{G}) = G$.* □

Corollary 1.13. *Let S be a semigroup of transformations of a finite set A and let G be a subgroup of S. Then there exists a subset Z of A such that the restrictions of the elements of G to Z are permutations forming a group isomorphic to G.* □

Corollary 1.14. *Let S be a semigroup of transformations of a finite set A, and assume that there exists a subset Z of A such that some elements of S when restricted to Z are permutations. Then there exists in S a subgroup $\widetilde{G}$ such that the permutation group G generated by these permutations of Z is a homomorphic image of $\widetilde{G}$.*

It is not difficult to verify the following useful fact.

Fact 1.15. *For all finite or infinite transformation semigroups (X, S), (X', S'), (Y, T), and (Y', T'), we have the following:*

(1) *$(Y, T) < (Y', T')$ and $(X, S) < (X', S')$, then $(Y, T) \wr (X, S) < (Y', T') \wr (X', S')$.*
(2) *If (X', S') is a permutation group and T' contains an idempotent, then $(Y, T) \hookrightarrow (Y', T')$ and $(X, S) \hookrightarrow (X', S')$ implies $(Y, T) \wr (X, S) \hookrightarrow (Y', T') \wr (X', S')$.*
(3) *For permutation groups, it always holds that if $(Y, T) \hookrightarrow (Y', T')$ and $(X, S) \hookrightarrow (X', S')$, then $(Y, T) \wr (X, S) \hookrightarrow (Y', T') \wr (X', S')$.*

Proof. Conclusion (1) is easily verified. It is also easy to verify that conclusion (2) holds more generally whenever $(X, S) \hookrightarrow (X', S')$ are any transformation semigroups if T' contains an idempotent $e^2 = e$ (e.g., when T' is finite or a group) and if the lifts $\tilde{X}$ of X and the lifts $\tilde{S}$ of S satisfy $(X' \setminus \tilde{X}) \cdot \tilde{S} \subseteq X' \setminus \tilde{X}$ by lifting a transformation (f, s) of $(Y, T) \wr (X, S)$ to $(\tilde{f}, \tilde{s})$, where $\tilde{s}$ is the unique lift of s in S', and

$$\tilde{f}(x') = \begin{cases} \widetilde{f(x)} & \text{if } x' \in \tilde{X} \text{ and } x' \text{ is a lift of } x, \\ e & \text{if } x' \in X' \setminus \tilde{X}. \end{cases}$$

From this observation, (2) and hence (3) follow. □

Lemma 1.16 (Lagrange coordinates). *If G is any group and N is a normal subgroup of G, then (G, G) embeds in the permutation group $(N, N) \wr (G/N, G/N)$.*

Proof. For each coset Ng choose a coset representative $\overline{g} \in Ng$. Map the states of the wreath product bijectively onto G via $\psi : N \times G/N \twoheadrightarrow G$ with $\psi(n, Ng) = n\overline{g}$. Given $g_1 \in G$ we choose a transformation of the wreath product $\tilde{g_1}$ covering g_1: $(n, Ng) \cdot$

$\tilde{g_1} = (n \cdot \overline{g}g_1\overline{gg_1}^{-1}, Ng \cdot Ng_1) = (n\overline{g}g_1\overline{gg_1}^{-1}, Ngg_1)$. Now since $N\overline{g}g_1 = Ngg_1 = N\overline{gg_1}$, it follows that $\overline{g}g_1\overline{gg_1}^{-1} \in N$. Moreover, this element of N is determined by Ng and g_1. Thus $\tilde{g_1}$ is an element of the wreath product. Now $\psi((n, Ng) \cdot \tilde{g_1}) = \psi(n\overline{g}g_1\overline{gg_1}^{-1}, Ngg_1) = n\overline{g}g_1\overline{gg_1}^{-1}\overline{gg_1} = n\overline{g}g_1 = \psi(n, Ng) \cdot g_1$. Furthermore, we claim that for $g_1, g_2 \in G$, one has $\widetilde{g_1g_2} = \tilde{g_1}\tilde{g_2}$. Indeed, for all $n \in N$ and $Ng \in G/N$, we compute $((n, Ng) \cdot \tilde{g_1}) \cdot \tilde{g_2} = (n\overline{g}g_1\overline{gg_1}^{-1}, Ngg_1) \cdot \tilde{g_2} = (n\overline{g}g_1\overline{gg_1}^{-1}\overline{gg_1}g_2\overline{gg_1g_2}^{-1}, Ngg_1g_2) = (n\overline{g}g_1g_2\overline{gg_1g_2}^{-1}, Ngg_1g_2) = (n, Ng) \cdot \widetilde{g_1g_2}$. It follows that (G, G) is isomorphically embedded inside $(N, N) \wr (G/N, G/N)$. □

Theorem 1.17 (Lagrange coordinate decomposition for groups). *If G is any nontrivial finite or infinite group, and $\{1\} = G_0 \lhd G_1 \lhd \cdots \lhd G_{n-1} \lhd G_n = G$ with each G_i a normal subgroup of G_{i+1} for all $0 \le i \le n-1$ (this is called a* subnormal chain*), then (G, G) embeds in the wreath product of permutation groups*

$$(G_1/G_0, G_1/G_0) \wr \cdots \wr (G_n/G_{n-1}, G_n/G_{n-1}).$$

Proof. The embedding is derived by induction on n, the length of the subnormal chain. If $n = 1$, then $G_1 = G$ and the embedding is just the identity on (G, G). Suppose that $n > 1$. Then by induction hypothesis (G_{n-1}, G_{n-1}) embeds in

$$(G_1/G_0, G_1/G_0) \wr \cdots \wr (G_{n-1}/G_{n-2}, G_{n-1}/G_{n-2}).$$

By applying Lemma 1.16 to $G_{n-1} \lhd G_n$,

$$(G, G) = (G_n, G_n) \hookrightarrow (G_{n-1}, G_{n-1}) \wr (G_n/G_{n-1}, G_n/G_{n-1}).$$

It then follows from Fact 1.15(3) that $(G_{n-1}, G_{n-1}) \wr (G_n/G_{n-1}, G_n/G_{n-1})$ embeds in

$$(G_1/G_0, G_1/G_0) \wr \cdots \wr (G_{n-1}/G_{n-2}, G_{n-1}/G_{n-2}) \wr (G_n/G_{n-1}, G_n/G_{n-1}),$$

and hence so does (G, G). □

We recall the following well-known theorem from finite group theory.

Theorem 1.18 (Jordan–Hölder theorem for finite groups). *Let G be a finite group, and let $\{1\} = G_0 \lhd G_1 \lhd \cdots \lhd G_{n-1} \lhd G_n = G$ be a composition series for G; i.e., each G_i is a maximal proper normal subgroup of G_{i+1} for all $0 \le i \le n-1$. Then if $\{1\} = H_0 \lhd H_1 \lhd \cdots \lhd H_{m-1} \lhd H_m = G$ is another composition series for G, then $n = m$ and there is a permutation π of the set $\{0, \ldots, n\}$ such that each G_{i+1}/G_i is isomorphic to $H_{\pi(i)+1}/H_{\pi(i)}$. Moreover, any proper subnormal chain of G can be completed to a composition series. That is, if $\{1\} = G'_0 \lhd \cdots \lhd G'_k = G$, $(G_i \neq G_{i+1}$ for $0 \le i < k)$, then this chain has a refinement which is a composition series.* □

For a composition series of a nontrivial finite group G, each G_{i+1}/G_i is a simple group since G_i is a maximal normal subgroup of G_{i+1} and is called a *Jordan–Hölder factor* of G. The theorem entails that the Jordan–Hölder factors G (with multiplicities) are uniquely determined up to isomorphism. Obviously each Jordan–Hölder factor G_{i+1}/G_i of G divides G, since it is a homomorphic image of the subgroup G_{i+1}. However, G may have other simple group divisors that are not Jordan–Hölder factors. For example, this is true if G is a simple nonabelian group (as is evident from considering the simple divisors of its cyclic subgroups).

We now have as a further corollary the following theorem.

Theorem 1.19 (Jordan–Hölder coordinate theorem for finite groups). *If G is a finite nontrivial group, and $\{1\} = G_0 \lhd G_1 \lhd \cdots \lhd G_{n-1} \lhd G_n = G$ with each G_i a maximal proper normal subgroup of G_{i+1} for all $0 \leq i \leq n-1$, then (G, G) embeds in the wreath product of permutation groups*

$$(G_1/G_0, G_1/G_0) \wr \cdots \wr (G_n/G_{n-1}, G_n/G_{n-1}),$$

where each G_{i+1}/G_i $(0 \leq i \leq n-1)$ is a simple group. Moreover, the components of this wreath product and their multiplicities (but not necessarily their sequence) are the same for all such decompositions.

Proof. The theorem is an immediate consequence of the Lagrange coordinate decomposition theorem and the Jordan–Hölder theorem for finite groups. □

1.4 Bibliographical Remarks

Section 1.1. Several books present various aspects of the basic notation and notions used in this book. Most are folklore going back to G. F. Frobenius at the turn of the nineteenth century. (See Frobenius's collected works, published [1968].)

Section 1.2. Theorem 1.3 is from J. Dénes and P. Hermann [1982]. Some other aspects of this result are given by Dénes [1986]. Except for this result and the new proof of Theorem 1.2, the other parts of this section should also be regarded as folklore.

Section 1.3. An important contribution to the theory of transformation semigroups by K. B. Krohn and J. L. Rhodes appeared in a book edited by M. A. Arbib [1968]. It is also presented in an elegant form by S. Eilenberg [1976]. A nice book on the generating systems of the finite symmetric groups is one by S. Picard [1946]. Theorem 1.4 is due to E. Galois [1832]. Proposition 1.5 is folklore. The formulation of position-contents duality (Fact 1.6 and Corollary 1.7) and the observation on its relation to time reversal in the following remark, although elementary, appear to be new. The main idea of the proof of Lemma 1.8 is well known in the literature; one can find various statements of the same flavor, although we have not seen a formulation suitable for a direct reference. For the reader's convenience we have included a short elementary proof. By some well-known ideas in Wielandt [1964], another proof can be created easily. This elementary proof is due to D. S. Ananichev; see Ananichev, Dömösi, and Nehaniv [2004]. Corollaries 1.12, 1.13, and 1.14 are derived as consequences of Proposition 1.11. They were formulated also by A. Ginzburg [1968]. Lemma 1.16 and Theorem 1.17 are folklore going back to Lagrange and Frobenius (and have been variously formulated by, e.g., L. Kaloujnine and M. Krasner [1950, 1951a, 1951b], H. Neumann [1967], K. B. Krohn and J. L. Rhodes [1962, 1965], S. Eilenberg [1976], and C. L. Nehaniv [1992, 1995, 1996]). Theorem 1.19 follows from the latter immediately using Theorem 1.18 due to C. Jordan [1869] and O. Hölder [1889]. The other statements are new but elementary.

Chapter 2

Directed Graphs, Automata, and Automata Networks

In this chapter we introduce the concepts of directed graphs (digraphs), automata, networks constructed from them, and related algebraic structures used for studying automata that communicate according to the links of an interconnection digraph. Techniques and concepts developed here will be used throughout the monograph. Various restrictions on the kind of digraph of interconnections lead important classes of automata networks, whose computational power (completeness of various types) and stability we will consider. Results proved here suggest why many networks are so stable even when a lot of links are omitted. We prove that a digraph (i.e., a network) with all loop edges and $m > n$ *vertices remains* n*-complete if it is strongly connected and has a branch. Therefore, even with a number of links omitted, the network is able to preserve its completeness. Such properties underline and yield insight into the well-known experimental results that real-world networks (for example, the Internet, neural networks, and genetic regulatory networks) can remain very stable even if many of their links are removed.*

Elementary relationships between automata and associated algebraic structures are introduced, as are important types of automata networks and the various notions of computational power of classes of automata under particular ways of constructing networks (i.e., products of automata over interconnection digraphs). Another important part of this chapter is the fact that certain semigroups of automata mappings have no basis, i.e., minimal generating system. By these negative results we know it is hopeless to seek such bases. In the last part of the chapter we show some simple but important properties of automata products which are also considered automata networks. These include presentations of the well-known classical decomposition theorems of Gluškov and Letichevsky that characterize minimal computational elements that are nevertheless powerful enough for different kinds of computational completeness.

2.1 Digraph Completeness

A (finite) *directed graph* (or *digraph*) $\mathcal{D} = (V, E)$ (of order $n > 0$) is a pair consisting of sets of *vertices* $V = \{v_1, \ldots, v_n\}$ and *edges* $E \subseteq V \times V$. Elements of V are sometimes called *nodes*. Moreover, if $(v, v') \in E$, then it is said that (v, v') is an *outgoing edge* of v, and, simultaneously, (v, v') is an *incoming edge* for v'. (In this way, a loop edge (v, v) has both of these properties concerning the vertex v.) An edge $(v, v') \in E$ is said to have *source* v and *target* v'. If $|V| = n$, then we also say that $\mathcal{D}$ is a digraph of order n. In addition, the *digraph associated to* $\mathcal{D}$ *with all loop edges* is the digraph $\mathcal{D}^{\ell} = (V, E \cup \{(v, v) \mid v \in V\})$ for every digraph $\mathcal{D} = (V, E)$.

A *walk* in $\mathcal{D} = (V, E)$ is a sequence of vertices $v_1, \ldots, v_n, n > 1$, such that $(v_i, v_{i+1}) \in E, i = 1, \ldots, n-1$. A walk is *closed* if $v_1 = v_n$. By a *(directed) path* from a vertex a to a vertex $b \neq a$ we shall mean a sequence $v_1 \ldots v_n, n > 1$, of pairwise distinct vertices such that $a = v_1, b = v_n$, and $(v_i, v_{i+1}) \in E$ for every $i = 1, \ldots, n-1$. The positive integer $n - 1$ is called the *length* of the path. Thus a path is a walk with all n vertices distinct. A closed walk with all vertices distinct except $v_1 = v_n$ is a *cycle* of length $n - 1$. If $n \geq 3$, then sometimes we speak about a *real cycle.* (Therefore, closed walks with just two distinct vertices and, moreover, loop edges are not considered real cycles.) Two cycles of a graph are called *disjoint* if they have no vertex in common. In the opposite case we say the cycles *intersect.* If $(v, v') \in E$ and $v = v'$, then (v, v') is called a *(self-)loop edge*. A *branch* in a digraph is a pair of nonloop edges $(v, v'), (v, v'')$ with $v' \neq v''$ (and $v \notin \{v', v''\}$). The *union* $\mathcal{D} \cup \mathcal{D}'$ of two digraphs $\mathcal{D} = (V, E)$ and $\mathcal{D}' = (V', E')$ is defined as a digraph $\mathcal{D}'' = (V \cup V', E \cup E')$. The digraph $\mathcal{D}' = (V', E')$ is a *subdigraph* of $\mathcal{D}$ if V' is a nonvoid subset of V, and $E' \subseteq E$. $\mathcal{D}$ *has a homomorphism onto* $\mathcal{D}' = (V', E')$ if $E' = \{(\psi(v), \psi(v')) \mid (v, v') \in E\}$ for some surjective $\psi : V \to V'$. If ψ is bijective, then we speak about an *isomorphism.* $\mathcal{D}$ is *connected for* $v \in V$ if for every vertex $v' \in V$ either $v = v'$ or there is a (directed) path from v to v'. $\mathcal{D}$ is called *strongly connected* if it is connected for all of its vertices. Moreover, $\mathcal{D}$ is *centralized* if there exists a $v \in V$ with $(V \setminus \{v\}) \times \{v\} \subseteq E$.

An *undirected graph* $\mathcal{G} = (V, E)$ is a set of vertices V ($|V| > 0$) and edges $E \subseteq \{\{v, v'\} \mid v, v' \in V\}$. An undirected graph is called, in short, a *graph.* For any directed graph $\mathcal{D} = (V, E)$, we consider the *associated undirected graph* $U_{\mathcal{D}} = (V, E')$ with $E' = \{\{v, v'\} \mid (v, v') \in E\}$. Then for every $\{v, v'\} \in E'$, the vertices v and v' are called the *endpoints* of the edge $\{v, v'\}$.

Like before, we define a *walk* in (an undirected) graph (V, E') to be a sequence of vertices $v_1, \ldots, v_n$, such that $\{v_i, v_{i+1}\} \in E', i = 1, \ldots, n-1$. A *path* is a walk with all n vertices distinct. A walk is *closed* if $v_1 = v_n$. A *cycle* in a graph is also a closed walk such that its $n - 1$ points are distinct. If $n \geq 3$, then sometimes we speak about (an undirected) *real cycle.* The concepts of *union, subgraph, homomorphism onto,* and *isomorphism* for graphs are also analogously defined. One may define *distance* $d(v, v')$ between two vertices v and v' in an undirected graph to be the least of all lengths among all possible paths from v to v', unless $v = v'$, in which case $d(v, v') = 0$; otherwise, if no such path exists, one sets $d(v, v') = \infty$.

We say that a graph $\mathcal{G} = (V, E)$ has the *ordered cycle property* if its nodes can be labeled with distinct positive integers such that if we identify each vertex with its label, then every cycle of length $k \geq 3$ can be arranged in the form $c_1 < \cdots < c_k$, where each $\{c_i, c_{i+1 (\mathrm{mod}\, k)}\}$ lies in E and is an edge of the cycle ($c_i \in V, 1 \leq i \leq k$).

A graph $\mathcal{G}$ is called *planar* if it can be represented on a plane by distinct points for vertices and simple curves for edges connecting the corresponding points in such a way that any two such curves do not meet anywhere other than possibly at their endpoints. In this case it is said that $\mathcal{G}$ has a planar embedding. If we modify $\mathcal{G}$ by replacing some edge

$\{v, v'\}$ of $\mathcal{G}$ by the edges $\{v, v_2\}$ and $\{v_2, v'\}$, where v_2 is a newly introduced vertex, we have subdivided an edge of $\mathcal{G}$. A graph $\mathcal{G}'$ is called a *subdivision* of the graph $\mathcal{G}$ if it can be obtained from $\mathcal{G}$ by a finite number of such operations. (Of course every graph is a subdivision of itself.) We say that a graph $\mathcal{G}$ *contains a subdivision* of a graph $\mathcal{G}'$ if $\mathcal{G}$ has a subgraph that is isomorphic to a subdivision of $\mathcal{G}'$.

Let n, k, ℓ be positive integers with $k \leq \ell$. Put $\mathcal{K}_n = (\{1, \ldots, n\}, E_n)$, $E_n = \{\{i, j\} \mid 1 \leq i < j \leq n\}$, $\mathcal{K}_{k,\ell} = (\{1, \ldots, k+\ell\}, E_{k,\ell})$, $E_{k,\ell} = \{\{i, j\} \mid 1 \leq i \leq k, k+1 \leq j \leq k+\ell\}$. Thus $\mathcal{K}_n$ is the *complete graph with n vertices* and $\mathcal{K}_{k,\ell}$ is the *complete bipartite graph with k and ℓ vertices* (in that order).

Theorem 2.1 (Kuratowski planar graph theorem). *A graph $\mathcal{G}$ is planar if and only if it contains no subdivision of $\mathcal{K}_5$, the complete graph with five vertices, nor of the complete bipartite graph $\mathcal{K}_{3,3}$.* □

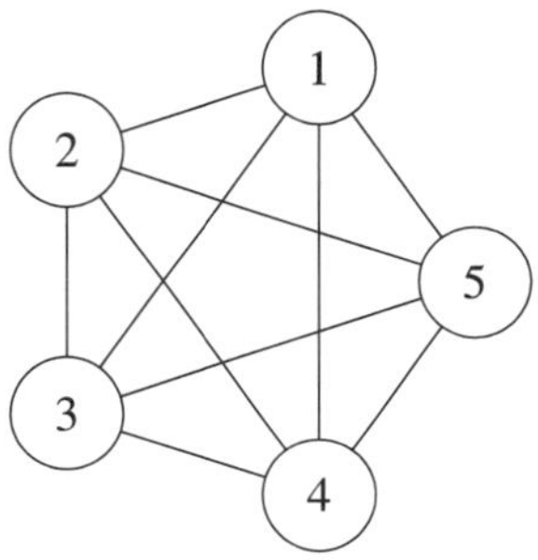

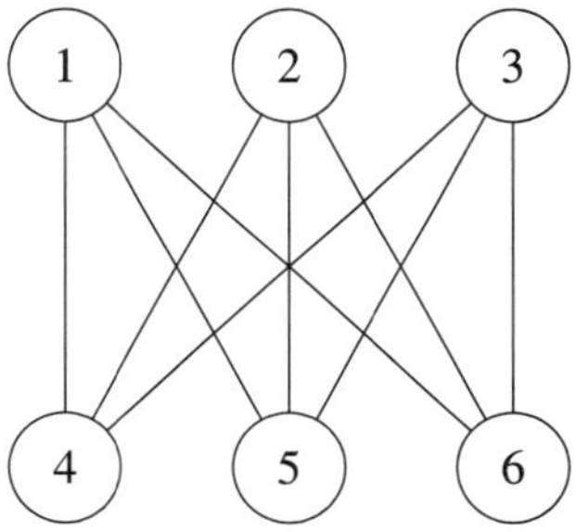

THE GRAPH $\mathcal{K}_5$ AND THE GRAPH $\mathcal{K}_{3,3}$

A realization of $\mathcal{G}$ on the plane, according to the conditions mentioned, is called a *topological planar graph* and is denoted $R(\mathcal{G})$. The connected portions (in the topology of the considered plane P) of $P \setminus R(\mathcal{G})$ are called *faces.* A face thus contains no vertex or point of any edge and its *boundary* is the set of edges and vertices in its closure. The (unique) unbounded component of $P \setminus R(\mathcal{G})$ is called the *outer face* or the *unbounded face.*

A graph is called *outerplanar* if it has a planar embedding so that all its vertices lie in the closure of the same face. In this case, this face may be taken to be the unbounded face. Outerplanarity is a strengthening of the notion of planarity, which has an analogous characterization in terms of forbidden subgraphs.

Theorem 2.2 (Chartrand–Harary outerplanarity theorem). *A graph $\mathcal{G}$ is outerplanar if and only if it contains no subdivision of $\mathcal{K}_4$, the complete graph with four vertices, nor of the complete bipartite graph $\mathcal{K}_{2,3}$.* □

A digraph $\mathcal{D}$ is *planar* (resp., *outerplanar*) if its associated graph $U_{\mathcal{D}}$ is. Similarly, a digraph has the *ordered cycle property* if its associated graph does.

Now we will study the situation in which, in a network of size n, an automaton at node i is to read a message of another automaton at node $\tau(i)$ where τ is a transformation over $\{1, \ldots, n\}$ such that τ is compatible with an *intercommunication digraph* $\mathcal{D}$ representing

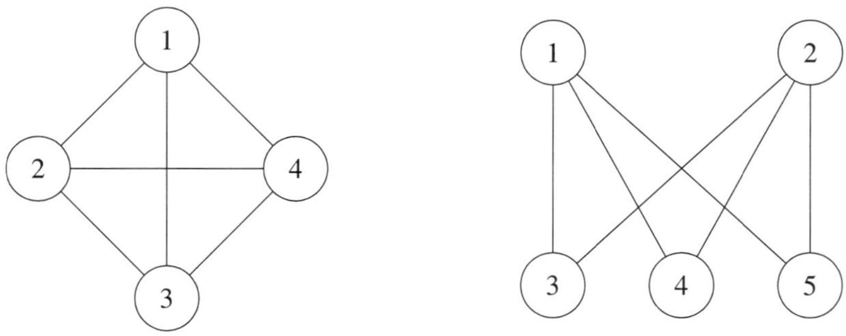

THE GRAPH $\mathcal{K}_4$ AND THE GRAPH $\mathcal{K}_{2,3}$

the communication links. (In this case each automaton can accept exactly one message from another automaton choosing one possibility from among the communication links.) Thus, $(v_{\tau(i)}, v_i) \in E$ for all i in $\{1, \ldots, n\}$. After synchronous update, each node $\tau(i)$ will contain the message transmitted from node i, $i \in \{1, \ldots, n\}$. It may also be important that we can send appropriate messages over the network using some consecutive applications of the various transformations compatible with the intercommunication digraph. Then the problem is to decompose any desired transformation into a product of transformations that are compatible with the communication digraph. In particular, if we restrict our investigations to the case in which the above transformations are permutations of $\{1, \ldots, n\}$, then the messages must be transmitted along the edges of the interconnection digraph $\mathcal{D}$, in such a way that, at any given moment, each of the nodes stores exactly one message; then the problem is to decompose any desired permutation into a product of permutations that are compatible with the communication digraph.[6]

Throughout this section we study digraphs in which every vertex has at least one incoming edge.

A transformation $f : V \to V$ is said to be *compatible* with a digraph $\mathcal{D} = (V, E)$ if $(f(v), v) \in E$ for every $v \in V$. The *semigroup* $S(\mathcal{D})$ *of the digraph* $\mathcal{D}$ is defined to be the semigroup generated by all $\mathcal{D}$-compatible transformations, and $T(\mathcal{D}) = (V, S(\mathcal{D}))$ is the *transformation semigroup* of $\mathcal{D}$. Then the minimal monoid $S^{\lambda}(\mathcal{D})$ containing $S(\mathcal{D})$ as a subsemigroup is the *monoid of* $\mathcal{D}$ and $\mathcal{M}(\mathcal{D}) = (V, S^{\lambda}(\mathcal{D}))$ is its *transformation monoid.* Moreover, the *group* $G(\mathcal{D})$ *of* $\mathcal{D}$ is generated by all $\mathcal{D}$-compatible permutations, and also $T_G(\mathcal{D}) = (V, G(\mathcal{D}))$ is the *permutation group of* $\mathcal{D}$. In addition, let $E(\mathcal{D}) = \{e : V \to V \mid$ there exists $v, v' \in V, (v, v') \in E, v \neq v' : e(v) = e(v') = v, e(w) = w, w \in V \setminus \{v, v'\}\}$. Then $E(\mathcal{D})$ is the set of all $\mathcal{D}^{(\ell)}$*-compatible elementary collapsings.*

Now consider a set C of possible contents that can sit at vertices V. The set C^V of all possible assignments of elements of C to vertices and is called the *configuration space* over $\mathcal{D}$ with contents from C. We write $(c_1, \ldots, c_n)$ for an element of C^V, with $c_i \in C$ $(1 \leq i \leq n = |V|)$, denoting that node v_i is assigned contents c_i (possibly $c_i = c_j$ for $i \neq j$). A *configuration map* is any function $F : C^V \to C^V$. The configuration map F is *induced* by $f : V \to V$ if $F(c_1, \ldots, c_n) = (c_{f(1)}, \ldots, c_{f(n)})$ for all $c_i \in C$ $(1 \leq i \leq n)$. (Note that $F = f^{\mathrm{I}}$ according to interpretation 1 of transformation in Section 1.3.) The configuration

[6]Note that we are interpreting the action of transformations according to what was called interpretation 1 in Section 1.3, i.e., so that the current contents of node $t(i)$ are copied to node i.

map F is said to be *$\mathcal{D}$-compatible* if f is.[7] Define the *configuration semigroup* $S_C^*(\mathcal{D})$ to be the semigroup generated by all $\mathcal{D}$-compatible maps of configurations. Then $(C^V, S_C^*(\mathcal{D}))$ is a transformation semigroup, the *configuration transformation semigroup of digraph $\mathcal{D}$* (for contents C). Similarly, we can define the *configuration permutation group* $(C^V, G_C^*(\mathcal{D}))$ and the *configuration group $G_C(\mathcal{D})$ of $\mathcal{D}$* generated by using configuration maps induced on C^V by compatible permutations of V.

Proposition 2.3. *Let $\mathcal{D} = (V, E)$ be a digraph and C a contents set with at least $|V|$ elements. Then the following hold:*

(1) *The configuration transformation semigroup $(C^V, S_C^*(\mathcal{D}))$ and the transformation semigroup $(V, S(\mathcal{D}))$ of $\mathcal{D}$ have anti-isomorphic semigroups. That is, there is a bijection $\psi : S_C^*(\mathcal{D}) \to S(\mathcal{D})$ with $\psi(F \circ G) = \psi(G) \circ \psi(F)$ for all $F, G \in S_C^*(\mathcal{D})$.*

(2) *If $F(c_1, \dots, c_n) = (c_{f(1)}, \dots, c_{f(n)})$, then*

$$F \in S_C^*(\mathcal{D}) \iff f \in S(\mathcal{D}) \text{ and}$$

$$F \in G_C^*(\mathcal{D}) \iff f \in G(\mathcal{D}).$$

(3) *Moreover, the configuration group $G_C^*(\mathcal{D})$ is isomorphic to the group $G(\mathcal{D})$ of the digraph.*

Proof. The first part of the proof can be seen from the position-contents duality lemma (Corollary 1.7). More explicitly, let F and G be induced by $\mathcal{D}$-compatible maps $f : V \to V$ and $g : V \to V$, respectively. Then

$$\begin{aligned} F \circ G(c_1, \dots, c_n) &= f^I \circ g^I(c_1, \dots, c_n) \\ &= g^I(c_{f(1)}, \dots, c_{f(n)}) \\ &= (c_{f(g(1))}, \dots, c_{f(g(n))}) \quad \text{(see discussion of interpretation I} \\ &\qquad\qquad\qquad\qquad\qquad\qquad \text{in Proposition 1.5)} \\ &= (c_{g\circ f(1)}, \dots, c_{g\circ f(n)}) \\ &= (g \circ f)^I(c_1, \dots, c_n). \end{aligned}$$

For any $H \in S_C^*(\mathcal{D})$, we may write $H = f_1^I \circ \cdots \circ f_m^I$ for some $\mathcal{D}$-compatible $f_i : V \to V$ $(1 \le i \le m)$. Then let $h = f_m \circ \cdots \circ f_1$ and $\psi(H) = h$. Using the assumption that C has at least n elements, there exist $(c_1, \dots, c_n)$ with pairwise distinct entries. Since $H(c_1, \dots, c_n) = (c_{h(1)}, \dots, c_{h(n)})$, it follows that h determines H and moreover that ψ is well defined. Clearly $\psi : S_C^*(\mathcal{D}) \to S(\mathcal{D})$ is also surjective as $\psi(h^I) = h$ for all $h \in S(\mathcal{D})$; hence it is bijective. The second part of the proposition is now clear.

The above calculation shows that $\psi(F \circ G) = g \circ f = \psi(G) \circ \psi(F)$ for any generators F and G of $S_C^*(\mathcal{D})$. Thus $\psi(H \circ H') = \psi(H') \circ \psi(H)$ holds for all $H, H' \in S_C^*(\mathcal{D})$, establishing that ψ is an anti-isomorphism.

Arguing as above for $\mathcal{D}$-compatible permutations, one constructs an anti-isomorphism ψ from $G_C^*(\mathcal{D})$ to $G(\mathcal{D})$. (This is just a restriction of the ψ constructed above.) However,

[7] We shall use a more general concept of compatibility in Section 2.4.

since every group G is anti-isomorphic to itself under the map $g \mapsto g^{-1}$ $(g \in G)$, composing ψ with the anti-isomorphism of $G(\mathcal{D})$ with itself yields an isomorphism of the two groups. □

Notation. *For $C = \{1, \ldots, n\}$, we shall write $F(1, \ldots, n) = (f(1), \ldots, f(n))$ as an abbreviation for $F(c_1, \ldots, c_n) = (c_{f(1)}, \ldots, c_{f(n)})$.*

Take a digraph $\mathcal{D} = (V, E)$ with an ordered set $V = \{v_1, \ldots, v_n\}, n > 1$ of vertices. Place a coin c_i onto v_i for every $i = 1, \ldots, n$ such that $c_i \neq c_j$ whenever $i \neq j$ for some $1 \leq i, j \leq n$. Let us say that a vertex is *free* if either it is covered by a coin c_n or there exists another vertex that is covered by the same type of coin. (The second case is also possible after we perform some moves explained below.)

Suppose that we are allowed to change the coins according to the following conditions:

(1) For every $i, j = 1, \ldots, n$, we can put a coin c_i onto the vertex v_j if we have one of the following properties:

(1a) v_k contains a coin c_i and $(v_k, v_j) \in E$;

(1b) v_j contains a coin c_i. (Then it may remain on the vertex v_j.)

(2) For every $j = 1, \ldots, n-1$, there exists a $k \in \{1, \ldots, n\}$, such that a coin c_j is placed onto v_k after the above procedure.

Of course, we can preserve the above properties if we are allowed to apply consecutively two types of rules, moving the coins according to them in the following manner:

(1) If $v_{i_1}, \ldots, v_{i_m}$, $1 < m \leq n$, form a cycle (i.e., $(v_{i_1}, v_{i_2}), (v_{i_2}, v_{i_3}), \ldots, (v_{i_{m-1}}, v_{i_m})$, $(v_{i_m}, v_{i_1}) \in E$), then we can move the coins such that after this step $v_{i_{j+1 \bmod m}}$ is covered by c_{k_j} whenever v_{i_j} was covered by c_{k_j}, and, in addition, $v_{i_{j+1 \bmod m}}$ is free whenever v_{i_j} was free before $(j = 1, \ldots, m)$. The rest of the coins do not move.

(2) A coin of the vertex v_j is changed for a coin c_i if there is an edge $(v_k, v_j) \in E$ such that v_k is covered by c_i and moreover v_j is free. After this move, we assume that v_j and v_k are each covered by one copy of c_i (and thus v_j and v_k have become free). Again, all other coins remain fixed.

We note, of course, that if there are two or more disjoint cycles of $\mathcal{D}$, then we will get the same result if we apply rule (1) for them simultaneously instead of with consecutive applications. In other words, we can also consider the application of $\mathcal{D}^{(\ell)}$-compatible permutations.

We should also take observations to rule (2).

Suppose that a vertex v_k is covered by a coin c_j and we apply rule (2) consecutively twice such that we change a coin c_j of the vertex v_k for c_ℓ and then immediately after change the coin c_ℓ of v_k for c_i. Of course, we have the same result of these two consecutive steps if we omit the first one and change the coin c_j of v_k for c_i directly.

Assume now that, applying rule (2), we change the coin c_j of the vertex v_k for c_ℓ, and after this, applying one or more consecutive rules of type (1), we move coin c_ℓ to a vertex v_u, and finally we change the coin c_ℓ of v_u for c_i, applying again rule (2). Observe that we have the same result for these consecutive steps if we omit the first application of rule (2) and after one or more consecutive rules of type (1) we change the coin c_j of the vertex v_u for c_i.

The above observations show that it is enough to apply rule (2) for a coin of a vertex v_k in the following cases:

(a) v_k is covered by c_n, and

(b) v_k is covered by c_j, $j \in \{1, \dots, n-1\}$, such that coin c_j did not arise by applying rule (2).

These rules are called allowed steps. We say that $\mathcal{D} = (V, E)$, $V = \{v_1, \dots, v_n\}$, *penultimately realizes* the permutation $p : \{1, \dots, n-1\} \to \{1, \dots, n-1\}$ with respect to $v_1, \dots, v_n$ if we can reach a configuration after one or more allowed steps such that $v_{p(i)}$ is covered by c_i, $i = 1, \dots, n-1$ (such that v_n should become free). In addition, if $\mathcal{D}$ penultimately realizes all permutations of the form $p : \{1, \dots, n-1\} \to \{1, \dots, n-1\}$, then we say that $\mathcal{D}$ is *penultimately permutation complete with respect to v_n*. $\mathcal{D}$ is called *penultimately permutation complete* if it is penultimately permutation complete with respect to every $v_i \in V$. More exactly, $\mathcal{D}$ is penultimately permutation complete if considering an arbitrary permutation $v_{P(1)}, \dots, v_{P(n)}$ of vertices, for every permutation $p : \{1, \dots, n-1\} \to \{1, \dots, n-1\}$ we can attain after one or more allowed steps that $v_{P(i)}$ is covered by $c_{p(i)}$, $i = 1, \dots, n-1$ (such that $v_{P(n)}$ should become free).

Lemma 2.4. *A digraph $\mathcal{D} = (V, E)$ is penultimately permutation complete with respect to vertex $v_0 \in V$ if and only if for each permutation p of the vertices $V \setminus \{v_0\}$, there is a transformation $p' \in S(\mathcal{D})$ with $p'(v) = p(v)$ for all $v \in V \setminus \{v_0\}$.*

Proof. This is an immediate consequence of Proposition 2.3(2). □

We note that we could also follow this interpretation: Let us say that a vertex is free if it is not covered by any coin. (Thus the last vertex is free before the coins are moved.) In this case we would always have exactly one free vertex. Then we should change the rule (2) as follows:

(2′) A coin c_i can be moved to a vertex v_j if there is an edge $(v_k, v_j) \in E$ such that v_k is covered by c_i and moreover v_j is free (i.e., v_j is not covered by any coin). The free (empty) vertex v_j is changed for a coin c_i if there is an edge $(v_k, v_j) \in E$ such that v_k is covered by c_i and moreover v_j is free. Again, all other coins remain fixed.

By the above interpretation this is evident: we can consider rule (2′) a special transposition when the empty space and coin c_i are changed. In each of these cases we have a configuration map $F = f^{\mathrm{I}}$ such that the contents of the position $f(j)$ move to position j. This observation implies that the two interpretations are equivalent from the point of view of penultimate completeness of digraphs.

In the most cases we will follow the first interpretation since it is closest to further discussions.

Now we will characterize the class of penultimately permutation complete digraphs. For simplicity, for every digraph we will identify the vertices with sequential numerical labels during this section. Therefore, we assume that a digraph of order n has the (ordered) set of vertices $V = \{1, \dots, n\}$.

We define the concept of an *allowed transformation* (with respect to $\mathcal{D}$) in the following way. Configuration map $F = f^{\mathrm{I}}$ is *allowed* if $f : V \to V$ is the composition of

$\mathcal{D}^{(\ell)}$-compatible permutations and elementary collapsings, i.e., of mappings from $G(\mathcal{D}^{(\ell)}) \cup E(\mathcal{D}^{(\ell)})$, and, moreover, either f is a permutation or f has rank $n-1$ with n not in the image of f. In this case, we also say f is allowed.

Note that for every allowed mapping $F(1,\dots,n) = (f(1),\dots,f(n))$, if $f(i) = f(j), i, j \in \{1,\dots,n\}$ for some $i \neq j$, then $f(1),\dots,f(i-1), f(i+1),\dots,f(n)$ is a permutation of $1,\dots,n-1$. Of course, the identity $F(1,\dots,n) = (1,\dots,n)$ is allowed, and every $\mathcal{D}^{(\ell)}$-compatible elementary collapsing $F = f^I$ with $f(i) \neq n, i \in \{1,\dots,n\}$ is allowed.[8]

Observe that F_2 is not necessarily allowed if F_1F_2 is allowed.[9] Additionally, F_1F_2 is not necessarily allowed if F_1 and F_2 are allowed.[10] However, we have the next fact.

Fact 2.5. *Let F_2 and F_1 be configuration transformations generated from compatible ones. If F_2 is not allowed, then F_2F_1 also is not.*

Proof. Write $F_i(1,\dots,n) = (f_i(1),\dots,f_i(n))$ for $i = 1, 2$. By the position-contents duality lemma (Corollary 1.7), we may work with f_1 and f_2 but must consider products in the reverse order. Thus, we will show that f_1f_2 is not allowed whenever f_2 is not allowed. It is clear that f_1f_2 is allowed only if $|\{f_i(j) : j = 1, 2, \dots, n\}| \geq n-1, i = 1, 2$. Of course, f_1f_2 is not allowed if f_2 has rank less than $n-1$. Since f_2 is not allowed, it cannot be a permutation, so we may assume f_2 has rank $n-1$. This implies that the rank of f_2f_1 is less than n. Then, since f_1 is not allowed, there are two cases:

Case 1. There are $u, v, w \in \{1,\dots,n\}$ with $u \neq v$ and $f_1(u) = f_1(v) \neq n$, $f_1(w) = n$. Of course, then $w \notin \{u, v\}$. If there exists $w' \in \{1,\dots,n\}$ having $f_2(w') = w$, then $f_1(f_2(w')) = n$ and $|f_2f_1| \leq n-1$ implies that f_2f_1 is not allowed. Now we assume that for every $w' \in \{1,\dots,n\}$, $f_2(w') \neq w$. Thus $|f_2| \leq n-1$, i.e., there are distinct $k, \ell \in \{1,\dots,n\}$ with $f_2(k) = f_2(\ell)$. Suppose that there are $u', v' \in \{1,\dots,n\}$ with $f_2(u') = u$, $f_2(v') = v$ $(u \neq v)$. Therefore, $\{k, \ell\} \neq \{u', v'\}$ with $f_1(f_2(u')) = f_1(f_2(v'))$, $f_1(f_2(k)) = f_1(f_2(\ell))$. Obviously, this implies $|\{f_1(f_2(u')), f_1(f_2(v')), f_1(f_2(k)), f_1(f_2(\ell))\}| \leq |\{k, \ell, u', v'\}| - 2$. But then $|f_2f_1| \leq n-2$, i.e., f_2f_1 is not allowed. Now, let us suppose that there exists no $w' \in \{1,\dots,n\}$ with $f_2(w') = u$ $(\neq w)$. Recall that for every $w' \in \{1,\dots,n\}$, $f_2(w') \neq w$. Hence $|f_2f_1| \leq n-2$, i.e., f_2f_1 is not allowed. Symmetrically, we have the same conclusion assuming that there exists no $w' \in \{1,\dots,n\}$ with $f_2(w') = v$ $(\neq w)$.

Case 2. There are $u, v \in \{1,\dots,n\}$ with $u \neq v$ and $f_1(u) = f_1(v) = n$.

Suppose that f_1 is a permutation. Then there are $u', v' \in \{1,\dots,n\}$, $u' \neq v'$ such that $f_1(u') = u$ and $f_1(v') = v$. Hence $f_2(f_1(u')) = f_2(f_1(v')) = n$ with $u' \neq v'$, which implies that f_1f_2 is not allowed.

Now we assume that $|\{f_1(j) : j = 1,\dots,n\}| = n-1$. Then there exists a $u' \in \{1,\dots,n\}$ with $f_1(u') \in \{u, v\}$, $u \neq v'$. Let, say, $f_1(u') = u$. Therefore, $f_2(f_1(u')) = n$, but $|\{f_1f_2(j) : j = 1,\dots,n\}| < n$. Hence f_1f_2 is not allowed. □

[8] Recall that $F_1F_2(1,\dots,n) = F_2(F_1(1,\dots,n)) = (f_1(f_2(1)),\dots,f_1(f_2(n)))$, where $F_i(1,\dots,n) = (f_i(1),\dots,f_i(n))$, $i = 1, 2$.

[9] $F_1(1,2,3) = (3,1,2)$, $F_2(1,2,3) = (3,2,3)$, $F_1F_2(1,2,3) = (2,1,2)$.

[10] $F_1(1,2,3) = (2,2,1)$, $F_2(1,2,3) = (1,2,2)$, but F_1F_2 (with $F_1F_2(1,2,3) = (2,2,2)$) is not allowed.

Lemma 2.6. *If a strongly connected digraph $\mathcal{D}$ is penultimately permutation complete with respect to a vertex v, it is also penultimately permutation complete with respect to any other vertex.*

Proof. Let $V = \{1, \dots, n\}$ denote the set of vertices of $\mathcal{D}$. Without any restriction, we may assume $v = n$.

Consider an arbitrary permutation $p : \{1, \dots, n-1\} \to \{1, \dots, n-1\}$. By our assumptions, we can reach a configuration after one or more allowed steps such that (i) is covered by $c_{p(i)}$, $i = 1, \dots, n-1$.

In other words, there exists an allowed configuration transformation T (generated by $\mathcal{D}^{(\ell)}$-compatible permutations and elementary collapsings) such that $T(1, \dots, n) = (t(1), \dots, t(n))$ with

$$t(i) = \begin{cases} p(i) & \text{if } i = 1, \dots, n-1, \\ \text{an appropriate element of } \{1, \dots, n\} & \text{if } i = n. \end{cases}$$

Take an arbitrary vertex u with $u \neq n$. To our statement, we prove that $\mathcal{D}$ is penultimately permutation complete with respect to u. In $\mathcal{D}$, there exists a path $uu_1 \cdots u_m n$ from u to n (because of the strong connectivity of $\mathcal{D}$). Let us fulfill the following procedure: Remove the coin of n and then move the coin of u_m to n. After that, in consecutive steps, move the coin of u_{i-1} to u_i, $i = m, m-1, \dots, 2$. Finally, duplicate the coin of u and put one of its copies to u_1 (leaving the other one on u). It is clear that all steps of our procedure are allowed. Formally, these consecutive allowed steps (as a product of $\mathcal{D}^{(\ell)}$-compatible elementary collapsings) result in an allowed transformation $T'(1, \dots, n) = (t'(1), \dots, t'(n))$ with

$$t'(i) = \begin{cases} i & \text{if } i \notin \{u_1, \dots, u_m, n\}, \\ u & \text{if } i = u_1, \\ u_{j-1} & \text{if } i = u_j,\ j \in \{2, \dots, m\}, \\ u_m & \text{if } i = n. \end{cases}$$

Then $\mathcal{D}$ penultimately realizes p with respect to $P(1), \dots, P(n)$, where $P : \{1, \dots, n\} \to \{1, \dots, n\}$ is a permutation having

$$P(i) = \begin{cases} i & \text{if } i \notin \{u, u_1, \dots, u_m, n\}, \\ u & \text{if } i = u_1, \\ u_{j-1} & \text{if } i = u_j,\ j \in \{2, \dots, m\}, \\ u_m & \text{if } i = n, \\ n & \text{if } i = u. \end{cases}$$

In other words, there exists an allowed configuration transformation T'' (generated by $\mathcal{D}^{(\ell)}$-compatible permutations and elementary collapsings) having $T''(1, \dots, n) = (t''(1), \dots, t''(n))$ with

$$t''(i) = \begin{cases} p(i) & \text{if } i \notin \{u, u_1, \dots, u_m, n\}, \\ p(u) & \text{if } i = u_1, \\ p(u_{j-1}) & \text{if } i = u_j,\ j \in \{2, \dots, m\}, \\ p(u_m) & \text{if } i = n, \\ \text{an appropriate element of } \{1, \dots, n\} & \text{if } i = u. \end{cases}$$

It is clear that we may assume $T'' = T'T$ (with $t'' = tt'$ having $t''(i) = t'(t(i)), i = 1, \ldots, n$).

But p was an arbitrary permutation over $\{1, \ldots, n-1\}$. Therefore, the digraph $\mathcal{D}' = (V, \{(P(j), P(k)) : (j, k) \in E\})$ is penultimately permutation complete with respect to n (where $n = P(u)$). Thus $\mathcal{D}$ also has this property with respect to u. □

Using the ideas of the proof of Lemma 2.6, we can derive the following statement.

Proposition 2.7. *$\mathcal{D}$ is penultimately permutation complete if and only if for every permutation $p : \{1, \ldots, n-1\} \to \{1, \ldots, n-1\}$ and vertex $i \in V$ there exists an allowed transformation F such that $F(1, \ldots, n) = (p(1), \ldots, p(i-1), k, p(i), \ldots, p(n-1))$ for a given $k \in \{1, \ldots, n\}$.* □

Lemma 2.8. *Let $\mathcal{D} = (V, E)$ be a digraph with vertices $V = \{1, \ldots, n\}$ and edges $E = \{(1, 2), \ldots, (n-1, n), (n, 1), (m, 1)\}$, $1 < m < n$. Then $\mathcal{D}$ is penultimately permutation complete.*

Proof. Obviously, $n \geq 3$. If $m = 2$, then our statement is a direct consequence of Proposition 1.5. Thus we may assume $n > m > 2$. We prove that $\mathcal{D}$ is penultimately permutation complete with respect to n. We will show that $\mathcal{D}$ penultimately realizes the $(n-1)$-cycle $\gamma_1'(i) = i - 1 \bmod (n-1)$ $(i \in \{1, \ldots, n-1\})$ and the transposition $\gamma_2'(1) = 2$, $\gamma_2'(2) = 1$, $\gamma_2'(i) = i$ for $2 < i \leq n-1$ with respect to $1, \ldots, n$.

Let us assume that every vertex i is covered by a coin c_i, $i = 1, \ldots, n$, and perform the following procedure: Change the coin c_n of the vertex n for a copy of c_{n-1}. (By the effect of this step, both vertices $n-1, n$ are covered by 1-1 copies of c_{n-1}.) Then shift all coins right cyclically such that we change the coin of the vertex i for the coin of the vertex $i - 1 (\bmod n)$, $i = 1, \ldots, n$. Finally, we obtain the configuration $(c_{n-1}, c_1, \ldots, c_{n-1})$. It is clear that all steps of our procedure are allowed and that the generated transformation is also allowed.

Formally, consider the mappings

$$F_{n-1}'(1, \ldots, n) = (1, 2, \ldots, n-1, n-1) \text{ [collapsing } n \text{ to } n-1],$$

$$F(1, \ldots, n) = (n, 1, \ldots, n-1) \text{ [shifting right cyclically]}.$$

Obviously, then $F_1 = FF_{n-1}'$ is allowed and $F_1(1, \ldots, n) = FF_{n-1}'(1, \ldots, n) = F(1, \ldots, n-1, n-1) = (n-1, 1, \ldots, n-2, n-1)$. This assures that $\mathcal{D}$ penultimately realizes γ_1' with respect to $1, \ldots, n$.

Now, let us consider the following procedure. Assume again that every vertex i is covered by a coin c_i, $i = 1, \ldots, n$. First apply a series of allowed steps as before such that we remove c_n and then, in consecutive steps, move coin c_{i-1} to the vertex i, $i = n, n-1, \ldots, 4$. Then duplicate the coin c_2 and put one of its copies to the vertex 3 (leaving the other one on the vertex 2). Then we get the configuration $(c_1, c_2, c_2, c_3, \ldots, c_{n-1})$. Shift the coins right cyclically $n-1$ times reaching $(c_2, c_2, c_3, \ldots, c_{n-1}, c_1)$. Now we can shift the first m coins right cyclically $m-1$ times, which results in $(c_2, \ldots, c_{m-1}, c_2, c_{m+1}, \ldots, c_{n-1}, c_1)$. We exchange the coin c_2 of the first vertex for a copy of c_1 covering the last vertex. Finally, shift again the first m coins right cyclically, obtaining the configuration $(c_2, c_1, c_3, \ldots, c_{n-1}, c_1)$.

Formally, let us give the mappings

$F(1, \ldots, n) = (n, 1, \ldots, n-1)$ [shifting right cyclically],

$F'(1, \ldots, n) = (m, 1, \ldots, m-1, m+1, \ldots, n)$ [shifting first m right cyclically],

$F_i'(1, \ldots, n) = (1, \ldots, i-1, i, i, i+2, \ldots, n), i = 1, \ldots, n-1$ [collapsing $i+1$ to i],

$F_n'(1, \ldots, n) = (n, 2, \ldots, m, \ldots, n-1, n)$ [collapsing 1 to n].

It is easy to check that $F_2 = F'F_n'(F')^{m-1}(F)^{n-1}F_2' \cdots F_{n-1}'$ is allowed and

$$\begin{aligned} F'F_n'(F')^{m-1}(F)^{n-1}F_2' \cdots F_{n-1}'(1, \ldots, n) &= F'F_n'(F')^{m-1}(F)^{n-1}(1, 2, 2, \ldots, n-1) \\ = F'F_n'(F')^{m-1}(2, 2, 3, \ldots, n-1, 1) &= F'F_n'(2, 3, 4, \ldots, m, 2, m+1, \ldots, n-1, 1) \\ = F'(1, 3, 4, \ldots, m, 2, m+1, \ldots, n-1, 1) &= (2, 1, 3, \ldots, n-1, 1). \end{aligned}$$

Therefore, $\mathcal{D}$ penultimately also realizes γ_2' with respect to n. Using Proposistion 1.5, $\mathcal{D}$ is penultimately permutation complete with respect to n. By Lemma 2.6 this ends the proof. □

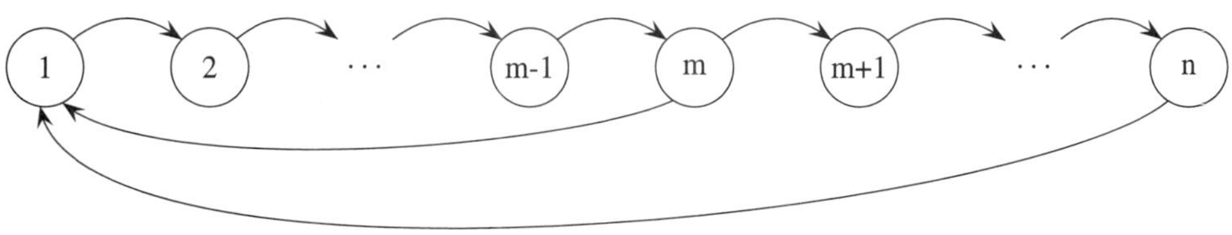

PENULTIMATELY PERMUTATION COMPLETE DIGRAPH (LEMMA 2.8)

Lemma 2.9. *Let $\mathcal{D} = (V, E)$ be a digraph with vertices $V = \{1, \ldots, n\}$ and edges $E = \{(1, 2), \ldots, (m, 1), (u, m+1), (m+1, m+2), \ldots, (n-1, n), (n, 1)\}, 1 \le u \le m, 1 < m < n$. Then $\mathcal{D}$ is penultimately permutation complete.*

Proof. First we prove that $\mathcal{D}$ is penultimately permutation complete with respect to n. Using Proposition 2.7, for this statement we show that $\mathcal{D}$ penultimately realizes the $(n-1)$-cycle γ_1' and the transposition γ_2', defined as above, with respect to n. (Then $1 < m < n$ implies $n > 2$.) If $u = m$, then by Lemma 2.8, we are done. Therefore, we may assume $1 \le u < m < n$.

Let us assume again that every vertex i is covered by a coin c_i, $i = 1, \ldots, n$, and consider the following composition of allowed configuration transformations: In consecutive steps, remove the coin c_i of i and then move a copy of the coin c_{i-1} of $i-1$ to i, $i = n, n-1, \ldots, m+2$. Then we get the configuration $(c_1, \ldots, c_m, c_{m+1}, c_{m+1}, \ldots, c_{n-1})$. Then shift cyclically the first m coins u times. Thus we reach $(c_{m-u+1}, \ldots, c_m, c_1, c_2, \ldots, c_{m-u}, c_{m+1}, c_{m+1}, \ldots, c_{n-1})$. Then u is covered by c_m. Therefore, removing c_{m+1} of $m+1$, we may put a copy of c_m to $m+1$. This results in $(c_{m-u+1}, \ldots, c_m, c_1, c_2, \ldots, c_{m-u}, c_m, c_{m+1}, \ldots, c_{n-1})$. Now we shift right cyclically the first m coins $m-u$ times reaching $(c_1, c_2, \ldots, c_m, c_m, \ldots, c_{n-1})$. The next treatment is that, in consecutive steps, remove the coin c_i of i and then move a copy of the coin c_{i-1} of $i - 1$ to $i, i = m, m - 1, \ldots, 2$. Finally, remove the coin c_1 of 1 and put a copy of the coin c_{n-1} of n to 1. Then we reach the configuration $(c_{n-1}, c_1, \ldots, c_{n-2}, c_{n-1})$.

Formally, define the mappings

$F'(1, \ldots, n) = (m, 1, \ldots, m - 1, m + 1, \ldots, n)$ [shifting first m right cyclically],

$F'_i(1, \ldots, n) = (1, \ldots, i - 1, i, i, i + 2, \ldots, n), i = 1, \ldots, m - 1, m + 1, \ldots, n - 1$ [collapsing $i + 1$ to i],

$F'_n(1, \ldots, n) = (n, 2, \ldots, n)$ [collapsing n to 1],

$F'_{u,m+1}(1, \ldots, n) = (1, \ldots, u, \ldots, m, u, m + 2, \ldots, n), i = m + 1, \ldots, n - 1$ [collapsing u to $m + 1$].

By an elementary computation we may check that

$$F_1 = F'_n F'_1 \cdots F'_{m-1} (F')^{m-u} F'_{m,u} (F')^u F'_{m+1} \cdots F'_{n-1}(1, \ldots, n)$$

is allowed and

$$\begin{aligned}
&F'_n F'_1 \cdots F'_{m-1} (F')^{m-u} F'_{m,u} (F')^u F'_{m+1} \cdots F'_{n-1}(1, \ldots, n) \\
&= F'_n F'_1 \cdots F'_{m-1} (F')^{m-u} F'_{m,u} (F')^u (1, \ldots, m, m + 1, m + 1, \ldots, n - 1) \\
&= F'_n F'_1 \cdots F'_{m-1} (F')^{m-u} F'_{m,u} (m - u + 1, \ldots, m, 1, \ldots, m - u, m + 1, \\
&\qquad m + 1, \ldots, n - 1) \\
&= F'_n F'_1 \cdots F'_{m-1} (F')^{m-u} (m - u + 1, \ldots, m, 1, \ldots, m - u, m, m + 1, \ldots, n - 1) \\
&= F'_n F'_1 \cdots F'_{m-1} (1, \ldots, m, m, \ldots, n - 1) \\
&= F'_n (1, 1, \ldots, m - 1, m, \ldots, n - 1) = (n - 1, 1, \ldots, m - 1, m, \ldots, n - 1).
\end{aligned}$$

This shows that $\mathcal{D}$ penultimately realizes γ'_1 with respect to $1, \ldots, n$.

Assuming again that every vertex i is covered by a coin c_i, $i = 1, \ldots, n$, we distinguish two cases.

Case 1. $m = n - 1$.

Repeat $u - 1$ times the above procedure of our proof resulting in F_1. Then we get the configuration $(c_{n-u+1}, \ldots, c_{n-1}, c_1, \ldots, c_{n-u}, c_{n-u+1})$ such that u is covered by c_1. Then, removing c_{n-u+1} of n, we can cover n by a copy of c_1. Hence, we obtain $(c_{n-u+1}, \ldots, c_{n-1}, c_1, \ldots, c_{n-u}, c_1)$. Now shift cyclically the first $n - 1$ coins $n - u - 1$ times. Thus we reach $(c_2, c_3, \ldots, c_{n-1}, c_1, c_1)$. Remove the coin c_1 of $n - 1$ and cover $n - 1$ by a copy of c_{n-1} of $n - 2$. In consecutive steps, remove the coin c_{i+1} of i and afterwards move a copy of the coin c_i of $i - 1$ to $i, i = n - 2, \ldots, 2$. Hence, we get $(c_2, c_2, c_3, \ldots, c_{n-1}, c_1)$. Then shift cyclically the first $n - 1$ coins $n - 2$ times. Thus we reach $(c_2, c_3, \ldots, c_{n-1}, c_2, c_1)$. Now we remove the coin c_2 of the first vertex and cover it by a copy of c_1 of the last vertex. This results in $(c_1, c_3, \ldots, c_{n-1}, c_2, c_1)$. Finally, cyclically shift the first $n - 1$ coins. This leads to $(c_2, c_1, c_3, \ldots, c_{n-1}, c_1)$.

In formulas, $F_2 = F' F'_n (F')^{n-2} F'_1 \cdots F'_{n-2} (F')^{n-u-1} F'_{u,n} (F_1)^{u-1}$ is allowed and

$$F' F'_n (F')^{n-2} F'_1 \cdots F'_{n-2} (F')^{n-u-1} F'_{u,n} (F_1)^{u-1} (1, \ldots, n)$$

$$= F'F'_n(F')^{n-2}F'_1 \cdots F'_{n-2}(F')^{n-u-1}F'_{u,n}(n-u+1,\ldots,n-1,1,2,\ldots,n-u,$$
$$n-u+1)$$
$$= F'F'_n(F')^{n-2}F'_1 \cdots F'_{n-2}(F')^{n-u-1}(n-u+1,\ldots,n-1,1,2,\ldots,n-u,1)$$
$$= F'F'_n(F')^{n-2}F'_1 \cdots F'_{n-2}(2,3,\ldots,n-1,1,1)$$
$$F'F'_n(F')^{n-2}(2,2,3,\ldots,n-1,1)$$
$$= F'F'_n(2,3,\ldots,n-1,2,1)$$
$$F'(1,3,\ldots,n-1,2,1) = (2,1,3,\ldots,n-1,1).$$

Case 2. $m < n-1$.

Consider the first procedure of our proof resulting in F_1. Repeating this procedure $n-1$ times, we obtain the configuration $(c_2, c_3, \ldots, c_{n-2}, c_{n-1}, c_1, c_2)$. Then, removing c_2 of n, we can cover n by a copy of c_1. Afterwards, remove the coin c_1 of $n-1$, cover $n-1$ by a copy of c_{n-1} covering $n-2$. Hence, we obtain $(c_2, c_3, \ldots, c_{n-2}, c_{n-1}, c_{n-1}, c_1)$. In consecutive steps, remove the coin c_{i+1} of i and then move a copy of the coin c_i of $i-1$ to i, $i = n-3, \ldots, m+2$. Hence, we get $(c_2, c_3, \ldots, c_{m+1}, c_{m+2}, c_{m+2}, \ldots, c_{n-1}, c_1)$. Now shift cyclically the first m coins u times. Thus we reach $(c_{m-u+2}, \ldots, c_{m+1}, c_2, c_3, \ldots, c_{m-u+1}, c_{m+2}, c_{m+2}, \ldots, c_{n-1}, c_1)$. Then u is covered by c_{m+1}. Remove c_{m+2} of $m+1$ and then cover $m+1$ by a copy of c_{m+1} covering u. Hence we have reached $(c_{m-u+2}, \ldots, c_{m+1}, c_2, c_3, \ldots, c_{m-u+1}, c_{m+1}, c_{m+2}, \ldots, c_{n-1}, c_1)$. Shifting the first m coins cyclically $m-u$ times, we obtain $(c_2, \ldots, c_{m+1}, c_{m+1}, c_{m+2}, \ldots, c_{n-1}, c_1)$. Now, in consecutive steps, remove the coin c_{i+1} of i and afterwards move a copy of the coin c_i of $i-1$ to i, $i = m, \ldots, 2$. This results in $(c_2, c_2, \ldots, c_m, c_{m+1}, c_{m+2}, \ldots, c_{n-1}, c_1)$. Shift cyclically the first m coins $m-1$ times, reaching $(c_2, c_3, \ldots, c_m, c_2, c_{m+1}, c_{m+2}, \ldots, c_{n-1}, c_1)$. Remove the coin c_2 of the first vertex and then cover the first position by a copy of c_1 covering the last vertex. This leads to $(c_1, c_3, \ldots, c_m, c_2, c_{m+1}, c_{m+2}, \ldots, c_{n-1}, c_1)$. Finally, shift again cyclically the first m coins (one time), reaching $(c_2, c_1, c_3, \ldots, c_m, c_{m+1}, \ldots, c_{n-1}, c_1)$.

Formally, it is easy to check that

$$F_1 = F'F'_n(F')^{m-1}F'_1 \cdots F'_{m-1}(F')^{m-u}F'_{u,m+1}(F')^u F'_{m+1} \cdots F'_{n-1}(F_2)^{n-1}$$

is allowed and

$$F'F'_n(F')^{m-1}F'_1 \cdots F'_{m-1}(F')^{m-u}F'_{u,m+1}(F')^u F'_{m+1} \cdots F'_{n-1}(F_2)^{n-1}(1,\ldots,n)$$
$$= F'F'_n(F')^{m-1}F'_1 \cdots F'_{m-1}(F')^{m-u}F'_{u,m+1}(F')^u F'_{m+1} \cdots$$
$$F'_{n-1}(2,3,\ldots,n-2,n-1,1,2)$$
$$= F'F'_n(F')^{m-1}F'_1 \cdots F'_{m-1}(F')^{m-u}F'_{u,m+1}(F')^u(2,3,\ldots,m+1,m+2,$$
$$m+2,\ldots,n-1,1)$$
$$= F'F'_n(F')^{m-1}F'_1 \cdots F'_{m-1}(F')^{m-u}F'_{u,m+1}(m-u+2,\ldots,m+1,2,3,\ldots,$$
$$m-u+1,m+2,m+2,\ldots,n-1,1)$$
$$= F'F'_n(F')^{m-1}F'_1 \cdots F'_{m-1}(F')^{m-u}(m-u+2,\ldots,m+1,2,3,\ldots,m-u$$
$$+1,m+1,m+2,\ldots,n-1,1)$$
$$= F'F'_n(F')^{m-1}F'_1 \cdots F'_{m-1}(2,\ldots,m+1,m+1,m+2,\ldots,n-1,1)$$
$$= F'F'_n(F')^{m-1}(2,2,\ldots,m,m+1,m+2,\ldots,n-1,1)$$

$$= F'F'_n(2, 3, \ldots, m, 2, m+1, m+2, \ldots, n-1, 1)$$
$$= F'(1, 3, \ldots, m, 2, m+1, m+2, \ldots, n-1, 1)$$
$$= (2, 1, 3, \ldots, m, m+1, m+2, \ldots, n-1, 1).$$

This shows that, for either $m = n-1$ or $m < n-1$, $\mathcal{D}$ penultimately realizes γ'_2 with respect to $1, \ldots, n$. Therefore, by Proposition 1.5, we obtain again that $\mathcal{D}$ is penultimately permutation complete with respect to n. By Lemma 2.6, the proof is complete. □

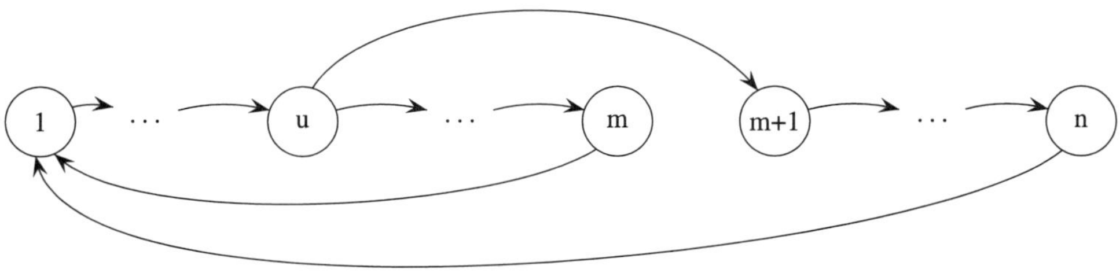

PENULTIMATELY PERMUTATION COMPLETE DIGRAPH (LEMMA 2.9)

Lemma 2.10. *Suppose that a strongly connected digraph $\mathcal{D}$ contains a strongly connected penultimately permutation complete subdigraph $\mathcal{D}'$ having at least three vertices. Then $\mathcal{D}$ is also penultimately permutation complete.*

Proof. Let $\mathcal{D}' = (V', E')$ be a maximal (with respect to the inclusion of vertex sets) strongly connected penultimately permutation complete subdigraph in $\mathcal{D}$. Suppose there exists a vertex v in $\mathcal{D}$ but not in $\mathcal{D}'$. Then there exist a shortest path from v to a vertex q_1 in $\mathcal{D}'$ and a shortest path from a vertex q_2 in $\mathcal{D}'$ to v. If there exists a vertex u in $\mathcal{D}$ but not in $\mathcal{D}'$ which occurs in both of these paths, then let u be the closest to $\mathcal{D}'$ having this property. If there does not exist any vertex u having the above property, then we identify u with the vertex v. If $q_1 = q_2$, then two paths $u \to q_1$ and $q_2 \to u$ form a cycle $\mathcal{C} = (C, E_C)$ where distinctness of the vertices follows from minimality. If $q_1 \neq q_2$, then in $\mathcal{D}'$ there is a path from q_1 to q_2 as $\mathcal{D}'$ is strongly connected. Then the three paths $u \to q_1$, $q_1 \to q_2$, and $q_2 \to u$ form a cycle $\mathcal{C} = (C, E_C)$.

Let p denote the (only) $\mathcal{C}$-compatible cyclic permutation and consider

$$p|^{V' \cup C}(a) = \begin{cases} p(a) & \text{if } a \in C, \\ a & \text{if } a \in V' \setminus C. \end{cases}$$

Moreover, for every allowed transformation φ with respect to $\mathcal{D}'$ define

$$\varphi|^{V' \cup C}(a) = \begin{cases} \varphi(a) & \text{if } a \in V', \\ a & \text{if } a \in C \setminus V'. \end{cases}$$

Let $v' \in V'$ be an arbitrary vertex of $\mathcal{D}'$. By Lemma 1.8, $\langle T_{G, V' \setminus \{v'\}} \cup \{p\} \rangle = T_{G, (V' \setminus \{v'\}) \cup C}$, where $T_{G, V' \setminus \{v'\}}$ and $T_{G, (V' \setminus \{v'\}) \cup C}$ denote the full permutation groups on $V' \setminus \{v'\}$ and $(V' \setminus \{v'\}) \cup C$, respectively. But all elements of $\{p|^{V' \cup C}\} \cup \{\varphi|^{V' \cup C} \mid$

φ is allowed with respect to $\mathcal{D}'\}$ are allowed with respect to $\mathcal{D}' \cup \mathcal{C}$. Thus,

$$T_{G,(V'\setminus\{v'\})\cup C} \subseteq \{\rho|_{(V'\setminus\{v'\})\cup C} \mid \rho = \rho_1 \cdots \rho_m, \rho_1, \ldots, \rho_m \text{ are allowed with respect to } \mathcal{D}' \cup \mathcal{C}\}.$$

In other words (by Lemma 2.4), $\mathcal{D}'\cup\mathcal{C}$ is penultimately permutation complete with respect to v'. But then, by Lemma 2.6, the (strongly connected) subdigraph $\mathcal{D}'\cup\mathcal{C}$ of $\mathcal{D}$ is penultimately permutation complete, a contradiction with the choice of $\mathcal{D}'$. □

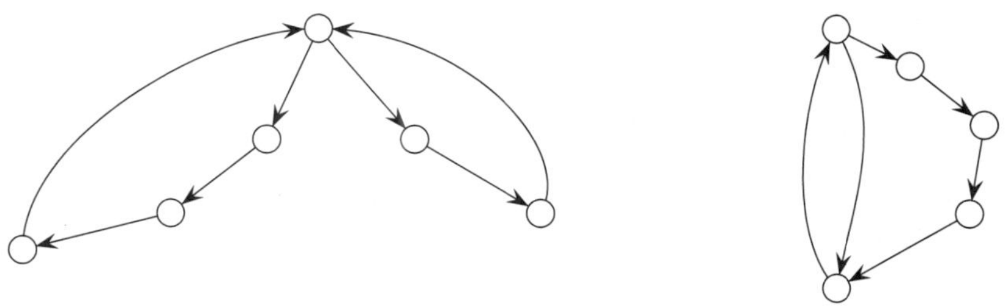

TWO PENULTIMATELY PERMUTATION COMPLETE DIGRAPHS

Theorem 2.11. *A digraph $\mathcal{D}$ with $n > 3$ vertices is penultimately permutation complete if and only if it is strongly connected and contains a branch.*

Proof. For the necessity, first we suppose that $\mathcal{D}$ is not strongly connected. Then there exists a pair $i, j, i \neq j$, of vertices such that there is no walk from i to j. But then $p : \{1, \ldots, n-1\} \to \{1, \ldots, n-1\}$ penultimately cannot be realized by $\mathcal{D}$ with respect to $P(1), \ldots, P(n)$ whenever $p(j) = i$ and P is a permutation of the vertices such that $P(n) \notin \{i, j\}$.

Now we assume that $\mathcal{D}$ is strongly connected but it does not have a branch. Then $\mathcal{D}$ consists of a cycle (up to the loop edges). Then for every allowed configuration transformation $F(1, \ldots, n) = (f(1), \ldots, f(n))$ we obtain either $F_1(1, \ldots, n) = (i, i+1 \pmod n), \ldots, n, 1, \ldots, i-1 \pmod n))$ (n-cyclic right shift) or a cyclic transformation applied to an elementary collapsing $F(1, \ldots, n) = (1, \ldots, k-1, k, k, k+2, \ldots, n)$, for some $k \in \{1, \ldots, n-1\}$, or $F(1, \ldots, n) = (n, 2, \ldots, n-1, n)$. Therefore, for example, no transposition can be penultimately realized by $\mathcal{D}$ with respect to any vertex $1, \ldots, n$. Thus $\mathcal{D}$ cannot be penultimately permutation complete. This ends the proof of necessity.

For the sufficiency, let us consider a strongly connected digraph $\mathcal{D}$ having a branch. Then it should have two intersecting cycles which form a strongly connected subdigraph $\mathcal{D}' = (V', E')$ having a branch. Using Lemma 2.8 or Lemma 2.9 we have that $\mathcal{D}'$ is penultimately permutation complete. (In this case it is understood that for every permutation $p : \{1, \ldots, m-1\} \to \{1, \ldots, m-1\}$ and $k_i \in V' = \{k_1, \ldots, k_m\}$ there exists a product F of $\mathcal{D}'^{(\ell)}$-compatible mappings having $F(k_1, \ldots, k_m) = (k_{p(1)}, \ldots, k_{p(i-1)}, \ell, k_{p(i)}, \ldots, k_{p(m-1)})$ with $\ell \in V'$.)

If $\mathcal{D} = \mathcal{D}'$, then we are done with the proof. Otherwise, by the strong connectivity of $\mathcal{D}$, we can apply Lemma 2.10. The proof is complete. □

Remark. *Observe that $n > 3$ is required in Theorem* 2.11. *The cyclic three-vertex digraph contains no branch, but allows $F_1(1, 2, 3) = (1, 2, 2)$ and $F_2(1, 2, 3) = (3, 2, 1)$,* *whose composition penultimately realizes a transposition, showing penultimate permutation completeness.*

Corollary 2.12. *If a digraph $\mathcal{D} = (V, E)$ is strongly connected and contains a branch, then it is penultimately permutation complete.*

Proof. Since $\mathcal{D}$ contains a branch, it must have pairwise distinct vertices v, w, w' with (v, w) and (v, w') in E. Thus $|V| \geq 3$. If $|V| > 3$, the result holds by Theorem 2.11. Otherwise $|V| = 3$. In this case strong connectivity implies that one of the following pairs of edges also must occur in E: (w, v) and (w', v); (w, w') and (w', v); or (w', w) and (w, v). In every case there exist distinct $x, y \in V$ with (x, y) and (y, x) both in E. Thus $\mathcal{D}$ can penultimately represent a transposition, whence penultimate permutation completeness follows. □

The next statement shows that even if a digraph contains all loop edges, its penultimately completeness does not imply that the degree $(|V| - 1)$ symmetric group can be embedded into its group.

Proposition 2.13. *There exists a penultimately complete digraph $\mathcal{D} = (V, E)$ such that the symmetric group of degree $(|V| - 1)$ cannot be embedded isomorphically into $G(\mathcal{D}^{(\ell)})$.*

Proof. Define $\mathcal{D} = (\{1, 2, 3, 4\}, \{(1, 2), (2, 4), (4, 1), (4, 3), (3, 2)\})$. It can be verified by a straightforward calculation that $\mathcal{D}$ is penultimately permutation complete, but this fact can also be derived from Theorem 2.11 since $\mathcal{D}$ is strongly connected and contains a branch. On the other hand, it is easy to see that every $\mathcal{D}^{(\ell)}$-compatible permutation p is even. Indeed, if $p(4) \neq 4$, then $p(4) = 1$ or $p(4) = 3$. In the former case, p must be the cycle (124) (i.e., $p(1) = 2, p(2) = 4, p(3) = 3, p(4) = 1$); in the latter, $p = (243)$ (i.e., $p(1) = 1, p(2) = 4, p(3) = 2, p(4) = 3$). Hence the group $G(\mathcal{D}^{(\ell)})$ is a subgroup in the alternating group A_4. The latter group is known to have no six-element subgroups; thus, the degree-3 symmetric group cannot embed in $G(\mathcal{D}^{(\ell)})$. □

Problem 2.14. *The following questions remain open problems:*

(1) *Characterize all penultimately permutation complete digraphs $\mathcal{D} = (V, E)$ for which the degree $(|V| - 1)$ symmetric group can be embedded isomorphically into $G(\mathcal{D}^{(\ell)})$.*

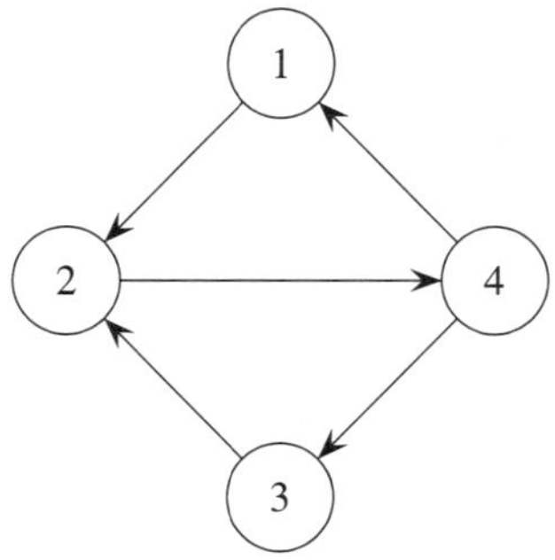

PENULTIMATELY PERMUTATION COMPLETE DIGRAPH $\mathcal{D}$ WITH FOUR VERTICES BUT WITH NO EMBEDDING OF THE SYMMETRIC GROUP OF DEGREE 3 INTO THE GROUP $G(\mathcal{D}^{\ell})$

(2) *Characterize all digraphs* $\mathcal{D} = (V, E)$ *for which the degree* $(|V| - 1)$ *symmetric group can be embedded isomorphically into* $G(\mathcal{D}^{(\ell)})$.

(3) *Characterize all digraphs* $\mathcal{D} = (V, E)$ *for which the degree* $(|V| - 1)$ *symmetric group can be embedded isomorphically into* $G(\mathcal{D})$.

(4) *Characterize all digraphs* $\mathcal{D} = (V, E)$ *for which the degree* $|V|$ *symmetric group can be embedded isomorphically into* $G(\mathcal{D}^{(\ell)})$.

(5) *Characterize all digraphs* $\mathcal{D} = (V, E)$ *for which the degree* $|V|$ *symmetric group can be embedded isomorphically into* $G(\mathcal{D})$.

Let $\mathcal{D}$ be a digraph. We say that $\mathcal{D}$ is *isomorphically n-complete* if the complete transformation semigroup on n letters embeds in the transformation semigroup of $\mathcal{D}$. $\mathcal{D}$ is *homomorphically n-complete* if the full transformation semigroup on n letters divides transformation semigroup of $\mathcal{D}$. $\mathcal{D}$ is *n-complete (with respect to its semigroup)* if the symmetric semigroup on n letters divides the semigroup of $\mathcal{D}$.

Now we prove the following statement.

Theorem 2.15. *Let* $\mathcal{D}$ *be a digraph containing all loop edges. Suppose that* $\mathcal{D}$ *has a strongly connected subdigraph with at least* $n + 1$ *vertices which contains a branch. Then* $\mathcal{D}$ *is isomorphically n-complete.*

Proof. By hypothesis, $\mathcal{D}$ contains a branch in a strongly connected subdigraph $\mathcal{D}' = (V', E')$ with $|V'| = m \geq n + 1$ vertices. By Corollary 2.12, $\mathcal{D}'$ is a penultimately permutation complete digraph. By definition of branch, there exist pairwise distinct vertices v_0, w, and w' in V' with $(v_0, w), (v_0, w') \in E'$ in $\mathcal{D}'$.

Now $\mathcal{D}'$ is also penultimately permutation complete with respect to $w \in V'$. In other words, for every bijection $p : V' \setminus \{w\} \to V' \setminus \{w\}$, the transformation semigroup $T(\mathcal{D}'^{(\ell)}) = (V', S(\mathcal{D}'^{(\ell)}))$ has an element p' such that $p'(u) = p(u)$ for every $u \in V' \setminus \{w\}$. Now p' is a product $f_1 \circ \cdots \circ f_k$ of some $\mathcal{D}'^{(\ell)}$-compatible maps $f_i : V' \to V'$. Since $\mathcal{D}$ contains all loop edges, we can extend each f_i to a $\mathcal{D}$-compatible map $\tilde{f}_i : V \to V$ by letting $\tilde{f}_i(v) = f_i(v)$ for all $v \in V'$ and $f_i(v) = v$ for all $v \in V \setminus V'$. Let $\tilde{p} = \tilde{f}_1 \circ \cdots \circ \tilde{f}_k$. Then $\tilde{p}$ is the identity on $V \setminus V'$ but agrees with p' on V'.

By $(v_0, w) \in E' \subseteq E$, the transformation $e : V \to V$ with

$$e(v) = \begin{cases} v_0 & \text{if } v \in \{v_0, w\}, \\ v & \text{otherwise} \end{cases}$$

is a $\mathcal{D}$-compatible collapsing that acts as the identity on $V \setminus V'$. Then $p'' = e\tilde{p}$ is also an element of $S(\mathcal{D})$ such that

$$p''(v) = \begin{cases} p(v_0) & \text{if } v = w, \\ p(v) & \text{if } v \in V' \setminus \{w\}, \\ v & \text{otherwise.} \end{cases}$$

In addition, if $q : V' \setminus \{w\} \to V' \setminus \{w\}$ is another arbitrary bijection, we have corresponding $q' : V' \to$ and $\tilde{q}$ in $S(\mathcal{D}^{(\ell)})$ constructed as for p. Thus $\tilde{q}(u) = q(u)$ for $u \in V' \setminus \{w\}$, $\tilde{q}(v_0) = \tilde{q}(w)$, and $\tilde{q}$ acts as the identity on $V \setminus V'$. Then we have for $p'' = e\tilde{p}$ and $q'' = e\tilde{q}$

that

$$p''q''(v) = \begin{cases} pq(v_0) & \text{if } v = w, \\ pq(v) & \text{if } v \in V' \setminus \{w\}, \\ v & \text{otherwise.} \end{cases}$$

Let P be the set of all functions $p'' : V \to V$ with

$$p''(v) = \begin{cases} p(v_0) & \text{if } v = w, \\ p(v) & \text{if } v \in V' \setminus \{w\}, \\ v & \text{otherwise} \end{cases}$$

such that $p : V' \setminus \{w\} \to V' \setminus \{w\}$ is a bijection. We have $P \subseteq S(\mathcal{D})$.

Now, for $p'', q'' \in P$, if $p''(v) = q''(v)$ holds for all $v \in V' \setminus \{w\}$, then $p'' = q''$ on all of V. Thus P acts faithfully on $V' \setminus \{w\}$. It follows that $(V' \setminus \{w\}, P)$ and (V', P) are isomorphically embedded in $T(\mathcal{D}) = (V, S(\mathcal{D}))$. Clearly, then, the former (but not the latter) is a permutation group isomorphic to the complete permutation group of degree $(|V'| - 1)$.

Then, since (v_0, w') and (v_0, w) are in E', the map $e' : V \to V$ is also a $\mathcal{D}$-compatible when

$$e'(v) = \begin{cases} v_0 & \text{if } v = w, \\ v_0 & \text{if } v = w', \\ v & \text{otherwise.} \end{cases}$$

Now $e'|_{V'} : V' \to V'$ is also $\mathcal{D}'^{(\ell)}$-compatible and $e'|_{V' \setminus \{w\}}$ is an elementary collapsing on $V' \setminus \{w\}$.

Let $S = \langle P \cup \{e'\} \rangle$. For every $f, g \in S$, it is clear that $fg(w) = fg(v_0)$ and fg acts as the identity on $V \setminus V'$. Also, if $f(v) = g(v)$ for all $v \in V \setminus \{w\}$, then $f = g$ on all of V. Thus S acts faithfully on $V' \setminus \{w\}$, so $(V' \setminus \{w\}, S)$ is a transformation semigroup containing all permutations of its states and an elementary collapsing. It follows by Proposition 1.5 that $(V' \setminus \{w\}, S)$ is isomorphic to the degree $(|V'| - 1)$ complete transformation semigroup. Thus the degree $(m - 1)$ full transformation semigroup is isomorphically embedded in $(V, S(\mathcal{D}))$; i.e., $\mathcal{D}$ is isomorphically $(m - 1)$-complete and hence also isomorphically n-complete. □

Let $\mathcal{D}$ be a digraph. We say that $\mathcal{D}$ is *isomorphically group n-complete* if the complete permutation group on n letters embeds into the transformation semigroup of $\mathcal{D}$. $\mathcal{D}$ is *homomorphically group n-complete* if the full permutation group on n letters divides the transformation semigroup of $\mathcal{D}$. $\mathcal{D}$ is *group n-complete (with respect to its semigroup)* if the symmetric group on n letters divides the semigroup of $\mathcal{D}$.

Of course, we have the following consequence of Theorem 2.15.

Theorem 2.16. *Let $\mathcal{D}$ be a digraph containing all loop edges. Suppose that $\mathcal{D}$ has a strongly connected subdigraph with at least $n + 1$ vertices which contains a branch. Then $\mathcal{D}$ is isomorphically group n-complete.* □

Lemma 2.17. *Let $\mathcal{E} = (V, E)$ be a digraph, possibly not containing some loop edges. Suppose no strongly connected subdigraph of $\mathcal{E}$ contains a branch. If G is a group and $G < S(\mathcal{E})$, then G is abelian.*

Proof. If a nontrivial group G divides $S(\mathcal{E})$, then, by Proposition 1.11, $S(\mathcal{E})$ has a subgroup $\tilde{G}$ mapping homomorphically onto G and $\tilde{G}$ acts faithfully by permutations on some subset $Z \subseteq V$. Now consider $h \in \tilde{G}$; h is a product $h = f_1 \dots f_K$ of $\mathcal{E}$-compatible maps $f_i : V \to V$ $(1 \leq i \leq K)$ for some $K \geq 1$. Since h permutes Z, each of the sets $h^\ell(Z) = Z$ for all $\ell > 0$, and, moreover, every $h^\ell f_1 \cdots f_i(Z)$ has $|Z|$ elements for all $1 \leq i \leq K$. Let f_i denote $f_{i (\mathrm{mod}\, K)}$, also for $i > K$.

Suppose G is not abelian; then neither is $\tilde{G}$. Thus there are $h_1, h_2 \in \tilde{G}$ with $h_1 h_2 \neq h_2 h_1$. Since $\tilde{G}$ acts faithfully, there exists $v_1 \in Z$ with $v_1 \cdot h_1 h_2 \neq v_1 \cdot h_2 h_1$. Since h_1 and h_2 are products of compatible maps, it follows that there is a path from v_1 to $v_1 \cdot h_1$ to $v_1 \cdot h_1 h_2$ and from v_1 to $v_1 \cdot h_2$ to $v_1 \cdot h_1 h_2$ in $\mathcal{E}$. Since $\tilde{G}$ acts by permutations there are also paths back so all the vertices mentioned lie within a strongly connected subdigraph $\mathcal{D}$ of $\mathcal{E}$. We may assume $\mathcal{D}$ is maximal with respect to inclusion of vertices. By hypothesis, $\mathcal{D}$ has no branch. Then strong connectedness implies that necessarily $\mathcal{D} = (V', E')$ is a cycle graph possibly with some loop edges. We may write its vertices as $\{v_1, \dots, v_{|V'|}\}$ and its edges as $\{(v_i, v_{i+1 (\mathrm{mod}\, |V'|)}) \mid 1 \leq i \leq |V'|\}$ plus zero or more loop edges (v_i, v_i).

Let $Z' = Z \cap V'$. Suppose, for a contradiction, that $v = f_1 \cdots f_i(z')$ lies outside of $\mathcal{D}$ for some $z' \in Z'$, $i \geq 1$; then, since $\mathcal{D}$ is a maximal strongly connected subdigraph and $f_{i+1}, \dots, f_N$ are $\mathcal{E}$-compatible, so does $f_1 \cdots f_i f_{i+1} \cdots f_N(z') = h^{N/K}(z')$ for every $N \geq i$ which is a multiple of K. It follows that $z' = (h^{N/K})^U(z')$ lies outside $\mathcal{D}$, when U is the order of the mapping $h^{N/K}$ as a permutation of Z. This contradicts $z' \in Z'$. Therefore, each $f_1 \cdots f_i$, $i \geq 1$, maps V' to V'. In particular, every $h \in \tilde{G}$ acts as a permutation of $Z' \subseteq Z$. Let $v_{i_1}, \dots, v_{i_{|Z'|}}$ with $1 \leq i_1 < \cdots < i_{|Z'|} \leq |V'|$ denote all the distinct elements of Z'. Observe that Z' is cyclically ordered in this way according to the structure of $\mathcal{D}'$ by defining the relation $v_{i_j} < v_{i_{j+1 (\mathrm{mod}\, |Z'|)}}$ for all $1 \leq j \leq |Z'|$. (Similarly, any subset of V' has a unique cyclic ordering structure arising from the structure of $\mathcal{D}$.)

As before, since h permutes Z', each of the sets $h^\ell(Z') = Z'$ for all $\ell > 0$, and, moreover, every $f_1 \cdots f_i(Z')$ has cardinality $|Z'|$. Thus, each $f_1 \cdots f_i$ $(i > 0)$ restricted to Z' is bijective. Since f_{i+1} is compatible, it follows that f_{i+1} preserves cyclic ordering among the members of $f_1 \cdots f_i(Z)'$ (without any collapsing). Thus $h = f_1 \cdots f_K$ restricted to Z' is a power of the cyclic permutation $c(v_{i_j}) = v_{i_{j+1 (\mathrm{mod}\, |Z'|)}}$, $1 \leq j \leq |Z'|$. Thus $h_1 h_2(v_1) = h_2 h_1(v_1)$ since h_1 and h_2 restricted to Z' are powers of c, a contradiction. Thus G is abelian. □

It is clear that isomorphic n-completeness implies homomorphic n-completeness, which leads to n-completeness with respect to the semigroup of the digraph. Similar to the general cases, isomorphic group n-completeness implies homomorphic group n-completeness, which implies group n-completeness with respect to the semigroup of the digraph.

Problem 2.18. *It is an open problem to characterize the six types of complete digraphs defined above (isomorphically n-complete, homomorphically n-complete, n-complete, isomorphically group n-complete, homomorphically group n-complete, group n-complete) for digraphs not necessarily containing all loop edges. It is also remains an open problem to determine which of these concepts are equivalent.*

We extend these concepts of digraph completeness to classes of digraphs as follows. Let Γ be a nonempty class of digraphs. Consider the following definitions. Γ is *isomorphically complete* if every transformation semigroup can be embedded in the transformation semigroup of a digraph in Γ. Γ is *homomorphically complete* if every transformation semigroup divides the transformation semigroup of a digraph in Γ. Γ is *complete* if every finite semigroup divides the semigroup of a digraph in Γ. Similarly, Γ is *isomorphically group complete* if every permutation group can be embedded in the transformation semigroup of a digraph in Γ. Γ is *homomorphically group complete* if every permutation group divides the transformation semigroup of a digraph in Γ. Finally, Γ is *group complete* if every finite group divides the semigroup of a digraph in Γ.

Obviously, we also have for digraph classes that isomorphic completeness implies homomorphic completeness, which implies completeness (with respect to the semigroup of the digraph). Similar to the general cases, isomorphic group completeness implies homomorphic group completeness, which implies group completeness (with respect to the semigroup of the digraph).

Now we characterize these digraph classes provided that all of their members contain all loop edges.

Corollary 2.19. *Let Γ be a nonempty class of digraphs containing all loop edges. The following conditions are equivalent:*

(1) Γ *is isomorphically group complete.*

(2) Γ *is homomorphically group complete.*

(3) Γ *is group complete.*

(4) *For every positive integer n there is a digraph in Γ which has a strongly connected subdigraph of order at least n and contains a branch.*

Corollary 2.20. *Let Γ be a nonempty class of digraphs containing all loop edges. The following conditions are equivalent:*

(1a) Γ *is isomorphically complete.*

(2a) Γ *is homomorphically complete.*

(3a) Γ *is complete.*

(4a) *For every positive integer n there is a digraph in Γ which has a strongly connected subdigraph of order at least n and contains a branch.*

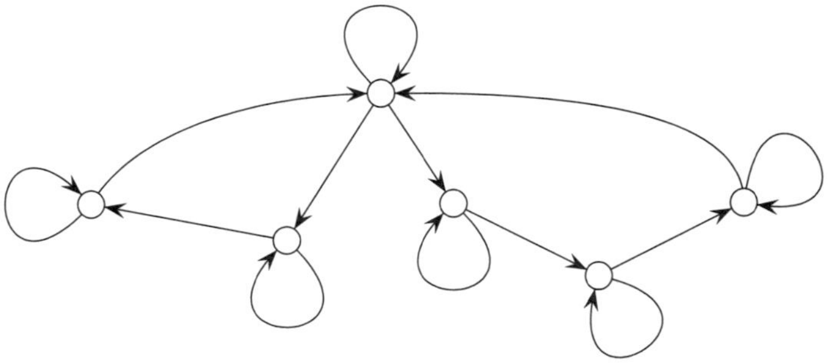

STRONGLY CONNECTED DIGRAPH WITH ALL LOOP EDGES CONTAINING A BRANCH

Note that $\mathcal{D} = \mathcal{D}^{(\ell)}$ holds for every $\mathcal{D} \in \Gamma$. Therefore, we could also consider $\mathcal{D}$ instead of $\mathcal{D}^{(\ell)}$ in the proofs of Corollaries 2.19 and 2.20 given below.

***Proof of Corollary* 2.19.** First we prove that (4) implies (1). Indeed, by (4) for every positive integer n, there exist a positive integer $m \geq n$ and a digraph in Γ having a strongly connected subdigraph $\mathcal{D} = (V, E)$ of order m containing all loop edges and a branch. Corollary 2.12 implies that this subdigraph is penultimately permutation complete. Therefore, by Theorem 2.16, the degree $(m-1)$ full permutation group can be embedded isomorphically in $S(\mathcal{D}^\ell)$. (See also Corollary 1.13.) Clearly then for every $1 \leq k < m$, the degree-k full permutation group can be embedded in the transformation semigroup of $\mathcal{D}^{(\ell)}$. (1) immediately implies (2), and (2) immediately implies (3). To end our proof we show that (3) implies (4). Consider the full symmetric group S_k of degree k for each $k > 0$. By hypothesis, S_k divides the semigroup of some digraph $\mathcal{E}_k$ in Γ. Suppose the contrary of (4) and assume that there exists a positive integer N such that for every strongly connected subdigraph $\mathcal{D} = (V, E)$ of an arbitrary digraph in $\mathcal{E}$, either $|V| < N$ or $\mathcal{D}$ does not have branch. Consider $k = \max(5, N)$. Then the alternating group A_k is well known to be a nonabelian simple subgroup of S_k for $k \geq 5$. (See Theorem 1.4.) If $|V| < N$, then A_k does not divide $S(\mathcal{D})$; indeed, every subgroup of $S(\mathcal{D})$ must obviously be a divisor of $S_{|V|}$, but the cardinality of A_k[11] is $\frac{k!}{2}$, which exceeds the cardinality ($|V|!$) of $S_{|V|}$ since $k \geq N > |V|$. On the other hand, if $\mathcal{D}$ is strongly connected but has no branch, then, by Lemma 2.17, the fact that A_k is nonabelian shows that A_k cannot divide $S(\mathcal{D})$. For every maximal strongly connected subdigraph $\mathcal{D}$ of $\mathcal{E}_k$, one or the other case applies, so A_k does not divide $S(\mathcal{D})$. Since S_k divides $S(\mathcal{E}_k)$, so does A_k, and then, by Proposition 1.11, a subgroup $\tilde{G}$ of $S(\mathcal{E}_k)$ acts by permutations on a subset $Z \subseteq E$ and maps onto A_k. Then, since $\tilde{G}$ cannot map nodes in Z between strongly connected components, we have that $\tilde{G}$ is isomorphic to a divisor of $S(\mathcal{D}_1) \times \cdots \times S(\mathcal{D}_m)$, where $\mathcal{D}_1, \ldots, \mathcal{D}_m$ are the maximal strongly connected subdigraphs of $\mathcal{E}_k$. Since A_k is simple and divides $\tilde{G}$, it must divide some $S(\mathcal{D}_i)$,[12] a contradiction. This proves (4) follows from (3). □

***Proof of Corollary* 2.20.** It is enough to prove that (4) (i.e., (4a)) implies (1a). But this statement coincides with Theorem 2.15. □

Since condition (4) of Corollary 2.20 coincides with condition (4a) of Corollary 2.19, we have the following.

Corollary 2.21. *For a class Γ of digraphs that contain all loop edges, the following are equivalent: isomorphic completeness, homomorphic completeness, completeness, group isomorphic completeness, group homomorphic completeness, and group completeness.* □

We finish this section with the following questions.

Problem 2.22. *The following questions remain open if we consider classes of digraphs not necessarily having all loop edges:*

(1) *Characterize the isomorphically group complete, homomorphically group complete, and group complete digraph classes. Decide whether the three concepts coincide.*

[11] It is a routine work to show that $|S_k| = k!\,(= k \cdot (k-1) \cdot \ldots \cdot 2 \cdot 1)$ and $|A_k| = \frac{k!}{2}$.

[12] We will see in Lemma 3.6 that a finite simple group that divides a direct product of finite semigroups must divide one of the factors.

(2) *Characterize the isomorphically complete, homomorphically complete, and complete digraph classes. Decide whether the three concepts coincide.*

The considerations of this section can be generalized by allowing more general types of compatible transformation and different nodes to have different types of contents and computing capacity. This leads to the notion of automata networks introduced in the next section. By considering digraphs in which the contents at each node must be a member of the same fixed finite set of possible contents and more general kinds of compatible mapping, one obtains the notion of a state-homogeneous automata network. Their power of simulation under projection is closely related to the notions of completeness that have been studied in this section and will be developed in Section 2.1.

2.2 Automata and Automaton Mappings

By an *automaton* $\mathcal{A} = (A, X, \delta)$ we mean a finite automaton without outputs. Here A is the (finite nonempty) *state set*, X is the *input alphabet*, and $\delta : A \times X \to A$ is the *transition function*. We also use δ in an extended sense, i.e., as a mapping $\delta^* : A \times X^* \to A$, where $\delta^*(a, \lambda) = a$ $(a \in A)$ and $\delta^*(a, px) = \delta(\delta^*(a, p), x)$ $(a \in A, p \in X^*, x \in X)$. In what follows, we shall simply write δ for δ^*. The *digraph* $D(\mathcal{A}) = (V, E)$ *of the automaton* $\mathcal{A}$ is defined as $V = A$ and $E = \{(a, b) \mid \text{ there exists } x \in X : \delta(a, x) = b\}$.

Let $\mathcal{A} = (A, X, \delta)$ be an automaton.

$\mathcal{A}$ is *trivial* if A is a singleton.

$\mathcal{A}$ is *discrete* if for every pair $a \in A, x \in X$ we have $\delta(a, x) = a$.

$\mathcal{A}$ is *monotone* if there is a partial ordering $\leq$ on A such that $a \leq \delta(a, x)$ for each $a \in A, x \in X$.

If for every triplet $a \in A, x, y \in X$ the equality $\delta(a, x) = \delta(a, y)$ holds, then we speak about an *autonomous automaton.*

$\mathcal{A}$ is a *reset automaton* if for all $x \in X$, the set $\{\delta(a, x) \mid a \in A\}$ is a singleton.

If every transformation $\delta_x : a \mapsto \delta(a, x)$ $(a \in A, x \in X)$ is a one-to-one mapping of A onto itself, then it is said that $\mathcal{A}$ is a *permutation automaton.* If every transformation δ_x is either a constant map or a permutation, then $\mathcal{A}$ is a *permutation-reset automaton.* If every transformation δ_x is either a constant map or the identity permutation, then $\mathcal{A}$ is an *identity-reset automaton.*

$\mathcal{A}$ *can be generated by the subset* B *of its states* if for every $a \in A$ there are $b \in B$ and $p \in X^*$ fulfilling $\delta(b, p) = a$. Then B is a *set of generators* in $\mathcal{A}$.

$\mathcal{A}$ is *connected* if it can be generated by one of its states. In other words, $\mathcal{A}$ is connected if it has a state a such that for every state b there exists an input word p having $\delta(a, p) = b$. Then we also say that $\mathcal{A}$ is connected for the state a.

$\mathcal{A}$ is said to be *strongly connected* if it can be generated by each of its states. In other words, $\mathcal{A}$ is strongly connected if for every pair $a, b \in A$ of states there is a word $p \in X^*$ with $\delta(a, p) = b$.

$\mathcal{A}$ is an *n-degree weakly nilpotent automaton* (or, in short, a weakly nilpotent automaton) if it has a state $a \in A$, called *dead state,* such that for every pair $b \in A, x \in X$ and positive integer $m \geq n$, $\delta(b, x^m) = a$. $\mathcal{A}$ is *n-degree nilpotent* (or, in short, *nilpotent*) if it has a state $a \in A$, called *dead state,* such that for every pair $b \in A, p \in X^*, |p| \geq n$, $\delta(b, p) = a$.

$\mathcal{A}$ is *directable* if there are a state $a \in A$ and an input word $p \in X^*$ such that $\delta(b, p) = a$ holds for every $b \in A$.

For an integer $k \geq 0$, the automaton $\mathcal{A}$ is called *weakly k-definite* if $\delta(a, p) = \delta(b, p)$ holds for every $a, b \in A$, $p \in X^*$, $|p| = k$. Moreover, it is said that $\mathcal{A}$ is definite if it is weakly k-definite for some integer $k \geq 0$.

For any integer $k \geq 0$, the automaton $\mathcal{A}$ is called *weakly reverse k-definite* if $\delta(a, px) = \delta(a, p)$ is valid for all $a \in A$, $p \in X^*$, $|p| = k$, $x \in X$. $\mathcal{A}$ is *reverse definite* if it is weakly reverse k-definite for some $k \geq 0$.

For any pair of integers $h, k \geq 0$, the automaton $\mathcal{A}$ is called *weakly* (h, k)*-definite* if $\delta(a, upv) = \delta(a, uv)$ is valid for all $a \in A$, $p, u, v \in X^*$, $|u| = h$, $|v| = k$. It is worthy of note that for every pair of integers $h' \geq h, k' \geq k$, the automaton $\mathcal{A}$ is weakly (h', k')-definite if it is weakly (h, k)-definite. We say that $\mathcal{A}$ is *generalized definite* if it is weakly (h, k)-definite for some integers $h, k \geq 0$.

In addition, $\mathcal{A}$ is called *commutative* if for any state $a \in A$ and input words $p, q \in X^*$, $\delta(a, pq) = \delta(a, qp)$.

We refer to the automaton $\mathcal{E}_2 = (\{0, 1\}, \{x_1, x_2\}, \delta_{\mathcal{E}_2})$, $\delta_{\mathcal{E}_2}(0, x_1) = 0$, $\delta_{\mathcal{E}_2}(0, x_2) = \delta_{\mathcal{E}_2}(1, x_1) = \delta_{\mathcal{E}_2}(1, x_2) = 1$ as the *(two-state) elevator.*

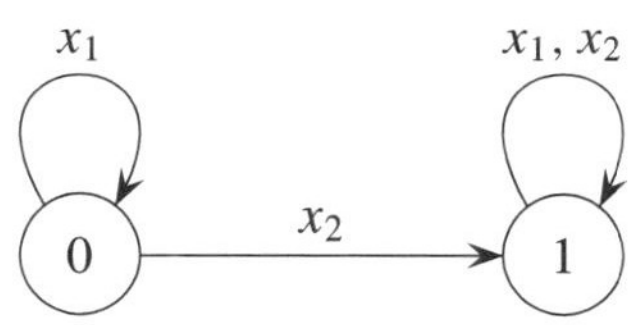

ELEVATOR $\mathcal{E}_2$

A *counter* of length $n \geq 1$ is an automaton with exactly one input letter $\mathcal{C}_n = (\{1, \ldots, n\}, \{x\}, \delta_{\mathcal{C}_n})$ with $\delta_{\mathcal{C}_n}(i, x) = i + 1 \pmod n$, so that $(\delta_{\mathcal{C}_n})_x$ induces a cyclic permutation of the state set. A counter of length $n \geq 1$ is also called a *modulo n counter*.

A *counter with identity* of length $n \geq 1$ is an automaton $\mathcal{C}_n^1 = (\{1, \ldots, n\}, \{x, y\}, \delta_{\mathcal{C}_n^1})$ with $\delta_{\mathcal{C}_n^1}(i, x) = i + 1 \pmod n$, and $\delta_{\mathcal{C}_n^1}(i, y) = i$, $i = 1, \ldots, n$. A counter with identity of length n is also called a *modulo n counter with identity*.

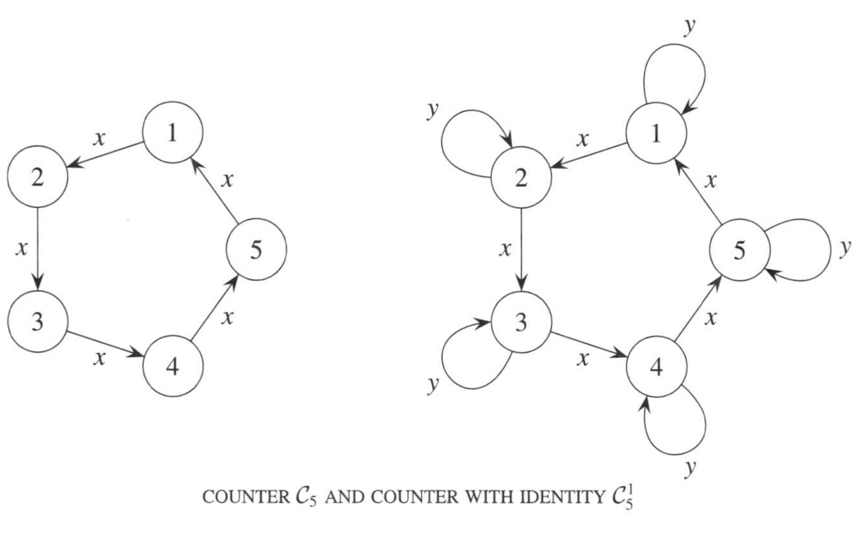

COUNTER $\mathcal{C}_5$ AND COUNTER WITH IDENTITY $\mathcal{C}_5^1$

The *two-state reset automaton* $\mathcal{A}_0 = (\{a_1, a_2\}, \{x_1, x_2\}, \delta_0)$ is defined by $\delta_0(a_i, x_j) = a_j$ for $i, j = 1, 2$. Finally, the *two-state identity-reset automaton* (also called the *flip-flop automaton*) $\mathcal{A}_0^1 = (\{a_1, a_2\}, \{x_0, x_1, x_2\}, \delta_0^1)$ is defined by

$$\delta_0^1(a_i, x_j) = \begin{cases} a_i & \text{if } j = 0, \\ a_j & \text{if } j \neq 0. \end{cases}$$

TWO-STATE RESET AUTOMATON $\mathcal{A}_0$ AND FLIP-FLOP AUTOMATON $\mathcal{A}_0^1$

Take an arbitrary automaton $\mathcal{A} = (A, X, \delta)$. A sequence $a_1, \ldots, a_n$ of pairwise distinct states of $\mathcal{A}$ is a *cycle* if there are input signs $x_1, \ldots, x_n$ such that $\delta(a_i, x_i) = a_{i+1 (\text{mod}\, n)}$ for every $i \in \{1, \ldots, n\}$. Then the positive integer n is the *length* of the cycle.

The following statement is obvious.

Proposition 2.23. *An automaton is nilpotent if and only if it has both of the following conditions:*

(1) *It has only one cycle.*

(2) *Its cycle is trivial (having only one element).* □

Each automaton can be considered as an algebra with unary operational symbols (corresponding to each input letter), or, alternatively, as an algebra with two sorts—states and input letters—and one binary operation—the transition function taking a state and a letter to a new state. Therefore, notions such as subautomaton, homomorphism, and isomorphism can be defined in the natural way. Thus $\mathcal{A}' = (A', X', \delta')$ is a *subautomaton* of the automaton $\mathcal{A} = (A, X, \delta)$ if $A' \subseteq A$, $X' \subseteq X$ and δ' is the restriction of δ to $A' \times X'$ (so that $\delta'(a', x') \in A'$ for any $a' \in A'$ and $x' \in X'$). In particular, if $A' = A$ or $X' = X$, then we speak about an *input-subautomaton* or a *state-subautomaton*, respectively. A pair $\psi = (\psi_1, \psi_2)$ of surjective mappings $\psi_1 : A \to A'$, $\psi_2 : X \to X'$ is a *homomorphism of* $\mathcal{A} = (A, X, \delta)$ onto $\mathcal{A}' = (A', X', \delta')$ if for every $a \in A, x \in X$, one has $\psi_1(\delta(a, x)) = \delta'(\psi_1(a), \psi_2(x))$. If ψ_1 and ψ_2 are bijective functions, then $\mathcal{A}$ is said to be *isomorphic to* $\mathcal{A}'$. In particular, if $X = X'$ and ψ_2 is the identity, then sometimes we will also refer to ψ_1 as a *state-homomorphism* (or a *state-isomorphism*) or, in short, as a homomorphism (or an isomorphism). In addition, if $A = A'$ and $\psi_1 : A \to A'$ is the identity mapping, then sometimes we will refer to ψ_2 as an *input-homomorphism* (or an *input isomorphism*). If $\mathcal{A}$ has a subautomaton which can be mapped homomorphically by (ψ_1, ψ_2) onto $\mathcal{A}'$, then we say that $\mathcal{A}$ *homomorphically represents* $\mathcal{A}'$ (under (ψ_1, ψ_2)). If $\mathcal{A}$ isomorphically represents $\mathcal{A}'$ (i.e., $\mathcal{A}$ has a subautomaton which can be mapped isomorphically onto $\mathcal{A}'$), then we also say that $\mathcal{A}'$ can be embedded isomorphically into $\mathcal{A}$.

Proposition 2.24. *Suppose that an automaton $\mathcal{A}$ has a homomorphism onto the strongly connected automaton $\mathcal{B}$. If $\mathcal{A}$ is minimal in the sense that $\mathcal{B}$ is not a homomorphic image of a proper state-subautomaton of $\mathcal{A}$, then $\mathcal{A}$ is also strongly connected.*

Proof. Suppose, for a contradiction, that $\mathcal{A} = (A, X_{\mathcal{A}}, \delta_{\mathcal{A}})$ is minimal in the above sense but not strongly connected. Then there exists a pair $a, b \in A$ of distinct states such that $b \notin \{\delta_{\mathcal{A}}(a, p) \mid p \in X_{\mathcal{A}}^*\}$. Let $\psi = (\psi_1, \psi_2)$ with $\psi_1 : A \to B, \psi_2 : X_{\mathcal{A}} \to X_{\mathcal{B}}$ be a homomorphism of $\mathcal{A}$ onto the (strongly connected) automaton $\mathcal{B} = (B, X_{\mathcal{B}}, \delta_{\mathcal{B}})$. Since $\mathcal{B}$ is strongly connected, for every $a', b' \in B$ there exists a $p' \in X_{\mathcal{B}}^*$ with $\delta_{\mathcal{B}}(a', p') = b'$. Thus, for every $a' \in B$, $\{\delta_{\mathcal{B}}(a', p') \mid p' \in X_{\mathcal{B}}^*\} = B$. Consider the state $b \in A$ with $b \notin \{\delta_{\mathcal{A}}(a, p) \mid p \in X_{\mathcal{A}}^*\}$. We have $\{\delta_{\mathcal{B}}(\psi_1(a), p') \mid p' \in X_{\mathcal{B}}^*\} = B$. Hence $\{\psi_1(\delta_{\mathcal{A}}(a, p)) \mid p \in X_{\mathcal{A}}^*\} = B$. Therefore, $\psi' = (\psi_1', \psi_2')$ with $\psi_1'|_{\{\delta_{\mathcal{A}}(a,p) \mid p \in X_{\mathcal{A}}^*\}}, \psi_2' = \psi_2$ is a homomorphism of the state-subautomaton $\mathcal{A}_a$ of $\mathcal{A}$ generated by the state $a \in A$. But $\mathcal{A}_a$ is a proper subautomaton of $\mathcal{A}$, a contradiction. □

Observe that the above proof also works if we assume $X_{\mathcal{A}} = X_{\mathcal{B}}$ and ψ_2 is the identity. In particular, we have the following.

Proposition 2.25. *Suppose that an automaton $\mathcal{A}$ has a state-homomorphism onto the strongly connected automaton $\mathcal{B}$. If $\mathcal{A}$ is minimal in the sense that $\mathcal{B}$ is not a state-homomorphic image of a proper state-subautomaton of $\mathcal{A}$, then $\mathcal{A}$ is also strongly connected.* □

The next statement is evident.

Proposition 2.26. *Given an automaton $\mathcal{A} = (A, X, \delta)$, let $a \in A$ be a state such that it is not a state of any strongly connected state-subautomaton of $\mathcal{A}$. Then $b \in A$ also has this property whenever $\delta(b, p) = a$ holds for some $p \in X^*$.* □

Take automata $\mathcal{A}_i = (A_i, X_i, \delta_i), i = 1, \dots, n$. The *direct product* $\mathcal{B} = \mathcal{A}_1 \times \cdots \times \mathcal{A}_n$ *of automata* $\mathcal{A}_1, \dots, \mathcal{A}_n$ is defined to be the automaton $\mathcal{B} = (B, Y, \delta_{\mathcal{B}})$, where $B = A_1 \times \cdots \times A_n, Y = X_1 \times \cdots \times X_n$, and $\delta_{\mathcal{B}}((a_1, \dots, a_n), (x_1, \dots, x_n)) = (\delta_1(a_1, x_1), \dots, \delta_n(a_n, x_n)), (a_1, \dots, a_n) \in B, (x_1, \dots, x_n) \in Y$. If $X_1 = \cdots = X_n$, then restricting to the input set in which all letters in the input n-tuple are equal, we have the subautomaton $\mathcal{B}' = \mathcal{A}_1 \Delta \cdots \Delta \mathcal{A}_n$, the *diagonal product of* $\mathcal{A}_1, \dots, \mathcal{A}_n$, whose input alphabet we may naturally identify with X. If $\mathcal{A}_1 = \cdots = \mathcal{A}_n$, then $\mathcal{B}$ is called the nth *direct power* $\mathcal{A}^n = (A^n, X^n, \delta^{(n)})$ of $\mathcal{A}$ and $\mathcal{B}'$ is called the nth *diagonal power* $\mathcal{A}^{\Delta n}$ *of* $\mathcal{A}$.

We shall use the following proposition.

Proposition 2.27. *Every directable automaton can be represented homomorphically by a diagonal product of its connected state-subautomata.*

Proof. Let $\mathcal{A} = (A, X, \delta)$ be an arbitrary directable automaton with an appropriate pair $d \in A, r \in X$ such that $\delta(b, r) = d$ holds for every $b \in A$. Consider its state set $A = \{a_1, \dots, a_m\}$ and let a subset $B = \{b_1, \dots, b_n\}$ of A be a set of generators in $\mathcal{A}$. In other words, suppose that $\{\delta(b, p) \mid b \in B, p \in X^*\} = A$. Let $\mathcal{A}_a$ denote the state-subautomaton of $\mathcal{A}$ generated by the state $a \in A$. Consider a diagonal product $\mathcal{M} = \mathcal{A}_{b_1} \Delta \cdots \Delta \mathcal{A}_{b_n} \Delta (\mathcal{A}_{a_1} \Delta \cdots \Delta \mathcal{A}_{a_m})^n$, where $(\mathcal{A}_{a_1} \Delta \cdots \Delta \mathcal{A}_{a_m})^n$ denotes the nth diagonal power of the diagonal product $\mathcal{A}_{a_1} \Delta \cdots \Delta \mathcal{A}_{a_m}$. Consider the following subset of the set of states of M. $H = \{b_1, \ldots, b_n, a_{1,1}, \ldots, a_{1,m}, \ldots, a_{n,1}, \ldots, a_{n,m}) \mid a_{1,1} = \cdots = a_{1,m} = \cdots = a_{j-1,1} = \cdots = a_{j-1,m} = a_{j+1,1} = \cdots = a_{j+1,m} = \cdots = a_{n,1} = \cdots = a_{n,m} = d, \{a_{j,1}, \ldots, a_{j,m}\} = A, j = 1, \cdots, n\}$. Clearly then for every $(b_1, \ldots, b_n, a_{1,1}, \ldots, a_{1,m}, \ldots, a_{n,1}, \ldots, a_{n,m}) \in H$ and $p \in X^*$, there are two possibilities: either all components of $(\delta(b_1, p), \ldots, \delta(b_n, p), \delta(a_{1,1}, p), \ldots, \delta(a_{n,m}, p))$ are the same or there exists exactly one $i \in \{1, \ldots, n\}$ such that $\delta(a_{i,j}, p) \neq \delta(a_{i,k}, p)$ holds for some $1 \leq j < k \leq m$. Therefore, it is clear that the following mapping $\psi : \{(\delta(b_1, p), \ldots, \delta(b_n, p), \delta(a_{1,1}, p), \ldots, \delta(a_{n,m}, p)) \mid (b_1, \ldots, b_n, a_{1,1}, \ldots, a_{n,m}) \in H, p \in X^*\} \to B$ is well defined and it is a state-homomorphism of a state-subautomaton of $\mathcal{M}$ onto $\mathcal{A}$:

$$\psi(\delta(b_1, p), \ldots, \delta(b_n, p), \delta(a_{1,1}, p), \ldots, \delta(a_{n,m}, p))$$
$$= \begin{cases} \delta(b_1, p) & \text{if } \delta(b_1, p) = \cdots = \delta(b_n, p) = \delta(a_{1,1}, p) = \cdots = \delta(a_{n,m}, p), \\ \delta(b_i, p) & \text{if } a_{i,j} \neq a_{i,k} \text{ holds for some } 1 \leq i \leq n, 1 \leq j < k \leq m. \end{cases}$$ □

Proposition 2.28. *Let $\mathcal{A} = (A, X, \delta)$ and $\mathcal{B} = (B, X, \delta')$ be arbitrary automata having the same input set. Suppose that $\mathcal{A}$ has two (distinct) states $a, b \in A$ such that $\delta(a, p) \neq \delta(b, p)$, $p \in X^*$. Then $\mathcal{B}$ can be represented homomorphically by a diagonal product of its connected state-subautomata and the automaton $\mathcal{A}$.*

Proof. The proof is similar to the proof of the above statement. Consider a set $B' = \{b_1, \ldots, b_n\}$ of generators in $\mathcal{B}$. In other words, suppose that $\{\delta(b, p) \mid b \in B', p \in X^*\} = B$. Let $\mathcal{B}_b$ again denote the state-subautomaton of $\mathcal{B}$ generated by the state $b \in B$. Consider a diagonal product $\mathcal{M} = \mathcal{B}_{b_1} \Delta \cdots \Delta \mathcal{B}_{b_n} \Delta \mathcal{A}^{2n}$ where $\mathcal{A}^{2n}$ denotes the $2n$th diagonal power of $\mathcal{A}$. Consider the following subset of the set of states of M. $H = \{b_1, \ldots, b_n, a_{1,1}, a_{1,2}, \ldots, a_{n,1}, a_{n,2}) \mid a_{1,1} = a_{1,2} = \cdots = a_{j-1,1} = a_{j-1,2} = a_{j+1,1} = a_{j+1,2} = \cdots = a_{n,1} = a_{n,2} = a, a_{j,1} = a, a_{j,2} = b, j = 1, \ldots, n\}$. Clearly then for every $(b_1, \ldots, b_n, a_{1,1}, a_{1,2}, \ldots, a_{n,1}, a_{n,2}) \in H$ and $p \in X^*$, there exists exactly one $i \in \{1, \ldots, n\}$ such that $\delta(a_{i,1}, p) \neq \delta(a_{i,2}, p)$. Therefore, it is clear that the mapping $\psi : \{(\delta'(b_1, p), \ldots, \delta'(b_n, p), \delta(a_{1,1}, p), \ldots, \delta(a_{n,2}, p)) \mid (b_1, \ldots, b_n, a_{1,1}, \ldots, a_{n,2}) \in H, p \in X^*\} \to B$ is well defined and it is a state-homomorphism of a state-subautomaton of $\mathcal{M}$ onto $\mathcal{B}$: $\psi(\delta'(b_1, p), \ldots, \delta'(b_n, p), \delta(a_{1,1}, p), \ldots, \delta(a_{n,2}, p)) = \delta'(b_i, p)$ whenever $\delta(a_{i,1}, p) \neq \delta(a_{i,2}, p)$ for some $i \in \{1, \ldots, n\}$. □

Now we turn to the automaton mappings. Given a pair of (not necessarily finite) nonempty sets X, Y, we say that $\varphi : X^* \to Y^*$ is an *automaton mapping* if it preserves the length of the words, and, moreover, an arbitrary initial part of a word is sent by φ into an initial part of the image. For example, if $\mathcal{A} = (A, X, \delta)$ is an automaton, fix $a \in A$; then we have an automaton mapping $\varphi_a : X^* \to A^*$ defined inductively by $\varphi_a(\lambda) = \lambda$, and $\varphi_a(px) = \varphi_a(p)\delta(a, px)$ $(x \in X, p \in X^*)$. Then $\varphi_a(p)$ records the trajectory of a in $\mathcal{A}$ as the successive letters of the word p are input.

Take an element p of X^*. Let φ_p denote a mapping $q \to \varphi_p(q)$ $(q \in X^*)$ having $\varphi(pq) = \varphi(p)\varphi_p(q)$. It is easy to show that φ_p is an automaton mapping. We say that φ_p is a *state* of φ. (Note that φ is a state of itself, namely, $\varphi = \varphi_\lambda$.)

If X and $\{\varphi_p \mid p \in X^*\}$ are finite sets, then we speak of a *finite automaton mapping*. Then $\mathcal{A}_\varphi = (\{\varphi_p : p \in X^*\}, X, \delta_\varphi)$ forms an automaton with $\delta_\varphi(\varphi_p, x) = \varphi_{px}$ $(\varphi_p \in \{\varphi_p : p \in X^*\}, x \in X)$ and then for every $p \in X^*, x_1, \ldots, x_k \in X$, $\varphi_p(x_1 \cdots x_k) = \varphi_p(x_1)\varphi_{px_1}(x_2) \cdots \varphi_{px_1 \cdots x_{k-1}}(x_k)$. Conversely, given an automaton $\mathcal{A} = (A, X, \delta)$, a nonempty finite set Y, let us consider arbitrary mappings $\varphi_a : X \to Y$ for every $a \in A$. Moreover, let a_0 be an arbitrary fixed element of A. Clearly then the mapping $\varphi : X^* \to Y^*$ is an automaton mapping whenever $\varphi(\lambda) = \lambda$ and $\varphi(x_1 x_2 \cdots x_k) = \varphi_{a_0}(x_1)\varphi_{a_1}(x_2) \cdots \varphi_{a_{k-1}}(x_k)$, $a_i = \delta(a_{i-1}, x_i), i \in \{1, \ldots, k-1\}$ $(x_1, \ldots, x_k \in X)$.

An *automaton transformation* is an automaton mapping of the form $\varphi : X^* \to X^*$, where X is a (not necessarily finite) nonempty set. By an easy proof we have that the set K_X of all automaton transformations over X^* forms a semigroup under the usual composition of mappings as multiplication. It is also known that the set A_X of all bijective transformations in K_X is a subgroup of the semigroup K_X. In addition, the set L_X of all finite automaton transformations in K_X is a subsemigroup of K_X, and, moreover, the set G_X of all finite bijective automaton transformations in K_X is a subgroup of A_X, K_X, and L_X.

The following statement is obvious.

Lemma 2.29. *For any $\varphi = \varphi^{(1)}\varphi^{(2)}, \varphi^{(1)}, \varphi^{(2)} \in K_X$, there exist a word $p \in X^*$ and distinct letters $x_1, x_2 \in X$ with $\varphi_p(x_1) = \varphi_p(x_2)$ if and only if there exist an $i \in \{1, 2\}$, a word $q \in X^*$ of length $|p|$, and distinct letters $y_1, y_2 \in X$ such that $\varphi_q^{(i)}(y_1) = \varphi_q^{(i)}(y_2)$.* □

Using this statement, we get a simple proof of the following.

Theorem 2.30. *Neither K_X nor L_X has a basis if X is not a singleton.*

Proof. Let X be a set having at least two elements. Given a nonnegative integer m, let $K_{m,X}$ and $L_{m,X}$ denote the subsets of K_X and L_X such that for every $\varphi \in K_{m,X} \cup L_{m,X}, p \in X^*$, and distinct $x_1, x_2 \in X$, $\varphi_p(x_1) \neq \varphi_p(x_2)$ whenever $|p| \neq m$. Moreover, let $K'_X = \bigcup_{i=0}^{\infty} K_{i,X}$ and $L'_X = \bigcup_{i=0}^{\infty} L_{i,X}$. Consider a fixed letter $x \in X$ and the mapping $\psi \in L_X$ for which $\psi_p(y) = x$, $p \in X^*$, $y \in X$. Let K and L denote generating systems of K_X and L_X, respectively. It is enough to prove that K is not a basis of K_X and L is not a basis of L_X.

It is clear that $\psi \neq \psi^{(1)} \cdots \psi^{(n)}$ on the condition that $\psi^{(1)}, \ldots, \psi^{(n)} \in K'_X$ (or $\psi^{(1)}, \ldots, \psi^{(n)} \in L'_X$). Therefore, $K \not\subseteq K'_X$ and $L \not\subseteq L'_X$. In other words, there exist $\alpha \in K$ and $\beta \in L$, nonnegative integers k_1, k_2, ℓ_1, ℓ_2 with $k_1 < k_2, \ell_1 < \ell_2$, words $p_1, p_2, q_1, q_2 \in X^*$ with $|p_i| = k_i, |q_i| = \ell_i, i = 1, 2$, and letters $x_{i,j}, y_{i,j} \in X, i, j = 1, 2$, with $x_{i,1} \neq x_{i,2}, y_{i,1} \neq y_{i,2}, i = 1, 2$ such that $\alpha_{p_i}(x_{i,1}) = \alpha_{p_i}(x_{i,2}), i = 1, 2$, and $\beta_{q_i}(y_{i,1}) = \beta_{q_i}(y_{i,2}), i = 1, 2$. Now we define the following mappings:

$$\alpha'_r(x) = \begin{cases} x & \text{if } |r| < |p_2|, \\ \alpha_r(x) & \text{otherwise,} \end{cases}$$

$$\alpha''_r(x) = \begin{cases} \alpha_r(x) & \text{if } |r| < |p_2|, \\ x & \text{otherwise,} \end{cases}$$

$$\beta'_r(x) = \begin{cases} x & \text{if } |r| < |q_2|, \\ \beta_r(x) & \text{otherwise,} \end{cases}$$

$$\beta_r''(x) = \begin{cases} \beta_r(x) & \text{if } |r| < |q_2|, \\ x & \text{otherwise.} \end{cases}$$

By Lemma 2.29 it is clear that for any $\mu, \nu \in K_X$, $\mu\alpha\nu \notin \{\alpha', \alpha''\}$, and $\mu\beta\nu \notin \{\beta', \beta''\}$. Therefore $K \setminus \{\alpha\}$ generates α' and α'', and $L \setminus \{\beta\}$ generates β' and β''. On the other hand, we have $\alpha = \alpha'\alpha''$ and $\beta = \beta'\beta''$. Thus $K \setminus \{\alpha\}$ generates K_X and $L \setminus \{\beta\}$ generates L_X. Therefore, K is not a basis of K_X and L is not a basis of L_X. This ends the proof. □

The next question remains an open problem.

Problem 2.31. *Do the group of all bijective automaton transformations and the group of all finite bijective automaton transformations over a fixed alphabet with at least two elements have any basis?*

Now we ask the following question, although it likely is hopeless to determine the correct answer.

Problem 2.32. *Is it decidable for every alphabet X with $|X| \geq 2$ and $\varphi, \varphi^{(1)}, \ldots, \varphi^{(n)} \in L_X$, whether or not $\varphi = \psi^{(1)} \cdots \psi^{(m)}$ for some $\psi^{(1)}, \ldots, \psi^{(m)} \in \{\varphi^{(1)}, \ldots, \varphi^{(n)}\}$? Is this problem decidable provided $\varphi, \varphi^{(1)}, \ldots, \varphi^{(n)} \in G_X$?*

Finally, we remark that Lemma 2.29 and the proof of Theorem 2.30 remain valid if K_X denotes the semigroup of all surjective automaton transformations and L_X denotes the semigroup of all surjective automaton transformations with finite number of states over an infinite set X.

2.3 Automata and Semigroups

Given an automaton $\mathcal{A} = (A, X, \delta)$, let us consider for every $p \in X^*$ the *transition* $\delta_p : A \to A$ *induced by the word* p as defined by $\delta_p(a) = \delta(a, p)$ for $a \in A$. For a given word $p \in X^*$, the *transition induced by* p is the function $\delta_p : A \to A$ that takes any state $a \in A$ to $\delta(a, p)$. We sometimes also say that the word p *represents* the transition δ_p and δ_p is the transformation of A corresponding to the word p. The *characteristic semigroup of* $\mathcal{A}$ is $S(\mathcal{A}) = \{\delta_p \mid p \in X^+\}$, where for every $\delta_p, \delta_q \in S(\mathcal{A})$, $\delta_p\delta_q = \delta_{pq}$; moreover, it is understood that $\delta_p = \delta_q$ if and only if $\delta(a, p) = \delta(a, q)$ holds for every $a \in A$. Sometimes we shall write $\delta(a, p)$ as $a \cdot p$. Then we have $(a \cdot p) \cdot q = a \cdot (pq)$ for all $a \in A$, $p, q \in X^*$. The set of all these mappings forms a monoid (i.e., a semigroup having identity element) $S_1(\mathcal{A})$ under composition as product operation and is called the *characteristic monoid* of $\mathcal{A}$. (Of course, the identity element of this monoid is δ_λ, where $\lambda \in X^*$ is the empty word.) If $S_1(\mathcal{A})$ is a group, then $\mathcal{A}$ is called a *permutation automaton.* A permutation automaton is said to be *trivial* if its characteristic monoid is a singleton. Consider a semigroup S. If S has no identity element, then let S^λ denote the semigroup with $S^\lambda = S \cup \{\lambda\}$, where λ is an arbitrary symbol with $\lambda \notin S$; moreover, the product operation in S is extended to S^λ by $\lambda s = s\lambda = s$ $(s \in S)$ and $\lambda\lambda = \lambda$. If S has identity element (i.e., if S is monoid), then let $S^\lambda = S$. In short, S^λ denotes the least monoid containing S as a subsemigroup.

A caveat: (1) Easily constructed examples show that it is possible for $S(\mathcal{A})$ to be a group even when $\mathcal{A}$ is not a permutation automaton. (2) It may happen that $S_1(\mathcal{A}) \neq$

$(S(\mathcal{A}))^{\lambda}$. For example, this occurs in all cases where $S(\mathcal{A})$ is a group but $\mathcal{A}$ is not a permutation automaton.

Example 2.33. *The automaton* $\mathcal{A} = (\{0, 1\}, \{x\}, \delta)$ *with* $\delta(0, x) = \delta(1, x) = 0$ *is a trivial example such that* $S(\mathcal{A})$ *is a group but* $\mathcal{A}$ *is not a permutation automaton.*

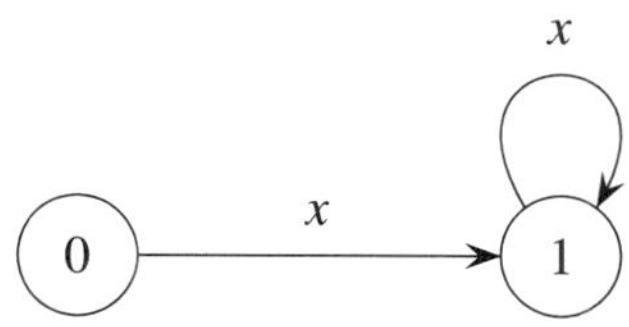

AUTOMATON WHICH IS NOT A PERMUTATION AUTOMATON
BUT WHOSE TRANSFORMATION SEMIGROUP IS A GROUP

Proposition 2.34. *Let* $\mathcal{A} = (A, X, \delta)$ *be an automaton and let* $\mathbf{F} = \{e, z_1, z_2\}$ *be a monoid having two right zero elements with identity e and distinct right zeros* z_1, z_2. *Moreover, let* $\mathbf{F}' = \{z_1, z_2\}$ *be a semigroup with two right zero elements.*

(1) $\mathbf{F}$ *is isomorphic to a submonoid of* $S(\mathcal{A})$ *if and only if there are states* $a_1, a_2 \in A$ *and nonempty words* $p_0, p_1, p_2 \in X^+$ *such that* $\delta(a_i, p_0) = a_i$ *and* $\delta(a_i, p_j) = a_j, i, j \in \{1, 2\}$.

(2) $\mathbf{F}'$ *is isomorphic to a subsemigroup of* $S(\mathcal{A})$ *if and only if there are states* $a_1, a_2 \in A$ *and nonempty words* $p_1, p_2 \in X^+$ *such that* $\delta(a_i, p_j) = a_j, i, j \in \{1, 2\}$.

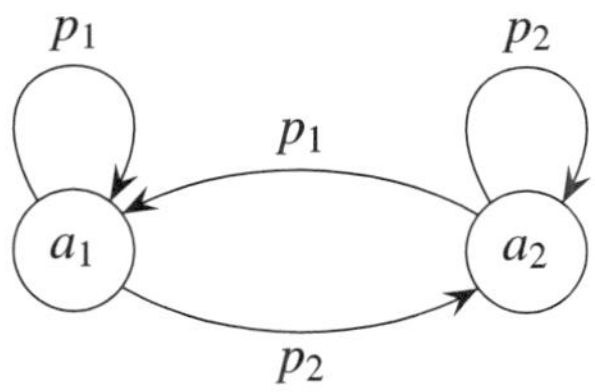

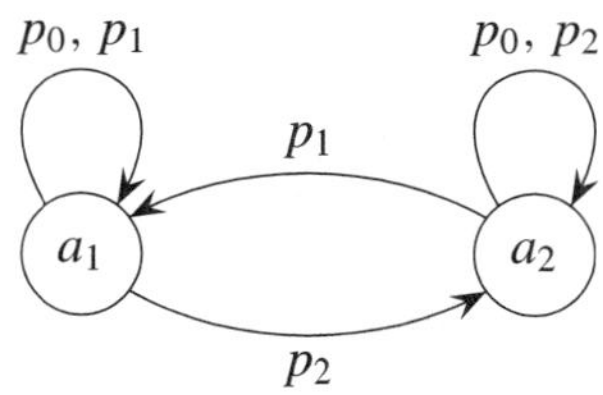

Proof. The sufficiency of (1) and (2) is clear. As regards necessity, first we consider an automaton $\mathcal{A} = (A, X, \delta)$ such that $S(\mathcal{A})$ has a submonoid to be isomorphic to $\mathbf{F}$. Then there are input words $p_0, p_1, p_2 \in X^+$, a subset B of the state set A such that $S' = \{\delta_{p_0}|_B, \delta_{p_1}|_B, \delta_{p_2}|_B\}$ is an isomorphic monoid to $\mathbf{F}$. Consider the isomorphism $\varphi : \mathbf{F} \to S'$ and assume that, say, $\varphi(e) = \delta_{p_0}|_B, \varphi(z_1) = \delta_{p_1}|_B, \varphi(z_2) = \delta_{p_2}|_B$. Then $(\delta_{p_0}|_B)(\delta_{p_0}|_B) = \delta_{p_0}|_B$ implies that $\delta_{p_0}(a) = a$ holds for every $a \in B$. By $\delta_{p_1}|_B \neq \delta_{p_2}|_B$, we obtain that there exists an $a_0 \in B$ such that $\delta_{p_1}(a_0) \neq \delta_{p_2}(a_0)$. Put $a_1 = \delta_{p_1}(a_0), a_2 = \delta_{p_2}(a_0)$. (Note that $a_0 \in \{a_1, a_2\}$ are possible.) Then $(\delta_{p_2}|_B)(\delta_{p_1}|_B) = \delta_{p_1}|_B$ implies $\delta_{p_1}(a_1) = \delta_{p_1}(a_2) = a_1$, and, similarly, $(\delta_{p_1}|_B)(\delta_{p_2}|_B) = \delta_{p_2}|_B$ implies $\delta_{p_2}(a_1) = \delta_{p_2}(a_2) = a_2$. Then we have two states a_1, a_2 satisfying condition (1). This completes the proof of necessity of (1).

Now we assume that $S(\mathcal{A})$ has a subsemigroup to be isomorphic to $\mathbf{F}'$. Then there are input words $p_1, p_2 \in X^*$, a subset B of the state set A such that $S' = \{\delta_{p_1}|_B, \delta_{p_2}|_B\}$ is an isomorphic semigroup to $\mathbf{F}'$. Using a treatment for $\delta_{p_1}|_B, \delta_{p_2}|_B$ as before, we obtain condition (2). This ends the proof. □

Proposition 2.35. *Let $\mathcal{A} = (A, X, \delta)$ be an automaton and let M be a submonoid of $S(\mathcal{A})$ or $S_1(\mathcal{A})$. There exists a nonempty set $B \subseteq A$ with the following properties:*

(1) *The elements of M map B into itself.*

(2) *The restriction of the identity of M to B is the identical mapping $e|_B : B \to B$.*

(3) *If m_1 and m_2 are distinct elements of M, then $m_1(b) \neq m_2(b)$ for at least one $b \in B$.*

Proof. Set $B = \{e(a) : a \in A\}$, where e denotes the identity of M. First we observe that whenever we have $e(a) = b$ for some $a, b \in A$, the equality $ee = e$ implies $e(b) = e(e(a)) = e(a) = b$. Thus $e(b) = b$ holds for every $b \in B$, in accordance with (2). On the other hand, for every $m \in M$, we have $m = em$. Therefore, by $m = em$ and $e(a) \in B, a \in A$, we get $m(b) = (em)(b) = e(m(b)) \in B, b \in B$. Hence, we obtain (1). Finally, let m_1 and m_2 be two elements of M such that $m_1(b) = m_2(b)$ for every $b \in B$. Of course, we have $m_1 = m_1 e, m_2 = m_2 e$ with $e(a) \in B, a \in A$. But then, for every $a \in A$, $m_1(a) = (m_1 e)(a) = m_1(e(a)) = m_2(e(a)) = (m_2 e)(a) = m_2(a)$. This implies $m_1 = m_2$. Therefore, if m_1 and m_2 are two distinct elements of M, then we should have (3). □

We also prove the following consequence of the above statement.

Proposition 2.36. *Let $\mathcal{A} = (A, X, \delta)$ be an automaton and M be a submonoid of $S(\mathcal{A})$ or $S_1(\mathcal{A})$. There exists a nonempty set $B \subseteq A$ with the following properties:*

(1) *The map $\varphi : M \to \{\delta_p|_B : \delta_p \in M, p \in X^+\}$ with $\varphi(\delta_p) = \delta_p|_B, \delta_p \in M, p \in X^+$ is a monoid isomorphism.*

(2) *If M is a group then every restriction $\delta_p|_B, \delta_p \in M, p \in X^+$ is a permutation of B; i.e., the monoid $\{\delta_p|_B : \delta_p \in M, p \in X^+\}$ is a permutation group over B.*

Proof. Consider an automaton $\mathcal{A} = (A, X, \delta)$ and let M be a submonoid of $S(\mathcal{A})$ or $S_1(\mathcal{A})$. Then A has a subset B having the properties of Proposition 2.35.

First we show (2). By conditions (1) and (2) of Proposition 2.35, $M' = \{\delta_p|_B : \delta_p \in M, p \in X^+\}$ is a monoid such that the restriction of the identity e of M to B is the identical mapping $e|_B : B \to B$. Then $e|_B$ is the identity element of M'. On the other hand, for every pair $f|_B, g|_B \in M'$, the mapping $f|_B g|_B$ is a permutation of B if and only if both mappings $f|_B$ and $g|_B$ have this property. In addition, $f|_B(b), g|_B(b), (fg)|_B(b) \in B, b \in B$ implies $f|_B g|_B = (fg)|_B$. Therefore, if g is the inverse of f in M, then $f|_B g|_B = e|_B$, i.e., both $f|_B$ and $g|_B$ should be a permutation of B. This completes the proof of (2).

Now we prove (1). By condition (3) of Proposition 2.35, we obtain that $\varphi : M \to \{\delta_p|_B : \delta_p \in M, p \in X^+\}$ with $\varphi(\delta_p) = \delta_p|_B, \delta_p \in M, p \in X^+$, is a one-to-one mapping. Thus it is enough to show $\delta_p|_B \delta_q|_B = \delta_{pq}|_B, \delta_p, \delta_q \in M, p, q \in X^+$. But this condition comes from $\delta_p \delta_q = \delta_{pq}$ and $\delta_p(b), \delta_q(b), \delta_{pq}(b) \in B, b \in B$. □

Given a semigroup S, let $\mathcal{A}_S = (S^\lambda, S, \delta_S)$ denote again the automaton with $\delta_S(s_1, s_2) = s_1 s_2, s_1 \in S^\lambda, s_2 \in S$. Moreover, for every automaton $\mathcal{A} = (A, X, \delta)$ and $p \in X^*$, let $\delta_p : A \to A$ be defined by $\delta_p(a) = \delta(a, p), a \in A$, as usual. Consider the automaton $\mathcal{A}^{(-)} = (A, \{\delta_x : x \in X\}, \delta^{(-)})$ having $\delta^{(-)}(a, \delta_x) = \delta(a, x), x \in X$.

Proposition 2.37. *For every automaton $\mathcal{A} = (A, X, \delta)$, $\mathcal{A}^{(-)}$ can be represented homomorphically by $\mathcal{A}_{S(\mathcal{A})}$.*

Proof. Define the automaton $\mathcal{A}' = (A, S(\mathcal{A}), \delta')$ such that $\delta'(a, s) = s(a), a \in A$, $s \in S(\mathcal{A})$. Let $|A| = n$ and consider the nth diagonal power $\mathcal{A}'^{\Delta n}$ of $\mathcal{A}'$. Give a fixed arrangement $a_1, \ldots, a_n$ of the elements of the state set A and let $B = \{(a_{s(1)}, \ldots, a_{s(n)}) : s \in S(\mathcal{A})\}$. Clearly, then $\mathcal{B} = (B, S(\mathcal{A}), \delta'')$ with $\delta''(b, s) = \delta(b, s), b \in B, s \in S(\mathcal{A})$, is a state subautomaton of $\mathcal{A}'^{\Delta n}$. On the other hand, it is obvious that $\tau : S \to B$ with $\tau(s) = (a_{s(1)}, \ldots, a_{s(n)}), s \in S$, is a state isomorphism of $\mathcal{A}_{S(\mathcal{A})}$ to $\mathcal{B}$. It is also clear that the mapping $\tau' : B \to A$ with $\tau'((a_{s(1)}, \ldots, a_{s(n)})) = a_{s(1)}, (a_{s(1)}, \ldots, a_{s(n)}) \in B$ is a state homomorphism of $\mathcal{B}$ onto $\mathcal{A}'$. In addition, $\mathcal{A}^{(-)}$ is an input subautomaton of $\mathcal{A}'$. Thus, of course, $\mathcal{A}'$ homomorphically represents $\mathcal{A}$. Therefore, using the transitive property of homomorphic representation, we obtain that $\mathcal{A}_{S(\mathcal{A})}$ homomorphically represents $\mathcal{A}^{(-)}$. □

Given a semigroup S, let $\mathcal{A}_S = (S^\lambda, S, \delta_S)$ be the automaton with $\delta_S(s_1, s_2) = s_1 s_2$ ($s_1 \in S^\lambda, s_2 \in S$), where $s_1 s_2$ denotes the product of s_1 by s_2 in S^λ. We call $\mathcal{A}_S$ the *semigroup automaton* (corresponding to S). If S is a simple group, then we will use the name *simple group automaton* too. Every finite transformation semigroup (A, S) can be identified with an automaton (A, S, δ_S), where $\delta_S(a, s) = s(a) = a \cdot s$ ($a \in A, s \in S$). Clearly, the characteristic semigroup of this automaton is S.

For an arbitrary automaton $\mathcal{A} = (A, X, \delta)$, the set $\{\delta_x : x \in X\}$ is a generating system of its characteristic semigroup $S(\mathcal{A})$. Similarly, $(A, \{\delta_x : x \in X\})$ is a generating system of the transformation semigroup $T(\mathcal{A}) = (A, S(\mathcal{A}))$, which is called *transformation semigroup of the automaton* $\mathcal{A}$. We have $a \cdot \delta_p = \delta_p(a) = \delta(a, p)$ for all $a \in A, \delta_p \in S(\mathcal{A})$. We can derive similarly the *transformation monoid of* $T_1(\mathcal{A}) = (A, S_1(\mathcal{A}))$ of $\mathcal{A}$. Of course, we can consider any automaton $\mathcal{A}$ as a generating system of the corresponding transformation semigroup $(A, S(\mathcal{A}))$ over the set A of its states (identifying x with y whenever $\delta_x = \delta_y$ $(x, y \in X)$). Finally, let us consider the semigroup automaton $\mathcal{A}_S$ corresponding to a semigroup S. Then we can identify this automaton with the transformation semigroup (S^λ, S), which we call the *transformation semigroup of the semigroup* S.

Our central notions are those of homomorphic and isomorphic simulations. An automaton $\mathcal{A} = (A, X, \delta)$ *homomorphically simulates* the automaton $\mathcal{A}' = (A', X', \delta')$ under a surjective mapping τ_1 of a subset B of A onto A' and a mapping τ_2 of X' into X^* with $\tau_1(\delta(b, \tau_2(x))) = \delta'(\tau_1(b), x)$ $(b \in B, x \in X')$. (It is understood that $\delta(b, \tau_2(x)) \in B$ holds for every pair $b \in B, x \in X'$.) If τ_1 is bijective, then we speak about *isomorphic simulation.*

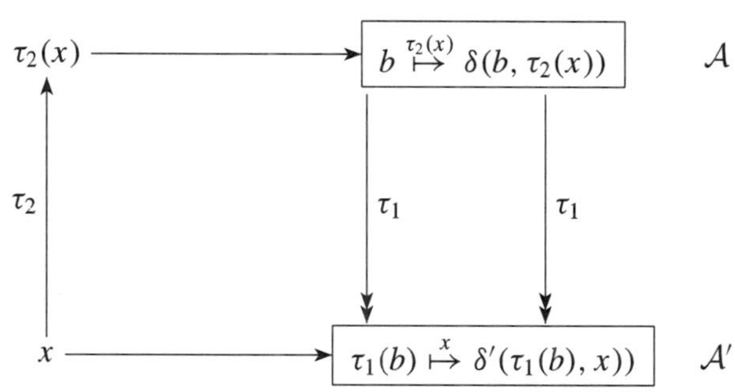

$\mathcal{A}$ HOMOMORPHICALLY SIMULATES $\mathcal{A}'$ UNDER (τ_1, τ_2)

If $\bigcup_{x' \in X'} \tau_2(x') \subseteq X^+$, then (τ_1, τ_2) is called a homomorphic (isomorphic) simulation by nonempty words. If we have a homomorphic simulation by nonempty words, then we also say that $\mathcal{A}'$ *divides* $\mathcal{A}$ (or $\mathcal{A}$ can be divided by $\mathcal{A}'$) and we will use the notation $\mathcal{A}' < \mathcal{A}$ to indicate this. In addition, if there is a positive integer n such that $|\tau_2(x')| = n$ for every $x' \in X'$, then we write $\mathcal{A}'||^{(n)}\mathcal{A}$, and, moreover, we say that $\mathcal{A}'$ *divides* $\mathcal{A}$ *into equal lengths*, denoting this by $\mathcal{A}'||\mathcal{A}$, if $\mathcal{A}'||^{(n)}\mathcal{A}$ holds for some positive integer n.

The above definition of division for automata (often encountered in the automata theoretic literatures) appears to be quite different from the definition of division for transformation semigroups of Section 1.2, which is most commonly encountered in the more algebraic literature. The latter involves two maps going in the same direction, while the former involves maps in opposite directions. This difference is superficial, as the following lemma shows. It establishes that for transformation semigroups the automata-theoretic notion of division is equivalent with the notion of division for transformation semigroups introduced in Chapter 1.

Now we prove that the next statement is a consequence of Proposition 2.37.

Corollary 2.38. *For every automaton* $\mathcal{A}$, *we have* $\mathcal{A}||\mathcal{A}_{S(\mathcal{A})}$.

Proof. Let $\mathcal{A}$ be an arbitrary automaton. Proposition 2.37 implies $\mathcal{A}^{(-)}||\mathcal{A}_{S(\mathcal{A})}$. On the other hand, $\mathcal{A}||\mathcal{A}^{(-)}$ is obvious. Hence, by the transitive property of division we obtain $\mathcal{A}||\mathcal{A}_{S(\mathcal{A})}$. □

Lemma 2.39. *Let* (X, S) *and* (X', S') *be transformation semigroups. Let* $\mathcal{X} = (X, S, \delta_X)$ *and* $\mathcal{X}' = (X', S', \delta')$ *be the corresponding automata. Then* (X, S) *divides* (X', S') *according to the definition of division for transformation semigroups if and only if* $\mathcal{X}$ *divides* $\mathcal{X}'$ *according to the definition of division for automata. That is,* $(X, S) < (X', S')$ *if and only if* $\mathcal{X} < \mathcal{X}'$.

Proof. If (X, S) divides (X', S'), by definition there is a subset $Y \subseteq X'$, subsemigroup T of S', with $Y \cdot T \subseteq Y$, an onto function $\psi_2 : Y \twoheadrightarrow X$, and a surjective semigroup homomorphism $\psi_1 : T \twoheadrightarrow S$ satisfying $\psi_1(y \cdot t) = \psi_2(y) \cdot \psi_1(t)$ for all $y \in Y$, $t \in T$. To establish the division of automata, let $\tau_1 = \psi_2$ and let $\tau_2 : S \to T \subseteq (S')^+$ be given by choosing $\tau_2(s)$ to be an arbitrary member of $\psi_1^{-1}(s)$. Now for all $y \in Y$, $s \in S$, $\tau_1(\delta'(y, \tau_2(s))) = \psi_2(y \cdot \tau_2(s)) = \psi_2(y) \cdot \psi_1(\tau_2(s)) = \psi_2(y) \cdot s = \delta(\tau_1(y), s)$. Also, $\delta'(y, \tau_2(s))$ always lies in Y since $Y \cdot T \subseteq Y$ and $\tau_2(s) \in T$. Thus (τ_1, τ_2) as constructed comprises a division of automata $\mathcal{X} < \mathcal{X}'$.

Conversely, suppose a division of these automata is given. By definition, this comprises a surjective mapping τ_1 of a subset Y of X' onto X and a mapping τ_2 of S into nonempty words $(S')^+$ with $\tau_1(\delta'(y, \tau_2(s))) = \delta(\tau_1(y), s)$ $(y \in Y, s \in S)$. Moreover, $\delta(y, \tau_2(s)) \in Y$ holds for every pair $y \in Y$, $s \in S$. We will use Proposition 1.10. We take the lifts $\tilde{x}$ of any x in X to be the elements of the nonempty set $\tau_1^{-1}(x)$. Distinct members of X never have a common lift, since τ_1 is a function. As a lift for $s \in S$, we take $\tilde{s}$ to be the element of S' represented by the nonempty word $\tau_2(s)$ in S'. Suppose two elements of S have a lift in common: if $\tau_2(s)$ and $\tau_2(s_1)$ represent the same element of S' for some $s, s_1 \in S$, then for all $y \in Y$, $\tau_1(y) \cdot s = \delta(\tau_1(y), s) = \tau_1(\delta'(y, \tau_2(s))) = \tau_1(\delta'(y, \tau_2(s_1))) = \delta(\tau_1(y), s_1) = \tau_1(y) \cdot s_1$. Since τ_1 maps Y onto X, this implies that $x \cdot s = x \cdot s_1$ for all $x \in X$. Since (X, S) is a transformation semigroup, it follows that $s = s_1$. Therefore, distinct elements of S have no lift in common. This establishes condition (1) of Proposition 1.10. For all lifts $\tilde{x} \in \tau_1^{-1}(x)$

of $x \in X$ and $\tilde{s}$ of $s \in S$, we have $\tau_1(\tilde{x} \cdot \tilde{s}) = \tau_1(\delta'(\tilde{x}, \tau_2(s))) = \delta(\tau_1(\tilde{x}), s) = \delta(x, s) = x \cdot s$, i.e., $\tilde{x} \cdot \tilde{s}$ is a lift of $x \cdot s$. This is condition (2) of Proposition 1.10; hence (X, S) divides (X', S'). □

The listed concepts of simulation and divisibility for automata are extended to transformation semigroups in such a way that we handle the transformation semigroups as automata in the manner just discussed. It is an easy exercise to verify the following.

Proposition 2.40. *Let $\mathcal{A} = (A, X, \delta)$ and $\mathcal{A}' = (A', X', \delta')$ be automata. Then the following are equivalent:*

(1) $\mathcal{A}' < \mathcal{A}$,
(2) $(A', S(\mathcal{A}')) < \mathcal{A}$,
(3) $(A', S(\mathcal{A}')) < (A, S(\mathcal{A}))$,
(4) $\mathcal{A}' < (A, S(\mathcal{A}))$. □

In addition, we note that, in general, $\mathcal{A}'||\mathcal{A}$ does not imply $(A', S(\mathcal{A}'))||\mathcal{A}$, but, of course, $\mathcal{A}'||\mathcal{A}$ implies $\mathcal{A}'||(A, S(\mathcal{A}))$. Finally, $(A', S(\mathcal{A}'))||\mathcal{A}$ implies $\mathcal{A}'||\mathcal{A}$ by definition.

Proposition 2.41. *Let $\mathcal{A} = (A, X, \delta)$ and $\mathcal{A}' = (A', X', \delta')$ be automata. Then the following hold:*

(1) *$\mathcal{A}'$ homomorphically simulates $\mathcal{A}$ if and only if $(A', S_1(\mathcal{A}'))$ homomorphically represents $\mathcal{A}$.*
(2) *$\mathcal{A}'$ homomorphically simulates $\mathcal{A}$ by nonempty words (i.e., $\mathcal{A} < \mathcal{A}'$) if and only if $(A', S(\mathcal{A}'))$ homomorphically represents $\mathcal{A}$.* □

Proposition 2.42. *Every automaton $\mathcal{A} = (A, X, \delta)$ is homomorphically represented by the direct product of a discrete automaton $(A, \{1_A\}, \delta_1)$ (with $\delta_1(a, 1_A) = a$ for all $a \in A$) and the characteristic semigroup automaton $(S^\lambda(\mathcal{A}), S(\mathcal{A}), \delta_{S(\mathcal{A})})$.*

Proof. Let $\psi_1(a, s) = a \cdot s$ for all $a \in A$, $s \in S^\lambda(\mathcal{A})$, and $\psi_2(1_A, s) = s$ for all $s \in S(\mathcal{A})$. These maps make up a homomorphic representation. □

Recall that if S' and S are semigroups, then $S' < S$ (S' divides S) means S' is a homomorphic image of a subsemigroup of S. It easy to check that this is equivalent to the division of transformation semigroups $(S'^\lambda, S') < (S^\lambda, S)$, recalling how division for transformation semigroups was defined in Section 1.2. Moreover, using Lemma 2.39, a division of the semigroup automata $(S'^\lambda, S', \delta_{S'}) < (S^\lambda, S, \delta_S)$ is equivalent to the corresponding division of transformation semigroups.

Lemma 2.43. *Let S and S' be finite semigroups. Then the following are equivalent:*

(1) $S' < S$,
(2) $(S'^\lambda, S') < (S^\lambda, S)$,
(3) $(S'^\lambda, S', \delta_{S'}) < (S^\lambda, S, \delta_S)$.

Proof. (2) $\Rightarrow$ (1) is trivial. Indeed, having (2), S has a subsemigroup which can be mapped homomorphically onto S', i.e., $S' < S$.
(1) $\Rightarrow$ (2). Suppose $S' < S$. Then S has a subsemigroup Y which can be mapped homomorphically onto S'. Obviously, then Y^λ is a submonoid of S^λ. Let $\psi : Y \to S'$ denote an appropriate homomorphism and consider the mapping $\psi_2 : Y^\lambda \to S'^\lambda$ with $\psi_2(\lambda) = \lambda$ and $\psi_2(y) = \psi(y)$, $y \in Y$. Moreover, let $\psi_1 = \psi$. Then $Y^\lambda \cdot Y \subseteq Y$ and $\psi_2(y \cdot t) = \psi_2(y) \cdot \psi_1(t)$, $y \in Y^\lambda$, $t \in Y$. Thus we get (2).

It remains to prove that (2) and (3) are equivalent. (2) means that there are a subset Y of S^λ, a subsemigroup T of S with $Y \cdot T \subseteq Y$, a surjective mapping $\psi_2 : Y \to S'^\lambda$, and a surjective semigroup homomorphism $\psi_1 : T \to S$ satisfying $\psi_2(y\dot{t}) = \psi_2(y)\dot{\psi}_1(t)$ for all $y \in Y, t \in T$. (1) means that there exists a nonempty set $Y \subseteq S^\lambda$, functions $h : Y \to X, \varphi : S' \to S$ such that h is surjective, $y \cdot \varphi(s) \in Y$, and $h(y \cdot \varphi(s)) = h(y) \cdot s$ for all $y \in Y$ and $s \in S$. But by Proposition 1.9, these properties are equivalent and thus, indeed, (2) and (3) are equivalent. The proof is complete. □

Proposition 2.44. *Let (X, S) be any transformation semigroup. Then*

(1) $(S^\lambda, S) < \prod_{x \in X}(X, S)$ *and*
(2) $(X, S) < (X, \{1_X\}) \times (S^\lambda, S)$.

Proof. (1) Lift each $s \in S^\lambda$ to the X-tuple whose x-component is $x \cdot s$. Lift each $s \in S$ to the X-tuple $(s, \dots, s)$. Distinct states and inputs have distinct lifts. Using Proposition 1.10 we have a division. (2) follows from the previous proposition. □

We say the semigroup S *divides* an automaton $\mathcal{A}$ ($S < \mathcal{A}$) if S divides the characteristic semigroup $S(\mathcal{A})$ of $\mathcal{A}$. We say $S||^{(n)}\mathcal{A}$ if $S \overset{\varphi}{\twoheadleftarrow} T \leq S(\mathcal{A})$ for T a subsemigroup of $S(\mathcal{A})$ such that for each $s \in S$, there exists $t \in S(\mathcal{A})$ induced by an input word in X^+ of length n (where X is the alphabet of $\mathcal{A}$) with $\varphi(t) = s$. We then say *S divides $\mathcal{A}$ in equal lengths n*, and we write $S||\mathcal{A}$ if this holds for some positive n. (We may also define $S||^{(n)}S(\mathcal{A}^\lambda)$ with $n = 0$ such that S' is trivial and so is S.) In particular, if φ is a one-to-one mapping, then we also say that *S embeds in $S(\mathcal{A})$ in equal lengths with respect to $\mathcal{B}$*, or, more precisely, *S embeds in $S(\mathcal{A})$ in equal lengths n with respect to $\mathcal{B}$*. Notice that for every monoid M and automaton $\mathcal{A}$, we have $M||A$ if and only if $M||S(\mathcal{A}^\lambda)$.

The following statement is obvious.

Proposition 2.45 (transitive property of simulation). *Let $\mathcal{A}, \mathcal{B}, \mathcal{C}$ be arbitrary automata. If $\mathcal{A}$ homomorphically (isomorphically) simulates $\mathcal{B}$ and $\mathcal{B}$ homomorphically (isomorphically) simulates $\mathcal{C}$, then $\mathcal{A}$ homomorphically (isomorphically) simulates $\mathcal{C}$.* □

(Of course, division and division in equal lengths also have the above transitive property.) Let $\mathcal{A} = (A, X, \delta)$ and $\mathcal{A}' = (A', X', \delta')$ be automata such that $\mathcal{A}$ homomorphically represents the automaton $\mathcal{A}'$ under an appropriate pair $\psi = (\psi_1, \psi_2)$, $\psi_1 : B \twoheadrightarrow A'$, $\psi_2 : Y \twoheadrightarrow X'$, $B \subseteq A, Y \subseteq X$. Clearly, then ψ_2 has a bijective restriction φ such that $\mathcal{A}$ homomorphically simulates $\mathcal{A}'$ under (ψ_1, φ^{-1}). Thus we have the following fact showing that homomorphic (or isomorphic) representation is more strict in a certain sense than homomorphic (or isomorphic) simulation. Similarly, the homomorphic representation is more strict in that sense than divisibility.

Proposition 2.46. *If the automaton $\mathcal{A}$ homomorphically (isomorphically) represents the automaton $\mathcal{A}'$, then $\mathcal{A}$ homomorphically (isomorphically) simulates $\mathcal{A}'$ (in equal lengths), and, simultaneously, then $\mathcal{A}'$ divides the automaton $\mathcal{A}$ (in equal lengths).* □

Proposition 2.47. *Given a semigroup S and an automaton $\mathcal{A}$ with n states, let $S < \mathcal{A}$. Then $\mathcal{A}_S < \mathcal{A}^n$, where $\mathcal{A}^n$ denotes the nth direct power of $\mathcal{A}$. Moreover, $S||\mathcal{A}$ implies $\mathcal{A}_S||\mathcal{A}^n$. In more detail, $\mathcal{A}_S$ divides the nth diagonal power $\mathcal{A}^{\Delta n}$ of $\mathcal{A}$ (and, respectively, $\mathcal{A}_S||\mathcal{A}^{\Delta n}$ provided $S||\mathcal{A}$) under appropriate mappings $\tau_1 : B \to S^\lambda$ ($B \subseteq A^n$), and $\tau_2 : S \to X^+$ (having a positive integer m with $|\tau_2(s)| = m$ ($s \in S$) provided $S||\mathcal{A}$).*

Proof. Since S divides $\mathcal{A} = (A, X, \delta)$, there is a surjective homomorphism $\psi : S' \to S$, where S' is a subsemigroup of $S(\mathcal{A})$. In particular, if $S||\mathcal{A}$, then ψ has the property for an appropriate positive integer m that we can correspond an m-length word p_s to every $s \in S$ satisfying $\psi(\delta_{p_s}) = s$, such that $\delta_{p_s} \in S'$.

Consider an arrangement $a_1, \ldots, a_n$ of the elements of A and identify every transformation $t : \{a_1, \ldots, a_n\} \to \{a_1, \ldots, a_n\}$ (including the identity transformation, too) with the state $(t(a_1), \ldots, t(a_n))$ of the direct power $\mathcal{A}^n$. By transitivity of division, it suffices to show $\mathcal{A}_S < \mathcal{A}^{\Delta n}$. (Or in the case of $S||\mathcal{A}$, by transitivity of division in equal lengths, it suffices to show that $\mathcal{A}_S||\mathcal{A}^{\Delta n}$.)

It is clear that, by this correspondence, $\mathcal{A}^{\Delta n}$ has a subautomaton isomorphic to the automaton $\mathcal{M} = (S_1(\mathcal{A}), X, \delta')$, where $\delta'(\delta_t, x) = \delta_{tx}$ ($\delta_t \in S_1(\mathcal{A}), x \in X$). Let us define for every $\delta_t \in B = S' \cup \{\delta_\lambda\} \subseteq S_1(\mathcal{A})$,

$$\tau_1(\delta_t) = \begin{cases} \psi(\delta_t) & \text{if } \delta_t \in S', \\ \lambda \in S^\lambda & \text{if } \delta_t = \delta_\lambda. \end{cases}$$

Moreover, for every $s \in S$, let $\tau_2(s) = p_s$ ($\in X^+$) such that $\psi(\delta'_p) = s$, and, simultaneously, $|p| = m$ provided $S||\mathcal{A}$. By an elementary computation we obtain that $\mathcal{A}^{\Delta n}$ homomorphically simulates the semigroup automaton $\mathcal{A}_S$ under (τ_1, τ_2); therefore, by the fact that τ_2 maps into X^+, $\mathcal{A}_S < \mathcal{A}^{\Delta n}$. Furthermore, we have $\mathcal{A}_S||\mathcal{A}^{\Delta n}$ provided $S||\mathcal{A}$. The proof is complete. □

We now prove the following.

Proposition 2.48. *Consider a pair $\mathcal{A}$, $\mathcal{B}$ of automata and let n denote the set of states of $\mathcal{B}$. Assume that $S(\mathcal{A}) < \mathcal{B}$ ($S(\mathcal{A})||\mathcal{B}$). If $\mathcal{B}$ is nontrivial (i.e., $n > 1$), then $\mathcal{A}$ divides (divides in equal length) the nth diagonal power of $\mathcal{B}$.*

Proof. Let $S(\mathcal{A}) < \mathcal{B}$ ($S(\mathcal{A})||\mathcal{B}$). By Proposition 2.47, this assumption implies $\mathcal{A}_{S(\mathcal{A})} < (\mathcal{B})^{\Delta n}$ ($\mathcal{A}_{S(\mathcal{A})}||(\mathcal{B})^{\Delta n}$), where $(\mathcal{B})^{\Delta n}$ denotes the nth diagonal power of $\mathcal{B}$. On the other hand, by Corollary 2.38, $\mathcal{A}||\mathcal{A}_{S(\mathcal{A})}$. Thus, by the transitive property of division, $\mathcal{A} < (\mathcal{B})^{\Delta n}$ ($\mathcal{A}||(\mathcal{B})^{\Delta n}$) . □

Now we are ready to prove the following statement.

Proposition 2.49. *Given an automaton $\mathcal{A}$ and a semigroup S, let $S < \mathcal{A}$. Suppose that S is either a noncyclic simple group or a subsemigroup of the monoid* **F** *with two right-zero elements. Then $S||\mathcal{A}$.*

Proof. By definition of division, since $S < \mathcal{A} = (A, X, \delta)$, we have a subsemigroup S' of the characteristic semigroup $S(\mathcal{A})$ and a surjective homomorphism $\psi : S' \twoheadrightarrow S$.

First we suppose that $S = \{g_1, \dots, g_n\}$ is a noncyclic simple group. Let $r_1, \dots, r_n \in X^+$ with $\psi(\delta_{r_i}) = g_i,\ i \in \{1, \dots, n\}$. Then, using Theorem 1.2, there exists a positive integer m such that for every $s \in S$ there are permutations $P_{s,1}, \dots, P_{s,m}$ over $\{1, \dots, n\}$, with $s = g_{P_{s,1}(1)} \cdots g_{P_{s,1}(n)} \cdots g_{P_{s,m}(1)} \cdots g_{P_{s,m}(n)}$. But then, of course, $\psi(\delta_{r_{P_{s,1}(1)}} \cdots \delta_{r_{P_{s,1}(n)}} \cdots \delta_{r_{P_{s,m}(1)}} \cdots \delta_{r_{P_{s,m}(n)}}) = \psi(\delta_{r_{P_{s,1}(1)} \cdots r_{P_{s,1}(n)} \cdots r_{P_{s,m}(1)} \cdots r_{P_{s,m}(n)}}) = s$. Consequently, there exists a positive integer t $(= m(|r_1| + \cdots + |r_n|))$ such that for every $s \in S$, there is a t-length word p with $\psi(\delta_p) = s$. Thus we have $S||A$ for the simple group S whenever $S < \mathcal{A}$.

Now we will study the case when S is a subsemigroup of the monoid having two right zero elements (for $S = \mathbf{F} = \{e, z_1, z_2\}$ with identity e and distinct right zeros z_1, z_2). Then we have input words, $p_0, p_1, p_2 \in X^+$ with $\psi(\delta_{p_0}) = e,\ \psi(\delta_{p_1}) = z_1,\ \psi(\delta_{p_2}) = z_2$. Take words $q_0, q_1, q_2 \in X^+$ having $q_0 = (p_0)^{|p_1||p_2|}$, $q_1 = (p_1)^{|p_0||p_2|}$, $q_2 = (p_2)^{|p_0||p_1|}$. It is clear that $|q_0| = |q_1| = |q_2|$, and, simultaneously, $\psi(\delta_{q_0}) = e,\ \psi(\delta_{q_1}) = z_1,\ \psi(\delta_{q_2}) = z_2$. We omit the easy proof for the subsemigroups of this monoid. The proof is complete. □

2.4 Automata Networks and Products of Automata

In what follows we also use the concept of compatibility in the following sense. This a broader concept than we encountered in Section 2.1, as it allows the new content of a node to be influence directly by several incoming messages.

A transformation $F : X^n \to X^n$ is said to be *compatible* with a digraph $\mathcal{D} = (V, E)$ (of order n) if F has the form $F(x_1, \dots, x_n) = (f_1(x_1, \dots, x_n), \dots, f_n(x_1, \dots, x_n))$ $((x_1, \dots, x_n) \in X^n)$, and $f_i : X^n \to X, i = 1, \dots, n$ may depend only on x_i and those x_j for which $(v_j, v_i) \in E$ (including the case $v_i = v_j$). If F is compatible with $\mathcal{D}$, then sometimes we also say that F is *$\mathcal{D}$-compatible*.

Given an automaton $\mathcal{A} = (A, X, \delta)$, let $A = A_1 \times \cdots \times A_n$ for some $|A_i| \geq 1$ and $n \geq 1$ (where $|A_i|$ denotes the cardinality, i.e., the number of elements in A_i). Then we say that $\mathcal{A}$ is a *finite automata network* of size n. Then the *underlying graph* $\mathcal{D}_\mathcal{A} = (V_\mathcal{A}, E_\mathcal{A})$ of $\mathcal{A}$ is defined by $V_\mathcal{A} = \{1, \dots, n\}$, $E_\mathcal{A} = \{(i, j) \mid$ there exists $x \in X : cp_j(\delta_x)$ really depends on its ith variable$\}$. $\mathcal{A}$ is a *$\mathcal{D}$-network* if $\mathcal{D} = (V, E)$ is a digraph with $V = V_\mathcal{A}$ and $E \supseteq E_\mathcal{A}$. In other words, $\mathcal{A}$ is a $\mathcal{D}$-network if every mapping $\delta_x : A \to A$ $(x \in X)$ is *compatible* with $\mathcal{D}$. Note that a size n automata network may be regarded as comprising n component automata $\mathcal{A}_i = (A_i, A_1 \times \cdots \times A_n \times X, \delta_i)$, $i \in \{1, \dots, n\}$, where the δ_i are defined by

$$\delta(a, x) = (\delta_1(a_1, (a, x)), \dots, \delta_n(a_n, (a, x)))$$

for $a = (a_1, \dots, a_n) \in A$, $a_i \in A_i$, $x \in X$. One may of course suppress the components of $A = A_1 \times \cdots \times A_n$ in the inputs to $\mathcal{A}_i$ on which δ_i does not really depend.

Let $\mathcal{A}_i = (A_i, X_i, \delta_i)$ be automata where $i \in \{1, \dots, n\}$, $n \geq 1$. Take a finite nonvoid set X and a *feedback function* $\varphi_i : A_1 \times \cdots \times A_n \times X \to X_i$ for every $i \in \{1, \dots, n\}$. The *general product*[13] (or *Gluškov-type product*) of the automata $\mathcal{A}_i$ with respect to the feedback functions φ_i $(i \in \{1, \dots, n\})$ is defined to be the automaton $\mathcal{A} = \mathcal{A}_1 \times \cdots \times$

[13]A natural extension of this concept is the so-called generalized product introduced by F. Gécseg (see also later in this monograph), when the feedback components map into the input monoids of the component automata. Several generalized product families, derived from the Gécseg-type generalization, have also been defined.

$\mathcal{A}_n(X, (\varphi_1, \ldots, \varphi_n))$ with state set $A = A_1 \times \cdots \times A_n$, input set X, transition function δ given by $\delta((a_1, \ldots, a_n), x) = (\delta_1(a_1, \varphi_1(a_1, \ldots, a_n, x)), \ldots, \delta_n(a_n, \varphi_n(a_1, \ldots, a_n, x)))$ for all $(a_1, \ldots, a_n) \in A$, and $x \in X$. In particular, if $\mathcal{A}_1 = \cdots = \mathcal{A}_n$, then we say that $\mathcal{A}$ is a *(general) power.* In the special case $n = 1$, then $\mathcal{A} = \mathcal{A}_1(X, \varphi_1)$, and we speak of a *single-factor product.*[14]

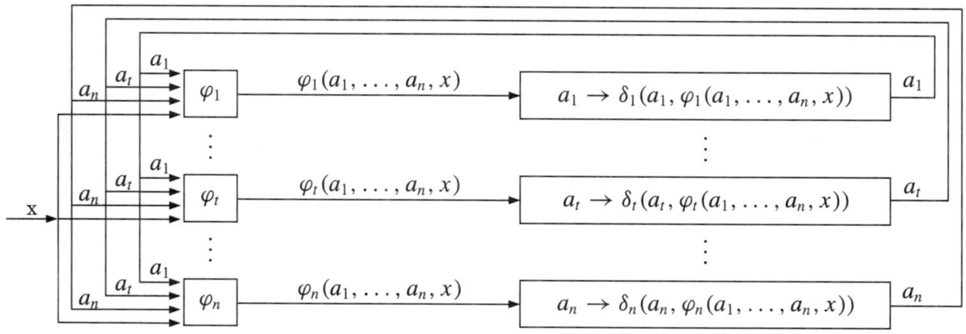

GENERAL PRODUCT

We shall use the feedback functions φ_i, $i = 1, \ldots, n$, in an extended sense as mappings $\varphi_i^* : A_1 \times \cdots \times A_n \times X^*$, where $\varphi_i^*(a_1, \ldots, a_n, \lambda) = \lambda$ and $\varphi_i^*(a_1, \ldots, a_n, px) = \varphi_i^*(a_1, \ldots, a_n, p)\varphi_i(\delta_1(a_1, \varphi_1^*(a_1, \ldots, a_n, p)), \ldots, \delta_n(a_n, \varphi_n^*(a_1, \ldots, a_n, p)), x)$, $a_i \in A_i$, $i = 1, \ldots, n$, $p \in X^*$, $x \in X$. In what follows, φ_i^*, $i \in \{1, \ldots, n\}$, will also be denoted by φ_i.

We can imagine this structure as a working model in the following way. The product is a collection of automata such that every member of this collection is supplied with a transformer which is a special type of an automaton. The transformers, realizing the feedback functions mentioned above, are able to get an input vector containing a common external input sign and the state of all component automata. They can each transform this input vector into an appropriate input sign for their component automaton. The product is at work along a discrete time scale in the following way: all transformers of the product get a common external input sign x, and, simultaneously, all transformers get the value of the instantaneous states $a_1, \ldots, a_n$ of all component-automata as input information. By the effect of this input vector $(a_1, \ldots, a_n, x)$, the transformers produce an input sign $x_t = \varphi_t(a_1, \ldots, a_n, x)$, $t = 1, \ldots, n$, for their component-automata. Then, by the effect of these (transformed) input signs, every component-automaton goes into a new (not necessarily different) $\delta_t(a_t, x_t) = \delta_t(a_t, \varphi_t(a_1, \ldots, a_n, x))$ state, and then, in the next period, this process happens again. We will use several generalizations and several restrictions of this concept. If the transformers, working as microprocessors for their component automata, can produce not only single input signs but entire input words (strings of input signs), so that by the effect of the inner input sign x and the value of the instantaneous states $a_1, \ldots, a_n$ they produce a (possibly empty) input word $\varphi_t(a_1, \ldots, a_n, x))$, then we get the model of the *generalized product*, which will be intensively studied in another volume. If we assume that transformers do not necessarily have access to all the instantaneous states of component automata, but only some restricted subset, then we will get the models of several special types of the products discussed in this volume.

[14]Note that a single-factor product is different from its factor in general.

We shall use the following statement.

Proposition 2.50. *Let $\mathcal{A} = \mathcal{A}_1 \times \cdots \times \mathcal{A}_n(X, \varphi_1, \ldots, \varphi_n)$ be a product of automata $\mathcal{A}_t = (A_t, X_t, \delta_t)$, $t = 1, \ldots, n$, and consider a permutation P over $\{1, \ldots, n\}$. Define the product $\mathcal{A}' = \mathcal{A}'_1 \times \cdots \times \mathcal{A}'_n(X, \varphi'_1, \ldots, \varphi'_n)$ such that $\mathcal{A}'_t = \mathcal{A}_{P(t)}$, and, moreover, for any state $(a_{P(1)}, \ldots, a_{P(n)}) \in A_{P(1)} \times \cdots \times A_{P(n)}$ and input letter $x \in X$, $\varphi'_t(a_{P(1)}, \ldots, a_{P(n)}, x) = \varphi_{P(t)}(a_1, \ldots, a_n, x)$, $t = 1, \ldots, n$. Then $\mathcal{A}$ is isomorphic to $\mathcal{A}'$.* □

It is obvious that $\mathcal{A}$ is a finite automata network if $X_i = A_1 \times \cdots \times A_n \times X$ and φ_i is identity for all $i \in \{1, \ldots, n\}$. Therefore, we can consider automata networks as a special type of products of automata. Conversely, we may assume that every feedback function is realized by special reset automata $\mathcal{R}_i = (X_i, A_1 \times \cdots \times A_n \times X, \delta_{\varphi_i})$, called the ith feedback automaton, such that $\delta_{\varphi_i}(x_i, (a_1, \ldots, a_n, x)) = \varphi_i(a_1, \ldots, a_n, x)$ $(a_1, \ldots, a_n) \in A, x \in X)$ for every $i \in \{1, \ldots, n\}$. Therefore, we can also consider the product of automata as a special type of automata network. (In this model, of course, every component automaton is directly connected to its feedback automaton and feedback automata can get all state components and the joint input letter in every (discrete) time point. Moreover, the component automata of the product do not have the same set of states in general.) Several families of products can be derived from the general product by defining restrictions on the feedback dependency. Thus, for example, A is called a *cascade product* or *α_0-product* if for every $i \in \{1, \ldots, n\}$, φ_i really may not depend on its jth state variable if $j \geq i$. In general, $\mathcal{A}$ is an *α_i-product* $(i = 0, 1, \ldots)$ if each φ_t $(t = 1, \ldots, n)$ is really independent of its jth state component $(j = 1, \ldots, n)$ whenever $j \geq t+i$. In particular, if $\mathcal{A}$ is an α_0-product, then we often give the system of feedback functions in the form $\varphi_1 : X \to X_1, \varphi_2 : A_1 \times X \to X_2, \ldots, \varphi_n : A_1 \times \cdots \times A_{n-1} \times X \to X_n$.[15] If i is a positive integer for which every φ_t $(t = 1, \ldots, n)$ really depends on not more than i state variables, then $\mathcal{A}$ is a *ν_i-product*. In addition, an *$\alpha_i - \nu_j$-product* $(i = 0, 1, \ldots, j = 1, 2, \ldots)$ is an α_i-product that is also a ν_j-product.

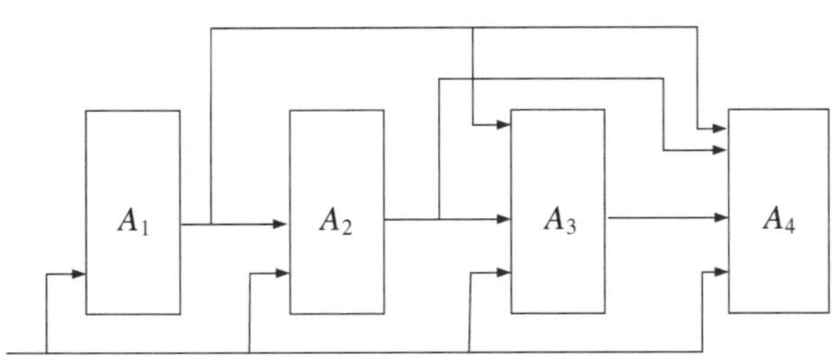

α_0-PRODUCT

[15]Feedback from a factor to itself is considered to be of length 1. Thus, in a sequence of automata, feedback of length 2 is understood to be to the preceding factor.

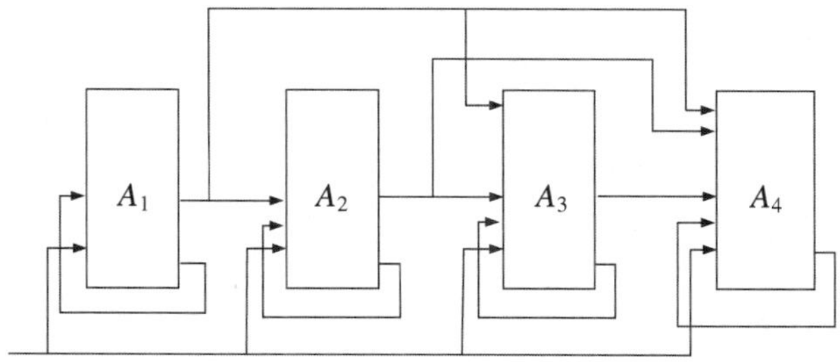

α_1-PRODUCT

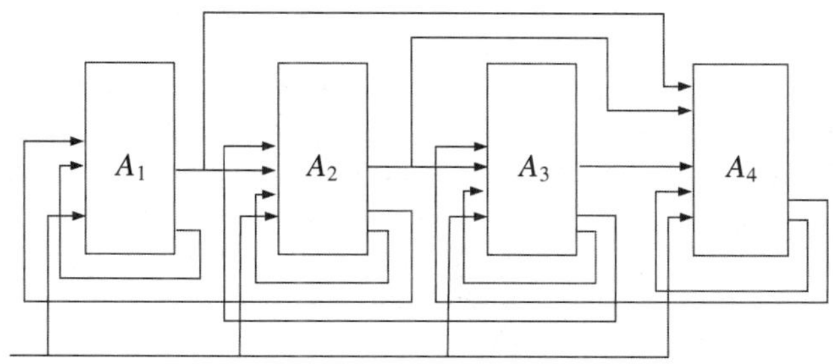

α_2-PRODUCT

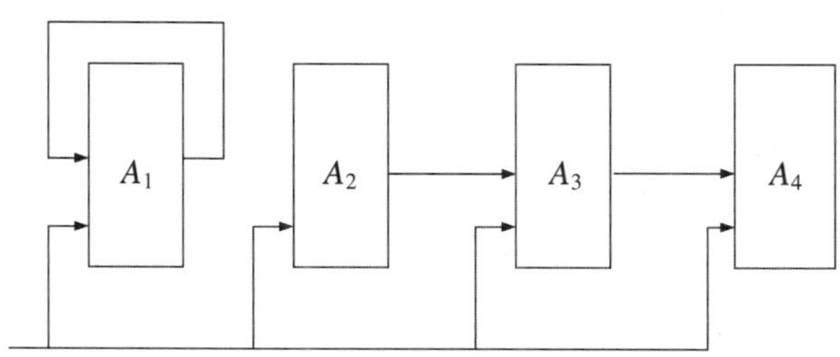

ν_1-PRODUCT

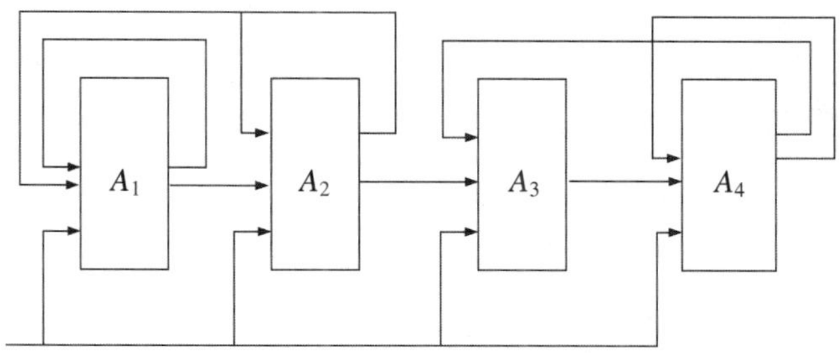

ν_2-PRODUCT

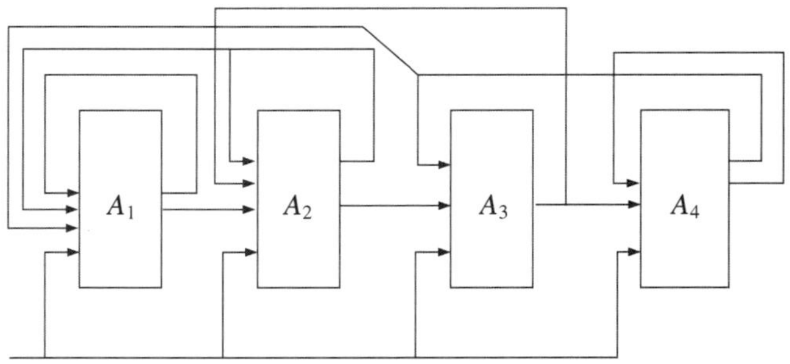

ν_3-PRODUCT

There is a close relationship between the α_0-product of automata and the wreath product of transformation semigroups. If we have an α_0-product as above, then consider the wreath product of the transformation semigroups of the $\mathcal{A}_i$, $(A_n, S(\mathcal{A}_n)) \wr \cdots \wr (A_1, S(\mathcal{A}_1))$; viewing this as an automaton, we see that the α_0-product embeds into it under $(a_1, \ldots, a_n) \mapsto (a_n, \ldots, a_1)$ and $x \mapsto (f_n, \ldots, f_1)$, where $f_1 \in S(\mathcal{A}_1)$ is defined by $f_1 = (\delta_1)_{x_1}$ using $x_1 = \varphi_1(x)$, and, similarly for $i > 1$, $f_i : A_{i-1} \times \cdots \times A_1 \to S(\mathcal{A}_i)$ is defined by $f_i(a_{i-1}, \ldots, a_1) = (\delta_i)_{x_i}$, using $x_i = \varphi_i(a_1, \ldots, a_{i-1}, x)$. Conversely, consider the wreath product $(A_n \times \cdots \times A_1, W)$ of transformation semigroups $\mathcal{A}_i = (A_i, S_i)$ as an automaton with input transformations $(f_n, \ldots, f_1) \in W$; then the α_0-product of the $\mathcal{A}_i$ considered as automata with feedback functions $\varphi_i(a_1, \ldots, a_{i-1}, (f_n, \ldots, f_1)) = f_i(a_{i-1}, \ldots, a_1)$ is isomorphic to the wreath product.

Proposition 2.51. *Given a nonnegative integer i, let $\mathcal{M} = \mathcal{M}_1 \times \cdots \times \mathcal{M}_n(X, \psi_1, \ldots, \psi_n)$, $n \geq 1$, be an α_0-product of α_i-products $\mathcal{M}_j = \mathcal{M}_{j,1} \times \cdots \times \mathcal{M}_{j,k_j}(X_j, \psi_{j,1}, \ldots, \psi_{j,k_j})$, $j = 1, \ldots, n$. Then $\mathcal{M}$ is isomorphic to an α_i-product of $\mathcal{M}_{j,\ell}$, $j = 1, \ldots, n$, $\ell = 1, \ldots, k_j$.* □

Let $\mathcal{D} = (V, E)$ be a digraph with $V = \{1, \ldots, n\}$ and, for every $v \in V$, let $\mathcal{A}_v = (A_v, X_v, \delta_v)$ be an automaton. A general product $\mathcal{A} = \mathcal{A}_1 \times \cdots \times \mathcal{A}_n(X, (\varphi_1, \ldots, \varphi_n))$ is a *$\mathcal{D}$-product* if each feedback function φ_v $(v \in V)$ is really independent of its uth $(u \in V)$ state variable whenever $(u, v) \notin E$. If Δ is a nonempty class of digraphs and $\mathcal{D} \in \Delta$, then it is also said that $\mathcal{A}$ is a *Δ-product.* (In what follows, by a class Δ of digraphs we always mean that Δ is a nonempty class.) Thus, if Δ is the class of all digraphs having neither cycles nor loop edges, then the Δ-product is just the *loop-free product*, which is further equivalent to the cascade product or, by another name, the α_0-product. If Δ consists of the cycles, then we obtain the concept of *loop product.* (Of course, if all factors are the same, then we speak about appropriate types of powers.) If Δ is the class of all digraphs having no edges, then the Δ-product is called a *parallel product* or a *quasi-direct product*, or, in short, a *q-product.*

In other words, we define the *underlying graph* $\mathcal{D} = (V, E)$ $(V = \{1, \ldots, n\}, E \subseteq V \times V)$ of $\mathcal{A}$ such that $(i, j) \in E$ if and only if the feedback function φ_j really depends on its ith-state variable. Thus, an underlying graph is a digraph which may contain loop edges.

We will use the following two facts throughout this monograph.

Proposition 2.52. *Given a digraph $\mathcal{D}$, suppose that an automaton $\mathcal{A}$ can be represented homomorphically (isomorphically) by a $\mathcal{D}$-product of automata $A_t, t = 1, \ldots, n$. Then there exists a $\mathcal{D}$-product of automata $A_t, t = 1, \ldots, n$, having a state-subautomaton which can be mapped state-homomorphically (state-isomorphically) onto $\mathcal{A}$.*

Proof. Let $\psi = (\psi_1, \psi_2), \psi_1 : B \to A, \psi_2 : X_{\mathcal{B}} \to X$ be a homomorphism (an isomorphism) of a subautomaton $B = (B, X_{\mathcal{B}}, \delta_{\mathcal{B}})$ of the $\mathcal{D}$-product $\mathcal{M}' = \mathcal{A}_1 \times \cdots \times \mathcal{A}_n(X', \varphi'_1, \ldots, \varphi'_n)$ onto $\mathcal{A} = (A, X, \delta)$. For every $x \in X$, let u_x be an arbitrary fixed letter in $X_{\mathcal{B}}$ with $\psi_2(u_x) = x$. Define the $\mathcal{D}$-product $\mathcal{M} = \mathcal{A}_1 \times \cdots \times \mathcal{A}_n(X, \varphi_1, \ldots, \varphi_n)$ such that for every $t \in \{1, \ldots, n\}$, $(a_1, \ldots, a_n) \in A_1 \times \cdots \times A_n$ and $x \in X$, $\varphi(a_1, \ldots, a_n, x) = \varphi'(a_1, \ldots, a_n, u_x)$. It is clear that $\mathcal{M}$ is also a $\mathcal{D}$-product. It is also obvious that ψ_1 is a state-homomorphism (a state isomorphism) of the state-subautomaton of $\mathcal{M}$ with the state set B onto A. The proof is complete. □

Proposition 2.53. *Given a digraph $\mathcal{D}$, suppose that an automaton $\mathcal{A}$ can be represented homomorphically (isomorphically) by a $\mathcal{D}$-product of automata $\mathcal{A}_t, t = 1, \ldots, n$. Consider automata $\mathcal{B}_t, t = 1, \ldots, n$, such that for every $t = 1, \ldots, n$, $\mathcal{B}_t$ homomorphically (isomorphically) represents $\mathcal{A}_t$. Then $\mathcal{A}$ can also be represented homomorphically (isomorphically) by a $\mathcal{D}$-product of $\mathcal{B}_t, t = 1, \ldots, n$.*

Proof. For every $t = 1, \ldots, n$, consider a homomorphism (isomorphism) $\psi_t = (\psi_{t,1}, \psi_{t,2})$, $\psi_{t,1} : B'_t \to A_t, \psi_{t,2} : Y'_t \to X_t$ of a subautomaton of $\mathcal{B}_t = (B_t, X'_t, \delta'_t)$ onto $\mathcal{A}_t = (A_t, X_t, \delta_t)$. Moreover, let $\mathcal{A}_1 \times \cdots \times \mathcal{A}_n(\varphi_1, \ldots, \varphi_n, X)$ be a $\mathcal{D}$-product of automata $\mathcal{A}_1, \ldots, \mathcal{A}_n$ which homomorphically (isomorphically) represents $\mathcal{A} = (A, X, \delta)$. Then, by Proposition 2.52, we can also assume that it has a state-subautomaton $\mathcal{M}$ which can be mapped homomorphically (isomorphically) onto $\mathcal{A}$ by a state-homomorphism (state-isomorphism) $\psi : M \to A$. Define the $\mathcal{D}$-product $\mathcal{B}_1 \times \cdots \times \mathcal{B}_n(\varphi'_1, \ldots, \varphi'_n, X)$ in the following manner.

For every $t = 1, \ldots, n$, consider a fixed element d_t of B'_t and define the mapping $\tau_t : B_t \to A_t$ such that

$$\tau_t(b) = \begin{cases} \psi_{t,1}(b) & \text{if } b \in B'_t, \\ \psi_{t,1}(d_t) & \text{otherwise.} \end{cases}$$

Moreover, let ρ_t denote an ordering on X'_t. In addition, define $\varphi'_t(b_1, \ldots, b_n, x) = x'$ such that $x' \in X'_t$ is the minimal input sign (with respect to ρ_t) having $\psi_{t,2}(x') = x''$ with $\varphi_t(\tau_1(b_1), \ldots, \tau_n(b_n), x) = x''$.

First we prove that $\mathcal{B}_1 \times \cdots \times \mathcal{B}_n(\varphi'_1, \ldots, \varphi'_n, X)$ is a $\mathcal{D}$-product. Consider $t \in \{1, \ldots, n\}$ and let $\{(t_1, t), \ldots, (t_m, t)\} = E \cap \{1, \ldots, n\} \times \{t\}$, where E denotes the set of edges of $\mathcal{D}$. Then for every $t \in \{1, \ldots, n\}$, $(b_1, \ldots, b_n) \in B_1 \times \cdots \times B_n, x \in X$, $\varphi_t(\tau_1(b_1), \ldots, \tau_n(b_n), x) = x''$ is unambiguously determined by the components $b_{t_1}, \ldots, b_{t_m}, x$. Therefore, these components unambiguously determine the set $\{z \in X'_t \mid \psi_{t,2}(z) = x''\}$. Then $\varphi'_t(b_1, \ldots, b_n, x) = x'$ is also unambiguously determined (by the components $b_{t_1}, \ldots, b_{t_m}, x$) because it is the minimal element of $\{z \in X'_t \mid \psi_{t,2}(z) = x'\}$ (with respect to ρ_t). Therefore, indeed, $\mathcal{B}_1 \times \cdots \times \mathcal{B}_n(\varphi'_1, \ldots, \varphi'_n, X)$ is a $\mathcal{D}$-product.

Now we prove that this $\mathcal{D}$-product homomorphically (isomorphically) represents $\mathcal{A}$. Let $M' = \{(b_1, \ldots, b_n) \in B'_1 \times \ldots \times B'_n \mid (\psi_{1,1}(b_1), \ldots, \psi_{n,1}(b_n)) \in M\}$; moreover, let $\psi' : M' \to A$ defined by $\psi'((b_1, \ldots, b_n)) = \psi((\psi_{1,1}(b_1), \ldots, \psi_{n,1}(b_n)))$. For every $(b_1, \ldots, b_n) \in M'$ and $t \in \{1, \ldots, n\}$, we have $\varphi'_t(b_1, \ldots, b_n, x) = x'$ such that $x' \in X'_t$ is the minimal input sign (with respect to ρ_t) having $\psi_{t,2}(x') = x''$ with $\varphi_t(\psi_{1,1}(b_1), \ldots, \psi_{n,2}(b_n), x) = x''$. This means that $\psi'((\delta'_1(b_1, \varphi'_1(b_1, \ldots, b_n, x)), \ldots, \delta'_n(b_n, \varphi'_n(b_1, \ldots, b_n, x)))) = \psi((\psi_{1,1}(\delta'_1(b_1, \varphi'_1(b_1, \ldots, b_n, x))), \ldots, \psi_{n,1}(\delta'_n(b_n, \varphi'_n(b_1, \ldots, b_n, x))))) = \psi((\delta_1(\psi_{1,1}(b_1), \psi_{1,2}(\varphi'_1(b_1, \ldots, b_n, x))), \ldots, \delta_n(\psi_{n,1}(b_n), \psi_{n,2}(\varphi'_n(b_1, \ldots, b_n, x))))) = \psi((\delta_1(\psi_{1,1}(b_1), \varphi_1(\psi_{1,1}(b_1), \ldots, \psi_{n,1}(b_n), x))), \ldots, \delta_n(\psi_{n,1}(b_n), \varphi_n(\psi_{1,1}(b_1), \ldots, \psi_{n,1}(b_n), x))) = \delta(\psi((\psi_{1,1}(b_1), \ldots, \delta_{n,1}(b_n))), x) = \delta(\psi'((b_1, \ldots, b_n)), x)$. Therefore, ψ' is a state-homomorphism (state-isomorphism) of a state-subautomaton of the $\mathcal{D}$-product $\mathcal{B}_1 \times \cdots \times \mathcal{B}_n(\varphi'_1, \ldots, \varphi'_n, X)$ onto $\mathcal{A}$. The proof is complete. □

We have the following direct consequence of the above statement.

Proposition 2.54. *Consider a class $\mathcal{K}$ of automata and two classes Δ, Λ of digraphs having the property that every Λ-product of Δ-products of automata from $\mathcal{K}$ is also a Δ-product of automata from $\mathcal{K}$. Suppose that an automaton $\mathcal{A}$ can be represented homomorphically (isomorphically) by a Λ-product of automata $\mathcal{A}_1, \ldots, \mathcal{A}_n$, and assume that for every $t = 1, \ldots, n$, $\mathcal{A}_t$ can be represented homomorphically (isomorphically) by a Δ-product of automata from $\mathcal{K}$. Then $\mathcal{A}$ can be represented homomorphically (isomorphically) by a Δ-product of automata from $\mathcal{K}$.* □

Here and throughout this monograph, if we are dealing with a *class $\mathcal{K}$ of automata,* we always assume that $\mathcal{K}$ is nonvoid. A class $\mathcal{K}$ of automata is called *complete with respect to homomorphic (isomorphic) simulations under the given type Θ of products* if every automaton can be simulated homomorphically (isomorphically) by a Θ-product of automata in $\mathcal{K}$.

$\mathcal{K}$ is called *finite* if it has a finite number of elements. Furthermore, it is said that $\mathcal{K}$ is *minimal* if for every $\mathcal{A} \in \mathcal{K}$, $\mathcal{K} \setminus \mathcal{A}$ is not a complete class of automata with respect to homomorphic (isomorphic) simulations under the Θ-product.

The complete (finite complete, minimal complete) classes of automata with respect to homomorphic (isomorphic) simulations *by nonempty words* are analogously defined.

The next statement is clear.

Proposition 2.55. *Given a digraph $\mathcal{D}$, let $\mathcal{M} = \mathcal{A}_1 \times \cdots \times \mathcal{A}_n(X, \varphi_1, \ldots, \varphi_n)$ be a $\mathcal{D}$-product of automata $\mathcal{A}_1, \ldots, \mathcal{A}_n$ such that $\mathcal{M}$ homomorphically (isomorphically) simulates an automaton $\mathcal{A}$ by some mappings $\tau_1 : B \to A$, $\tau_2 : Y \to X^*$. Then there exists a $\mathcal{D}$-product $\mathcal{M}' = \mathcal{A}_1 \times \cdots \times \mathcal{A}_n(X', \varphi'_1, \ldots, \varphi'_n)$ with the same factors such that $\mathcal{M}'$ homomorphically (isomorphically) simulates $\mathcal{A}$ by $\tau_1 : B \to A$, $\tau'_2 : Y \to X'^*$ having the following properties: $|\tau_2(y)| = |\tau'_2(y)|$, $y \in Y$; moreover, for every positive integer k, ℓ and $y_1, y_2 \in Y$, the kth letter of $\tau'_2(y_1)$ and the ℓth letter of $\tau'_2(y_2)$ coincide only if $k = \ell$ and $y_1 = y_2$.* □

Given a digraph $\mathcal{D} = (V, E)$, let $D^\ell = (V, E')$ be a digraph with $E' = E \cup \{(i, i) \mid i \in V\}$. Similarly, if Δ is a (nonempty) class of digraphs, then we put $\Delta^\ell = \{D^\ell \mid D \in \Delta\}$. Thus, for every digraph $\mathcal{D}$, a $\mathcal{D}^\ell$-product of automata is a general product having an underlying graph $\mathcal{D}^\ell$. Similarly, if $\mathcal{M}$ is a $\mathcal{D}^\ell$-product of automata and $\mathcal{D} \in \Delta$ holds for a class of digraphs, then $\mathcal{M}$ is also said to be Δ^ℓ-product. In this sense we will speak about α_i^ℓ-product, ν_i^ℓ-product, q^ℓ-product, etc.

The following statement is obvious.

Proposition 2.56. *Given a digraph $\mathcal{D}$, an automaton is a $\mathcal{D}^\ell$-product of automata $\mathcal{A}_t, t = 1, \ldots, n$, if and only if it is a $\mathcal{D}$-product of products $\mathcal{A}_t(X_t, \varphi_t)$ of automata $\mathcal{A}_t, t = 1, \ldots, m$, each having a single factor.* □

Direct consequences of Proposition 2.56 include the following three statements.

Proposition 2.57. *Every q^ℓ-product of automata $\mathcal{A}_1, \ldots, \mathcal{A}_n$ coincides with a diagonal product of automata $\mathcal{B}_1, \ldots, \mathcal{B}_n$, where $\mathcal{B}_i$ is a single-factor product of $\mathcal{A}_i$ for every $i = 1, \ldots, n$. Moreover, every q^ℓ-product of automata $\mathcal{A}_i$, $i = 1, \ldots, n$, coincides with a quasi-direct product of automata $\mathcal{B}_i$ which are each single-factor products of $\mathcal{A}_i$.* □

Proposition 2.58. *Every quasi-direct product of automata $\mathcal{A}_1, \ldots, \mathcal{A}_n$ coincides with a diagonal product of automata $\mathcal{B}_1, \ldots, \mathcal{B}_n$, where $\mathcal{B}_i$ is a loop-free product of a single factor $\mathcal{A}_i$ for every $i = 1, \ldots, n$.* □

Proposition 2.59. *Every α_1-product of automata $\mathcal{A}_1, \ldots, \mathcal{A}_n$ coincides with an α_0-product of automata $\mathcal{B}_1, \ldots, \mathcal{B}_n$, where $\mathcal{B}_i$ is a product of a single factor $\mathcal{A}_i$ for every $i = 1, \ldots, n$.* □

The next six statements are also obvious.

Proposition 2.60. *The α_0^ℓ-product coincides with the α_1-product. In addition, if $i > 0$, then the α_i^ℓ-product coincides with the α_i-product.* □

Proposition 2.61. *Every q^ℓ-product is a ν_1-product. Furthermore, every ν_i^ℓ-product is a ν_{i+1}-product.* □

Proposition 2.62. *Given a cycle digraph $\mathcal{D}$ with $\mathcal{D} = (\{1, \ldots, m\}, \{(1, m), (2, 1), (3, 2), \ldots, (m, m-1)\})$, let $\mathcal{M}$ be a $\mathcal{D}^\ell$-product of automata. Then $\mathcal{M}$ also is an α_2-product of its factors.* □

Proposition 2.63. *Suppose that the automaton $\mathcal{M}$ is an α_0-product of factors $\mathcal{M}_1$ and $\mathcal{M}_2$, where $\mathcal{M}_1$ is an α_i-product of $\mathcal{A}_1, \ldots, \mathcal{A}_m$ (having m factors for a given positive integer m); moreover, $\mathcal{M}_2$ is an α_j-ν_k-product (α_j-ν_k^ℓ-product) of $\mathcal{A}_{m+1}, \ldots, \mathcal{A}_n$. Then $\mathcal{M}$ is an $\alpha_{\max(i,j)}$-ν_{m+k}-product ($\alpha_{\max(i,j)}$-ν_{m+k}^ℓ-product) of $\mathcal{A}_1, \ldots, \mathcal{A}_n$.* □

Proposition 2.64. *Suppose that an automaton $\mathcal{A}$ can be represented homomorphically by a general product of nilpotent automata. Then $\mathcal{A}$ is a nilpotent automaton.* □

Proposition 2.65. *Given a monotone automaton $\mathcal{A}$, suppose that $\mathcal{B}$ is a single factor general product of $\mathcal{A}$. Then $\mathcal{B}$ is a monotone automaton.* □

Next we prove the following.

Proposition 2.66. *Let* $\mathcal{A} = (A_1 \times \cdots \times A_n, X, \delta_{\mathcal{A}}) = \mathcal{A}_1 \times \cdots \times \mathcal{A}_n(X, \varphi_1, \ldots, \varphi_n)$ *be a product of automata having an underlying graph* $\mathcal{D} = (\{1, \ldots, n\}, E)$*, vertices* i, j, k *with* $i < j, k$ *such that* $(i, j), (i, k) \in E$*. Suppose that for any pair* ℓ, m*,* $\ell \leq i < m$ *implies* $(m, \ell) \notin E$*. Then there exists a product* $\mathcal{A}' = (A_1 \times \cdots \times A_i \times A_1 \times \cdots \times A_n, X, \delta_{\mathcal{A}'}) = \mathcal{A}_1 \times \cdots \times \mathcal{A}_i \times \mathcal{A}_1 \times \cdots \times \mathcal{A}_n(X, \varphi'_1, \ldots, \varphi'_{i+n})$ *having the underlying graph with nodes* $\{1, \ldots, i+n\}$ *and edges* $(\{(i+u, i+v) \mid (u, v) \in E\} \setminus \{(2i, i+j)\}) \cup \{(i, i+j)\} \cup \{(u, v) \mid (u, v) \in E, u, v \leq i\})$ *such that for any* $a_1 \in A_1, \ldots, a_n \in A_n, x \in X$,

$$\delta_{\mathcal{A}'}((a_1, \ldots, a_i, a_1, \ldots, a_n), x) = (a'_1, \ldots, a'_i, a'_1, \ldots, a'_n)$$

whenever

$$\delta_{\mathcal{A}}((a_1, \ldots, a_n), x) = (a'_1, \ldots, a'_n).$$

Proof. By the condition on edges, $\varphi_1, \ldots, \varphi_i$ do not depend on their $(i+1)$th, ..., nth-state components.

Fix any arbitrary $a'_{i+1} \in A_{i+1}, \ldots, a'_n \in A_n$. We construct the following feedback functions:

$$\varphi'_t(a_1, \ldots, a_{i+n}, x) = \begin{cases} \varphi_t(a_1, \ldots, a_i, a'_{i+1}, \ldots a'_n, x) & \text{if } t \leq i, \\ \varphi_{t-i}(a_{i+1}, \ldots, a_{i+n}, x) & \text{if } t > i \text{ and } t \neq i+j, \\ \varphi_j(a_{i+1}, \ldots, a_{2i-1}, a_i, & \\ \quad a_{2i+1}, \ldots, a_{i+n}, x) & \text{if } t = i+j \end{cases}$$

(where $t = 1, \ldots, i+n$, $(a_1, \ldots, a_{i+n}) \in A_1 \times \cdots \times A_i \times A_1 \times \cdots \times A_n, x \in X$).

It is easy to check that the product $\mathcal{A}'$ having the above feedback function components satisfies the required conditions. □

Corollary 2.67. *Every cascade of automata can be isomorphically represented by a cascade of (copies of the same) automata such that for each* i*, at most one feedback function* φ_j *really depends on the state of* $\mathcal{A}_i$*. Also, the analogous statement holds for the* α_1*-product.* □

Completeness problems are investigated intensively for several families of products of automata.

A class $\mathcal{K}$ of automata is called *complete with respect to homomorphic (isomorphic) representations/simulations under the given type* Θ *of products* if every automaton can be represented/simulated homomorphically (isomorphically) by a Θ-product of automata in $\mathcal{K}$. We also say $\mathcal{K}$ is *homomorphically (isomorphically) complete* under the Θ-product if every finite automaton can be homomorphically (isomorphically) represented by a Θ-product of automata from $\mathcal{K}$. Homomorphic (isomorphic) completeness under any of various other products is defined analogously.

$\mathcal{K}$ is called *finite* if it has a finite number of elements. Furthermore, it is said that $\mathcal{K}$ is *minimal* if for every $\mathcal{A} \in \mathcal{K}$, $\mathcal{K} \setminus \mathcal{A}$ is not a complete class of automata with respect to homomorphic (isomorphic) representations under the Θ-product.

Theorem 2.68 (Gluškov decomposition theorem). *A class $\mathcal{K}$ of automata is complete with respect to isomorphic representations under the general product if and only if there exists an automaton $\mathcal{A} = (A, X, \delta)$ in $\mathcal{K}$ which has input letters $x_1, x_2, x_3, x_4 \in X$ and distinct states $a_0, a \in A$ such that $\delta(a_0, x_1) = a_0, \delta(a_0, x_2) = a, \delta(a, x_3) = a$, and $\delta(a, x_4) = a_0$ hold.*

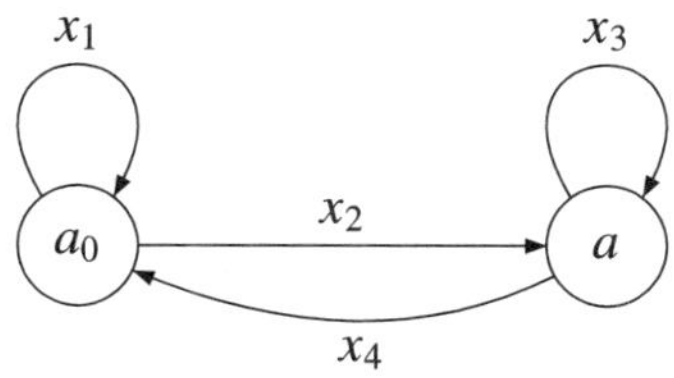

GLUŠKOV CRITERION

Proof. For the proof of necessity, let $\mathcal{B} = \mathcal{B}_1 \times \cdots \times \mathcal{B}_n(X, \varphi_1, \ldots, \varphi_n)$ be a product of automata $\mathcal{B}_t = (B_t, x_t, \delta_t), t = 1, \ldots, n$, such that a subautomaton of $\mathcal{B}$ has an isomorphism $\tau = (\tau_1, \tau_2)$ onto $\mathcal{A}$. For appropriate states $(b_1, \ldots, b_n), (b'_1, \ldots, b'_n)$ and input letters $y_i, i = 1, 2, 3, 4$, let $\tau_1((b_1, \ldots, b_n)) = a_0, \tau_1((b'_1, \ldots, b'_n)) = a, \tau_2(y_i) = x_i, i = 1, 2, 3, 4$. Because of $a \neq a_0$, there exists a $t \in \{1, \ldots, n\}$ with $b_t \neq b'_t$. Let $z_u = \varphi_t(b_1, \ldots, b_n, y_u), u = 1, 2$, and $z_v = \varphi_t(b'_1, \ldots, b'_n, y_v), v = 3, 4$.

Then $\delta_t(b_t, z_1) = b_t, \delta_t(b_t, z_2) = b'_t, \delta_t(b'_t, z_3) = b'_t, \delta_t(b', z_4) = b_t$. Thus $\mathcal{B}_t$ has the conditions of necessity.

Conversely, prove that an arbitrary automaton $\mathcal{M} = (M, X, \delta_{\mathcal{M}})$ can be represented isomorphically by a power of $\mathcal{A}$.

Using the transitive property of isomorphic representation, without loss of generality we may assume that $M = \{a{a_0}^{n-1}, a_0 a {a_0}^{n-2}, \ldots, {a_0}^{n-1}a\}, n > 1$. Define the functions $\varphi_t : \{a_0, a\}^n \times X \to \{x_1, x_2, x_3, x_4\}, t = 1, \ldots, k$, such that for every $b_1 \ldots b_k, c_1 \ldots c_k \in M, x \in X, \delta_{\mathcal{M}}(b_1 \ldots b_k, x) = c_1 \ldots c_k$ implies $\varphi_t(b_1, \ldots, b_n, x) = x_i$ whenever $\delta(b_t, x_i) = c_t$. Clearly, then the power $\mathcal{A}^n(X, \varphi_1, \ldots, \varphi_n)$ isomorphically represents $\mathcal{M}$, where the appropriate isomorphism is $\tau = (\tau_1, \tau_2)$ having $\tau_1(a, a_0, \ldots, a_0) = a{a_0}^{n-1}, \tau_1(a_0, a, a_0, \ldots, a_0) = a_0 a {a_0}^{n-2}, \ldots, \tau_1(a, a_0, \ldots, a_0) = a{a_0}^{n-1}$, and $\tau_2(x) = x, x \in X$.

The proof is complete. □

Note that in the proof of sufficiency of the above theorem, the automaton $\mathcal{M}$ can also be represented by a power of $\mathcal{A}$ having $k \geq \log_2 n$ factors. (We leave to the reader the proof of this statement.)

For homomorphic representations by automata networks, the minimal computational elements required to achieve an arbitrary finite state computation are characterized in the Letichevsky decomposition theorem by a simple criterion. Necessity of the criterion will be shown in Proposition 2.71. Although giving a proof of sufficiency is not difficult, we delay this until the end of Chapter 5. There two proofs of sufficiency are derived from much stronger independently proved results (e.g., Theorem 5.26).

Theorem 2.69 (Letichevsky decomposition theorem). *A class $\mathcal{K}$ of automata is complete with respect to homomorphic representations under the general product if and only if there* *exists an automaton $\mathcal{A} = (A, X, \delta)$ which has a state $a_0 \in A$, two input letters $x, y \in X$, and two input words $p, q \in X^*$, under which*

$$(*) \qquad \delta(a_0, x) \neq \delta(a_0, y) \ \ \textit{and} \ \ \delta(a_0, xp) = \delta(a_0, yq) = a_0.$$

It is said that an automaton $\mathcal{A}$ *satisfies Letichevsky's criterion* if it has the above property $(*)$. If $\mathcal{A} = (A, X, \delta)$ does not satisfy Letichevsky's criterion but we have

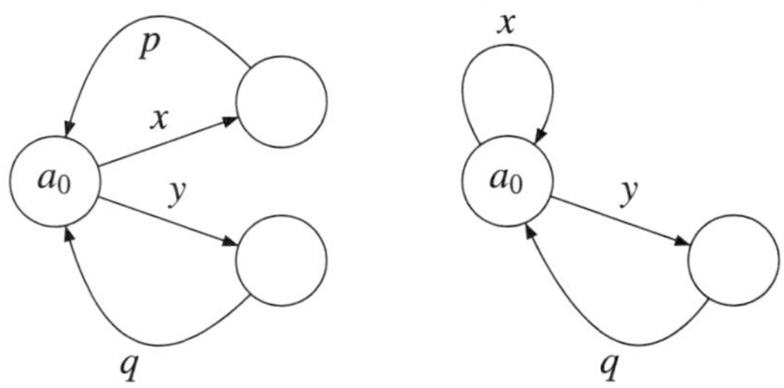

LETICHEVSKY CRITERION

$\delta(a_0, x) \neq \delta(a_0, y)$, and $\delta(a_0, xp) = a_0$ for some $a_0 \in A$, $x, y \in X$, and $p \in X^*$, then $\mathcal{A}$ *satisfies the semi-Letichevsky criterion.* Otherwise we say that $\mathcal{A}$ *does not satisfy any Letichevsky criteria* or *is without Letichevsky criteria.*

Proposition 2.70. *Let there be given an automaton $\mathcal{A} = (A, X, \delta)$, a state $a_0 \in A$, four input words $u, v, p, q \in X^*$ with $|p| = |q|$ under which $\delta(a_0, u) \neq \delta(a_0, v)$, and $\delta(a_0, up) = \delta(a_0, vq) = a_0$. Then $\mathcal{A}$ satisfies Letichevsky's criterion.*

Proof. We shall use the following simple fact. Assume that there are $w_1, w_2, w_1', w_2' \in X^*, x, y \in X$ $w_1xw_2, w_1'yw_2' \in \{up, vq\}$ such that $\delta(a_0, w_1) = \delta(a_0, w_1'), \delta(a_0, w_1x) \neq \delta(a_0, w_1'y)$. Then we obtain Letichevsky's criterion by setting a_0, u, v, p, q to $\delta(a_0, w_1)(= \delta(a_0, w_1')), x, y, w_2w_1, w_2'w_1$, respectively. Therefore, it remains to study the case when for every $w_1, w_2, w_1', w_2' \in X^*, x, y \in X$ with $w_1xw_2, w_1'yw_2' \in \{up, vq\}$ and $\delta(a_0, w_1) = \delta(a_0, w_1')$, it holds that $\delta(a_0, w_1x) = \delta(a_0, w_1'y)$. In this case, there are $x_1, \dots x_n \in X$ having $u = x_1 \cdots x_i, p = x_{i+1} \cdots x_n(x_1 \cdots x_n)^s, v = x_1 \cdots x_j, q = x_{j+1} \cdots x_n(x_1 \cdots x_n)^t$ for appropriate $s, t \geq 0$. But $\delta(a_0, u) \neq \delta(a_0, v)$ implies $i \neq j$. Hence $|p| \neq |q|$, a contradiction. □

If a class $\mathcal{K}$ of automata contains an automaton satisfying Letichevsky's criterion, then we also say that $\mathcal{K}$ *satisfies Letichevsky's criterion.* Otherwise, we say that $\mathcal{K}$ *does not satisfy it.* If $\mathcal{K}$ does not satisfy Letichevsky's criterion but there exists $\mathcal{A} \in \mathcal{K}$ such that $\mathcal{A}$ satisfies the semi-Letichevsky criterion, then it is also said that $\mathcal{K}$ *satisfies the semi-Letichevsky criterion.* Otherwise, we also say that $\mathcal{K}$ *does not satisfy Letichevsky criteria* or *is without any Letichevsky criteria.*

As already mentioned, necessity of the Letichevsky criterion in proving the Letichevsky decomposition theorem follows by the next statement.

Proposition 2.71. *Suppose that a product of automata satisfies Letichevsky's criterion. Then it has a factor with this property.*

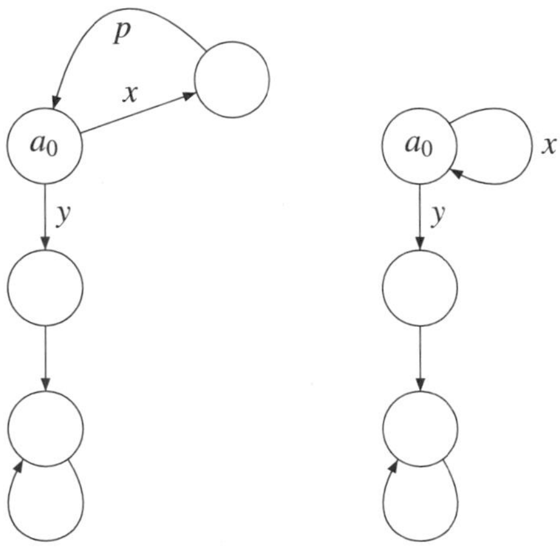

SEMI-LETICHEVSKY CRITERION

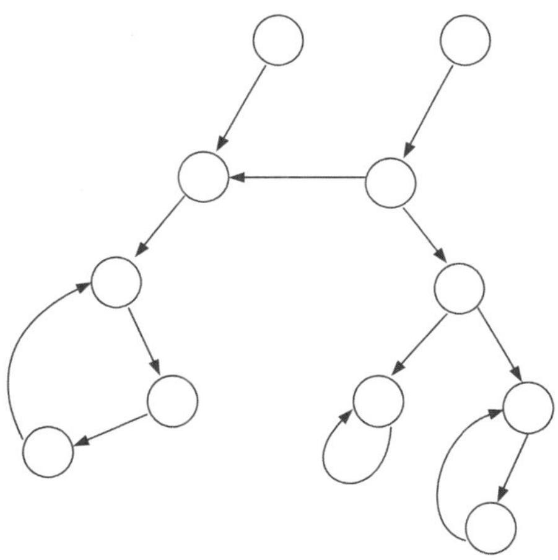

WITHOUT LETICHEVSKY CRITERIA

Proof. Let $\mathcal{A} = \mathcal{A}_1 \times \cdots \times \mathcal{A}_n(X, \varphi_1, \ldots, \varphi_n)$ be a product of automata $\mathcal{A}_t = (A_t, X_t, \delta_t)$, $t = 1, \ldots, n$. Suppose that $\mathcal{A}$ satisfies Letichevsky's criterion. Then there are a state $(a_1, \ldots, a_n) \in A_1 \times \cdots \times A_n$, input letters $x, y \in X$, input words $p, q \in X^*$ such that $(\delta_1(a_1, \varphi_1(a_1, \ldots, a_n, x)), \ldots, \delta_n(a_n, \varphi_n(a_1, \ldots, a_n, x))) \neq (\delta_1(a_1, \varphi_1(a_1, \ldots, a_n, y)), \ldots, \delta_n(a_n, \varphi_n(a_1, \ldots, a_n, y)))$, and, simultaneously, $(\delta_1(a_1, \varphi_1(a_1, \ldots, a_n, xp)), \ldots, \delta_n(a_n, \varphi_n(a_1, \ldots, a_n, xp))) = (\delta_1(a_1, \varphi_1(a_1, \ldots, a_n, yq)), \ldots, \delta_n(a_n, \varphi_n(a_1, \ldots, a_n, yq))) = (a_1, \ldots, a_n)$. But then there are $t \in \{1, \ldots, n\}, x_t = \varphi_t(a_1, \ldots, a_n, x), y_t = \varphi_t(a_1, \ldots, a_n, y), p_t = \varphi_t(a'_1, \ldots, a'_n, p), q_t = \varphi_t(a''_1, \ldots, a''_n, q)$ with $(a'_1, \ldots, a'_n) = (\delta_1(a_1, \varphi_1(a_1, \ldots, a_n, x)), \ldots, \delta_n(a_n, \varphi_n(a_1, \ldots, a_n, x)))$ and $(a''_1, \ldots, a''_n) = (\delta_1(a_1, \varphi_1(a_1, \ldots, a_n, y)), \ldots, \delta_n(a_n, \varphi_n(a_1, \ldots, a_n, y)))$ such that $\delta_t(a_t, x_t) \neq \delta_t(a_t, y_t)$, and, simultaneously, $\delta_t(a_t, x_t p_t) = \delta_t(a_t, y_t q_t) = a_t$. But then $\mathcal{A}_t$ has Letichevsky's criterion.

□

Proposition 2.72. *Suppose that a product of automata satisfies the semi-Letichevsky criterion. Then it has either a factor with Letichevsky's criterion or a factor with the semi-Letichevsky criterion.*

Proof. Let $\mathcal{A} = \mathcal{A}_1 \times \cdots \times \mathcal{A}_n(X, \varphi_1, \ldots, \varphi_n)$ be a product of automata $\mathcal{A}_t = (A_t, X_t, \delta_t)$, $t = 1, \ldots, n$. Suppose that $\mathcal{A}$ satisfies the semi-Letichevsky criterion. Then there are a state $(a_1, \ldots, a_n) \in A_1 \times \cdots \times A_n$, input letters $x, y \in X$, input words $p, q \in X^*$ such that $(\delta_1(a_1, \varphi_1(a_1, \ldots, a_n, x)), \ldots, \delta_n(a_n, \varphi_n(a_1, \ldots, a_n, x))) \neq (\delta_1(a_1, \varphi_1(a_1, \ldots, a_n, y)), \ldots, \delta_n(a_n, \varphi_n(a_1, \ldots, a_n, y)))$, and, simultaneously, $(\delta_1(a_1, \varphi_1(a_1, \ldots, a_n, xp)), \ldots, \delta_n(a_n, \varphi_n(a_1, \ldots, a_n, xp))) = (a_1, \ldots, a_n)$. But then there are $t \in \{1, \ldots, n\}$, $x_t = \varphi_t(a_1, \ldots, a_n, x)$, $y_t = \varphi_t(a_1, \ldots, a_n, y)$, $p_t = \varphi_t(a'_1, \ldots, a'_n, p)$, with $(a'_1, \ldots, a'_n) = (\delta_1(a_1, \varphi_1(a_1, \ldots, a_n, x)), \ldots, \delta_n(a_n, \varphi_n(a_1, \ldots, a_n, x)))$ and $(a''_1, \ldots, a''_n) = (\delta_1(a_1, \varphi_1(a_1, \ldots, a_n, y)), \ldots, \delta_n(a_n, \varphi_n(a_1, \ldots, a_n, y)))$ such that $\delta_t(a_t, x_t) \neq \delta_t(a_t, y_t)$, and, simultaneously, $\delta_t(a_t, x_t p_t) = a_t$. It does not matter whether there exists a $q_t = \varphi_t(a''_1, \ldots, a''_n, q)$ with $\delta_t(a_t, y_t q_t) = a_t$ because then $\mathcal{A}_t$ has either Letichevsky's criterion or the semi-Letichevsky criterion. □

The next two observations show that a statement analogous to Proposition 2.71 does not hold for the semi-Letichevsky criterion.

Proposition 2.73. *There exists an automaton having Letichevsky criterion such that it has a single-factor product satisfying the semi-Letichevsky criterion.*

Proof. Let $\mathcal{A} = (\{0, 1, 2\}, \{x_0, x_1, x_2\}, \delta)$ be defined by $\delta(0, x_i) = i$, $\delta(1, x_i) = 0$, $\delta(2, x_i) = 2$, $i = 0, 1, 2$. Let $\mathcal{A}(\{x, y\}, \varphi)$ be given with $\varphi(i, x) = x_1$, $\varphi(i, y) = x_2$, $i = 0, 1, 2$. Then $\delta(0, \varphi(0, x)) = 0$, $\delta(0, \varphi(0, y)) = 2$, $\delta(2, \varphi(2, x)) = \delta(2, \varphi(2, y)) = 2$, and $\delta(1, \varphi(1, x)) = \delta(1, \varphi(1, y)) = 0$. It is easy to check that $\mathcal{A}$ satisfies Letichevsky's criterion, but the product $\mathcal{A}(\{x, y\}, \varphi)$ does not satisfy it. However, $\mathcal{A}(\{x, y\}, \varphi)$ satisfies the semi-Letichevsky criterion. □

Proposition 2.74. *There exists an automaton having Letichevsky criterion such that it has a single-factor product which is without any Letichevsky criteria.*

Proof. Let us consider again the automaton $\mathcal{A} = (\{0, 1, 2\}, \{x_0, x_1, x_2\}, \delta)$ with $\delta(0, x_i) = i$, $\delta(1, x_i) = 0$, $\delta(2, x_i) = 2$, $i = 0, 1, 2$. Define $\mathcal{A}(\{x, y\}, \varphi)$ with $\varphi(i, x) = \varphi(i, y) = x_2$, $i = 0, 1, 2$. Then for every $i \in \{0, 1, 2\}$ and $p \in \{x, y\}^*$ with $|p| \geq 2$, we obtain $\delta(i, \varphi(i, p) = 2$. Thus the product $\mathcal{A}(\{x, y\}, \varphi)$ does not satisfy Letichevsky's criteria. □

Proposition 2.75. *Given a product of automata, it is without Letichevsky criteria if all its factors have this property.*

Proof. Consider a product of automata such that all of its factors are without Letichevsky criteria. If this product satisfies Letichevsky's criterion, then, by Proposition 2.71, one of its factors has this property, which is a contradiction. If this considered product has the semi-Letichevsky criterion, then, by Proposition 2.72, one of its factors has either Letichevsky's criterions or the semi-Letichevsky criterion, which is a contradiction. Therefore, the considered product is without Letichevsky criteria. □

Proposition 2.76. *Let $\mathcal{A}$ be an arbitrary noncommutative strongly connected automaton. Suppose that an automaton $\mathcal{B}$ homomorphically simulates $\mathcal{A}$. Then $\mathcal{B}$ satisfies Letichevsky's criterion.*

Proof. Consider a noncommutative strongly connected automaton $\mathcal{A} = (A, X, \delta)$. Suppose that $\mathcal{A}$ can be simulated homomorphically by an automaton $\mathcal{B} = (B, X_{\mathcal{B}}, \delta_{\mathcal{B}})$ under $\tau_1 : B' \to A, \tau_2 : X \to X_{\mathcal{B}}{}^*$. Suppose that B' is minimal such that it has no proper subset B'' for which $\mathcal{B}$ homomorphically simulates $\mathcal{A}$ under some $\tau_1' : B'' \to A, \tau_2 : X \to X_{\mathcal{B}}{}^*$. Then, by the strong connectivity of $\mathcal{A}$, for every $b_1, b_2 \in B'$ there exist $x_1, \ldots, x_k \in X$ with $\delta_{\mathcal{B}}(b_1, \tau_2(x_1) \ldots \tau_2(x_k)) = b_2$. In addition, because of the noncommutativity of $\mathcal{A}$, there are a state $a \in A$ and input words $p, q \in X^*$ having $\delta(a, pq) \neq \delta(a, qp)$. Consider the natural extension $\tau_2 : X^* \to X_{\mathcal{B}}{}^*$ of τ_2 (with $\tau_2(\lambda) = \lambda, \tau_2(x_1 \ldots x_n) = \tau_2(x_1) \ldots \tau_2(x_n)$). Then for every $b \in \tau_1{}^{-1}a$, $\delta_{\mathcal{B}}(b, \tau_2(p)\tau_2(q)) \neq \delta_{\mathcal{B}}(b, \tau_2(q)\tau_2(p))$. In addition, by the minimality of B', there exist words $r, s \in X^*$ with $\delta_{\mathcal{B}}(b, \tau_2(p)\tau_2(q)\tau_2(r)) = \delta_{\mathcal{B}}(b, \tau_2(q)\tau_2(p))$, $\delta_{\mathcal{B}}(b, \tau_2(q)\tau_2(p)\tau_2(s)) = b$. On the other hand, $\delta_{\mathcal{B}}(b, \tau_2(p)\tau_2(q)) \neq \delta_{\mathcal{B}}(b, \tau_2(q)\tau_2(p))$ implies that $\lambda \notin \{\tau_2(p)\tau_2(q), \tau_2(q)\tau_2(p)\}$. (Indeed, $\lambda \in \{\tau_2(p)\tau_2(q), \tau_2(q)\tau_2(p)\}$ leads to $\lambda = \tau_2(p)\tau_2(q) = \tau_2(q)\tau_2(p)$, resulting in $\delta_{\mathcal{B}}(b, \tau_2(p)\tau_2(q)) = \delta_{\mathcal{B}}(b, \tau_2(q)\tau_2(p))$ $(= b)$, a contradiction.) Therefore, there are $u, v, v' \in X_{\mathcal{B}}{}^*, x, y \in X_{\mathcal{B}}$ such that $uxv = \tau_2(p)\tau_2(q)), uyv' = \tau_2(q)\tau_2(p))$, and, moreover, $\delta_{\mathcal{B}}(b, ux) \neq \delta_{\mathcal{B}}(b, uy)$. Let $b' = \delta_{\mathcal{B}}(b, u), p' = v\tau_2(r)\tau_2(s)u, q' = v'\tau_2(s)u$. Obviously, then $\delta_{\mathcal{B}}(b', x) \neq \delta_{\mathcal{B}}(b', y)$ and $b' = \delta_{\mathcal{B}}(b', xp') \neq \delta_{\mathcal{B}}(b', yq')$. The proof is complete. □

2.5 Bibliographical Remarks

Section 2.1. Extensive treatment of graph theory is given by F. Harary [1969]. Theorem 2.1 is from K. Kuratowski [1930]. A nice presentation of this result was developed by G. A. Dirac and S. Schuster [1954]. Theorem 2.2 was found by G. Chartrand and F. Harary [1967]. Lemmas 2.6, 2.7, 2.8, 2.9, and 2.10, Theorem 2.11, Proposition 2.13, and Theorems 2.15 and 2.16 can be found in Ananichev, Dömösi, and Nehaniv [in press]. Lemmas 2.8 and 2.9 can also be derived from Z. Ésik [1989b]. The concept of digraph completeness is from Z. Ésik [1991a]. Corollaries 2.19 and 2.20 are in Ésik [1991a]. The other parts of this section are essentially new.

Section 2.2. Many books have given accounts of various aspects of the algebraic theory of automata—for example, Bavel [1983], Eilenberg [1974], Gécseg and Peák [1972], Gécseg [1986], Ginzburg [1968], Hartmanis and Stearns [1966], Holcombe [1982], Nelson [1968], Salomaa [1969], and Simon [1999]. The concept of automaton mapping is given and intensively studied in Raney [1958] and Hořejš [1963]. The generating systems of semigroups and groups of automaton mappings are intensively studied in Csákány and Gécseg [1965], Gécseg [1965], and Zarovnyĭ [1965]. Theorem 2.30 is due to S. V. Alešin [1970a] and P. Dömösi [1972]. (In the present book a new and simpler proof of this statement has been produced.) S. V. Alešin [1970b] stated that the answer to Problem 2.31 is in the affirmative. But, unfortunately, there is a gap in the proof of his Lemma 3, and so the validity of his results may be questionable. (See also Csákány, Mathematical Reviews 45#1687.) The other results are new but elementary.

Section 2.3. Proposition 2.34 issued from Eilenberg [1974]. Proposition 2.47 is new. Proposition 2.49 was proved by Z. Ésik [1991a]. All the other statements are folklore.

Section 2.4. Investigation of finite automata networks goes back to W. S. McCulloch and W. Pitts [1943], J. von Neumann [1966], E. F. Codd [1968], M. Minsky and S. Papert [1969], A. W. Burks [1970], and C. Choffrut [1986]. An extensive algebraic treatment of automata networks was given by M. Tchuente [1979,1982,1983,1985,1986] and by F. Fogelman-Soulié, Y. Robert, and M. Tchuente [1987]. Structural and behavioral equivalence relations in automata networks were studied by T. Saito and H. Nishio [1989]. A verification tool for distributed systems using reduction of finite automata networks was described by E. Madelaine and D. Vergamini [1989]. Finite (and infinite) automata systems as parallel communicating finite (and infinite) automata networks were intensively investigated by Z. Fülöp [1991], C. Martin-Vide and Gh. Păun [1999], C. Martin-Vide and V. Mitrana [2000, 2001], C. Martin-Vide, A. Mateescu, and V. Mitrana [2002], and I. Babcsányi and A. Nagy [2004]. Product and completeness of automata were intensively investigated by F. Gécseg and I. Peák [1972], S. Eilenberg [1974, 1976], J. Dassow [1981], and F. Gécseg [1986]. The concept of the Gluškov-type product is introduced by V. M. Gluškov [1961]. Several specialized types of the Gluškov-type product were defined. The concepts of $\mathcal{D}$-product and Δ-product were proposed by Z. Ésik [1991b]. The quasi-direct product was given by F. Gécseg and I. Peák [1972]. The cascade product is from M. Yoeli [1961]. The loop product was defined by Z. Ésik [1987a]. The family of α_i-products was introduced and intensively studied by F. Gécseg [1974, 1976a, 1986]. The family of ν_i-products is due to P. Dömösi and B. Imreh [1989]. The family of α_i-ν_j-products was given by F. Gécseg and H. Jürgensen [1991]. Products of automata with identity are from Z. Ésik and J. Virágh [1986]. Theorem 2.68 was proved by V. M. Gluškov [1961]. Theorem 2.69 can be found in Letichevsky [1961]. The proof of Proposition 2.76 is new.

Chapter 3

Krohn–Rhodes Theory and Complete Classes

While the fundamental information concerning complete classes with respect to homomorphic representations under the general (Gluškov-type) product is concentrated in the celebrated classical criterion of A. A. Letichevsky, the well-known Krohn–Rhodes decomposition theorem is the basis for studying the cascade product of automata. The cascade product of automata is a general model of automata networks without feedback, and the theorem describes how to synthesize any finite state automaton using such a cascade, and, moreover, it describes the necessary irreducible components in detail. We shall derive the Krohn–Rhodes decomposition theorem from a sophisticated result called the holonomy decomposition theorem, which generally yields much more efficient decompositions than in the original proofs of the former.

Characterization of homomorphic representation is important since one of the major tools for representations is homomorphism. While it is not too general, it is powerful enough. We study homomorphic representation in networks of automata with no feedback (cascade and quasi-direct products) and with low bounds on feedback length (α_i-products for $i \leq 2$) here.

3.1 Krohn–Rhodes and Holonomy Decomposition Theorems

Theorem 3.1 (Krohn–Rhodes decomposition theorem). *Given a finite automaton $\mathcal{A}$, let $\mathbf{F}$ be the flip-flop monoid (the smallest monoid with two right-zero elements); moreover, let $G_1, \ldots, G_n$ denote all simple groups that divide the characteristic semigroup $S(\mathcal{A})$. Then $\mathcal{A}$ can be represented homomorphically by a cascade product of components from $\{\mathcal{A}_{\mathbf{F}}, \mathcal{A}_{G_1}, \ldots, \mathcal{A}_{G_n}\}$. Moreover, if $\mathcal{A}$ is a nontrivial permutation automaton, then the factor $\mathcal{A}_{\mathbf{F}}$ may be excluded.*

Conversely, let $\mathcal{B} = \mathcal{B}_1 \times \cdots \times \mathcal{B}_n(X, (\varphi_1, \ldots, \varphi_n))$ be a cascade product of automata which homomorphically represents the automaton $\mathcal{A}$. If a subsemigroup S of the flip-flop monoid or a simple group S is a homomorphic image of a subsemigroup of $S(\mathcal{A})$, then S is a homomorphic image of a subsemigroup of $S(\mathcal{B}_t)$ for some component automaton $\mathcal{B}_t$ $(t \in \{1, \ldots, n\})$. In addition, a subsemigroup S of the monoid $\mathbf{F}$ with two right-zero elements *can be embedded isomorphically into a finite semigroup T whenever T has a subsemigroup that can be mapped homomorphically onto S.*

Irreducibility. A finite semigroup S is *irreducible* if whenever S divides a wreath product of finite transformation semigroups $(X_2, S_2) \wr (X_1, S_1)$, then S divides S_2 or S divides S_1. Equivalently, if S divides the cascade product of finite automata $\mathcal{A}$ and $\mathcal{B}$, then either $S < \mathcal{A}$ or $S < \mathcal{B}$. Or again, equivalently, "S is irreducible" means that if S divides the α_0-product of finite automata, then it divides one of the factors.

For any finite automaton $\mathcal{A}$, let $PRIMES(\mathcal{A})$ denote the set of simple groups that divide $S(\mathcal{A})$. This notation also applies to transformation semigroups viewed as automata.

We will derive Theorem 3.1 from the following, for whose proof we shall rely on the holonomy decomposition theorem and a series of lemmas.

Theorem 3.2. (Krohn–Rhodes prime decomposition theorem for transformation semigroups). *Let (A, S) be a finite transformation semigroup. Then (A, S) divides a wreath product of transformation monoids (M_i, M_i) such that each $M_i \in PRIMES(S)$ or M_i is the flip-flop monoid. In the case that (A, S) is a permutation group, then the flip-flop monoid is not required and the division can be chosen to be an embedding.*

Conversely, for every wreath product of finite transformation semigroups (B_i, S_i) which (A, S) divides, we have that $G \in PRIMES(S)$ implies that G divides S_i for some i. Moreover, if M is any divisor of the flip-flop monoid which divides S, then M embeds into some S_i.

The subsemigroups of flip-flop monoid **F** are the monoid **F** with two-right-zero elements itself, the two-element monoid with zero elements, the semigroup with two right-zero elements, and the trivial monoid. It is clear that any divisor of **F** is actually isomorphic to one of these. Thus the second part of this theorem characterizes the irreducible finite semigroups as the finite simple groups and the subsemigroups of the flip-flop monoid.

We will say that the class $\mathcal{K}$ of automata *satisfies Krohn and Rhodes' criterion* if for every finite simple group S there is an automaton $\mathcal{A} \in \mathcal{K}$ having that S is a homomorphic image of a subsemigroup of $S(\mathcal{A})$, and, moreover, following the last part of the Krohn–Rhodes decomposition theorem, a class $\mathcal{K}$, satisfying the Krohn–Rhodes criterion, must contain an automaton $\mathcal{A}$ such that a subsemigroup of $S(\mathcal{A})$ is isomorphic to the monoid **F** with two right-zero elements.

These two forms of the Krohn–Rhodes theorem yield an immediate corollary.

Corollary 3.3. *A finite transformation semigroup (resp., finite automaton) divides the wreath product (resp., cascade product) of flip-flops if and only if it has no nontrivial group divisors, or, equivalently, if and only if it has no simple group divisors.*

Such transformation semigroups (resp., automata) all of whose group divisors are trivial are called *aperiodic*, and the result shows that they can be represented homomorphically by cascades of flip-flops (resp., divide a wreath product of flip-flops).

We begin with last part of Theorem 3.2.

Lemma 3.4. *If S is a subsemigroup of the flip-flop monoid* **F** *and S divides a finite semigroup T, then S is isomorphic to a subsemigroup of T.*

Proof. $\mathbf{F} = \{e, \ell, r\}$ with identity e and $xy = y$ for any $x \in \mathbf{F}$, $y \in \{\ell, r\}$. We have $T \supseteq T' \overset{\varphi}{\twoheadrightarrow} S$. Take $\tilde{e}, \tilde{\ell}, \tilde{r} \in T'$ such that φ maps them to e, ℓ, and r, respectively. By replacing $\tilde{e}$ by its unique idempotent power and $\tilde{r}$ and $\tilde{\ell}$ by the unique idempotent powers of $\tilde{e}\tilde{r}\tilde{e}$ and $\tilde{e}\tilde{\ell}\tilde{e}$, we may assume that $\tilde{e}\tilde{r}\tilde{e} = \tilde{r}$ and $\tilde{e}\tilde{\ell}\tilde{e} = \tilde{\ell}$. Let $f = (\tilde{\ell}\tilde{r})^m$ be the unique idempotent power of $\tilde{\ell}\tilde{r}$. Let $g = (f\tilde{\ell})^n$ be the unique idempotent power of $f\tilde{\ell}$. Then $\tilde{e}f = f = f\tilde{e}$, $\tilde{e}g = g = g\tilde{e}$, and $fg = f(f\tilde{\ell})^n = (f\tilde{\ell})^n = g$ and $gf = (f\tilde{\ell})^n(\tilde{\ell}\tilde{r})^m = ((\tilde{\ell}\tilde{r})^m\tilde{\ell})^n(\tilde{\ell}\tilde{r})^m = ((\tilde{\ell}\tilde{r})^{nm}\tilde{\ell})(\tilde{\ell}\tilde{r})^m = ((\tilde{\ell}\tilde{r})^{nm})(\tilde{\ell}\tilde{r})^m = ((\tilde{\ell}\tilde{r})^m)^n(\tilde{\ell}\tilde{r})^m = (f^m)^n f = f$. Clearly $\varphi(\tilde{e}) = e$, $\varphi(f) = r$, and $\varphi(g) = \ell$, so $\tilde{e}$, f, and g are distinct. Thus $\{\tilde{e}, f, g\}$ makes up a subsemigroup of T isomorphic to **F**. Simplifications of this proof establish the result for the other subsemigroups of **F**. □

Next we do some work toward establishing which semigroups are irreducible.

Lemma 3.5. *The flip-flop monoid* **F** *and all its subsemigroups are irreducible.*

Proof. Suppose **F** divides a wreath product of transformation semigroups $(X \times Y, W) = (X, S) \wr (Y, T)$; then by the above lemma $\mathbf{F} \hookrightarrow W$. Let (E, e) be the identity and (L, ℓ) and (R, r) be the two right-zeros in the embedded copy of the flip-flop.

Case 1. $r \neq \ell$. Since $(R, r)(L, \ell) = (L, \ell)$, we have $r\ell = \ell$ and, similarly, $re = r$, whence ℓ cannot equal e. Similarly one shows r cannot equal e. Thus $e, r, \ell \in T$ are pairwise distinct, so the projection from W to T is injective on the embedded copy of **F**. Thus **F** embeds in T.

Case 2. $r = \ell$. We examine the right coordinates of the embedded flip-flop monoid. We have $E, R, L : Y \to S$. Since (L, ℓ) and $(R, r) = (R, \ell)$ are right-zeros in the embedded copy, for all $y \in Y$, $L(y)L(y \cdot \ell) = L(y)$, $R(y)R(y \cdot \ell) = R(y)$, $L(y)R(y \cdot \ell) = R(y)$, $R(y)L(y \cdot \ell) = L(y)$. Similarly, since (E, e) is the identity in this copy of **F**, also $E(y)F(y \cdot e) = F(y)$ and $F(y)E(y \cdot e) = F(y)$ for each $F \in \{E, L, R\}$, $y \in Y$. Since $r = \ell$, R cannot equal L, so there exists a $y_0 \in Y$ with $R(y_0) \neq L(y_0)$. We have $R(y_0)R(y_0 \cdot \ell) = R(y_0)$ and $R(y_0)L(y_0 \cdot \ell) = L(y_0)$, whence $R(y_0 \cdot \ell) \neq L(y_0 \cdot \ell)$. Let $y_1 = y_0 \cdot \ell$. Since $\ell e = \ell = e\ell$, we have $y_1 \cdot e = y_1 = y_1 \cdot \ell$. Thus, the above equations applied to $y = y_1$ show that $\{L(y_1), R(y_1), E(y_1)\}$ has the multiplication table of **F**. We already know that $L(y_1) \neq R(y_1)$, so $L(y_1)E(y_1) = L(y_1)$ and $L(y_1)R(y_1) = R(y_1)$ imply $E(y_1) \neq R(y_1)$. Similarly $E(y_1) \neq L(y_1)$. Thus we have an embedded copy of **F** in S.

The proofs of irreducibility of other subsemigroups of **F** are simplifications of this one. □

Lemma 3.6. *A nontrivial finite group is irreducible if and only if it is a simple group.*

Proof. Let G be a nontrivial finite group. If a group G is not simple, then it has a proper normal subgroup N, so by Lemma 1.16 G is not irreducible. Conversely, if G is a simple group and G divides $(X, S) = (X_2, S_2) \wr (X_1, S_1)$ for some finite transformation semigroups (X_i, S_i) $(i = 1, 2)$, then by Proposition 1.11 there exists a permutation group $(Z, \tilde{G})$ with $Z \subseteq X_2 \times X_1$ and subgroup $\tilde{G}$ of S mapping homomorphically onto G. $\varphi : \tilde{G} \twoheadrightarrow G$.

Consider the projection homomorphism π_1 from S to S_1 restricted to $\tilde{G}$. Let $N = \ker \pi_1$. Let $1_{\tilde{G}} = (i, e)$ denote the identity of $\tilde{G}$. Clearly, $e^2 = e \in S_1$. Define a homomorphism $\psi : \ker \pi_1 \to S_2^{X_1} = S_2 \times \cdots \times S_2$ ($|X_1|$ times) by $\psi(h, e)(x_1) = h(x_1 \cdot e)$ for all $x_1 \in X_1$. (This defines each x_1-coordinate of the $|X_1|$-tuple.) Of course $\psi(h, e)$ can be considered a function from X_1 to S_2. Now, given $(h, e), (h', e) \in N = \ker \pi_1$, if $\psi(h, e) = \psi(h', e)$, then for all $(x_2, x_1) \in X_2 \times X_1$ we have $(x_2, x_1) \cdot (h, e) = (x_2, x_1) \cdot (i, e)(h, e) = (x_2 \cdot i(x_1)h(x_1 \cdot e), x_1 \cdot e) = (x_2 \cdot i(x_1)h'(x_1 \cdot e), x_1 \cdot e) = (x_2, x_1) \cdot (i, e)(h', e) = (x_2, x_1) \cdot (h', e)$, whence $(h, e) = (h', e)$, and so ψ is injective. We also have, for each $x_1 \in X_1$, $\psi((h, e)(h', e))(x_1) = h(x_1 \cdot e)h'(x_1 \cdot e \cdot e) = h(x_1 \cdot e)h'(x_1 \cdot e) = (\psi(h, e)(x_1))(\psi(h', e)(x_1))$. Thus ψ is an injective homomorphism on N. Thus $\tilde{G}/N$ is isomorphic to the subgroup $\pi_1(\tilde{G})$ of S_1 and N is isomorphic to the subgroup $\psi(N)$ of $S_2^{X_1}$. Therefore every Jordan–Hölder factor of $\tilde{G}$ occurs as a Jordan–Hölder factor of $\tilde{G}/N$ or of N, then hence divides $\tilde{G}/N$ or N, and hence divides S_1 or $S_2^{X_1}$. But since $\tilde{G}/\ker \varphi = G$ is simple, G is a Jordan–Hölder factor of $\tilde{G}$. This proves G divides $S_2^{X_1}$ or S_1. In the latter case, we are done. In the former case, G divides $S_2^{X_1} = S_2 \times \cdots \times S_2$. Take H a subgroup of the direct product mapping onto G, say, $\eta : H \twoheadrightarrow G$, and let H_i denote the subsemigroup of S_2 consisting of those elements s_i of S_2 such that s_i occurs as the ith component of some element of H. Consider the homomorphisms $p_i : H_i \to G$ given by $p_i(s_i) = \eta(1, \ldots, 1, s_i, 1, \ldots, 1)$, $s_i \in H_i$; 1 in each coordinate $j \neq i$ denotes the idempotent of S_2 occurring in that position in the identity element of the group H. Let $k = |X_1|$. Then $\eta(s_1, \ldots, s_k) = p_1(s_1) \cdots p_k(s_k)$ for all $(s_1, \ldots, s_k) \in H$, and since the $p_i(s_i)p_j(s_j) = p_j(s_j)p_i(s_i)$ ($s_i \in H_i, s_j \in H_j$) always holds for $i \neq j$, it follows that $\eta(s_1, \ldots, s_k)p_i(s_i')\eta(s_1, \ldots, s_k)^{-1} = p_i(s_i)p_i(s_i')p_i(s_i)^{-1} \in p_i(H_i)$. Thus each $p_i(H_i)$ is a normal subgroup of G. Since G is not trivial and η is surjective, these subgroups cannot all be trivial. Therefore by simplicity of G there is an i such that $p_i(H_i) = G$, whence G divides S_2. This proves that G is irreducible. □

Lemma 3.7. *If* $(X, \overline{G})$ *is a permutation-reset transformation semigroup, then* $(X, \overline{G}) < (X, \overline{\{1_X\}}) \wr (G, G)$.

Proof. Recall that $(X, \overline{\{1_X\}})$ has a semigroup consisting of the identity permutation 1_X and all constant maps $c_x : X \to X$, with $c_x(x') = x$ for all $x' \in X$. Define $\psi : X \times G \twoheadrightarrow X$ by $(x, g) \mapsto x \cdot g$. For $g \in G \subseteq \overline{G}$, define $\tilde{g}$ by $(x_1, g_1) \cdot \tilde{g} = (x_1 \cdot 1_X, g_1 \cdot g) = (x_1, g_1 g)$. For $c_x \in \overline{G}$, define $\tilde{c_x}$ by $(x_1, g_1) \cdot \tilde{c_x} = (x_1 \cdot c_{x \cdot g_1^{-1}}, g_1 \cdot 1) = (x \cdot g_1^{-1}, g_1)$. Then $\psi((x_1, g_1) \cdot \tilde{g}) = x_1 \cdot g_1 g = \psi(x_1, g_1) \cdot g$, and $\psi((x_1, g_1) \cdot \tilde{c_x}) = \psi(x \cdot g_1^{-1}, g_1) = x \cdot g_1^{-1} g_1 = x = \psi(x_1, g_1) \cdot c_x$. This proves the lemma. □

Lemma 3.8. *If* $(X, \overline{\{1_X\}})$ *is a finite transformation semigroup with transformations consisting of the identity transformation and all constant maps on* X*, then* $(X, \overline{\{1_X\}})$ *embeds into the direct product of copies of the flip-flop.*

Proof. Let $A = \{0, 1\}$; then the flip-flop is $(A, \overline{\{1_A\}})$. Let n be such that $2^n \geq |X|$. Let $f : X \to A^n$ be any injective function. Let $f_i : X \to A$ be the ith component of f. Let $x' \in X$ and $s \in \overline{\{1_X\}}$. Then $s = c_x$ for some $x \in X$, or $s = 1_X$. Let $\psi(c_x) = (c_{f_1(x)}, \ldots, c_{f_n(x)})$ and $\psi(1_X) = (1_A, \ldots, 1_A)$. Then ψ is an injective homomorphism.

We have $f(x' \cdot 1_X) = f(x') = (f_1(x'), \ldots, f_n(x')) = (f_1(x') \cdot 1_A, \ldots, f_n(x') \cdot 1_A) = f(x') \cdot \psi(1_X)$. Also $f(x' \cdot c_x) = f(x) = (f_1(x), \ldots, f_n(x)) \in A^n$. On the other hand, $f(x') \cdot \psi(c_x) = f(x') \cdot (c_{f_1(x)}, \ldots, c_{f_n(x)}) = (f_1(x'), \ldots, f_n(x')) \cdot (c_{f_1(x)}, \ldots, c_{f_n(x)}) = (f_1(x') \cdot c_{f_1(x)}, \ldots, f_n(x') \cdot c_{f_n(x)}) = (f_1(x), \ldots, f_n(x))$. This establishes the embedding. □

Holonomy. We introduce the holonomy groups and related notions. First, fix a finite transformation semigroup (A, S). The holonomy decomposition theorem (Theorem 3.9), from which we shall derive the Krohn–Rhodes prime decomposition theorem, is proved by a detailed study of how S acts on subsets of A. Recall that if $q \subseteq A$ and $s \in S$, then

$$q \cdot s = \{a \cdot s \in A \mid a \in q\}.$$

Let $Q = \{A \cdot s \mid s \in S\} \cup \{A\} \cup \{\{a\} \mid a \in A\}$. In this section, we write λ for the identity transformation on A. Then clearly $S^\lambda = S \cup \{\lambda\}$ acts on Q as just described.

We have a reflexive and transitive relation on Q given by

$$p \le q \iff \text{there exists } s \in S^\lambda,\ p \subseteq q \cdot s \quad (p, q \in Q).$$

Consequently, we have an equivalence relation on Q,

$$p \equiv q \iff p \le q \text{ and } q \le p.$$

For each equivalence class in $Q/\!\equiv$, choose a unique representative $\overline{q} \in q/\!\equiv$. Observe that $p \cdot s \le p$ always holds, since $p \cdot s \subseteq p \cdot s$. If $p \le q$, then for appropriate $s, s', s'' \in S^\lambda$, we have $p \subseteq q \cdot s$, $\overline{p} \subseteq p \cdot s'$, and $q \subseteq \overline{q} \cdot s''$, implying $\overline{p} \subseteq \overline{q} \cdot s''ss'$, whence $\overline{p} \le \overline{q}$. By symmetry, it follows that $p \le q \iff \overline{p} \le \overline{q}$. Thus, the set of representatives of the equivalence classes $Q/\!\equiv$ is partially ordered by $\le$. We also write $p < q$ if $p \le q$ but not $q \le p$. Thus, $p < q \iff \overline{p} < \overline{q}$. We say p is a *tile* of q and write $p \prec q$, if $p \subsetneq q$ and for all $z \in Q$, $p \le z \le q$ implies $z = p$ or $z = q$. The set of tiles of q for any $|q| > 1$ is denoted by $B_q = \{p \in Q \mid p \prec q\}$. Since Q contains the singletons, for $|q| > 1$, q equals the union of its tiles, i.e., $q = \bigcup_{p \in B_q} p$.

Define H_q to be the set of permutations of B_q induced by elements of $s \in S^\lambda$. That is, if for $s \in S^\lambda$, the function $s_q : Q \to Q$ defined by $s_q(z) = z \cdot s = \{a \cdot s \mid a \in z\}$ $(z \in Q)$ restricts to $s_q : B_q \to B_q$ and permutes the elements of B_q, then $s_q \in H_q$. H_q is called the *holonomy group* of q in (A, S), and clearly H_q divides S,[16] and (B_q, H_q) is a permutation group called the *holonomy permutation group* of q.

Suppose $q \equiv p$ and $|q| > 1$; then we can write $p \subseteq q \cdot u_{q,p}$ and $q \subseteq p \cdot v_{q,p}$ (for some $u_{q,p}, v_{q,p} \in S^\lambda$). Let $s = u_{q,p}v_{q,p}$; then we claim $s_q \in H_q$. Since q is finite, $q \subseteq p \cdot v_{q,p} \subseteq (q \cdot u_{q,p}) \cdot v_{q,p} = q \cdot s$ implies $q = q \cdot s$. This shows that s permutes the elements of q. Let $z \in B_q$; we have $|z| < |q|$ and so $z \cdot s \subsetneq q$. Suppose $z' \in Q$ with $z \cdot s \subseteq z' \subseteq q$. Let $n > 1$ be such that s^n is the identity permutation on q, and let $z'' = z' \cdot s^{n-1} \in Q$. Then $z' = z'' \cdot s$. Since $z \cdot s \subseteq z' = z'' \cdot s$, we have $z \cdot ss^{n-1} \subseteq z'' \cdot ss^{n-1}$,

[16] In the exceptional case $H_q = \{\lambda_q\}$, we have that division of S is guaranteed since S contains an idempotent. In all other cases, the identity transformation on B_q is, of course, represented by the idempotent power of any nontrivial $s_q \in H_q$, so that H_q is a quotient of the subsemigroup of S consisting of those elements s for which s_q permutes B_q.

and so $z \subseteq z''$. We have $z \subseteq z'' \subseteq q$, and, since z is a tile of q, it follows that $z'' = z$ or $z'' = q$, whence $z' = z \cdot s$ or $z' = q \cdot s = q$. This proves if z is a tile of q, then $z \cdot s$ is also a tile of q. Moreover, if $z_1, z_2 \in B_q$ and $z_1 \cdot s = z_2 \cdot s$, then $z_1 \cdot s^n = z_2 \cdot s^n$, i.e., $z_1 = z_2$. Since B_q is finite, this proves s permutes the elements of B_q. Thus $s_q \in H_q$ as claimed.

Furthermore, $q \equiv p$ implies that, say, (B_q, H_q) is isomorphic to (B_p, H_p). We saw $s = u_{q,p}v_{q,p}$ permutes the elements of q and, similarly, $v_{q,p}u_{q,p}$ permutes those of p. Take $n > 1$ such that $(u_{q,p}v_{q,p})^n$ is the identity permutation of q. Let $\overline{u}_{q,p} = v_{q,p}(u_{q,p}v_{q,p})^{n-1}$. Then $u_{q,p}\overline{u}_{q,p}$ acts as the identity on q and also $\overline{u}_{q,p}u_{q,p}$ acts as the identity on p. For $z \in B_q$, then $z \mapsto z \cdot u_{q,p} \in B_p$ is bijective with inverse $z' \mapsto z' \cdot \overline{u}_{q,p}$. For $s_q \in H_q$, the map $s_q \mapsto \overline{u}_{q,p}s_q u_{q,p} \in H_p$ is an isomorphism of groups with $(z \cdot u_{q,p}) \cdot \overline{u}_{q,p}s_q u_{q,p} = (z \cdot s_q) \cdot u_{q,p}$. Hence, we have an isomorphism of permutation groups.

Now for each q let $u_q = u_{q,\overline{q}}$ and $\overline{u}_q = \overline{u}_{q,\overline{q}}$ be elements of S^λ determining isomorphisms as above between the holonomy permutation group of q and that of its unique representative $\overline{q}$. Observe that since A is equivalent only to itself, we can take u_A and $\overline{u}_A$ to be the identity.

Define the *height* of a member q of Q by $h(q) = 0$ if q is a singleton and otherwise inductively by $h(q) = i$ when $q_0 < q_1 < \cdots < q_i = q$ is a longest such chain in Q ending in q with $h(q_0) = 0$ and $h(q_j) = j$ $(j = 0, \ldots, i-1)$. The singletons are exactly the elements of height zero, and we call $h = h(A)$ the height of (A, S). Note that $h(q) = h(A)$ implies $A = q$. Note also that if q is equivalent to $\overline{q}$, then we have $h(q) = h(\overline{q})$ (using the fact that $p < p'$ if and only if $\overline{p} < \overline{p'}$ applied to the elements in the maximal chains up to q and $\overline{q}$). It is important to note that if $h(p) = h(p') \geq 1$ and $p \subseteq p' \cdot s$, then actually p' is equivalent to p: if $h(p) = i$, we have $q_0 < \cdots < q_i = p \leq p'$, and so by $h(p') = i$ the last inequality cannot be strict, i.e., $p \equiv p'$. Clearly, for each i with $0 \leq i \leq h$, Q contains at least one element having height i.

For each i $(1 \leq i \leq h)$, define $(\mathcal{B}_i, \mathcal{H}_i)$ to be the direct product of the holonomy permutation groups of the height i representatives in Q. Then $\mathcal{B}_i = \prod_{h(\overline{p})=i} B_{\overline{p}}$ and $\mathcal{H}_i = \prod_{h(\overline{p})=i} H_{\overline{p}}$. Then $(\mathcal{B}_i, \mathcal{H}_i)$ is a permutation group and $(\mathcal{B}_i, \overline{\mathcal{H}_i})$ is the associated holonomy permutation-reset transformation semigroup obtained by adjoining all constant maps taking values in $\mathcal{B}_i$. Denote elements of $\mathcal{B}_i$ by boldface variables $\mathbf{b}$ or $\mathbf{b}_i$.

Notation. Suppose that b is a tile of some $p \in Q$ with height i and the p represents its equivalence class. That is, $b \prec p$ and $p = \overline{p}$. Then we denote by $[b]_p$ any arbitrary element of $\mathcal{B}_i$ containing tile b in the p-position.

Also, if $g \in H_p$, then we write $[g]_p$ for any arbitrary element of $\mathcal{H}_i$ containing g in the $\overline{p}$-position and identity elements in all other positions. Observe that $\mathcal{B}_h = B_A$ and $\mathcal{H}_h = H_A$ since A is the only set of height h.

Thus $\mathbf{b}_h = [b_h]_A = b_h$ for all tiles b_h of A and $[g]_A = g$ for all permutations g in the holonomy group of $A = \overline{A}$.

Theorem 3.9 (holonomy decomposition theorem). *Let (A, S) be a finite transformation semigroup; then (A, S) divides a wreath product of its holonomy permutation-reset transformation semigroups $(\mathcal{B}_1, \overline{\mathcal{H}_1}) \wr \cdots \wr (\mathcal{B}_h, \overline{\mathcal{H}_h})$.*

Proof. Let $*$ be a new symbol and define $\eta_i : \mathcal{B}_i \times \cdots \times \mathcal{B}_h \to Q \cup \{*\}$ inductively as i goes from h to 1 by

$$\eta_h(\mathbf{b}_h) = b_h,$$

and, letting $p = \eta_{i+1}(\mathbf{b}_{i+1}, \ldots, \mathbf{b}_h)$, which we suppose has already been well defined for $h \geq i+1 > 1$, we define

$$\eta_i(\mathbf{b}_i, \ldots, \mathbf{b}_h) = \begin{cases} p & \text{if } h(p) < i, \\ b_i \cdot \overline{u}_p & \text{if } h(p) = i,\ b_i \text{ appears in the } \overline{p}\text{-position of } \mathbf{b}_i, \\ * & \text{otherwise.} \end{cases}$$

(It is understood that the last case will apply also if $p = *$.) Observe that in the second case $p \cdot u_p = \overline{p}$ is the representative of the equivalence class of p, and so $b_i \cdot \overline{u}_p \subseteq \overline{p} \cdot \overline{u}_p = p$, and in fact $b_i \cdot \overline{u}_p \prec p$.

Observe that if $\eta_i(\mathbf{b}_i, \ldots, \mathbf{b}_h)$ is not $*$, then $h(\eta_i(\mathbf{b}_i, \ldots, \mathbf{b}_h)) < i$. Thus $\eta_1(\mathbf{b}_1, \ldots, \mathbf{b}_h)$ is either $*$ or a singleton. If $h(p) < i$, then $\mathbf{b}_i$ can have no effect on the value of $\eta_i(\mathbf{b}_i, \ldots, \mathbf{b}_h)$ in the above definition. If $h(p) = i$, then only the $\overline{p}$-position of $\mathbf{b}_i$ may affect the value of η_i. In all cases at most one position of $\mathbf{b}_i$ can affect the value of η_i.

Let Ψ be the set of $(\mathbf{b}_1, \ldots, \mathbf{b}_h)$ such that $\eta_1(\mathbf{b}_1, \ldots, \mathbf{b}_h)$ is a singleton. The η_i's will give our covering map η by "successive approximation." Indeed, an easy induction establishes that $\eta_i(\mathbf{b}_i, \ldots, \mathbf{b}_h) \subseteq \eta_{i+1}(\mathbf{b}_{i+1}, \ldots, \mathbf{b}_h)$ for every $(\mathbf{b}_1, \ldots, \mathbf{b}_h) \in \Psi$, $1 \leq i < h$. Let $\eta : \Psi \to A$ be given by letting $\eta(\mathbf{b}_1, \ldots, \mathbf{b}_h)$ be the unique element of the singleton $\eta_1(\mathbf{b}_1, \ldots, \mathbf{b}_h)$ for $(\mathbf{b}_1, \ldots, \mathbf{b}_h) \in \Psi$.

We show that $\eta : \Psi \twoheadrightarrow A$ is a surjective function: given an arbitrary $a \in A$, we choose $\mathbf{b}_h = [b_h]_A = b_h \prec A$ containing a. Then $a \in \eta_h(\mathbf{b}_h)$ and $\eta_h(\mathbf{b}_h) < h$. Now assuming $\mathbf{b}_{i+1}, \ldots, \mathbf{b}_h$ are defined and $a \in p = \eta_{i+1}(\mathbf{b}_{i+1}, \ldots, \mathbf{b}_h)$ with $h(p) < i+1$, we proceed by induction for $i \geq 1$ and let

$$\mathbf{b}_i = \begin{cases} \mathbf{b}_* & \text{if } h(p) < i, \text{ for fixed but arbitrary } \mathbf{b}_* \in \mathcal{B}_i, \\ [b]_{\overline{p}} & \text{if } h(p) = i, b \prec p \cdot u_p, \text{ and } b \cdot \overline{u}_p \text{ contains } a \\ & \text{for some fixed but arbitrary } b \in B_{\overline{p}}. \end{cases}$$

In the case $h(p) < i$, existence of elements of height i in Q guarantees that an arbitrary fixed $\mathbf{b}_*$ exists. In the case $h(p) = i$, since $i > 0$ the fact that p is the union of its tiles guarantees the existence of a tile b' of p containing a, so b may be taken to be $b' \cdot u_p$.

It is clear that $a \in \eta_i(\mathbf{b}_i, \ldots, \mathbf{b}_h)$ and $h(\eta_i(\mathbf{b}_i, \ldots, \mathbf{b}_h)) < i$ since either (1) $h(p) < i$, and so $\eta_i(\mathbf{b}_i, \ldots, \mathbf{b}_h) = p$ which contains a, or (2) $h(p) = i$, and then $\eta_i(\mathbf{b}_i, \ldots, \mathbf{b}_h)$ is a tile $b \cdot \overline{u}_p$ of p containing a. It follows by induction that $a \in \eta_1(\mathbf{b}_1, \ldots, \mathbf{b}_h)$ and $h(\eta_1(\mathbf{b}_1, \ldots, \mathbf{b}_h)) < 1$, whence $\eta_1(\mathbf{b}_1, \ldots, \mathbf{b}_h)$ is the singleton $\{a\}$. This proves that η is surjective.

We shall use Proposition 1.10 to establish the division. In the terminology of Proposition 1.10, surjectivity of η gives an element $(\mathbf{b}_1, \ldots, \mathbf{b}_h) \in \Psi$ as a lift of state a.

Recall that an element of the wreath product is given by describing its component actions. (See the discussion following the definition of wreath product in Section 1.3.) Thus to specify lift $\tilde{s}$ of an $s \in S$ to the wreath product we need to give appropriate functions $\tilde{s}_i : \mathcal{B}_{i+1} \times \cdots \times \mathcal{B}_h \to \overline{\mathcal{H}_i}$ for $i = h, \ldots, 1$. (For $i = h$, $\tilde{s}_h$ is just an element of $\overline{H_A} = \overline{\mathcal{H}_h}$.) Such an h-tuple $(\tilde{s}_1, \ldots, \tilde{s}_h)$ of functions determines a transformation in the wreath product.

We define a lift for a member s of S to the wreath product by defining $\tilde{s}_i(\mathbf{b}_{i+1}, \ldots, \mathbf{b}_h)$ for $i = h, \ldots, 1$ inductively. First,

$$\tilde{s}_h = \begin{cases} \text{constant } b & \text{if } A \cdot s \subsetneq A \text{ and } b \text{ is fixed but arbitrary such that} \\ & \quad A \cdot s \subseteq b \prec A, \\ s_A & \text{if } A \cdot s = A. \end{cases}$$

We record that $\eta_h(\mathbf{b}_h) \cdot s \subseteq \eta_h(\mathbf{b}_h \cdot \tilde{s}_h)$, which has height less than $h = h(A)$. (This is just the observation that $\mathbf{b}_h \cdot s \subseteq \mathbf{b}_h \cdot \tilde{s}_h \prec A$.)

Fixing $(\mathbf{b}_{i+1}, \ldots, \mathbf{b}_h)$ for the moment, write $\tilde{s}_j$ for $\tilde{s}_j(\mathbf{b}_{j+1}, \ldots, \mathbf{b}_h)$ for $j = i + 1, \ldots, h$. Then, writing p for $\eta_{i+1}(\mathbf{b}_{i+1}, \ldots, \mathbf{b}_h)$ and q for $\eta_{i+1}(\mathbf{b}_{i+1} \cdot \tilde{s}_{i+1}, \ldots, \mathbf{b}_h \cdot \tilde{s}_h)$, define $\tilde{s}_i(\mathbf{b}_{i+1}, \ldots, \mathbf{b}_h)$ inductively as i goes down from $h - 1$ to 1, by

$$\tilde{s}_i(\mathbf{b}_{i+1}, \ldots, \mathbf{b}_h) = \begin{cases} \text{constant } [b \cdot u_q]_{\overline{q}} & \text{if } h(p) \le i \text{ and } h(q) = i, \text{ and } p \cdot s \subsetneq q, \\ & \quad \text{where } b \prec q \text{ is a fixed arbitrary tile with} \\ & \quad p \cdot s \subseteq b, \\ [(\overline{u}_p s u_q)_{\overline{q}}]_{\overline{q}} & \text{if } h(p) = i \text{ and } h(q) = i \text{ and } p \cdot s = q, \\ 1_{\mathcal{H}_i} & \text{otherwise.} \end{cases}$$

Of course $h(p)$ and $h(q)$ are necessarily less than $i + 1$. It is understood that the third case applies also if $p = *$. In the first case, constant $[b \cdot u_q]_{\overline{q}}$ is a constant map taking a value in $\mathcal{B}_i$ whose $\overline{q}$-position is a particular tile of q. Clearly this constant map lies in $\overline{\mathcal{H}_i}$. In the second case, we have $h(p) = i = h(q) = h(p \cdot s) \ge 1$, whence $q = p \cdot s \equiv p$. Therefore $\overline{p} = \overline{q}$. This implies that $\overline{u}_p s u_q$ represents an element of $H_{\overline{p}}$, so that $[(\overline{u}_p s u_q)_{\overline{p}}]_{\overline{p}} \in \mathcal{H}_i$. In all cases, $\tilde{s}_i(\mathbf{b}_{i+1}, \ldots, \mathbf{b}_h) \in \overline{\mathcal{H}_i}$ as required.

Now we are ready to establish the division. Fixing $(\mathbf{b}_1, \ldots, \mathbf{b}_h) \in \Psi$, we again write $p = \eta_{i+1}(\mathbf{b}_{i+1}, \ldots, \mathbf{b}_h)$ and suppose by induction for $i + 1 \le h$ that for $\tilde{s}_h, \ldots, \tilde{s}_{i+1}(\mathbf{b}_{i+2}, \ldots, \mathbf{b}_h)$ as defined above we have

$$p \cdot s = \eta_{i+1}(\mathbf{b}_{i+1}, \ldots, \mathbf{b}_h) \cdot s \subseteq \eta_{i+1}(\mathbf{b}_{i+1} \cdot \tilde{s}_{i+1}, \ldots, \mathbf{b}_h \cdot \tilde{s}_h) = q.$$

We have already observed that this holds with $i + 1 = h$, and now assuming inductively that it holds for $i + 1$, we establish it for i ($i > 0$). We must show that

$$\eta_i(\mathbf{b}_i, \ldots, \mathbf{b}_h) \cdot s \subseteq \eta_i(\mathbf{b}_i \cdot \tilde{s}_i, \ldots, \mathbf{b}_h \cdot \tilde{s}_h) \text{ holds.}$$

Now we shall consider four cases.

Case 1. $h(p) < i$ and $h(q) = i$. We have $p \cdot s \subsetneq q$—lest $q = p \cdot s$ imply $q \le p$, contradicting $h(p) < h(q)$. $\eta_i(\mathbf{b}_i, \ldots, \mathbf{b}_h) = p$ since by definition of η_i since $h(p) < i$. Now

$$\begin{aligned} \eta_i(\mathbf{b}_i \cdot \tilde{s}_i, \ldots, \mathbf{b}_h \cdot \tilde{s}_h) &= (\mathbf{b}_i \cdot \tilde{s}_i)_{\overline{q}} \cdot \overline{u}_q \text{ since } h(q) = i, \\ &= (b \cdot u_q)\overline{u}_q, \text{ where } p \cdot s \subseteq b \prec q \text{ according to the definition} \\ &\quad \text{of } \tilde{s}_i, \\ &= b. \end{aligned}$$

Therefore,

$$\eta_i(\mathbf{b}_i, \ldots, \mathbf{b}_h) \cdot s \subseteq b = \eta_i(\mathbf{b}_i \cdot \tilde{s}_i, \ldots, \mathbf{b}_h \cdot \tilde{s}_h).$$

Case 2. $h(p) = i$, $h(q) = i$, $p \cdot s \subsetneq q$. Now $\eta_i(\mathbf{b}_i, \ldots, \mathbf{b}_h) = (\mathbf{b}_i)_{\overline{p}} \cdot \overline{u}_p \prec p$ by definition of η_i since $h(p) = i$. And since $h(q) = i$, by definition of η_i, we have

$$\begin{aligned}\eta_i(\mathbf{b}_i \cdot \tilde{s}_i, \ldots, \mathbf{b}_h \cdot \tilde{s}_h) &= (\mathbf{b}_i \cdot \tilde{s}_i)_{\overline{q}} \cdot \overline{u}_q,\\ &= (b \cdot u_q) \cdot \overline{u}_q, \text{ where } p \cdot s \subseteq b \prec q \text{ by definition of } \tilde{s}_i,\\ &= b.\end{aligned}$$

So

$$\eta_i(\mathbf{b}_i, \ldots, \mathbf{b}_h) \cdot s = (\mathbf{b}_i)_{\overline{p}} \cdot \overline{u}_p \cdot s \subseteq p \cdot s \subseteq b = \eta_i(\mathbf{b}_i \cdot \tilde{s}_i, \ldots, \mathbf{b}_h \cdot \tilde{s}_h).$$

Case 3. $h(p) = i$, $h(q) = i$, $p \cdot s = q$. We have $p \equiv q$, so $\overline{p} = \overline{q}$. This implies that s maps B_p bijectively onto B_q. Again $\eta_i(\mathbf{b}_i, \ldots, \mathbf{b}_h) = (\mathbf{b}_i)_{\overline{p}} \cdot \overline{u}_p \prec p$ and

$$\begin{aligned}\eta_i(\mathbf{b}_i \cdot \tilde{s}_i, \ldots, \mathbf{b}_h \cdot \tilde{s}_h) &= (\mathbf{b}_i \cdot \tilde{s}_i)_{\overline{q}} \cdot \overline{u}_q,\\ &= (\mathbf{b}_i)_{\overline{p}} \cdot (\overline{u}_p s u_q)_{\overline{p}} \overline{u}_q \text{ since } \overline{p} = \overline{q} \text{ and by definition of } \tilde{s}_i,\\ &= (\mathbf{b}_i)_{\overline{p}} \cdot \overline{u}_p s u_q \overline{u}_q,\\ &= (\mathbf{b}_i)_{\overline{p}} \cdot \overline{u}_p s (u_q \overline{u}_q),\\ &= (\mathbf{b}_i)_{\overline{p}} \cdot \overline{u}_p s \text{ since } u_q \overline{u}_q \text{ acts as the identity on } B_q.\end{aligned}$$

Therefore $\eta_i(\mathbf{b}_i, \ldots, \mathbf{b}_h) \cdot s = (\mathbf{b}_i)_{\overline{p}} \cdot \overline{u}_p \cdot s = \eta_i(\mathbf{b}_i \cdot \tilde{s}_i, \ldots, \mathbf{b}_h \cdot \tilde{s}_h)$.

Case 4. $h(p) < i$ and $h(q) < i$. Then by definition, $\eta_i(\mathbf{b}_i, \ldots, \mathbf{b}_h) \cdot s = p$ and $\eta_i(\mathbf{b}_i \cdot \tilde{s}_i, \ldots, \mathbf{b}_h \cdot \tilde{s}_h) = q$, so the conclusion holds by induction hypothesis.

By induction we conclude that $\eta_i(\mathbf{b}_i, \ldots, \mathbf{b}_h) \cdot s \subseteq \eta_i(\mathbf{b}_i \cdot \tilde{s}_i, \ldots, \mathbf{b}_h \cdot \tilde{s}_h)$ for all i $(1 \leq i \leq h)$, all $s \in S$, and all $(\mathbf{b}_1, \ldots, \mathbf{b}_h) \in \Psi$. In particular, it follows from $* \neq \eta_i(\mathbf{b}_i \cdot \tilde{s}_i, \ldots, \mathbf{b}_h \cdot \tilde{s}_h)$ that the height of the latter is less than i and from the case $i = 1$ that $(\mathbf{b}_1, \ldots, \mathbf{b}_h) \cdot \tilde{s}$ lies in Ψ and that $\eta(\mathbf{b}_1, \ldots, \mathbf{b}_h) \cdot s = \eta((\mathbf{b}_1, \ldots, \mathbf{b}_h) \cdot \tilde{s})$ holds for all $s \in S$ and $(\mathbf{b}_1, \ldots, \mathbf{b}_h) \in \Psi$.

Moreover, lifts of distinct members of A are distinct since η is a function, and lifts of distinct members of S are distinct. If $s_1 \neq s_2$ $(s_1, s_2 \in S)$, then there is an $a \in A$ such that $a \cdot s_1 \neq a \cdot s_2$. Taking a lift $\tilde{a}$ of a we have $\eta(\tilde{a} \cdot \tilde{s_i}) = \eta(\tilde{a}) \cdot s_i = a \cdot s_i$, but these are distinct for $i = 1, 2$, and therefore the lifts $\tilde{s_1}$ and $\tilde{s_2}$ are also distinct. By Proposition 1.10, this establishes the division and proves the holonomy theorem. □

We now derive the Krohn–Rhodes theorem as a consequence of the holonomy theorem. Using the lemmas above, the following theorem implies the Krohn–Rhodes decomposition theorem for transformation semigroups.

Theorem 3.10. *A finite semigroup is irreducible if and only if it is a nontrivial simple group or a subsemigroup of the flip-flop monoid.*

Proof. By Lemmas 3.5 and 3.6, the nontrivial simple groups and subsemigroups of the flip-flop monoid are irreducible. Let S be any irreducible finite semigroup; (S^λ, S) divides a wreath product of permutation-reset transformation semigroups of the form $(\mathcal{B}_i, \overline{\mathcal{H}_i})$ by the holonomy theorem. Since S is irreducible it divides one of the $\overline{\mathcal{H}_i}$. Now $(\mathcal{B}_i, \overline{\mathcal{H}_i}) <$ $(\mathcal{B}_i, \overline{\{1_{\mathcal{B}_i}\}}) \wr (\mathcal{H}_i, \mathcal{H}_i)$ by Lemma 3.7. So either S divides $\overline{\{1_{\mathcal{B}_i}\}}$ and by Lemma 3.8 embeds in the flip-flop monoid, or S divides the group $\mathcal{H}_i$. In the latter case S is a group and, by Lemma 3.6, S is a simple group or the trivial group. This completes the proof. □

From the holonomy decomposition theorem, the Jordan–Hölder coordinate theorem for finite groups, the preceding lemmas, and Fact 1.15, we obtain the Krohn–Rhodes prime decomposition theorem for transformation semigroups. The corresponding result (Theorem 3.1) for a finite automaton $\mathcal{A} = (A, X, \delta)$ follows by taking the transformation semigroup $(A, S(\mathcal{A}))$ and using the result for transformation semigroups.

3.2 Some Results Related to the Krohn–Rhodes Decomposition Theorem

In this section we characterize the complete classes of finite automata with respect to the homomorphic representation under the α_i products for $i = 0, 1, 2$. First we show an application of the Krohn–Rhodes theory.

Theorem 3.11. *Let $\mathcal{A} = \mathcal{B} \times \mathcal{C}(X, \varphi_1, \varphi_2)$ be an α_0-product. If G is a simple group with $G||^{(n)}\mathcal{A}$, then either $G||^{(n)}\mathcal{B}$ or $G||^{(n)}\mathcal{C}$.*

Proof. First, we note that if we restrict ourselves to a noncyclic simple group G, then our statement is a direct consequence of the Krohn–Rhodes theorem and Proposition 2.49. We consider a direct proof of our statement which does not use this direct consequence. Let $\mathcal{A} = (A, X, \delta)$, $\mathcal{B} = (B, Y, \delta_1)$, and $\mathcal{C} = (C, Z, \delta_2)$. Obviously, the map $\beta : S(\mathcal{A}) \to S(\mathcal{B})$ with $\beta(\delta_p) \to (\delta_1)_{\varphi_1(p)}$, and $p \in X^+$ is a well-defined semigroup homomorphism. For each $b \in B$ let $S_b \subseteq S(\mathcal{A})$ be the collection of those transformations δ_p, $p \in X^+$, with $\delta_1(b, \varphi_1(p)) = b$. If S_b is nonempty, then it is a subsemigroup of $S(\mathcal{A})$ and the mapping $\beta_b : S_b \to S(\mathcal{C})$ with $\beta_b(\delta_p) = (\delta_2)_{\varphi_2(b,p)}$ is a homomorphism. Since $G||^{(n)}S(\mathcal{A})$, there is a subgroup H of $S(\mathcal{A})$ such that G is a homomorphic image of H and $H \subseteq \{\delta_p : p \in X^+, |p| = n\}$. Since H is a subgroup of $S(\mathcal{A})$, by Proposition 2.36, there is a nonempty set $W \subseteq B \times C$ with the following properties:

(a) The restriction $\delta_p|_w$ of each $\delta_p \in H$, $p \in X^+$, to W is a permutation of W.

(b) The mapping $\gamma : H \to \{\delta_p|_W : \delta_p \in H, p \in X^+\}$ with $\gamma(\delta_p) = \delta_p|_W$, $\delta_p \in H$, $p \in X^+$, is an isomorphism of H onto a permutation group over W.

Thus for $\delta_p, \delta_q \in H$, $p, q \in X^+$, we have $\delta_p = \delta_q$ if and only if $\delta((b, c), p) = \delta((b, c), q)$ for all $(b, c) \in W$. Let B_1 be the set of the first components of the elements of W. For each $\delta_p \in H$, $p \in X^+$ define $\psi(\delta_p)$ to be the permutation of B_1 obtained by taking the restriction of $(\delta_1)_{\varphi_1(p)}$ to B_1, i.e., $\psi(\delta_p) = (\delta_1)_{\varphi_1(p)}|_{B_1}$. Then $\kappa : H \to \{\psi(\delta_p) : \delta_p \in H, p \in X^+\}$ with $\kappa(\delta_p) = \psi(\delta_p)$, $\delta_p \in H$, $p \in X^+$, is a well-defined homomorphism of H onto a permutation group $P = \{\psi(\delta_p) : \delta_p \in H, p \in X^+\}$ over B_1. Set $N = \ker \psi$, so that N is a normal subgroup of H and $H/N = P$. Let $\psi' : H \to S(\mathcal{B})$ be the homomorphism taking $\delta_p \in H$, $p \in X^+$, to $(\delta_1)_{\varphi_1(p)}$. Clearly, $H_1 = \{\psi'(\delta_p) : \delta_p \in H, p \in X^+\}$ is a subgroup of $S(\mathcal{B})$. We can consider ψ' as a homomorphism of H onto H_1. If $(\delta_1)_{\varphi_1(p)} = (\delta_1)_{\varphi_1(q)}$ for some $\delta_p, \delta_q \in H$, $p, q \in X^+$, then also $(\delta_1)_{\varphi_1(p)}|_{B_1} = (\delta_1)_{\varphi_1(q)}|_{B_1}$. Therefore $\ker \psi' \subseteq N$. Moreover, ψ factors through ψ'. It follows that H/N is a homomorphic image of H_1. From $H \subseteq \{\delta_p : p \in X^+, |p| = n\}$ we also have $H_1 \subseteq \{(\delta_1)_q : q \in Y^+, |q| = n\}$. Since the simple group G is a homomorphic image of H and N is a normal subgroup of H, either G is a homomorphic image of H/N or N maps homomorphically onto G. In the former case G is a homomorphic image of H_1 and therefore $G||^{(n)}\mathcal{B}$. From now on we assume that G is a homomorphic image of N. Let $b \in B_1$ be any fixed state. If $\delta_p \in N$ for a word $p \in X^+$, then $\delta_1(b, \varphi_1(p)) = b$, i.e., $N \subseteq S_b$. Define $\psi_b : N \to S(\mathcal{C})$ by $\psi_b(\delta_p) = (\delta_2)_{\varphi_2(b,p)}$ for all $\delta_p \in N$ with $p \in X^+$. We already know that ψ_b is a well-defined homomorphism of N into $S(\mathcal{C})$. Therefore $H_b = \{\psi_b(\delta_p) : \delta_p \in N, p \in X^+\}$ must be a group, a subgroup of $S(\mathcal{C})$. We also view ψ_b as a homomorphism of N onto H_b If $p, q \in X^+$ with $\delta_p, \delta_q \in N$, and $\delta_p \neq \delta_q$, then there is a pair $(b, c) \in W$ with $\delta((b, c), p) \neq \delta((b, c), q)$. But $\delta((b, c), p) = (b, \delta_2(c, \varphi_2(b, p)))$ and $\delta((b, c), q) = (b, \delta_2(c, \varphi_2(b, q)))$, so that $\psi_b(\delta_p) = (\delta_2)_{\varphi_2(b,p)} \neq (\delta_2)_{\varphi_2(b,q)} = \psi_b(\delta_q)$. Thus $\cap(\ker \psi_b : b \in B_1)$ is the trivial normal subgroup consisting of the identity of N. Since $\cap(\ker \psi_b : b \in B_1)$ is trivial, N is isomorphic to a subdirect product (i.e., a subgroup of the direct product) of the groups $H_b, b \in B_1$. Since the simple group G is a homomorphic image of N, it is also a homomorphic image of a subgroup of some H_b, i.e., G divides H_b. The group H_b consists of the transformations of the form $(\delta_2)_{\varphi_2(b,p)}$, where $p \in X^+$ and $\delta_p \in N$. Since each member of N is induced by some word of length n, the same holds for H_b, i.e., $H_b \subseteq \{(\delta_2)_q : q \in Z^+, |q| = n\}$. Because G divides H_b, we obtain $G||^{(n)}\mathcal{C}$. □

Corollary 3.12. *Let G be a nontrivial simple group. If $G||\mathcal{A}$ for an α_0-product $\mathcal{A}$ of automata in $\mathcal{K}$, where $\mathcal{K}$ is any class of automata, then $G||\mathcal{B}$ for some $\mathcal{B} \in \mathcal{K}$.* □

Theorem 3.13. *Let S be an irreducible semigroup. If $S||\mathcal{A}$ for an α_0-product $\mathcal{A}$ of automata in $\mathcal{K}$, where $\mathcal{K}$ is any class of automata, then $S||\mathcal{B}$ for some $\mathcal{B} \in \mathcal{K}$. In addition, if $S||\mathcal{B}$ for some $\mathcal{B} \in \mathcal{K}$ and S is a subsemigroup of the flip-flop monoid, then S embeds in $S(\mathcal{B})$ in equal lengths with respect to $\mathcal{B}$.*

Proof. If S is an irreducible semigroup, then by Theorem 3.10, S is either a simple group or a subsemigroup of the flip-flop monoid. If S is either a noncyclic simple group or a subsemigroup of the flip-flop monoid, then we can apply Proposition 2.49. If S is a (cyclic or noncyclic) simple group, then we can apply Corollary 3.12. This completes the first part of the proof. For the second part, assume that $S||\mathcal{B}$ for some $\mathcal{B} = (B, X, \delta) \in K$ such that S is a subsemigroup of the flip-flop monoid. First we suppose that $S = \{e, \ell, r\}$ is the flip-flop monoid. Using Lemma 3.4, there exists a subsemigroup $S' = \{\delta_u, \delta_v, \delta_w\}$ of $S(\mathcal{B})$ with $u, v, w \in X^+$ having an isomorphism κ onto S. Let, say, $\kappa(\delta_u) = e, \kappa(\delta_v) = \ell, \kappa(\delta_w) = r$ such that e is the identity of S. We shall use the obvious facts that for every pair i, j of positive integers, $\delta_u = (\delta_u)^i = \delta_{u^i}, \delta_v = (\delta_u)^i(\delta_v)^j = \delta_{u^iv^j} = (\delta_w)^i(\delta_v)^j = \delta_{w^iv^j}, \delta_w = (\delta_u)^i(\delta_w)^j = \delta_{u^iw^j} = (\delta_v)^i(\delta_w)^j = \delta_{v^iw^j}$. Let $u' = u^{|v|+|w|}, v' = w^{|u|}v^{|u|}, w' = v^{|u|}w^{|u|}$. Then $|u'| = |v'| = |w'|$ and $\delta_u = \delta_{u'}, \delta_v = \delta_{v'}, \delta_w = \delta_{w'}$. Therefore, $\kappa(\delta_{u'}) = e, \kappa(\delta_{v'}) = \ell, \kappa(\delta_{w'}) = r$ such that u', v', w' have equal lengths and κ is an isomorphism. This completes the proof if S is the flip-flop semigroup. We can give similar treatment for the subsemigroups of S. □

Proposition 3.14. *Every reset automaton can be represented isomorphically by a direct power of the two state reset automaton.*

Proof. Let $\mathcal{A} = (A, X, \delta)$ be an arbitrary reset automaton with state set $A = \{a_1, \ldots, a_n\}$. Consider the nth direct power $\mathcal{A}_0^n$ of the two state reset automaton $\mathcal{A}_0=(\{0, 1\}, \{x_0, x_1\}, \delta_0)$, for which $\delta(i, x_j) = j, i, j \in \{0, 1\}$. Let $\varphi_1 : A \to \{0, 1\}^n$, $\varphi_2 : X \to \{x_0, x_1\}^n$ be mappings such that for every $a_i \in A$ and $x \in X$, $\varphi_1(a_i) = 0^{i-1}10^{n-i}, i = 1, \ldots, n$; furthermore, $\varphi_2(x) = x_0^{j-1}x_1x_0^{n-j}$ if and only if $\{\delta(a_i, x) \mid a_i \in A\} = \{a_j\}$. Clearly then $\mathcal{A}$ can be embedded isomorphically into $\mathcal{A}_0^n$ under (φ_1, φ_2). □

Take two alphabets X and Y. For a fixed positive integer n, consider a mapping $\tau : \{p \in X^+ : |p| = n\} \to \{p \in Y^+ : |p| = n\}$. Moreover, let $\mathcal{R}_\tau = (X, R_\tau, \delta_\tau)$ be the automaton, where $R_\tau = \{(p, q) \in X^* \times Y^* \mid 1 \le |p|, |q| \le n, |p| + |q| = n + 1\}$ and, for arbitrary $(p, yq) \in R_\tau$ $(y \in Y)$ and $x \in X$,

$$\delta_\tau((p, yq), x) = \begin{cases} (px, q) & \text{if } |p| < n, \\ (x, \tau(p)) & \text{if } |p| = n. \end{cases}$$

Lemma 3.15. *For every $\tau : \{p \in X^+ : |p| = n\} \to \{p \in Y^+ : |p| = n\}$ $(n > 0)$, $\mathcal{R}_\tau$ can be represented homomorphically by a cascade product of an n-state counter and two-state reset automata.*

Proof. Of course, a direct product of automata can be considered as a special type of their cascade product. Therefore, using Proposition 3.14, every reset automaton can be represented homomorphically by a cascade product of two-state reset automata. To prove our lemma, therefore, it is enough to show that $\mathcal{R}_\tau$ can be represented homomorphically by a cascade product of an n-state counter and certain reset automata.

Let $\mathcal{A}_1 = (\{1, \ldots, n\}, \{x_c\}, \delta_1)$ be a counter. Moreover, take the automata $\mathcal{A}_i = (X \cup \{\{*\}, \{1, \ldots, n\} \times X, \delta_i)(i \in \{2, \ldots, n + 1\})$, and $\mathcal{A}_{n+1+i} = (Y \cup \{*\}, (X \cup \{*\})^n \times (Y \cup \{*\}), \delta_{n+1+i})$ $(i \in \{1, \ldots, n\})$, where $*$ is an arbitrary symbol with $* \notin X \cup Y$; moreover, for arbitrary $a \in X \cup \{*\}$, $a', b \in Y \cup \{*\}$, $k \in \{1, \ldots, n\}$, $x \in X$, $p \in \{p' \in (X \cup \{*\})^* : |p'| = n\}$, and $i \in \{1, \ldots, n\}$,

$$\delta_{1+i}(a, (k, x)) = \begin{cases} * & \text{if } i > 1 \text{ and } k = n, \\ x & \text{otherwise,} \end{cases}$$

$$\delta_{n+1+i}(a', (b, p)) = \begin{cases} b & \text{if } p \notin \{p' \in X^+ : |p'| = n\}, \\ y & \text{if } p \in \{p' \in X^+ : |p'| = n\} \text{ and } y \text{ is the } (n - i + 1)\text{th} \\ & \quad \text{letter of } \tau(p). \end{cases}$$

Clearly, all the $\mathcal{A}_i$ $(i \in \{1, \ldots, n\})$ are reset automata. Take the cascade product

$$\mathcal{B} = (B, X, \delta) = \prod_{i=1}^{2n+1} \mathcal{A}_i(X, \varphi_1, \ldots, \varphi_n)$$

with

$$\varphi_i(k, a_2, \ldots, a_{2n+1}, x) = \begin{cases} x_c & \text{if } i = 1, \\ (k, x) & \text{if } i = 2, \\ (k, a_{i-1}) & \text{if } i \in \{3, \ldots, n + 1\}, \\ (a_{n+1} \ldots a_2, *) & \text{if } i = n + 2, \\ (a_{n+1} \ldots a_2, a_{i-1}) & \text{if } i \in \{n + 3, \ldots, 2n + 1\}, \end{cases}$$

where $i \in \{1, \ldots, 2n + 1\}$, $k \in \{1, \ldots, n\}$, $a_j \in A_j$ $(j \in \{2, \ldots, 2_{n+1}\})$, and $x \in X$.

Now we define a subautomaton $\mathcal{B} = (B', X', \delta)$ with $X' = X$ of $\mathcal{B}$ and that of a mapping $\psi : B' \to R_\tau$ under which $\mathcal{R}_\tau$ is a homomorphic image of $\mathcal{B}'$. Let B' consist of all $b \in B$ for which there are words $p = x_1 \dots x_k$ $(x_1, \dots, x_k \in X)$ and $q = y_1 \dots y_{n+1-k} \in Y)$ with $1 \leq |p|, |q| \leq n$ such that

$$pr_i(b) = \begin{cases} k & \text{if } i = 1, \\ x_{i-1} & \text{if } i \in \{2, \dots, k+1\}, \\ y_{2n+2-i} & \text{if } i \in \{n+k+1, \dots, 2n+1\}, \\ * & \text{otherwise.} \end{cases}$$

Moreover, let $\psi(b) = (\bar{p}, q)$, where $\bar{p}$ denotes the mirror image of p. It is routine work to show that $\mathcal{B}'$ is a subautomaton of $\mathcal{B}$, and ψ is a homomorphism of $\mathcal{B}'$ onto $\mathcal{R}_\tau$. □

We will also use the following lemma.

Lemma 3.16. *Let $\mathcal{A} = (A, X, \delta)$ and $\mathcal{B} = (B, Y, \delta')$ be automata. Assume that for an integer $n > 0$, there exist mappings $\sigma : B \to A$ and $\tau : \{p \in X^+ : |p| = n\} \to \{p \in Y^+ : |p| = n\}$ such that the following two conditions are satisfied:*

(1) $\delta(\sigma(b), p) = \sigma(\delta'(b, \tau(p)))$ *for arbitrary* $b \in B$ *and* $p \in \{p' \in X^+ : |p'| = n\}$.
(2) $\{\delta(\sigma(\delta'(b, q)), p) : b \in B, (p, q) \in R_\tau\} = A$.

Then a cascade product of $\mathcal{R}_\tau$ by $\mathcal{B}$ homomorphically represents $\mathcal{A}$.

Proof. Form the cascade product $\mathcal{D} = (D, X, \delta'') = \mathcal{R}_\tau \times \mathcal{B}(X, \varphi_1, \varphi_2)$, where, for arbitrary $(p, yq) \in \mathcal{R}_\tau$ $(y \in Y)$, $b \in B$, $\varphi_1((p, yq), b, x) = x$ and $\varphi_2((p, yq), b, x) = y$.

To an arbitrary state $d = ((p, yq), b)$ of $\mathcal{D}$ we correspond the state $\delta(\sigma(\delta'(b, yq)), p)$ of $\mathcal{A}$. Assume that $\mathcal{D}$ receives an input signal x in this state d. If $|p| < n$, then $\delta''(d, x) = ((px, q), \delta'(b, y))$, to which the state $\delta(\sigma(\delta'(\delta'(b, y), q)), px) = \delta(\delta(\sigma(\delta'(b, yq)), p), x)$ of $\mathcal{A}$ is corresponded. In the opposite state, i.e., if $|p| = n$, then $\delta''(d, x) = ((x, \tau(p)), \delta'(b, y))$. The state of $\mathcal{A}$ corresponding to this $\delta''(d, x)$ is $\delta(\sigma(\delta'(\delta'(b, y), \tau(p))), x)$ which is equal to $\delta(\delta(\sigma(\delta'(b, y)), p), x)$ since, by (1), $\sigma(\delta'(b, \tau(p))) = \delta(\sigma(b), p)$ for arbitrary $b \in A$. (Observe that in the second case $q = \lambda$.) In both cases we have that the mapping ψ given by $\psi(((p, q), b)) = \delta(\sigma(\delta'(b, q)), p)$ $(((p, q), b) \in D)$ is a homomorphism of $\mathcal{D}$ into $\mathcal{A}$. By (2), ψ is a mapping onto A. □

We are ready to give a proof of the next statement.

Theorem 3.17. *A class $\mathcal{K}$ of automata is complete with respect to homomorphic representations under the cascade product if and only if*

(1) *the two-state reset automaton can be represented homomorphically by a cascade product of automata from $\mathcal{K}$,*
(2) *every counter (of prime power length) can be represented homomorphically by a cascade product of automata from $\mathcal{K}$,*

(3a) *there is an automaton* $\mathcal{A} \in K$ *such that a subsemigroup* S *of* $S(\mathcal{A})$ *is isomorphic to the monoid with two right-zero elements, and*

(3b) *for every simple group* G *there is an automaton* $\mathcal{A} \in \mathcal{K}$ *with* $G < S(\mathcal{A})$.

Proof. The first two conditions are obviously necessary.[17] The necessity of (3a) and (3b) comes directly from the Krohn–Rhodes decomposition theorem. (Actually, (3a) and (3b) constitute Krohn–Rhodes' criterion.) For the converse, first note that every counter can be represented homomorphically by a cascade product of counters with prime power length. (We omit the easy proof of this statement.) Thus we may assume by the conditions (1), (2), and Lemma 3.15 that for every $\tau : \{p \in X^+ : |p| = n\} \to \{p \in Y^+ : |p| = n\}$, the automaton $\mathcal{R}_\tau$ can be represented homomorphically by a cascade product of automata from $\mathcal{K}$. Let S be an arbitrary noncyclic, irreducible monoid. By our conditions (3a), (3b), and Proposition 2.49 we have that $S||\mathcal{A}$ for some $\mathcal{A} \in \mathcal{K}$. Using Proposition 2.47, we then get $\mathcal{A}_S||\mathcal{B}^{\Delta n}$, where $\mathcal{B}^{\Delta n}$ denotes the nth diagonal power of an appropriate subautomaton $\mathcal{B} = (B, Y, \delta_\mathcal{B})$ of the automaton $\mathcal{A}$ satisfying the conditions of Proposition 2.47 and n is the number of states of $\mathcal{A}$. Thus, let $\mathcal{A}_S||\mathcal{B}^{\Delta n}$ under the mappings $\tau_1 : B^n \to S$, $\tau_2 : S \to Y^+$. Then, denoting by e the identity element of S, we get $\{\delta_S(\tau_1(\delta'(\mathbf{b}, \tau_2(e))), e) : \mathbf{b} \in B^n\} = \{\tau_1(\delta'(\mathbf{b}, \tau_2(e))) : \mathbf{b} \in B^n\} = \{\tau_1(\mathbf{b}) : \mathbf{b} \in B^n\} = S$. Therefore (taking $\sigma, \tau, \mathcal{A}, \mathcal{B}$ of the lemma to be $\tau_1, \tau_2, \mathcal{A}_S, \mathcal{B}$, respectively), we have the conditions (1) and (2) of Lemma 3.16. Then $\mathcal{A}_S$ can be represented by a cascade product of the automaton $\mathcal{R}_{\tau_2}$ by the direct power $\mathcal{B}^n$. Therefore, combining this with (2), for every irreducible semigroup S, we conclude that $\mathcal{A}_S$ can be represented homomorphically by a cascade product of automata from $\mathcal{K}$ for every irreducible semigroup S. Using the first part of the Krohn–Rhodes decomposition theorem, this implies that $\mathcal{K}$ is complete with respect to homomorphic representations under the cascade product. The proof is complete. □

Theorem 3.18. *None of conditions* (1), (2), (3a), *and* (3b) *of Theorem* 3.17 *can be omitted.*

Proof. For (3a) and (3b), the statement comes directly from the Krohn–Rhodes decomposition theorem. (See Theorem 3.1.)

For (1), we now show that there exists a class $\mathcal{K}$ satisfying (2), (3a), and (3b), which is not complete with respect to homomorphic representations under the cascade product. Let $\mathcal{K}_0 = \{\mathcal{A}_n = (\{1, \ldots, n+1\}, \{x, y\}, \delta_n) \mid n = 2, 3, \ldots\}$ be a (countable) system of automata with

$$\delta_n(i, z) = \begin{cases} i + 1 \pmod{n+1} & \text{if } z = x, \\ 1 & \text{if } i = 2, z = y, \\ 2 & \text{if } i = 1, z = y, \\ i & \text{if } 2 < i \leq n+1, z = y, \end{cases}$$

$n = 2, 3, \ldots$; moreover, let $\mathcal{A}_1 = (\{1, 2\}, \{x, y\}, \delta_1)$ be defined by $\delta_1(1, x) = \delta_1(2, x) = 2$, $\delta_1(1, y) = 2$, $\delta_1(2, y) = 1$.

Then $\delta_1(1, x) = \delta_1(2, x) = 2$, $\delta_1(1, xy) = \delta_1(2, xy) = 1$, $\delta_1(1, yy) = 1$, and $\delta_1(2, yy) = 2$. Thus a subsemigroup of $S(\mathcal{A}_1)$ is isomorphic to the monoid with two

[17] Although it may be counterintuitive to those used to transformation semigroups rather than automata, to obtain (2), it is not sufficient to homomorphically represent all prime-length counters with single-letter input sets. For example, using such counters, the single-letter-input modulo-four counter cannot be homomorphically represented.

right-zero elements. On the other hand, by Proposition 1.5, for every $n = 2, 3, \ldots,$ the degree-$n+1$ symmetric group is isomorphic to $S(\mathcal{A}_n)$. Since the degree-two symmetric group can be embedded isomorphically into a larger symmetric group, and, furthermore, every simple group can be embedded isomorphically into a symmetric group, then $\mathcal{K}_0$ satisfies (3b). Therefore, $\{\mathcal{A}_0\} \cup \mathcal{K}_0$ satisfies (3a) and (3b).

Let $\mathcal{A} = (\{0, 1\}, \{x, y\}, \delta)$ be an automaton having $\delta(i, x) = 1$ and $\delta(i, y) = 2, i = 1, 2$. Then $\mathcal{A}$ is a two-state reset automaton. For every positive integer n, set $\mathcal{B}_n = (A \times \{1, \ldots, n+1\} \cup \{*\}, A \times \{x, y\}, \delta'_n)$, where $*$ is a new symbol and

$$\delta'_n((a, b), (c, z)) = \begin{cases} (\delta(a, z), \delta_n(b, z) & \text{if } a = c, \\ * & \text{otherwise,} \end{cases}$$

$$\delta'_n(*, (c, z)) = *$$

with $(a, b) \in \{0, 1\} \times \{1, \ldots, n+1\}$, $(c, z) \in \{0, 1\} \times \{x, y\}$.

Let $\mathcal{K}$ consist of the counters and all automata $\mathcal{B}_n, n = 1, 2, \ldots$. It is now obvious that $\mathcal{K}$ satisfies (2). For (3a) and (3b), first note that for every n, $\mathcal{A}_n$ is a homomorphic image of a subautomaton of a cascade product $\mathcal{A} \times \mathcal{B}_n(\{x, y\}, \varphi_1, \varphi_2)$. (Hint: $\varphi_1(z) = z, \varphi_2(a, z) = (a, z), a \in \{0, 1\}, z \in \{x, y\}$, and the state set of the subautomaton is just $\{(a, (a, b)) \mid a \in A, b \in A_n\}$.) But the monoid **F** with two right-zero elements can be embedded isomorphically into $S(\mathcal{A}_1)$. Thus, from the Krohn–Rhodes theorem, **F** is isomorphic to a subsemigroup of $S(\mathcal{A})$ or to a subsemigroup of $S(\mathcal{B}_1)$. The first case is impossible by the choice of $\mathcal{A}$; hence, $\mathcal{K}$ satisfies (3a). Similarly, let G be a noncommutative simple group. Since $\mathcal{K}_0$ satisfies (3b), we may assume $G < \mathcal{A}_n$ for some $\mathcal{A}_n \in \mathcal{K}_0, n > 1$. But then we have again that $\mathcal{A}_n$ is a homomorphic image of a subautomaton of a cascade product $\mathcal{A} \times \mathcal{B}_n(\{x, y\}, \varphi_1, \varphi_2)$. Hence $G < \mathcal{A} \times \mathcal{B}_n(\{x, y\}, \varphi_1, \varphi_2)$ and $G < \mathcal{A}$ is impossible. Then $G < \mathcal{B}_n$ follows from the Krohn–Rhodes theorem. We have seen that $\mathcal{K}$ satisfies all of the conditions (2), (3a), and (3b).

To see that $\mathcal{K}$ is not complete we prove the following: every strongly connected subautomaton of a cascade product of factors in $\mathcal{K}$ is autonomous. Of course, this also shows that $\mathcal{K}$ does not satisfy condition (1).

Assume the contrary. Let $\mathcal{D}_1 \times \cdots \times \mathcal{D}_m(X, \varphi_1, \ldots, \varphi_m)$ be a cascade product of automata of factors $\mathcal{K}$ having a strongly connected nonautonomous subautomaton and such that m is minimal with this property. First we assume that either $m = 1$ or $m > 1$ and $\mathcal{D}_1, \ldots, \mathcal{D}_{m-1}$ are counters. Then a cascade product $\mathcal{C} \times \mathcal{D}_m(X, \psi_1, \psi_2)$ also has a nonautonomous strongly connected subautomaton, where $\mathcal{C}$ is a counter $\mathcal{C}_k$. ($\mathcal{C}$ can be a trivial counter having only one state if $m = 1$). Let us denote this subautomaton by $\mathcal{D} = (D, X, \delta'')$. The automaton $\mathcal{D}_m$ cannot be a counter, so it is $\mathcal{B}_n$ for a particular n. Since $\mathcal{D}$ is a nonautonomous strongly connected automaton, no state in D has the symbol $*$ as its second component. Let $(i, (a_0, b_0)) \in D$ and $z_1, z_2 \in X$ be fixed with $\delta''((i, (a_0, b_0)), z_1) \neq \delta''((i, (a_0, b_0)), z_2)$. Put $\delta''((i, (a_0, b_0)), z_1) = (j, (a_1, b_1))$. $\delta''((i, (a_0, b_0)), z_2) = (j, (a_2, b_2))$. Thus, $(a_1, b_1) \neq (a_2, b_2)$. If $a_1 = a_2$, then, by $\mathcal{A}$ and the definition of $\mathcal{B}_n$, we have $\psi_2(i, z_1) = \psi_2(i, z_2)(\in \{(a_0, x), (a_0, y)\}$. But then $b_1 = b_2$ follows. Therefore, $a_1 \neq a_2$ and, consequently, one of $\delta''((j, (a_1, b_1)), z_1)$ and $\delta''((j, (a_2, b_2)), z_1)$ has the symbol $*$ as its second component.

Hence, $m > 1$ and there exists a $k, 1 \leq k < m$, such that $\mathcal{D}_k = \mathcal{B}_\ell$ for a particular ℓ. Denote again $\mathcal{D} = (D, X, \delta'')$ a nonautonomous strongly connected subautomaton of $\mathcal{D}_1 \times \cdots \times \mathcal{D}_m(X, \varphi_1, \ldots, \varphi_m)$. If $\mathcal{D}$ has a state for which its ℓth component is $*$, then all state of $\mathcal{D}$ has this property. Clearly, then $\mathcal{D}$ can also be represented by a cascade product $\mathcal{D}_1 \times \cdots \times \mathcal{D}_{\ell-1} \times \mathcal{D}_{\ell+1} \times \cdots \times \mathcal{D}_m(X, \varphi'_1, \ldots, \varphi'_m)$, which is impossible by the minimality of m. Then every state $(d_1, \ldots, d_m)$ of $\mathcal{D}$ has the property $d_\ell \in A \times A_\ell$. Suppose that there are $(d_1, \ldots, d_m) \in D_1 \times \cdots \times D_m, z_1, z_2 \in X$ with $\delta''((d_1, \ldots, d_m), z_1) \neq \delta''((d_1, \ldots, d_m), z_2)$. Similar to the first case, it can be easily seen that one of $\delta''((d_1, \ldots, d_m), z_1z_1)$ and $\delta''((d_1, \ldots, d_m), z_2z_1)$ has the symbol $*$ as its ℓth component. But then all states of $\mathcal{D}$ have this property and thus $\mathcal{D}$ can also be represented by a cascade product $\mathcal{D}_1 \times \cdots \times \mathcal{D}_{\ell-1} \times \mathcal{D}_{\ell+1} \times \cdots \times \mathcal{D}_m(X, \varphi'_1, \ldots, \varphi'_m)$, which is impossible by the minimality of m. Hence we then obtain for every $(d_1, \ldots, d_m) \in D_1 \times \cdots \times D_m, z_1, z_2 \in X$, $\delta''((d_1, \ldots, d_m), z_1) = \delta''((d_1, \ldots, d_m), z_2)$. But then, contrary to our assumptions, $\mathcal{D}$ is autonomous.

It remains to show that there exist a class $\mathcal{K}$ satisfying (1), (3a), and (3b), which is not complete with respect to homomorphic representations under the cascade product.

Now we consider a (countable) system of automata $\mathcal{K}_1 = \{\mathcal{A}_n = (\{1, \ldots, n\}, \{x, y, w\}, \delta_n) \mid n = 1, 2, \ldots\}$ with

$$\delta_n(i, z) = \begin{cases} i + 1 \pmod n & \text{if } z = x, \\ 1 & \text{if } i = 2, z = y, \\ & \text{or } i \in \{1, 2\}, z = w, \\ 2 & \text{if } i = 1, z = y, \\ i & \text{if } 2 < i \le p_n, z \in \{y, w\}, \end{cases}$$

$n = 1, 2, \ldots$. Moreover, let $\mathcal{A}_0 = (\{1, 2\}, \{x, y\}, \delta_1)$ be defined by $\delta_0(i, x) = 1, \delta_0(i, y) = 2, i \in \{1, 2\}$.

By Proposition 1.5, for every $n \geq 1$, $S(\mathcal{A}_n)$ is isomorphic to the degree n symmetric semigroup. Thus both of the monoid **F** with two right-zero elements and the degree-n symmetric group divide $S(\mathcal{A}_n)$ if $n > 1$. On the other hand, $\mathcal{A}_0$ is a two-state reset automaton. Therefore, $\{\mathcal{A}_0\} \cup K_1$ satisfies the conditions (1), (3a), and (3b).

Let m be an arbitrary positive integer with $m > 1$, and for every positive integer n set $\mathcal{B}_n = (\{1, \ldots, m\} \times \{1, \ldots, n\} \cup \{*\}, \{1, \ldots, m\} \times \{x, y, w\}, \delta'_n)$, where $*$ is a new symbol and

$$\delta'_n((a, b), (c, z)) = \begin{cases} (a + 1 \pmod m), \delta_n(b, z)) & \text{if } a = c, \\ * & \text{otherwise}, \end{cases}$$

$$\delta'_n(*, (c, z)) = *$$

with $(a, b) \in \{1, \ldots, m\} \times \{1, \ldots, n\}, (c, z) \in \{1, \ldots, m\} \times \{x, y, w\}$.

Let $\mathcal{K}_2$ consist of all automata $\mathcal{B}_n, n = 1, 2, \ldots$. It is obvious that $\mathcal{K}' = \{\mathcal{A}_0\} \cup \mathcal{K}_2$ satisfies (1). For (3a) and (3b), first note that for every $n \geq 1$, $\mathcal{A}_n$ is a homomorphic image of a subautomaton of a cascade product $\mathcal{C}_m \times \mathcal{B}_n(\{x, y, w\}, \varphi_1, \varphi_2)$, where $\mathcal{C}_m$ is a counter with m states. (Hint: $\varphi_1(z) = z, \varphi_2(a, z) = (a, z), a \in \{1, \ldots, m\}, z \in \{x, y, w\}$, and the state set of the subautomaton is just $\{(a, (a, b)) \mid a \in \{1, \ldots, m\}, b \in \{1, \ldots, n\}\}$). But for every $n > 1$ the monoid **F** with two right-zero elements is a submonoid of $S(\mathcal{A}_n)$. Thus, from the Krohn–Rhodes theorem, **F** is isomorphic to a subsemigroup of $S(\mathcal{C}_m)$ or to a subsemigroup of $S(\mathcal{B}_n)$. The first case is impossible; hence, $\mathcal{K}'$ satisfies (3a). Similarly, let G be a nontrivial simple group. Since every group can be embedded in a larger noncommutative simple group, we may assume without loss of generality that G is noncommutative (and simple). We may assume that $G < \mathcal{A}_n$ for some $n > 1$. On the other hand, we have again that $\mathcal{A}_n$ is a homomorphic image of a subautomaton of a cascade product $\mathcal{C}_m \times \mathcal{B}_n(\{x, y, w\}, \varphi_1, \varphi_2)$. Hence, $G < \mathcal{C}_m \times \mathcal{B}_n(\{x, y, w\}, \varphi_1, \varphi_2)$ and, using that G is noncommutative, $G < \mathcal{C}_m$ is impossible. Then $G < \mathcal{B}_n$ follows from the Krohn–Rhodes theorem. We have seen that $\mathcal{K}' = \{\mathcal{A}_0\} \cup \mathcal{K}_2$ satisfies all of the conditions (1), (3a), and (3b).

To see that $\mathcal{K}'$ is not complete we prove that none of the nontrivial counters can be represented homomorphically by a cascade product of factors in $\mathcal{K}'$. This also shows that $\mathcal{K}'$ does not satisfy condition (2).

Let us first note that if a counter $\mathcal{C}$ is a homomorphic image of an automaton $\mathcal{A}$, then $\mathcal{A}$ has a subautomaton $\mathcal{B}$ isomorphic to a counter which can be mapped homomorphically onto $\mathcal{C}$. Moreover, the number of states in $\mathcal{C}$ always divides the number of states in $\mathcal{B}$. Therefore, it is enough to show that whenever $\mathcal{B} = (\{x\}, B, \delta')$ is a subautomaton of a cascade product $\mathcal{A} = (A, \{x\}, \delta) = \prod_{j=1}^{k} \mathcal{D}_j(\{x\}, \varphi_1, \ldots, \varphi_k)$, $\mathcal{D}_j = (D_j, X'_j, \delta''_j) \in K'$, $j = 1, \ldots, k$, and $\mathcal{B}$ is isomorphic to a counter, then the number of states in $\mathcal{B}$ is not divisible by m. For this, take a $b = (b_1, \ldots, b_k) \in B$. For every $j \in \{1, \ldots k\}$, denote by ℓ_j the least positive integer such that $\delta''_t(b_t, \varphi_t(b_1, \ldots, b_k, x^{\ell_j})) = b_t$, $t = 1, \ldots, j$. We prove that m does not divide ℓ_j, $j = 1, \ldots, k$, which in the case $j = k$ means that m does not divide $|B|$. Obviously, if $j = 1$, then either $\mathcal{D}_1 = \mathcal{A}_0$ or $b_1 = *$ and thus $\ell_1 = 1$. We prove that $\ell_j = 1$ whenever $\ell_{j-1} = 1$ and $j > 1$. Indeed, $\ell_{j-1} = 1$ implies that $\varphi_t(b_1, \ldots, b_k, x) = \varphi_t(b'_1, \ldots, b'_k, x)$, $t = 1, \ldots, \ell_j$, $(b'_1, \ldots, b'_k) = \delta'((b_1, \ldots, b_k), x^n)$, $n \geq 1$. But then either $\mathcal{D}_j = \mathcal{A}_0$ or $b_j = *$. Thus, we get $\ell_k = 1$ by induction. Therefore, if $\mathcal{B}$ homomorphically represents a counter $\mathcal{C}$, then $\mathcal{C}$ is trivial (having only one state). The proof is complete. □

Proposition 3.19. *There exists a class $\mathcal{K}$ of automata which is complete with respect to homomorphic representations under the cascade product such that all elements of $\mathcal{K}$ have two input letters.*

Proof. We consider again a two-state reset automaton $\mathcal{A}_0$ (having two input letters) and define $\mathcal{K}_0$ as the above proof of Theorem 3.18.

Let $\mathcal{K}_0 = \{\mathcal{A}_n = (\{1, \ldots, n+1\}, \{x, y\}, \delta_n) \mid n = 2, 3, \ldots\}$ with

$$\delta_n(i, z) = \begin{cases} i + 1 \pmod{n+1} & \text{if } z = x, \\ 1 & \text{if } i = 2, z = y, \\ 2 & \text{if } i = 1, z = y, \\ i & \text{if } 2 < i \leq n+1, z = y, \end{cases}$$

$n = 2, 3, \ldots$; moreover, let $\mathcal{A}_1 = (\{1, 2\}, \{x, y\}, \delta_1)$ be defined by $\delta_1(1, x) = \delta_1(2, x) = 2$, $\delta_1(1, y) = 2$, $\delta_1(2, y) = 1$. We prove the completeness of $\mathcal{K}'' = \{\mathcal{A}_0\} \cup \mathcal{K}_0$. Of course, it trivially satisfies conditions (1) and (2) of Theorem 3.17. On the other hand, by $\delta_1(1, x) = \delta_1(2, x) = 2$, $\delta_1(1, xy) = \delta_1(2, xy) = 1$, $\delta_1(1, yy) = 1$, and $\delta_1(2, yy) = 2$. The monoid $\mathbf{F}$ with two right-zero elements can be embedded isomorphically into $\mathcal{A}_1$, as we proved in the previous proof of Theorem 3.18. Thus we obtain (3a). Finally, Proposition 1.5 implies that $S(\mathcal{A})_n$ is isomorphic to the degree-$n+1$ symmetric group (as we have also already mentioned in the previous proof of Theorem 3.18). Since every simple group can be embedded into an appropriate symmetric group and every symmetric group can be embedded into a larger symmetric group, we also have condition (3b). The proof is complete. □

By an elementary computation we obtain the next observation.

Proposition 3.20. *Let C_n be a nontrivial counter, i.e., $n > 1$, and let $\mathcal{K}$ be a class of automata. C_n can be represented homomorphically by a cascade product of factors from $\mathcal{K}$ if there is a multiple of m such that C_m can be embedded isomorphically by a cascade product of factors from $\mathcal{K}$. Further, C_m can be embedded isomorphically into a cascade product of factors from $\mathcal{K}$ if and only if there are automata $\mathcal{A}_i = (A_i, X_i, \delta_i) \in \mathcal{K}$, $i = 1, \ldots, k$, $k > 0$, and integers $1 = m_0 < m_1 < \cdots < m_k = m$ such that*

(1) *m_{i-1} is a divisor of m_i $(i \in \{1, \ldots, k\})$, and*

(2) *for every $i \in \{1, \ldots, k\}$ there are distinct states $a_1, \ldots, a_{m_i/m_{i-1}} \in A_i$ and a word $u \in X_i^*$ with $|u| = m_{i-1}$ and*

$$\delta_i(a_1, u) = a_2, \ldots, \delta_i(a_{m_i/m_{i-1}-1}, u) = a_{m_i/m_{i-1}}, \delta_i(a_{m_i/m_{i-1}}, u) = a_1. \qquad \square$$

Let us call $\mathcal{K}$ of automata *precomplete* if it satisfies the following conditions:

(i) There is an automaton $\mathcal{A}$ such that $\mathbf{F}$ is isomorphic to a subsemigroup of $S(\mathcal{A})$.

(ii) For every simple group G there is an $\mathcal{A} \in \mathcal{K}$ with $G < S(\mathcal{A})$.

The Krohn–Rhodes decomposition theorem readily implies that every complete class for the cascade product is precomplete, but the converse fails in general. It may well happen that although $\mathcal{K}$ is precomplete, any strongly connected automaton is trivial, i.e., a one-state automaton if it can be represented homomorphically by a cascade product of factors in $\mathcal{K}$.

Proposition 3.21. *Let $q = p^\ell$ be a prime power. There exists a precomplete class $\mathcal{K}'$ such that for any $\mathcal{K}_0$, C_q cannot be represented homomorphically by a cascade product of factors from $\mathcal{K}' \cup \mathcal{K}_0$ unless C_q can be represented homomorphically by a cascade product of factors from $\mathcal{K}_0$.*

Proof. Define the automata $\mathcal{A}_n$, $n \geq 3$, as follows: $\mathcal{A}_n = (\{1, \ldots, qn\} \cup \{*\}, \{x_1, \ldots, x_q, y\}, \delta_n)$, where, for every $i \in \{1, \ldots, qn\}$ and $j \in \{1, \ldots, q\}$,

$$\delta_n(i, x_j) = \begin{cases} i + 1 \,(\mathrm{mod}\, qn) & \text{if } i \,(\mathrm{mod}\, q) = j, \\ * & \text{otherwise;} \end{cases}$$

$$\delta_n(i, y) = \begin{cases} 2 & \text{if } i = 1, \\ q(n-1) + 2 & \text{if } i = q + 1, \\ i - q + 1 & \text{if } i \,(\mathrm{mod}\, q) = 1 \text{ and } i \notin \{1, q+1\}, \\ * & \text{otherwise;} \end{cases}$$

$$\delta_n(*, x_j) = \delta_n(*, y) = *.$$

Observe that the word $x_1, \ldots, x_q$ induces a cyclic permutation of the states in $A' = \{1, q+1, \ldots, q(n-1)+1\}$. Moreover, $yx_2 \cdots x_q$ induces the transposition $\gamma(1) = q+1$, $\gamma(q+1) = 1$, $\gamma(a') = a'$, $a' \in A' \backslash \{1, q+1\}$ of A'. Consequently, by Proposition 1.5, the degree-n symmetric group divides $S(\mathcal{A}_n)$. In addition to the automata $\mathcal{A}_n$ define $\mathcal{A} = (\{1, \ldots, 2q\} \cup \{*\}, \{x, y, z, x_2, \ldots, x_q, \}, \delta)$ by the following rules: $\delta(1, x) = \delta(q+1, x) = \delta(q+1, z) = 2$, $\delta(1, y) = \delta(1, z) = \delta(q+1, y) = q+2$, $\delta(i, x_i) = i + 1 \,(\mathrm{mod}\, 2q)$, $\delta(q+i, x_i) =$

$q+i+1 \pmod{2q}$, $i = 2, \ldots, q$, and the value of the transition function δ is $*$ for the rest of the cases. We have $\delta(1, xx_2 \cdots x_q) = \delta(q+1, xx_2 \cdots x_q) = q+1$, $\delta(1, yx_2 \cdots x_q) = \delta(q+1, yx_2 \cdots x_q) = 1$; further, $\delta(1, zx_2 \cdots x_q) = 1$, $\delta(q+1, zx_2 \cdots x_q) = q-1$. Consequently, by Proposition 2.34, $S(\mathcal{A})$ has a subsemigroup isomorphic to $\mathbf{F}$.

Besides those mentioned above, the automata $\mathcal{A}_n$ have the following property: for every $i \in \{1, \ldots, qn\}$ and $u \in \{x_1, \ldots, x_q, y\}^+$ with $q \not| \, |u|$, it holds that $\delta_n(i, uu) = *$. Similarly, $\delta(i, uu) = *$ in $\mathcal{A}$ whenever $i \in \{1, \ldots, 2q\}$ and $u \in \{x, y, z, x_2, \ldots, x_q\}^+$ with $q \not| \, |u|$. Set $\mathcal{K}' = \{\mathcal{A}\} \cup \{\mathcal{A}_n \mid n \geq 3\}$. $\mathcal{K}'$ is a precomplete class. Let $\mathcal{K}_0$ be any class of automata such that $\mathcal{C}_q$ cannot be represented homomorphically by a cascade composition of factors from $\mathcal{K}_0$. The above property of the automata in $\mathcal{K}'$ and Proposition 3.20 jointly imply that for every integer $m > 1$, $\mathcal{C}_m$ can be represented isomorphically by a cascade product of factors from $\mathcal{K}' \cup \mathcal{K}_0$ only if $\mathcal{C}_m$ can be represented isomorphically by a cascade product of factors from $\mathcal{K}_0$. It follows that $\mathcal{C}_q$ cannot be represented homomorphically by any cascade product of factors from $\mathcal{K}' \cup \mathcal{K}_0$. □

Now we need some auxiliary results and definitions. Let m, n be positive integers with $m \geq 2$ or $n \geq 2$. We denote by $\mathcal{K}(m, n)$ the class of all strongly connected automata $\mathcal{A} = (A, \{\bar{x}, \bar{y}\}, \delta)$ satisfying the following condition: there are distinct states $a_1, \ldots, a_m \in A$ such that

(i) $\delta(a_i, \bar{x}) \neq \delta(a_i, \bar{y})$, $\delta(a_i, \bar{y}^n) = a_i$, $\delta(a_i, \bar{x}^n) = a_{i+1 \pmod{m}}$ for all $i \in \{1, \ldots, m\}$, and

(ii) for every $i \in \{1, \ldots, m\}$, $z \in \{\bar{x}, \bar{y}\}$, and $u, v \in \{\bar{x}, \bar{y}\}^*$ with $|u|, |v| < n$, we have $\delta(a_i, zu) = \delta(a_i, zv)$ if and only if $|u| = |v|$.

An example of an automaton in $\mathcal{K}(m, n)$ if $m \geq 2$ is $\mathcal{C}(m, n) = (\{1, \ldots, n\} \times \{1, \ldots, m\}, \{\bar{x}, \bar{y}\}, \delta)$, where

$$\delta(i, j), \bar{x}) = \begin{cases} (i+1 \pmod{n}, j+1 \pmod{m}) & \text{if } i = 1, \\ (i+1 \pmod{n}, j) & \text{if } i \neq 1, \end{cases}$$

$$\delta((i, j), \bar{y}) = (i+1 \pmod{n}, j)$$

for all $i \in \{1, \ldots, n\}$ and $j \in \{1, \ldots, m\}$. For later use we remark that $\mathcal{C}(m, n)$ is just the cascade product $\mathcal{C}_n \times \mathcal{C}_m^1(\{\bar{x}, \bar{y}\}, \varphi)$ with $\varphi_1(i, j, \bar{x}) = \varphi_1(i, j, \bar{y}) = \bar{x}$ and

$$\varphi_2(i, j, \bar{x}) = \begin{cases} \bar{x} & \text{if } i = 1, \\ \bar{y} & \text{if } i \neq 1, \end{cases}$$

$$\varphi_2(i, j, \bar{y}) = \bar{y}, i \in \{1, \ldots, n\}, j \in \{1, \ldots, m\},$$

where $\mathcal{C}_n$ denotes a counter (with n states) and $\mathcal{C}_m^1$ denotes a counter with identity (having m states). In particular, $\mathcal{C}(m, 1)$ is isomorphic to $\mathcal{C}_m^1$. To include the case $m = 1$ (so that $n \geq 2$), we define $\mathcal{C}(1, n) = (\{1, \ldots, n\} \cup \{2'\}, \{\bar{x}, \bar{y}\}, \delta)$, where $\delta(1, \bar{x}) = 2$, $\delta(1, \bar{y}) = 2'$, $\delta(i, \bar{x}) = \delta(i, \bar{y}) = i + 1 \pmod{n}$, $i = 2, \ldots, n$, $\delta(2', \bar{x}) = \delta(2', \bar{y}) = 3 \pmod{n}$. We see that $\mathcal{C}(1, n) \in \mathcal{K}(1, n)$. The proof of the following statement is omitted.

Lemma 3.22. *For every pair of integers m, n with $m \geq 2$ or $n \geq 2$ and automaton $\mathcal{A} \in \mathcal{K}(m, n)$ we have that $\mathcal{C}(m, n)$ is a homomorphic image of $\mathcal{A}$.*

Take an automaton $\mathcal{C}(1, n)$ so that $n \geq 2$. For technical reasons we treat the two-state reset automaton $\mathcal{A}_0$ as being equipped with the fixed input signs $\bar{x}, \bar{y}$, i.e., $\mathcal{A}_0 = (\{0, 1\}, \{\bar{x}, \bar{y}, \delta\})$ with $\delta(i, \bar{x}) = 0$, $\delta(i, \bar{y}) = 1$, $i = 0, 1$. It is easy to see that for every pair of words $u, v \in \{\bar{x}, \bar{y}\}^*$ if $\delta_u = \delta_v$ (i.e., u and v induce the same transition in $\mathcal{C}(1, n)$), then either $u = v = \lambda$ or $u, v \neq \lambda$, and the last letter of u coincides with that of v. On the other hand, $\mathcal{A}_0$ also has this property.

Lemma 3.23. *For all $\mathcal{A} \in \mathcal{K}(1, n)$, $\mathcal{A}_0$ can be represented homomorphically by an nth-diagonal power of $\mathcal{A}$.*

Proof. Let $\mathcal{A} = (A, \{\bar{x}, \bar{y}\}, \delta) \in \mathcal{K}(1, n)$ be given. Then

(i) $\delta(a_1, \bar{x}) \neq \delta(a_1, \bar{y})$, $\delta(a_1, \bar{x}^n) = \delta(a_1, \bar{y}^n) = a_1$, and

(ii) for every $z \in \{\bar{x}, \bar{y}\}$ and $u, v \in \{\bar{x}, \bar{y}\}^*$ with $|u|, |v| < n$, we have $\delta(a_1, zu) = \delta(a_1, zv)$ if and only if $|u| = |v|$.

By $\delta(a_1, \bar{x}) \neq \delta(a_1, \bar{y})$, we may assume without loss of generality that $a_1 \neq \delta(a_1, \bar{y})$. Consider the nth-diagonal power $\mathcal{B} = (A^n, X, \delta') = \mathcal{A}_1 \Delta \cdots \Delta \mathcal{A}_n$ of $\mathcal{A}$ with $\mathcal{A}_1 = \cdots = \mathcal{A}_n = \mathcal{A}$ and let $\mathcal{B}' = (B', X, \delta'')$ be a state-subautomaton of $\mathcal{B}$ generated by its state $(a_1, \delta(a_1, \bar{x}), \ldots, \delta(a_1, \bar{x}^{n-1}))$. Then none of the states $a_1, \delta(a_1, \bar{x}), \ldots, \delta(a_1, \bar{x}^{n-1})$ coincides with $\delta(a_1, \bar{y})$; moreover, for every $p \in \{\bar{x}, \bar{y}\}^*$, $z \in \{\bar{x}, \bar{y}\}$, $\delta(a_1, \bar{y}) \in \{\delta(a_1, pz), \delta(a_1, \bar{x}pz), \ldots, \delta(a_1, \bar{x}^{n-1}pz)$ if and only if $z = \bar{y}$. Let $\psi : B \to \{0, 1\}$ be given by

$$\psi((a_1, \delta(a_1, \bar{x}), \ldots, \delta(a_1, \bar{x}^{n-1}))) = 0;$$

moreover, let

$$\psi((\delta(a_1, pz), \delta(a_1, \bar{x}pz), \ldots, \delta(a_1, \bar{x}^{n-1}pz))) = \begin{cases} 0 & \text{if } \delta(a_1, \bar{x}) \in \{\delta(a_1, pz), \\ & \quad \delta(a_1, \bar{x}pz), \ldots, \delta(a_1, \bar{x}^{n-1}pz)\}, \\ 1 & \text{otherwise} \end{cases}$$

for every $p \in X^*$, $z \in \{\bar{x}, \bar{y}\}$. By the definition of $\mathcal{A}$, ψ is well defined and it is a state-homomorphism of $\mathcal{B}'$ onto $\mathcal{A}_0$. □

Lemma 3.24. *For all $\mathcal{A} \in \mathcal{K}(m, n)$, $\mathcal{C}_m^1$ can be represented homomorphically by an nth-diagonal power of $\mathcal{A}$.*

Proof. We may assume that $m \geq 2$ since otherwise the statement is trivial. By Lemma 3.22 it suffices to prove that $\mathcal{C}_m^1$ is a homomorphic image of a state-subautomaton of the nth-diagonal power of $\mathcal{C}(m, n)$.

Consider the nth-diagonal power $\mathcal{B} = (A^n, X, \delta') = \mathcal{A}_1 \Delta \cdots \Delta \mathcal{A}_n$, of $\mathcal{C}(m, n)$ with $\mathcal{A}_1 = \cdots = \mathcal{A}_n = \mathcal{C}(m, n)$ and let $\mathcal{B}' = (B', X, \delta'')$ be a state-subautomaton of $\mathcal{B}$ generated by its state $((1, 1), (2, 1), \ldots, (n, 1))$. Put $((i_1, j_1), \ldots, (i_n, j_n)) = (\delta((1, 1), p), \delta((2, 1), p), \ldots, \delta((n, 1), p))$ and $((i_1', j_1'), \ldots, (i_n', j_n')) = (\delta((1, 1), pz), \delta((2, 1), pz), \ldots, \delta((n, 1), pz))$ for a given pair $p \in \{\bar{x}, \bar{y}\}^*$, $z \in \{\bar{x}, \bar{y}\}$. Then, by the definition of $\mathcal{C}(m, n)$, $\{i_1, \ldots, i_n\} = \{i_1', \ldots, i_n'\} = \{1, \ldots, n\}$, and simultaneously,

$$j_1' + j_2' + \cdots + j_n' \pmod m = \begin{cases} \sum_{k=1}^n j_k + 1 \pmod m & \text{if } z = \bar{x}, \\ \sum_{k=1}^n j_k \pmod m & \text{if } z = \bar{y}. \end{cases}$$

Define $\psi : B \to \{1, \dots, m\}$ such that $\psi(((1,1), \dots, (n,1))) = 1$; moreover, for every $p \in \{\bar{x}, \bar{y}\}$, $\psi((\delta((1,1), p), \delta((2,1), p), \dots, \delta((n,1), p))) = p(\bar{x}) + 1 \pmod{m}$, where $p(\bar{x})$ denotes the number of occurrences of the letter $\bar{x}$ in the word p.

Clearly, then ψ is well defined and it is a state-homomorphism of the state-subautomaton $\mathcal{B}'$ of the nth-diagonal product of $\mathcal{C}(m, n)$ onto $\mathcal{C}_m^1$. □

Lemma 3.25. *Given a class $\mathcal{K}$ of automata, suppose that all counters and a strongly connected nonautonomous automaton can be represented homomorphically by a cascade product of factors from $\mathcal{K}$. Then either $\mathcal{A}_0$ or $\mathcal{C}_m^1$ can be represented homomorphically by a cascade product of factors from $\mathcal{K}$ for an integer $m \geq 2$.*

Proof. Suppose that a strongly connected nonautonomous automaton $\mathcal{A} = (A, X, \delta)$ can be represented homomorphically by a cascade product of factors from $\mathcal{K}$. For every $x \in X$, let A_x denote the set of all states $a \in A$ such that $a = \delta(a, x^r)$ for some $r \geq 1$. Since $\mathcal{A}$ is a strongly connected nonautonomous automaton, there are $x_1, x_2 \in X$ and $a_1 \in A_{x_1}$ with $\delta(a_1, x_1) \neq \delta(a_1, x_2)$. Let $n \geq 1$ be any integer with the property that $\delta(a, x_1^{n-1}) \in A_{x_1}$ for all $a \in A$, and $\delta(a, x_1^n) = a$ whenever $a \in A_{x_1}$. Starting with a_1, successively compute the states $a_1, \dots, a_t, a_i = \delta(a_{i-1}, x_2 x_1^{n-1}), i \geq 2$, until repetition occurs. Thus the states $a_1, \dots, a_t$ are pairwise distinct and $\delta(a_t, x_2 x_1^{n-1}) = a_s$ for some $s \in \{1, \dots, t\}$. From the choice of n and a_1 we have $a_1, \dots, a_t \in A_{x_1}$ and $\delta(a_i, x_1^n) = a_i, i \in \{1, \dots, t\}$.

If $\delta(a_i, x_1) = \delta(a_i, x_2)$ for some $i \in \{1, \dots, t\}$, then take a word $v \in X^+$ with $\delta(a_i, v) = a_1$ and define $u = x_1^{n-1}(x_2 x_1^{n-1})^{i-1} v$. We see that $\delta(a_1, x_1) \neq \delta(a_1, x_2)$ and $\delta(a_1, x_1 u) = \delta(a_1, x_2 u) = a_1$. Let $k = |u|$. It is easy to prove that a cascade product of $\mathcal{C}_{k+1}$ with $\mathcal{A}$ has a subautomaton belonging to $\mathcal{K}(1, k+1)$. (Hint: define $\mathcal{C}_{k+1} \times \mathcal{A}(\{\bar{x}, \bar{y}\}, \varphi_1, \varphi_2)$ by $\varphi_1(c, a, \bar{x}) = \varphi_1(c, a, \bar{y}) = \bar{x}, \varphi_2(1, a, \bar{x}) = x_1, \varphi_2(1, a, \bar{y}) = x_2, \varphi_2(i, a, \bar{x}) = \varphi_2(i, a, \bar{y}) = u_{i-1}$ for all $a \in A, c, i \in \{1, \dots, k+1\}, i \neq 1$, where u_{i-1} denotes the $(i-1)$th letter of u; then take the state-subautomaton of this cascade product generated by its state $(1, a_1)$.) Since $\mathcal{C}_{k+1}$ can be represented homomorphically by a cascade product of factors from $\mathcal{K}$, we obtain that at least one element of $\mathcal{K}(1, k+1)$ can be represented homomorphically by a cascade product of factors from $\mathcal{K}$. From Lemma 3.23 it follows that $\mathcal{A}_0$ can be represented homomorphically by a cascade product of factors from $\mathcal{K}$.

Suppose now that $\delta(a_i, x_1) \neq \delta(a_i, x_2)$ for all $i \in \{1, \dots, t\}$. Define $m = t - s + 1, b_1 = a_s, \dots, b_m = a_t$. If $m = 1$, then $n \geq 2$ and we again have that $\mathcal{A}_0$ can be represented homomorphically by a cascade product of factors from $\mathcal{K}$ because of $\delta(b_1, x_1) \neq \delta(b_1, x_2)$ and $\delta(b_1, x_1 x_1^{n-1}) = \delta(b_1, x_2 x_1^{n-1}) = b_1$. Let $m > 1$. Form the cascade product $\mathcal{C}_n \times \mathcal{A}(\{\bar{x}, \bar{y}\}, \varphi_1, \varphi_2)$, where $\varphi_1(c, a, \bar{x}) = \varphi_1(c, a, \bar{y}) = \bar{x}, \varphi_2(1, a, \bar{x}) = x_2, \varphi_2(1, a, \bar{y}) = x_1, \varphi_2(i, a, \bar{x}) = \varphi_2(i, a, \bar{y}) = x_1$ for all $a \in A, c, i \in \{1, \dots, n\}, i \neq 1$. The state-subautomaton generated by the state $(1, b_1)$ of this cascade product contains each of the states $(1, b_1), \dots, (1, b_m)$ and belongs to $\mathcal{K}(m, n)$. From Lemma 3.24 we obtain that $\mathcal{C}_m^1$ can be represented homomorphically by a cascade product of factors from $\mathcal{K}$. □

Lemma 3.26. *Suppose that all counters can be represented homomorphically by appropriate cascade products of factors from $\mathcal{K}$. Then the two-state reset automaton $\mathcal{A}_0$ can be represented homomorphically by a cascade product of factors from $\mathcal{K}$ if and only if the* *following hold:*

(1) *There is a strongly connected automaton which can be represented homomorphically by a cascade product of factors from* $\mathcal{K}$.

(2) *There is an automaton* $\mathcal{A} \in \mathcal{K}$ *such that* $S(\mathcal{A})$ *is isomorphic to the right-zero semigroup with two elements.*

Proof. The necessity of (1) is obvious, while the necessity of (2) can be derived from the Krohn–Rhodes decomposition theorem. By Lemma 3.25, either $\mathcal{A}_0$ or $\mathcal{C}_m^1$ can be represented homomorphically by a cascade product of factors from $\mathcal{K}$. In the first case the proof is done. Supposing that $\mathcal{C}_m^1$ can be represented homomorphically by a cascade product of factors from $\mathcal{K}$, choose $\mathcal{A} = (A, X, \delta)$ with property (2). By Proposition 2.34, there are states $a_1, a_2 \in A$ and words $v_1, v_2 \in X^+$ with $\delta(a_i, v_1) = a_1$ and $\delta(a_i, v_2) = a_2, i = 1, 2$. We may as well suppose that $|v_1| = |v_2| = mn$ for some $n \geq 2$ (for v_1 can be replaced by $(v_2v_1)^m$ and v_2 by $(v_1v_2)^m$). Observe that there exists an nth-cascade power of $\mathcal{C}_m^1$ which is isomorphic to the counter $\mathcal{C}_{mn}$ having mn states. Indeed, let $\mathcal{C}_m^1 = (\{1, \dots, m\}, \{\bar{x}, \bar{y}\}, \delta_{\mathcal{C}_m^1})$ be given with

$$\delta_{\mathcal{C}_m^1}(i, z) = \begin{cases} i + 1 \pmod m & \text{if } z = \bar{x}, \\ i & \text{if } z = \bar{y}, \end{cases}$$

$i \in \{1, \dots, m\}$; moreover let $(\mathcal{C}_m^1)^n(\{\bar{x}\}, \varphi_1, \dots, \varphi_n)$ be given with

$$\varphi_t(c_1, \dots, c_n, \bar{x}) = \begin{cases} \bar{x} & \text{if } t = 1 \\ & \text{or } c_1 = \cdots = c_{t-1} = 1, t > 1, \\ \bar{y} & \text{otherwise}, \end{cases}$$

$(c_1, \dots, c_n) \in \{1, \dots, m\}^n$. An easy technical computation shows that this cascade power is isomorphic to $\mathcal{C}_{mn}$.

Now we consider a cascade product $\mathcal{B} = \mathcal{C}_{mn} \times \mathcal{A}(\{\bar{x}, \bar{y}\}, \varphi_1, \varphi_2)$ such that $\varphi_1(c, a, z) = \bar{x}$ and

$$\varphi_2(c, a, z) = \begin{cases} c\text{th letter of } v_1 & \text{if } z = \bar{x}, \\ c\text{th letter of } v_2 & \text{if } z = \bar{y}, \end{cases}$$

$c \in \{1, \dots, mn\}, a \in A$. Let $\mathcal{B}'$ be a state-subautomaton of the cascade product $\mathcal{B}$ generated by the state $(1, a_1)$. It is obvious that $\mathcal{B}' \in \mathcal{K}(1, n)$. By Lemma 3.23 this completes the proof. □

Let $m \geq 1$ and $n \geq 2$ be integers. We call the automaton $\mathcal{A} = (A, X, \delta)$ an (m, n)-*automaton* if there are $a \in A$, sets $X_1, \dots, X_m \subseteq X$, and signs $x_1, x_2 \in X_1$ with the following conditions, where L denotes the language $X_1 \cdots X_m$:

(1) $1 \leq |X_1|, \dots, |X_m| \leq n$,

(2) $\delta(a, x_1) \neq \delta(a, x_2)$, and

(3) for every $u \in L^+$ there is a $v \in L^*$ with $\delta(a, uv) = a$.

Moreover, we say that A is an m-*automaton* if it is an (m, n)-automaton for some $n \geq 2$. Obviously, $\mathcal{A}$ is an m-automaton if and only if $|X| \geq 2$ and it is an (m, n)-automaton for $n = |X|$.

Proposition 3.27. *If A is an m-automaton and $\mathcal{A}$ is a homomorphic image of the automaton $\mathcal{B}$, then $\mathcal{B}$ is also an m-automaton. Also, if none of $\mathcal{A}_1, \dots, \mathcal{A}_n$ is an m-automaton, then no cascade product with these components is an m-automaton.*

Proof. For the first part of this statement, consider an m-automaton $\mathcal{A} = (A, X, \delta)$ with $a \in A$, sets $X_1, \dots, X_m \subseteq X$, and signs $x_1, x_2 \in X_1$ having the above properties (1) through (3). Let $\mathcal{B} = (B, X', \delta')$ be an automaton having a homomorphism $\psi = (\psi_1, \psi_2)$ onto $\mathcal{A}$. Then for every $a' \in \psi_1^{-1}, x_1' \in \psi_2^{-1}(x_1), x_2' \in \psi_2^{-1}(x_2)$, and $u' \in \psi_2^{-1}(u), u \in L^+, L = X_1 \cdots X_m$ we obtain $\delta'(a', x_1') \neq \delta'(a', x_2')$ and that $\delta'(a', u'v') \in \psi_1^{-1}(a)$ holds for some $v' \in \psi_2^{-1}(v), v \in L^*$. By the finiteness of the state set B of $\mathcal{B}$, there exists a state $a_0' \in \psi_1^{-1}(a)$ such that for every $u' \in \psi_2^{-1}(u), u \in L^+$, there is a $v' \in \psi_2^{-1}(v), v \in L^*$, having $\delta'(a', u'v') = a'$.

For the second part of our statement, consider a cascade product $\mathcal{A} = (A, X, \delta) = \mathcal{A}_1 \times \cdots \times \mathcal{A}_n(X, \varphi_1, \dots, \varphi_n)$ such that none of the automata $\mathcal{A}_t = (A_t, Y_t, \delta_t), t = 1, \dots, n$, is an m-automaton for some m. Assume that contrary to our statement, $\mathcal{A}$ is an m-automaton with $a = (a_1, \dots, a_n) \in A, a_t \in A_t, t = 1, \dots, n$, sets $X_1, \dots, X_m \subseteq X$, and signs $x_1, x_2 \in X_1$ having the above properties (1) through (3). But then for every $u \in L^+, L = X_1 \cdots X_n$, there exists a $v \in L^*$ with $\delta(a, uv) = a$.

Suppose that $\delta_1(a_1, \varphi_1(a_1, \dots, a_n, z_1)) \neq \delta_1(a_1, \varphi_1(a_1, \dots, a_n, z_2))$ for some $z_1, z_2 \in X_1$ and put $X_i' = \{\varphi_1(a_1', \dots, a_n', x') \mid (a_t' \in A_t, t = 1, \dots, n, x' \in X_i\}, i = 1, \dots, m$. But φ_1 may not really depend on its state components and thus for every $x_1' \dots x_m' \in X_1' \cdots X_m'$, there exists a word $x_1 \dots x_m \in L$ such that $\varphi_1(a_1, \dots, a_n, x_1 \dots x_m) = x_1' \dots x_m'$. Then we have that for every $u' \in (X_1' \cdots X_m')^+$ there exists a $v' \in (X_1' \cdots X_n')^*$ such that $\delta_1(a_1, u'v') = a_1$. Thus we obtain that $\mathcal{A}_1$ is an m-automaton, which is a contradiction. Therefore, $\delta_1(a_1, \varphi_1(a_1, \dots, a_n, z_1)) = \delta_1(a_1, \varphi_1(a_1, \dots, a_n, z_2))$ necessarily holds. We get in a similar way that $\delta_1(a_1, \varphi_1(a_1, \dots, a_n, uz_1)) = \delta_1(a_1, \varphi_1(a_1, \dots, a_n, uz_2))$ for every $u \in X_1 \cdots X_t, z_1, z_2 \in X_{t+1 \,(\mathrm{mod}\, m)}, 1 \leq t < m$. This means that for every $u_1, u_2 \in L^+, |u_1| = |u_2|$ implies $\delta_1(a_1, \varphi_1(a_1, \dots, a_n, u_1)) = \delta_1(a_1, \varphi_1(a_1, \dots, a_n, u_2))$.

Suppose that $n \geq 2$ $\delta_2(a_2, \varphi_2(a_1, \dots, a_n, z_1)) \neq \delta_2(a_2, \varphi_2(a_1, \dots, a_n, z_2))$ for some $z_1, z_2 \in X_1$ and put $X_i' = \{\varphi_2(a_1', \dots, a_n', x') \mid (a_1', \dots, a_n') \in \delta((a_1, \dots, a_n), u),$ $u \in L^*$ if $i = 1, u \in L^* X_1 \cdots X_{i-1}$ if $i > 1, x' \in X_i\}, i = 1, \dots, m$. Recall that for arbitrary $u_1, u_2 \in (L^*)X_1 \cdots X_t, t = 1, \dots, n$, with $|u_1| = |u_2|$ and $x \in X_{t+1 \,(\mathrm{mod}\, n)}$, $\varphi_2(a_1', \dots, a_n', x) = \varphi_2(a_1'', \dots, a_n'', x)$ whenever $(a_1' \dots, a_n') = \delta((a_1, \dots, a_n), u_1)$, and $(a_1'' \dots, a_n'') = \delta((a_1, \dots, a_n), u_2)$. But then, for every $x_1' \dots x_m' \in X_1' \cdots X_m'$, there exists a word $x_1 \dots x_m \in L$ such that $\varphi_2(a_1, \dots, a_n, x_1 \dots x_m) = x_1' \dots x_m'$. Hence we obtain that for every $u' \in (X_1' \cdots X_m')^+$ there can be found a $v' \in (X_1' \cdots X_n')^*$ having $\delta_2(a_2, u'v') = a_2$. But then $\mathcal{A}_2$ is an m-automaton, which is a contradiction. By a similar method we obtain $\delta_2(a_2, \varphi_2(a_1, \dots, a_n, u_1 z_1)) = \delta_2(a_2, \varphi_2(a_1, \dots, a_n, u_2 z_2))$ for every $u_1, u_2 \in X_1 \cdots X_t$, $z_1, z_2 \in X_{t+1 \,(\mathrm{mod}\, m)}, 1 \leq t < m$.

Repeating this procedure for $\mathcal{A}_t, t = 3, 4, \dots, n$, finally, we obtain that $\delta(a, x_1) = \delta(a, x_2)$, a contradiction. Therefore, $\mathcal{A}$ is not an m-automaton. □

The following result, which can be derived from the previous theorem, shows the complete structure of the complete classes of automata with respect to homomorphic representations under the cascade product.

Theorem 3.28. *A class $\mathcal{K}$ of automata is complete with respect to homomorphic representations under the cascade product if and only if*

(1) *there is an m-automaton in $\mathcal{K}$ for some $m \geq 1$,*

(2) *for every prime power $n > 1$, there is a multiple m of n, automata $\mathcal{A}_i = (A_i, X_i, \delta_i) \in \mathcal{K}$, $i \in \{1, \ldots, k\}$, $k > 0$, and integers $1 = m_0 < m_1 < \cdots < m_k = m$ such that*

(2a) *m_{i-1} is a divisor of m_i ($i \in \{1, \ldots, k\}$),*

(2b) *for every $i \in \{1, \ldots, k\}$ there are distinct states $a_1, \ldots, a_{m_i/m_{i-1}} \in A_i$ and a word $u \in X_i^*$ with $|u| = m_{i-1}$ and*

$$\delta_i(a_1, u) = a_2, \ldots, \delta_i(a_{m_i/m_{i-1}-1}, u) = a_{m_i/m_{i-1}}, \delta_i(a_{m_i/m_{i-1}}, u) = a_1,$$

(3a) *there is an automaton $\mathcal{A} \in K$ such that a subsemigroup S of $S(\mathcal{A})$ is isomorphic to the monoid with two right-zero elements, and*

(3b) *for every simple group G there is an automaton $\mathcal{A} \in \mathcal{K}$ with $G < S(\mathcal{A})$.* □

Proof. The necessity of condition (1) can be derived from Proposition 3.27. Indeed, if none of the automata $\mathcal{A}_t$, $t = 1, \ldots, n$, is an m-automaton for some m, then, by Proposition 3.27, all of their cascade products preserve this property. In addition, if $\mathcal{B}$ is a subautomaton of $\mathcal{A}$ and $\mathcal{B}$ is an m-automaton for some m, then $\mathcal{A}$ is also an m-automaton by definition. Thus, applying again Proposition 3.27, none of the cascade products of the above automata $\mathcal{A}_t$, $t = 1, \ldots, n$, can represent homomorphically an m-automaton. This ends the proof of the necessity of condition (1).

As regards the necessity of condition (2) of the above result, observe that all counters can be represented homomorphically by a cascade product of automata from $\mathcal{K}$ if and only if $\mathcal{K}$ has property (2). (See also Proposition 3.20.) By Theorem 3.17, this establishes the necessity of condition (2).

The necessity of conditions (2a) and (2b) comes directly from the Krohn–Rhodes decomposition theorem.

As to sufficiency, condition (2) is equivalent of condition (2) of Theorem 3.17. In addition, conditions (3a) and (3b) are the same as conditions (3a) and (3b) of Theorem 3.17. Thus, applying Theorem 3.17, it remains to show that, by our conditions, we can ensure condition (1) of Theorem 3.17.

Let $\mathcal{A} = (A, X, \delta)$ be again an m-automaton with a state $a \in A$, sets $X_1, \ldots, X_m \subseteq X$, and signs $x_1, x_2 \in X_1$ with the following conditions, where L denotes the language $X_1 \cdots X_m$: (a) $1 \leq |X_1|, \ldots, |X_m| \leq n$; (b) $\delta(a, x_1) \neq \delta(a, x_2)$; (c) for every $u \in L^+$ there is a $v \in L^*$ with $\delta(a, uv) = a$.

Consider a cascade product $\mathcal{B} = (\{1, \ldots, m\} \times A, X, \delta') = \mathcal{C}_m \times \mathcal{A}(X, \varphi_1, \varphi_2)$ such that $\mathcal{C}_m = (\{1, \ldots, m\}, \{x_{\mathcal{C}_m}\}, \delta_{\mathcal{C}_m})$, $\delta_{\mathcal{C}_m}(c, x_{\mathcal{C}_m}) = c + 1 \pmod m$, $\varphi_1(c, a, x) = x_{\mathcal{C}_m}$, $c \in C$, $a \in A$, $x \in X$, $\varphi_2(i, a, x) = x$ if $x \in X_i$, $\varphi_2(i, a, x) \in X_i$ if $x \notin X_i$, $i = 1, \ldots, m$.

Obviously, then $\delta'((1, a), x_1) \neq \delta'((1, a), x_2)$. On the other hand, for every $y_1 \ldots y_k \in (X_1 \cdots X_m)^+$, there exists a $y_1 \ldots y_k \in (X_1 \cdots X_m)^+$ such that $\delta(a, y_1 \ldots y_k) = b$ if and only if $\delta'((1, a), x'_1 \ldots x'_k) = (m, b)$. Conversely, for every $x'_1 \ldots x'_k \in L^+$, there exists an $y_1 \ldots y_k \in (X_1 \cdots X_m)^+$ such that $\delta'((1, a), x'_1 \ldots x'_k) = (m, b)$ if and only if $\delta(a, y_1 \ldots y_k) = b$. By condition (c) of the m-automaton $\mathcal{A}$, this implies that for every

$p \in X^*$ there exists a $q \in X^*$ such that $\delta'((1, a)) = (1, a)$. In other words, the state $(1, a)$ of $\mathcal{B}$ generates a strongly connected nonautonomous state-subautomaton of $\mathcal{B}$. By Lemma 3.26, this completes the proof. □

Proposition 3.29. *None of conditions* (1), (2), (3a), *or* (3b) *of Theorem* 3.28 *can be omitted.*

Proof. By Proposition 3.27 we cannot omit condition (1). The rest of the statement is a direct consequence of Theorem 3.18. □

A well-known open problem is whether the following natural question is undecidable in general.

Problem 3.30. *Given a finite class $\mathcal{K}$ of automata and a finite automaton $\mathcal{A}$, decide whether $\mathcal{A}$ can be represented by a cascade product of automata from $\mathcal{K}$.*

Lemma 3.31. *Let $\mathcal{A} = (A, X, \delta)$ be an automaton having states $a, b \in A, a \neq b$, words $p, q, r \in X^+, |p| = |q|$, with $\delta(a, p) = a, \delta(b, p) = b, \delta(a, q) = b, \delta(b, r) = a$. Then there exists a single-factor product $\mathcal{A}(X, \varphi)$ such that* **F** *can be embedded isomorphically into $S(\mathcal{A}(X, \varphi))$, where* **F** *denotes the monoid with two right-zero elements.*

Proof. Consider the automaton $\mathcal{A}$ having the conditions of Lemma 3.31. Furthermore, let $\mathcal{B} = (\{a_1, a_2\}, \{x_0, x_1, x_2\}, \delta_{\mathcal{B}})$ be an automaton with $\delta_{\mathcal{B}}(a_i, x_0) = a_i, \delta_{\mathcal{B}}(a_i, x_j) = a_j, i, j \in \{1, 2\}$. Clearly then $S(\mathcal{B})$ is isomorphic to **F**. Thus, it is enough to find a single-factor product $\mathcal{A}(X, \varphi)$ which isomorphically simulates $\mathcal{B}$ by nonempty words.

By our conditions in Lemma 3.31, there are words $p, v'(= qr), v''(= pr), w'(= q), w''(= p) \in X^+, |v'| = |v''|, |w'| = |w''|$ with $\delta(a, p) = a, \delta(b, p) = b, \delta(a, v') = \delta(b, v'') = a, \delta(a, w') = \delta(b, w'') = b$.

It can be seen that in this case there exists an unambiguously defined $\varphi : A \times Y \to X$ such that for appropriate words $u, v, w \in Y^+$ $\delta(a, \varphi(a, u)) = a, \delta(b, \varphi(b, u)) = b, \delta(a, \varphi(a, v)) = \delta(b, \varphi(b, v)) = a, \delta(a, \varphi(a, w)) = \delta(b, \varphi(b, w)) = b$.

Indeed, let Y be an arbitrary nonempty set having at least $3|p| + |r|$ elements. Thus there are words $u, v, w \in Y^+$ with $|u| = |p|, |v| = |v'| = |v''| = |pr|, |w| = |w'| = |w''| = |p|$ such that uvw does not contain letters in Y^+ with double occurrences.

Define $\varphi : A \times Y \to X$ such that

$$\begin{aligned}
&\varphi(\delta(a, p_1), z) = \varphi(\delta(b, p_1), z) = x \text{ if } u = u_1 z u_2,\ p = p_1 x p_2,\ |u_1| = |p_1|,\\
&\varphi(\delta(a, v_1'), z) = x',\ \varphi(\delta(b, v_1''), z) = x'' \text{ if } v = v_1 z v_2,\ v' = v_1' x' v_2',\ v'' = v_1'' x'' v_2'',\\
&\qquad |v_1| = |v_1'| = |v_1''|,\ \delta(a, v_1') \neq \delta(b, v_1''),\\
&\varphi(\delta(a, v_1'), z) = \varphi(\delta(b, v_1''), z) = x' \text{ if } v = v_1 z v_2,\ v' = v_1' x' v_2',\ v'' = v_1'' x'' v_2'',\\
&\qquad |v_1| = |v_1'| = |v_1''|,\ \delta(a, v_1') = \delta(b, v_1''),\\
&\varphi(\delta(a, w_1'), z) = y',\ \varphi(\delta(b, w_1''), z) = y'' \text{ if } w = w_1 z w_2,\ w' = w_1' y' w_2',\\
&\qquad w'' = w_1'' y'' w_2'',\ |w_1| = |w_1'| = |w_1''|,\ \delta(a, w_1') \neq \delta(b, w_1''),\\
&\varphi(\delta(a, w_1'), z) = \varphi(\delta(b, w_1''), z) = y' \text{ if } w = w_1 z w_2,\ w' = w_1' y' w_2',\\
&\qquad w'' = w_1'' y'' w_2'',\ |w_1| = |w_1'| = |w_1''|,\ \delta(a, w_1') = \delta(b, w_1'').
\end{aligned}$$

Then $\mathcal{A}$ isomorphically simulates $\mathcal{B}$ by nonempty words under $\tau_1 : \{a, b\} \to \{a_1, a_2\}$, $\tau_2 : \{x_0, x_1, x_2\} \to \{u, v, w\}$ with $\tau_1(a) = a_1$, $\tau_1(b) = a_2$, $\tau_2(x_0) = u$, $\tau_2(x_1) = v$, $\tau_2(x_2) = w$. This is the end of the proof. □

Lemma 3.32. *Let $\mathcal{A} = (A, X, \delta)$ be an automaton such that $G < S(\mathcal{A})$ for some noncommutative group G. There exists a single-factor product $\mathcal{A}(X, \varphi)$ such that the monoid with two right-zero elements can be embedded isomorphically into the semigroup $S(\mathcal{A}(X, \varphi))$.*

Proof. By Proposition 1.11, there exists a subgroup $\tilde{G}$ of $S(\mathcal{A})$ which acts on a subset of $Z \subseteq A$ by permutations so that $(Z, \tilde{G})$ is a permutation group and $\tilde{G}$ maps homomorphically onto G. Since G is noncommutative, so is $\tilde{G}$. Thus there exist words $x, y \in X^+$ such that x and y represent members of $\tilde{G}$ that correspond to noncommuting permutations of the states Z. That is, $\delta_x, \delta_y \in \tilde{G}$ but $\delta_x\delta_y \neq \delta_y\delta_x$. Hence there exists a state $a_0 \in Z$ such that $a \neq b$ for $a = \delta(a_0, xy)$, $b = \delta(a_0, yx)$. Recall that $o(g)$ denotes the order of a group element. By definition of order, $x^{o(\delta_x)}$ acts as the identity permutation on Z, and similarly for $y^{o(\delta_y)}$. Now define the following words in X^+:

$$p = x^{o(\delta_x)}y^{o(\delta_y)},$$

$$q = y^{o(\delta_y)-1}x^{o(\delta_x)-1}yx,$$

$$r = x^{o(\delta_x)-1}y^{o(\delta_y)-1}xy.$$

(Note that the orders of δ_x and δ_y are each more than 1, since these group elements do not commute.) Observe that each of these words is of the same length, namely, of length $o(\delta_x)|x| + o(\delta_y)|y|$. We compute that $a \cdot q = a_0 \cdot xyq = a_0 \cdot xyy^{o(\delta_y)-1}x^{o(\delta_x)-1}yx = a_0 \cdot xy^{o(\delta_y)}x^{o(\delta_x)-1}yx = a_0 \cdot xx^{o(\delta_x)-1}yx = a_0 \cdot x^{o(\delta_x)}yx = a_0 \cdot yx = b$. It is trivial to check that $a \cdot p = a$, $b \cdot p = b$, and $b \cdot r = a$. But then the states a, b and the words p, q, r of $\mathcal{A}$ satisfy the conditions of Lemma 3.31. This ends the proof. □

Lemma 3.33. *Let $\mathcal{A} = (A, X, \delta)$ be an automaton such that $G < S(\mathcal{A})$ for some noncommutative group G. Then $\mathcal{A}$ satisfies Letichevsky's criterion.*

Proof. Given an automaton $\mathcal{A}$ with n states, let $G < S(\mathcal{A})$ for a noncommutative group G. By Proposition 2.47 we obtain $\mathcal{A}_G < \mathcal{A}^{\Delta n}$, where $\mathcal{A}^{\Delta n}$ denotes the nth-diagonal power of $\mathcal{A}$. It is clear that $\mathcal{A}_G$ is strongly connected. In addition, G is noncommutative. Thus $\mathcal{A}_G$ is a noncommutative strongly connected automaton. Therefore, by Proposition 2.76, $\mathcal{A}^{\Delta n}$ satisfies Letichevsky's criterion. Obviously, then $\mathcal{A}$ also has this property. This ends the proof. □

Lemma 3.34. *Let $\mathcal{A} = (A, X, \delta)$ be an automaton having Letichevsky's criterion with a state $a_0 \in A$, input letters $x, y \in X$, and words $p, q \in X^*$ such that $\delta(a_0, x) \neq \delta(a_0, y)$ and $\delta(a_0, xp) = \delta(a_0, yq)$. There exists a single-factor product $\mathcal{B} = \mathcal{A}(X, \varphi)$ and a counter $\mathcal{C}_k$ such that the two-state reset automaton can be homomorphically represented by an α_0-product $\mathcal{C}_k \times \mathcal{B}^{|pq|+2}(\{x, y\}, \varphi_1, \dots, \varphi_{|pq|+2})$.*

Proof. Let $\mathcal{A} = (A, X, \delta)$ be an automaton having Letichevsky's criterion with a state $a_0 \in A$, input letters $x, y \in X$, and words $p, q \in X^*$ such that $\delta(a_0, x) \neq \delta(a_0, y)$ and

$\delta(a_0, xp) = \delta(a_0, yq)$. Suppose that $\delta(a_0, xp') = \delta(a_0, yq')$ holds for some $p', q' \in X$ having $p = p'p'', q = q'q''$. Then $\mathcal{A}$ will also have Letichevsky's criterion for the state a_0, input letters $x, y \in X$, and words $p'q'', q'q''$. Therefore, we may assume $p'' = q''$ whenever $\delta(a_0, xp') = \delta(a_0, yq')$ holds with $p = p'p'', q = q'q''$. We also assume the minimality of p and q such that $\delta(a_0, xp') = \delta(a_0, xp'p'')$ implies $p'' = \lambda$, and, similarly, $\delta(a_0, yq') = \delta(a_0, yq'q'')$ implies $q'' = \lambda$ for every $p', p'', p''', q', q'', q'''$ with $p = p'p''p'''$ and $q = q'q''q'''$. Thus the following feedback function is unambiguously defined.

$$\varphi(a, z) = \begin{cases} x & \text{if } a = a_0, z = x, \\ y & \text{if } a = a_0, z = y, \\ x' & \text{if } a = \delta(a_0, xp'), p = p'x'p'', \\ y' & \text{if } a = \delta(a_0, yq'), q = q'y'q'', \\ \text{fixed element} & \text{otherwise} \end{cases}$$

(for $z \in \{x, y\}, a \in A$).

Consider the single-factor product $\mathcal{B} = (A, \{x, y\}, \delta_{\mathcal{B}}) = \mathcal{A}(\{x, y\}, \varphi)$ and nonnegative integers m, n with $m = |p|, n = |q|$. Let, say, $m \leq n$. Observe that there exists a word $r \in \{x, y\}^+$ having $\delta_{\mathcal{B}}(a_0, xr) = \delta_{\mathcal{B}}(a_0, yr)$. (Indeed, for example, let $r = x^m y^{n-m} x^{m+1}$.) Put $r = z_1 \dots z_{k-1}, z_1, \dots, z_{k-1} \in \{x, y\}$.

Take the counter $\mathcal{C}_k = (\{1, \dots, k\}, \{x\}, \delta_{\mathcal{C}_k})$ (with $k = |r| + 1$) and define the α_0-product $\mathcal{M} = (M, \{x, y\}, \delta_{\mathcal{M}}) = \mathcal{C}_k \times \mathcal{B}^k(\{x, y\}, \varphi_1, \dots, \varphi_{k+1})$ such that

$$\varphi_i(c, b_1, \dots, b_k, z) = \begin{cases} x & \text{if } i = 1, \\ z & \text{if } 1 < i \leq k+1, c = i - 1 \,(\text{mod}\, k), \\ z_s & \text{if } 1 < i \leq k+1, c = i + s - 1 \,(\text{mod}\, k), \end{cases}$$

$(c, b_1, \dots, b_k) \in \{1, \dots, k\} \times A^k, z \in \{x, y\}$.

Put $H = \{(c, b_1, \dots, b_k) \mid 1 \leq c \leq k, b_{c-1(\text{mod}\, k)} \in \{\delta_{\mathcal{B}}(a_0, x), \delta_{\mathcal{B}}(a_0, y)\}, b_{c+s-1(\text{mod}\, k)} \in \{\delta(a_0, xz_1 \dots z_s), \delta(a_0, yz_1 \dots z_s)\}, 1 \leq s \leq k-1\}$. (Note that $\delta(a_0, xz_1 \dots z_{k-1}) = \delta(a_0, yz_1 \dots z_{k-1}) = a_0$.)

Let the product $\mathcal{M}$ receive the input signal z in its state $(c, b_1, \dots, b_k) \in H$. The next state $\delta_{\mathcal{M}}((c, b_1, \dots, b_k), z)$ is obtained in the following way. The first component c is set to $c + 1 \,(\text{mod}\, k)$, the $c \,(\text{mod}\, k) + 1$-th component $b_{c\,(\text{mod}\, k)} = a_0$ is replaced by $\delta_{\mathcal{B}}(a_0, z)$, the $c + 1 \,(\text{mod}\, k) + 1$-th component $b_{c+1\,(\text{mod}\, k)} = \delta(a_0, z'), z' \in \{x, y\}$ assumes the value $\delta(a_0, z'z_1)$, and each of the components $b_{c+s\,(\text{mod}\, k)+1} = \delta(a_0, z''z_1 \dots z_{s-1}), z'' \in \{x, y\}$, $s = 2, \dots, k-1$ is changed for $b_{s+i+1\,(\text{mod}\, k)} = \delta(a_0, z''z_1 \dots z_s)$.

Thus, for every $(c, b_1, \dots, b_k)$ the $c - 1 \,(\text{mod}\, k) + 1$-th component $b_{c-1\,(\text{mod}\, k)}$ shows the value of the last incoming input letter for $\mathcal{M}$.

Define $\mathcal{A}_0 = (\{a_1, a_2\}, \{x_1, x_2\}, \delta_0)$ with $\delta_0(a_i, x_j) = a_j,\ i, j \in \{1, 2\}$. Then $\mathcal{A}_0$ is the two-state reset automaton.

For every $(i, b_1, \dots, b_k) \in M$ we represent the state of $\mathcal{A}_0$ by $b_{i-1\,(\text{mod}\, k)}$ such that, say,

$$\psi(i, b_1, \dots, b_k) = \begin{cases} a_1 & \text{if } b_{i-1(\text{mod}\ k)} = \delta_{\mathcal{B}}(a_0, x), \\ a_2 & \text{if } b_{i-1(\text{mod}\ k)} = \delta_{\mathcal{B}}(a_0, x). \end{cases}$$

It is obvious that $\psi : H \to \{a_1, a_2\}$ is a homomorphism of a state-subautomaton of $\mathcal{M}$ onto $\mathcal{A}_0$. This ends the proof. $\square$

We have the following direct consequence of Theorem 3.17.

Theorem 3.35. *A class $\mathcal{K}$ of automata is complete with respect to homomorphic representations under the α_1-product if and only if*

(1) *the two-state reset automaton can be represented homomorphically by an α_1-product of automata from $\mathcal{K}$,*

(2) *every counter (of prime power length) can be represented homomorphically by an α_1-product of automata from $\mathcal{K}$,*

(3a) *there is a single-factor product $\mathcal{B}$ of an automaton $\mathcal{A} \in K$ such that a subsemigroup S of $S(\mathcal{B})$ is isomorphic to the monoid with two right-zero elements, and*

(3b) *for every simple group G there is a single-factor product $\mathcal{B}$ of an automaton $\mathcal{A} \in \mathcal{K}$ with $G < S(\mathcal{B})$.* □

Now we are ready to prove the following well-known result.

Theorem 3.36. *A class $\mathcal{K}$ is complete with respect to the homomorphic representations under the α_1-product if and only if*

(1) *every counter (of prime power length) can be represented homomorphically by an α_1-product of automata from $\mathcal{K}$, and*

(2) *for every simple group G there is an automaton $\mathcal{A} \in \mathcal{K}$ having a single-factor product $\mathcal{B}$ with $G < S(\mathcal{B})$.*

Proof. The necessity of (1) and (2) are obvious. To show sufficiency it is enough to prove that by our conditions (1) and (2) we obtain the conditions of Theorem 3.35. Indeed, using conditions (1) and (2) of our statement, condition (1) of Theorem 3.35 comes from Lemmas 3.33 and 3.34. Furthermore, using again conditions (1) and (2) of our statement, condition (3a) of Theorem 3.35 is a direct consequence of Lemma 3.32. The proof is complete. □

Proposition 3.37. *Neither condition* (1) *nor* (2) *of Theorem* 3.36 *can be omitted.*

Proof. It is trivial that none of the counters satisfies Letichevsky's criterion. Thus the class of all counters is not complete with respect to the homomorphic representations under the general product and thus it is not complete for the homomorphic representations under the α_1-product. Then it is obvious that we cannot omit condition (2). Now we prove that we cannot omit condition (1). For every $n \geq 2$, let $\mathcal{A}_n = (A_n, X_n, \delta_n)$ be the automaton where $A_n = \{0, 1, \ldots, n, 1', \ldots, n'\}$, $X_n = \{x_1, \ldots, x_n\}$, $\delta_n(0, x_i) = i$, $\delta_n(i, x_1) = 0$, $\delta_n(i, x_j) = i'$, $\delta_n(i', x_k) = i$ for every $i, k \in \{1, \ldots, n\}$, $j \in \{2, \ldots, n\}$. To see that $\mathcal{K} = \{\mathcal{A}_n \mid n \geq 2\}$ satisfies (2) of Theorem 3.36, we show that the degree-$(n-1)$ symmetric group can be embedded isomorphically into the semigroup of a single factor product of $\mathcal{A}_n$. Obviously, this holds if and only if the degree $(n-1)$ symmetric group can be embedded isomorphically into the semigroup of the digraph $\mathcal{D}_n = (A_n, \{(a, b) \mid a, b \in A_n,$ there exists $x \in X_n : \delta_n(a, x) = b\})$ which has the structure $\mathcal{D}_n = (A_n, \{(0, i), (i, 0), (i, i'), (i', i) \mid i = 1, \ldots, n\})$.

To prove this fact, we consider a game similar to that in the first part of Section 2.1: let us place $n-1$ coins c_i onto the vertices $1', \ldots, (n-1)'$ so that c_i is placed onto i'. It is sufficient to prove that every pair of coins can be interchanged so that the others get back to their initial locations. We can restrict ourselves to the case that c_1 and c_2 are the coins to be interchanged. Let us first move c_1 to n' in four steps along the path $1'10nn'$; meanwhile all other coins are rotated around the cycles $i'ii'$, $i = 2, \ldots, n-1$, since there are no loop edges. After this transformation we see that c_1 is on n', and for every $i = 2, \ldots, n-1$, c_i is back on vertex i'. Next, in a similar way, move c_2 to $1'$ along the path $2'2011'$ and rotate the coins $c_1, c_3, \ldots, c_{n-1}$ around the cycles $n'nn', 3'33', \ldots, (n-1)'(n-1)(n-1)'$, respectively. Now the placement of the coins is this: c_1 is on the vertex n', c_2 is on $1'$, and for $i = 3, \ldots, n-1$, c_i is on i'. Finally, with a similar procedure, move c_1 to $2'$ so that all the coins c_i get back to i', $i = 3, \ldots, n-1$, and c_2 gets back $1'$. This completes the proof that $\mathcal{K}$ satisfies (2).

To see that $\mathcal{K} = \{\mathcal{A}_n \mid n \geq 2\}$ does not satisfy (1) we show that for every α_1-product $\mathcal{B}$ of factors from $\mathcal{K}$, none of the counters of length greater than 2 can be represented homomorphically by $\mathcal{B}$. Assume to the contrary that there is such a counter. By Propositions 2.59 and 3.20, there are a single-factor product $\mathcal{A} = (A_n, X, \delta) = \mathcal{A}_n(X, \varphi)$, distinct states $a_1, \ldots, a_k \in A_n$, $k \geq 3$, and a word $u \in X^*$, $|u| = 2$, with $\delta(a_1, u) = a_2, \ldots, \delta(a_{k-1}, u) = a_k$, $\delta(a_k, u) = a_1$. Put $u = xy$. If $a_1 = i'$, then $a_2 = 0$ and $a_3 = j'$ with $j \neq i$. But then $\delta(a_3, u) \in \{0, j'\}$, a contradiction. If $a_1 = i$ with $i \neq 0$, then $a_2 = j$ so that $j \neq i$, $j \in \{1, \ldots, n\}$. We see that $\delta(i, x) = 0$, and whether $\delta(j, x) = 0$ or $\delta(j, x) = j'$, we have $a_2 = a_3$, a contradiction. The last case is that $a_1 = 0$. But then clearly $a_2 = i'$ for some $i \in \{1, \ldots, n\}$ and either $a_3 = a_1$ or $a_3 = a_2$, completing the proof. □

The following problem remains open.

Problem 3.38. *Does there exist a minimal homomorphically complete class of automata under the α_1-product?*

3.3 Homomorphically Complete Classes Under the Quasi-Direct Product

Recall that a quasi-direct product of automata is a special type of the general product such that the feedback functions of the component automata are really independent of the state components. Therefore, a quasi-direct product of automata $\mathcal{A}_t = (A_t, X_t, \delta_t)$, $t = 1, \ldots, n$, can be given as an automaton $\mathcal{A} = (A, X, \delta) = \prod_{t=1}^{n} \mathcal{A}_t(X, \varphi_1, \ldots, \varphi_n)$ with $A = A_1 \times \cdots \times A_n$ and $\delta((a_1, \ldots, a_n), x) = (\delta_1(a_1, \varphi_1(x)), \ldots, \delta_n(a_n, \varphi_n(x)))$, $(a_1, \ldots, a_n) \in A$, $x \in X$ (where the feedback functions $\varphi_t : X \to X_t$, $t = 1, \ldots, n$, do not depend on the state variables). The following statement is obvious.

Proposition 3.39. *The quasi-direct product of automata $\mathcal{A}_t = (A_t, X_t, \delta_t)$ given by X and $\varphi_t : X \to X_t$, $t = 1, 2$, is isomorphic to the quasi-direct product of automata $\mathcal{A}_2$ and $\mathcal{A}_1$ given by X and φ_2, φ_1.* □

Next we prove the following theorem.

Theorem 3.40. *There exists no minimal complete class of automata with respect to homomorphic representations under the quasi-direct product.*

Proof. Consider an arbitrary complete class $\mathcal{K}$ of automata with respect to homomorphic representations under the quasi-direct product. It is enough to show that for every automaton $\mathcal{A} = (A, X', \delta_{\mathcal{A}})$, the class $\mathcal{K} \setminus \{\mathcal{A}\}$ also is a complete class of automata with respect to homomorphic representations under the quasi-direct product.

Let $A = \{a_1, \ldots, a_n\}$ and take two distinct symbols $x_0, x_1 \notin X'$. Consider an automaton $\mathcal{B} = (B, X, \delta_{\mathcal{B}})$ for which $B = \{1, \ldots, n^2 + 1\}$, $X = X' \cup \{x_0, x_1\}$; moreover, for all $b \in B$ and $x \in X$,

$$\delta_{\mathcal{B}}(b, x) = \begin{cases} c & \text{if } 1 \le b \le n, x \in X', \delta_{\mathcal{A}}(a_b, x) = a_c, \\ b+1 & \text{if } 1 \le b \le n, x \in \{x_0, x_1\}, \\ b+1 & \text{if } n < b < n^2 + 1, \\ 1 & \text{if } b = n^2 + 1,\ x \in X' \cup \{x_0\}, \\ 2 & \text{if } b = n^2 + 1,\ x = x_1. \end{cases}$$

Consider a subautomaton $\mathcal{B}'' = (\{1, \ldots, n\}, X', \delta''_{\mathcal{B}})$ of $\mathcal{B}$. It is clear that $\mathcal{B}''$ is isomorphic to $\mathcal{A}$. Therefore, for our theorem it is enough to prove that $\mathcal{B}$ can be represented homomorphically by an appropriate quasi-direct product of automata from $\mathcal{K} \setminus \{\mathcal{A}\}$. It can be seen that $\mathcal{B}$ satisfies the following condition:

(1) For every $b, c \in B$ there exists a $p \in X^+$ with $\delta_{\mathcal{B}}(a, p) = \delta_{\mathcal{B}}(b, p)$.

Indeed, in case $b = c$ this is obvious. Now let $b \neq c$. Using the definition of $\delta_{\mathcal{B}}$, we have that by $b < c$, $p = x_0^{n^2+1-b}(x_1 x_0^{n^2-1})^{c-b}$ and, similarly, by $b > c$, $p = x_0^{n^2+1-c}(x_1 x_0^{n^2-1})^{b-c}$ satisfy condition (1).

We show that for all distinct states $a_i, a_j \in A$ of $\mathcal{A}$ and also for every word $p'_0 = x'_1 \ldots x'_t, x'_1, \ldots, x'_t \in X'$ satisfying $\delta_{\mathcal{B}}(a_i, p') = \delta_{\mathcal{B}}(a_j, p')$, there exists a word $p \in \{x'_1, \ldots, x'_t\}^*$ such that $\delta_{\mathcal{B}}(a_i, p) = \delta_{\mathcal{B}}(a_j, p)$ and $|p| \le n(n-1)$.

Indeed, assume that for an appropriate index r, $1 \le r \le t$, there exists an s, $1 \le s \le t - r$, such that $\delta_{\mathcal{B}}(a_i, x'_1 \ldots x'_{r+s}) = \delta_{\mathcal{B}}(a_j, x'_1 \ldots x'_{r+s}) = \delta_{\mathcal{B}}(a_j, x'_1 \ldots x'_r)$. Then, by the input word $x'_{r+s+1} \ldots x'_t$, $\delta_{\mathcal{B}}(a_i, x'_1 \ldots x'_{r+s} x'_{r+s+1} \ldots x'_t) = \delta_{\mathcal{B}}(a_i, x'_1 \ldots x'_r x'_{r+s+1} \ldots x'_t)$ and $\delta_{\mathcal{B}}(a_j, x'_1 \ldots x'_{r+s} x'_{r+s+1} \ldots x'_t) = \delta_{\mathcal{B}}(a_j, x'_1 \ldots x'_r x'_{r+s+1} \ldots x'_t)$. Consequently, the word p can be constructed such that its length does not exceed the number of pairs $(a_r, a_s) \in A \times A$, $a_r \neq a_s$. Thus, $|p| \le n(n-1)$ holds.

Let $\mathcal{D}_t = (D_t, X_t, \delta_t)$, $t = 1, \ldots, m$, be finite automata having an index $t \in \{1, \ldots, m\}$ with $\mathcal{D}_t = \mathcal{A}$. Suppose that a quasi-direct product $\mathcal{M} = \prod_{t=1}^m \mathcal{D}_t(Y, \varphi_1, \ldots, \varphi_m)$ can be given such that it homomorphically represents $\mathcal{B}$. By Proposition 2.52, we may also assume that $\mathcal{M}$ has a state-subautomaton $\mathcal{N} = (N, X, \delta_{\mathcal{N}})$ with a state-homomorphism $\psi : N \to B$ onto $\mathcal{B}$. (And then $Y = X$.) Assume that $\mathcal{N}$ is minimal in the sense that $\mathcal{B}$ is not a state-homomorphic image of a proper state-subautomaton of $\mathcal{N}$. By the definition of $\delta_{\mathcal{B}}$, $\mathcal{B}$ is strongly connected. Thus, using Proposition 2.25, $\mathcal{N}$ is also strongly connected. Because of Proposition 3.39, we can assume that either every $\mathcal{D}_t$, $t = 1, \ldots, m$, is equal to $\mathcal{A}$ or there exists an index $k \in \{1, \ldots, m\}$ such that $\mathcal{A} \notin \{\mathcal{D}_1, \ldots, \mathcal{D}_k\}$, but $\mathcal{A} = \mathcal{D}_{k+1} = \cdots = \mathcal{D}_m$. Especially, thus $\mathcal{D}_m = \mathcal{A}$ necessarily holds. Since $|A| < |B|$, by the assumption that $\mathcal{M}$ state-homomorphically represents $\mathcal{B}$, it follows that $m > 1$.

Consider the quasi-direct product $\mathcal{M}_1 = \prod_{t=1}^{m-1} \mathcal{D}_t(X, \varphi_1, \ldots, \varphi_{m-1})$. We now show that for all pairs $(d_1, \ldots, d_{m-1}, a_i), (d_1, \ldots, d_{m-1}, a_j) \in N, \psi((d_1, \ldots, d_{m-1}, a_i)) = \psi((d_1, \ldots, d_{m-1}, a_j))$, i.e., the homomorphism ψ is independent of the last factor of the quasi-direct product $\mathcal{M}$.

To this, first we assume the contrary of this statement, i.e., for a suitable pair of states $(d_1, \ldots, d_{m-1}, a_i), (d_1, \ldots, d_{m-1}, a_j) \in N$, suppose that $\psi((d_1, \ldots, d_{m-1}, a_i)) \neq \psi((d_1, \ldots, d_{m-1}, a_j))$. Then, by the definition of $\delta_\mathcal{B}$, there exists a positive integer t such that $\delta_\mathcal{B}(\psi((d_1, \ldots, d_m, a_i)), x_0^t) = n + 1$ and, using again the definition of $\delta_\mathcal{B}$, and taking into consideration $\psi((d_1, \ldots, d_{m-1}, a_i)) \neq \psi((d_1, \ldots, d_{m-1}, a_j))$, we obtain

$$\delta_\mathcal{B}(\psi((d_1, \ldots, d_m, a_i)), x_0^t) = n + 1 \neq \delta_\mathcal{B}(\psi((d_1, \ldots, d_m, a_j)), x_0^t).$$

However, using that for every $b, c \in B$ there exists a $p \in X^+$ having $\delta_\mathcal{B}(b, p) = \delta_\mathcal{B}(c, p)$, we obtain the existence of a word $q \in X^+$ with $\delta_\mathcal{B}(\psi((d_1, \ldots, d_m, a_i)), x_0^t q) = \delta_\mathcal{B}(\psi((d_1, \ldots, d_m, a_j)), x_0^t q)$. Since $\mathcal{N}$ is strongly connected, an $r \in X^+$ can be found such that $\delta_\mathcal{N}((d_1, \ldots, d_m, a_i), x_0^t qr) = \delta_\mathcal{N}((d_1, \ldots, d_m, a_i), x_0^t)$.

Hence, $\delta_{\mathcal{M}_1}((d_1, \ldots, d_m), x_0^t qr) = \delta_{\mathcal{M}_1}((d_1, \ldots, d_m), x_0^t)$, where $\delta_{\mathcal{M}_1}$ denotes the transition function of the quasi-direct product $\mathcal{M}_1$.

On the other hand, using, in order,

$$\delta_\mathcal{B}(\psi((d_1, \ldots, d_m, a_i)), x_0^t) = n + 1,$$
$$\delta_\mathcal{B}(\psi((d_1, \ldots, d_m, a_i)), x_0^t q) = \delta_\mathcal{B}(\psi((d_1, \ldots, d_m, a_j)), x_0^t q),$$

and

$$\delta_\mathcal{N}((d_1, \ldots, d_m, a_i), x_0^t qr) = \delta_\mathcal{N}((d_1, \ldots, d_m, a_i), x_0^t),$$

we get

$$\delta_\mathcal{B}(\psi((d_1, \ldots, d_m, a_j)), x_0^t qr) = \delta_\mathcal{B}(\psi((d_1, \ldots, d_m, a_i)), x_0^t) = n + 1.$$

For every $u = 1, \ldots, m$ we extend the function $\varphi_u : X \to X_u$ to X^+ such that for all $x_1', \ldots, x_k' \in X$, $\varphi_u(x_1' \ldots x_k') = \varphi_u(x_1') \ldots \varphi_u(x_k')$.

Continuing the proof, that $\psi((d_1, \ldots, d_{m-1}, a_i)) = \psi((d_1, \ldots, d_{m-1}, a_j))$ holds for all pairs $(d_1, \ldots, d_{m-1}, a_i)$, $(d_1, \ldots, d_{m-1}, a_j) \in N$, we show that

(2) for every positive integer z and $r' \in X^+$, $\delta_\mathcal{A}(a_j, \varphi_m(x_0^t r')) \neq \delta_\mathcal{A}(a_j, \varphi_m(x_0^t (qr)^z r'))$.

Indeed, by virtue of $\delta_\mathcal{B}(\psi((d_1, \ldots, d_m, a_j)), x_0^t qr) = \delta_\mathcal{B}(\psi((d_1, \ldots, d_m, a_i)), x_0^t) = n + 1$ and $\delta_\mathcal{N}((d_1, \ldots, d_m, a_i), x_0^t qr) = \delta_\mathcal{N}((d_1, \ldots, d_m, a_i), x_0^t)$, we get $\delta_\mathcal{B}(\psi((d_1, \ldots, d_m, a_j)), x_0^t (qr)^z) = \delta_\mathcal{B}(\psi((d_1, \ldots, d_m, a_i)), x_0^t (qr)^{z-1}) = \cdots = \delta_\mathcal{B}(\psi((d_1, \ldots, d_m, a_i)), x_0^t) = n + 1$.

Consequently, because of $\delta_\mathcal{B}(\psi((d_1, \ldots, d_m, a_i)), x_0^t) = n + 1 \neq \delta_\mathcal{B}(\psi((d_1, \ldots, d_m, a_j)), x_0^t)$, $\delta_\mathcal{B}(\psi((d_1, \ldots, d_m, a_j)), x_0^t) \neq n + 1 = \delta_\mathcal{B}(\psi(d_1, \ldots, d_m, a_j), x_0^t (qr)^z)$.

Thus, according to the definition of $\delta_\mathcal{B}$, for every word $r' \in X^+$ with $|r'| \leq n(n-1)$ we have $\delta_\mathcal{B}(\psi((d_1, \ldots, d_m, a_j)), x_0^t r') \neq \delta_\mathcal{B}(\psi((d_1, \ldots, d_m, a_j)), x_0^t (qr)^z r')$.

Simultaneously, by $\delta_{\mathcal{M}_1}((d_1, \ldots, d_m), x_0^t qr) = \delta_{\mathcal{M}_1}((d_1, \ldots, d_m), x_0^t)$ (where $\delta_\mathcal{M}$ denotes the transition function of the quasi-direct product $\mathcal{M}_1$), it follows that for every input word $r'' \in X^+$ of $\mathcal{M}_1$, $\delta_{\mathcal{M}_1}((d_1, \ldots, d_m), x_0^t r'') = \delta_{\mathcal{M}_1}((d_1, \ldots, d_m), x_0^t (qr)^z r'')$.

From this, by the substitution $r'' = r'$, we obtain $\delta_u(d_u, \varphi_u(x_0^t r')) = \delta_u(d_u, \varphi_u(x_0^t (qr)^z r'))$, $u = 1, \dots, m-1$, where δ_t denotes the transition function of the tth factor $\mathcal{D}_t$ of the quasi-direct product $\mathcal{M}_1$. Hence, by $\delta_\mathcal{B}(\psi((d_1, \dots, d_m, a_j)), x_0^t r') \neq \delta_\mathcal{B}(\psi((d_1, \dots, d_m, a_j)), x_0^t (qr)^z r')$ we have $\delta_\mathcal{A}(a_j, \varphi_m(x_0^t r')) \neq \delta_\mathcal{A}(a_j, \varphi_m(x_0^t (qr)^z r'))$, that is, for every input word $r' \in X^+$ with $|r'| \leq n(n-1)$, $\delta_\mathcal{A}(a_j, \varphi_m(x_0^t)\varphi_m(r')) \neq \delta_\mathcal{A}(a_j, \varphi_m(x_0^t)(\varphi_m(qr))^z \, \varphi_m(r'))$ holds.

Now we suppose that there exists a word $r'' \in X^+$ for which $\delta_\mathcal{A}(a_j, \varphi_m(x_0^t)\varphi_m(r'')) = \delta_\mathcal{A}(a_j, \varphi_m(x_0^t)(\varphi_m(qr))^z \varphi_m(r''))$. As we have already established, then there exists such a word $r'' \in X^+$ having $|r''| \leq n(n-1)$, contradicting $\delta_\mathcal{A}(a_j, \varphi_m(x_0^t)\varphi_m(r')) \neq \delta_\mathcal{A}(a_j, \varphi_m(x_0^t)(\varphi_m(qr))^z \varphi_m(r'))$, $r' \in X^*$, $|r'| \leq n(n-1)$.

Herewith it is proved that $\delta_\mathcal{A}(a_j, \varphi_m(x_0^t)\varphi_m(r')) \neq \delta_\mathcal{A}(a_j, \varphi_m(x_0^t)(\varphi_m(qr))^z \varphi_m(r'))$ and $\delta_\mathcal{A}(a_j, \varphi_m(x_0^t r')) \neq \delta_\mathcal{A}(a_j, \varphi_m(x_0^t (qr)^z r'))$ for every $r' \in X^+$ hold and this implies (2).

Because of $\delta_\mathcal{A}(a_j, \varphi_m(x_0^t r')) \neq \delta_\mathcal{A}(a_j, \varphi_m(x_0^t (qr)^z r'))$, by choosing $r' = (qr)^k$, we have that for every pair $z, k > 0$, $\delta_\mathcal{A}(a_j, \varphi_m(x_0^t (qr)^k)) \neq \delta_\mathcal{A}(a_j, \varphi_m(x_0^t (qr)^{z+k}))$. Thus the elements $\delta_\mathcal{A}(a_j, \varphi_m(x_0^t qr)), \delta_\mathcal{A}(a_j, \varphi_m(x_0^t (qr)^2)), \dots, \delta_\mathcal{A}(a_j, \varphi_m(x_0^t (qr)^{n+1}))$ are distinct states of $\mathcal{A}$ contradicting the assumption $|A| = n$. This contradiction shows that for every pair $(d_1, \dots, d_{m-1}, a_i), (d_1, \dots, d_{m-1}, a_j) \in N$, $\psi((d_1, \dots, d_{m-1}, a_i)) = \psi((d_1, \dots, d_{m-1}, a_j))$, that is, ψ is independent of the last factor of the quasi-direct product $\mathcal{M}$.

By this fact, we prove that $\mathcal{M}_1$ homomorphically represents $\mathcal{B}$. To this, take the subset $N' \subseteq D_1 \times \cdots \times D_{m-1}$ of the state set of $\mathcal{M}_1$ consisting of all elements $(d_1, \dots, d_{m-1}) \in D_1 \times \cdots \times D_{m-1}$ for which there exists an $a_t \in A$ such that $(d_1, \dots, d_{m-1}, a_t) \in N$ holds. Consider the mapping ψ' defined by $\psi'((d_1, \dots, d_{m-1})) = \psi((d_1, \dots, d_{m-1}, a_t))$ for all $(d_1, \dots, d_{m-1}) \in N'$, where $a_t \in A_t$ is an arbitrary state such that $(d_1, \dots, d_{m-1}, a_t) \in N$. It is clear that the mapping ψ' is well defined according to the fact that for every pair $(d_1, \dots, d_{m-1}, a_i), (d_1, \dots, d_{m-1}, a_j) \in N$, $\psi((d_1, \dots, d_{m-1}, a_i)) = \psi((d_1, \dots, d_{m-1}, a_j))$. By an easy computation we obtain that ψ' is s homomorphism of a subautomaton $\mathcal{N}' = (N', X, \delta_{N'})$ of $\mathcal{M}$ onto $\mathcal{B}$.

Therefore, if the automaton $\mathcal{B}$ can be represented homomorphically by a quasi-direct product $\mathcal{M}$ having ℓ factors equal to $\mathcal{A}$, then $\mathcal{B}$ can be represented homomorphically by a quasi-direct product $\mathcal{M}_1$ having $\ell - 1$ factors equal to $\mathcal{A}$, too. If $\mathcal{D}_{m-1} = \mathcal{A}$, then, similar to $\mathcal{M}_1$, we can construct a quasi-direct product $\mathcal{M}_2 = \prod_{t=1}^{m-2} \mathcal{D}_t(X, \varphi_1, \dots, \varphi_{m-2})$ such that $\mathcal{B}$ can be represented homomorphically by $\mathcal{M}_2$ and $\mathcal{M}_2$ has $\ell - 2$ factors equal to $\mathcal{A}$. Repeating this procedure, finally we get a quasi-direct product $\mathcal{M}_{m-k} = \prod_{t=1}^{k} \mathcal{D}_t(X, \varphi_1, \dots, \varphi_k)$ which homomorphically represents $\mathcal{B}$ and $\mathcal{D}_1, \dots, \mathcal{D}_k \in \mathcal{K} \setminus \{\mathcal{A}\}$. Consequently, $\mathcal{K} \setminus \{\mathcal{A}\}$ is also a complete class with respect to the homomorphic representation under the quasi-direct product and this completes the proof of our theorem. □

3.4 Homomorphically Complete Classes Under the Cascade Product

The following statement is a direct consequence of Theorem 3.1.

Lemma 3.41. *Let $\mathcal{K}$ be a class of automata having fewer states than an appropriate prime number p and let $\mathcal{C}$ be a counter with p number of states. Then $\mathcal{C}$ cannot be represented homomorphically by a cascade product of factors from $\mathcal{K}$.* □

The next statement is obvious.

Lemma 3.42. *Let $\mathcal{C}$ be a nontrivial counter and take a cascade product $\mathcal{M} = (M_1 \times \cdots \times M_n, X, \delta)$ of automata $\mathcal{M}_t = (M_t, X_t, \delta_t)$, $t = 1, \ldots, n$, with the following properties:*

(1) *$\mathcal{C}$ can be represented homomorphically by $\mathcal{M}$.*
(2) *There exists a positive integer k and a fixed-state m of an appropriate $\mathcal{M}_s$, $s \in \{1, \ldots, n\}$, such that $\delta((m_1, \ldots, m_n), x^{k+\ell}) \in M_1 \times \cdots \times M_{s-1} \times \{m\} \times M_{s+1} \times \cdots \times M_n$, $(m_1, \ldots, m_n) \in M_1 \times \cdots \times M_n$, $x \in X$, $\ell = 0, 1, \ldots$.*

Then $\mathcal{C}$ can be represented homomorphically by a cascade product of factors $\mathcal{M}_1, \ldots, \mathcal{M}_{s-1}, \mathcal{M}_{s+1}, \ldots, \mathcal{M}_n$. □

Now we prove the following.

Lemma 3.43. *Let $\mathcal{C} = (C, \{x_\mathcal{C}\}, \delta_\mathcal{C})$ be a nontrivial counter and let $\mathcal{K}$ be a class of automata such that*

(1) *$\mathcal{C}$ can be represented homomorphically by a cascade product of factors from $\mathcal{K}$,*
(2) *$\mathcal{N} \in \mathcal{K}$ implies that $\mathcal{C}$ cannot be represented homomorphically by a cascade product of factors from $\mathcal{K} \setminus \{\mathcal{N}\}$.*

Then for every automaton $\mathcal{A}=(A, X_\mathcal{A}, \delta_\mathcal{A})$ there exists an automaton $\mathcal{B}=(B, X_\mathcal{B}, \delta_\mathcal{B})$ and a nontrivial counter $\mathcal{D} = (D, \{x_\mathcal{D}\}, \delta_\mathcal{D})$ as follows:

(3) *$\mathcal{B}$ is a two-degree weakly nilpotent automaton.*
(4) *$\mathcal{A}$ and $\mathcal{D}$ can be represented homomorphically by a cascade product of components from $\mathcal{K} \cup \{\mathcal{B}\}$.*
(5) *$\mathcal{D}$ cannot be represented homomorphically by a cascade product of factors from $\mathcal{K}$.*
(6) *If any nontrivial counter $\mathcal{E}$ can be represented homomorphically by a cascade product of factors from $(\mathcal{K} \setminus \{\mathcal{N}\}) \cup \{\mathcal{B}\}$ and $\mathcal{N} \in \mathcal{K}$, then $\mathcal{E}$ can be represented homomorphically by a cascade product of factors from $\mathcal{K} \setminus \{\mathcal{N}\}$.*

Proof. Denote p a prime number for which every element of $\mathcal{K}$ has fewer states than p. Construct automata $\mathcal{D} = (\{1, \ldots, p\}, \{x_\mathcal{D}\}, \delta_\mathcal{D})$ and $\mathcal{B} = (B, X_\mathcal{B}, \delta_\mathcal{B})$ in the following way. Set

$$\delta_\mathcal{D}(d, x_\mathcal{D}) = \begin{cases} d+1 & \text{if } 1 \le d < p, \\ 1 & \text{if } d = p; \end{cases}$$

moreover, let $B = C \times A \times \{1, \ldots, p\} \cup \{(a_0, a_0, a_0)\}$ with $a_0 \notin C$, $(a_0, a_0, a_0) \notin C \times A \times \{1, \ldots, p\}$, $X_\mathcal{B} = C \times X_\mathcal{A}$, and

$$\delta_\mathcal{B}((c, a, d), (r, x)) = \begin{cases} (\delta_\mathcal{C}(c, x_\mathcal{C}), \delta_\mathcal{A}(a, x), d+1) & \text{if } r = c \text{ and } d < p, \\ (\delta_\mathcal{C}(c, x_\mathcal{C}), \delta_\mathcal{A}(a, x), 1) & \text{if } r = c \text{ and } d = p, \\ (a_0, a_0, a_0) & \text{otherwise.} \end{cases}$$

Obviously, $\delta_\mathcal{B}((a_0, a_0, a_0), (r, x)) = (a_0, a_0, a_0)$ and $\delta_\mathcal{B}(\delta_\mathcal{B}((c, a, d), (r, x)), (r, x)) = (a_0, a_0, a_0)((c, a, d) \in B, (r, x) \in X_\mathcal{B})$. Thus, (3) holds.

To (4), take a cascade product $\mathcal{N}' = \prod_{t=1}^{n} M_t(\{x_C\}, \varphi_1', \dots, \varphi_n')$ of factors from $\mathcal{K}$ and let $\mathcal{N} = (N, \{x_C\}, \delta_{\mathcal{N}})$ be a (state-)subautomaton of $\mathcal{N}'$ having a (state-)homomorphism $\psi : N \to C$ onto $\mathcal{C}$. Because of (1) there exists such an $\mathcal{N}$, obviously. Denote (c_0, x_0) a fixed element of $C \times X_{\mathcal{A}}$ and construct the α_0-product $\mathcal{M} = \prod_{t=1}^{n+1} M_t(X_{\mathcal{A}}, \varphi_1, \dots, \varphi_{n+1})$ with $\mathcal{M}_{n+1} = \mathcal{B}$ as follows:

For any state $(m_1, \dots, m_n, (c, a, d))$ and input letter x of $\mathcal{M}$ let

$$\varphi_t(m_1, \dots, m_n, (c, a, d), x) = \varphi_t'(m_1, \dots, m_n, x), \quad t = 1, \dots, n,$$

and

$$\varphi_{n+1}(m_1, \dots, m_n, (c, a, d), x)) = \begin{cases} (\psi((m_1, \dots, m_n)), x) & \text{if } (m_1, \dots, m_n) \in N, \\ (c_0, x_0) & \text{otherwise.} \end{cases}$$

Take the subset M' of state set of $\mathcal{M}$ with $M' = \{(m_1, \dots, m_n, (c, a, d)) \mid (m_1, \dots, m_n, (c, a, d)) \in N, (c, a, d) \in B \setminus \{(a_0, a_0, a_0)\}, \psi(m_1, \dots, m_n) = c\}$. By an easy computation we get that $\psi' : M' \to A$ with $\psi'((m_1, \dots, m_n, (c, a, d))) = a$ is a state-homomorphism of a suitable state-subautomaton of $\mathcal{M}$ (with state set M') onto $\mathcal{A}$.

Take an arbitrary fixed element $x_{\mathcal{A}}$ of $X_{\mathcal{A}}$ and let $\mathcal{M}' = \prod_{t=1}^{n+1} M_t\left(\{x_C\}, \varphi_1'', \dots, \varphi_{n+1}''\right)$ be a cascade product, for which $\varphi_t''(m_1, \dots, m_n, (c, a, d), x_C) = \varphi_t(m_1, \dots, m_n, (c, a, d), x_{\mathcal{A}})$, $t = 1, \dots, n+1$, holds for every state $(m_1, \dots, m_n, (c, a, d))$ of $\mathcal{M}'$. It can be easily shown that $\psi'' : M' \to D$ with $\psi''((m_1, \dots, m_n, (c, a, d))) = d$ is a state-homomorphism of an appropriate state-subautomaton of $\mathcal{M}'$ (with state set M') onto $\mathcal{D}$. This ends the proof of (4).

In consequence of Lemma 3.41 we receive (5).

To (6) let us take a cascade product $\mathcal{M} = (M_1 \times \cdots \times M_n, X, \delta_{\mathcal{M}}) = \prod_{t=1}^{n} M_t(X_{\mathcal{A}}, \varphi_1, \dots, \varphi_n)$ of factors $\mathcal{M}_t = (M_t, X_t, \delta_t), t = 1, \dots, n$, from $(\mathcal{K} \setminus \{\mathcal{N}\}) \cup \{\mathcal{B}\}$ and suppose that a nontrivial counter $\mathcal{C}'$ can be represented homomorphically by $\mathcal{M}$. Obviously, by $\mathcal{M}_1 = \mathcal{B}$ the first components of states $\delta_{\mathcal{M}}((m_1, \dots, m_n), x^{2+\ell}), (m_1, \dots, m_n) \in M_1 \times \cdots \times M_n, x \in X, \ell = 0, 1, \dots,$ are equal to (a_0, a_0, a_0). Thus, by Lemma 3.42, we can suppose $\mathcal{M}_1 \neq \mathcal{B}$.

Let s, $2 \le s \le n$, be the first index (if it exists) with $\mathcal{M}_s = \mathcal{B}$. By (2) we obtain that $\mathcal{C}$ cannot be represented homomorphically by a cascade product of factors $\mathcal{M}_1, \dots, \mathcal{M}_{s-1}$. Consequently, for every pair $c \in C, m \in M_1 \times \cdots \times M_n$ and input letter $x \in X$ there exist positive integers k_1, k_2 such that the first $s-1$ factors of $\delta_{\mathcal{M}}(m, x^{k_1})$ and $\delta_{\mathcal{M}}(m, x^{k_2})$ coincide but $\delta_{\mathcal{C}}(c, x_{\mathcal{C}}^{k_1}) \neq \delta_{\mathcal{C}}(c, x_{\mathcal{C}}^{k_2})$. Then either the sth factor of $\delta_{\mathcal{M}}(m, x^{k_1+1})$ or the sth factor of $\delta_{\mathcal{M}}(m, x^{k_2+1})$ coincides with (a_0, a_0, a_0). Obviously, then the sth factor of $\delta_{\mathcal{M}}(m, x^{\max(k_1,k_2)+\ell}), \ell = 1, 2, \dots,$ coincides with (a_0, a_0, a_0). Therefore, there exists a positive integer k, such that for every state m and input letter of $\mathcal{M}$, the sth factor of $\delta_{\mathcal{M}}(m, x^k)$ coincides with (a_0, a_0, a_0). Then, using Lemma 3.42, we can suppose $\mathcal{M}_s \neq \mathcal{B}$.

Repeating this procedure we have that $\mathcal{C}'$ can be represented homomorphically by a cascade product of factors from $\mathcal{K} \setminus \mathcal{N}$. This ends the proof of Lemma 3.43. □

Lemma 3.44. *Assume that for a given finite list $\mathcal{C}_0, \dots, \mathcal{C}_j$ of nontrivial counters there can be found a finite list $\mathcal{B}_1, \dots, \mathcal{B}_i, \mathcal{B}_{i+1}, \dots, \mathcal{B}_{i+j}$ of automata having the following properties:*

(1) *For arbitrary $k \in \{0, \dots, j\}$, $\mathcal{C}_k$ can be represented homomorphically by a cascade product of factors from $\{\mathcal{B}_1, \dots, \mathcal{B}_{i+k}\}$.*

(2) *By* $\mathcal{N} \in \{\mathcal{B}_1, \ldots, \mathcal{B}_{i+k}\}$, $k = 0, \ldots, j$, *none of* $\mathcal{C}_k, \ldots, \mathcal{C}_j$ *can be represented homomorphically by a cascade product of factors from* $\{\mathcal{B}_1, \ldots, \mathcal{B}_{i+j}\} \setminus \mathcal{N}$.

Then for every automaton $\mathcal{A}$ *there exists an automaton* $\mathcal{B}_{i+j+1}$ *and a nontrivial counter* $\mathcal{C}_{j+1}$ *as follows:*

(3) $\mathcal{B}_{i+j+1}$ *is a two-degree weakly nilpotent automaton.*

(4) $\mathcal{A}$ *and* $\mathcal{C}_{j+1}$ *can be represented homomorphically by a cascade product of factors from* $\{\mathcal{B}_1, \ldots, \mathcal{B}_{i+j+1}\}$.

(5) *If* $\mathcal{N} \in \{\mathcal{B}_1, \ldots, \mathcal{B}_{i+k}\}$, $k = 0, \ldots, j+1$, *then none of* $\mathcal{C}_k, \ldots, \mathcal{C}_{j+1}$ *can be represented homomorphically by a cascade product of factors from* $\{\mathcal{B}_1, \ldots, \mathcal{B}_{i+j+1}\} \setminus \{\mathcal{N}\}$.

Proof. Considering $\mathcal{K} = \{\mathcal{B}_1, \ldots, \mathcal{B}_{i+j}\}$ and $\mathcal{C} = \mathcal{C}_j$, we obtain conditions (1) and (2) of Lemma 3.43. Then, using the notation $\mathcal{B}_{i+j+1} = \mathcal{B}$ and $\mathcal{C}_{j+1} = \mathcal{D}$, it can be assumed that properties (3)–(6) of Lemma 3.43 are satisfied. Because of (3) and (4) of Lemma 3.43, we can see the validity of (3) and (4) directly. It remains to show (5).

Suppose that to the contrary of our statement, for an appropriate k, $0 \le k \le j+1$, and $\mathcal{N} \in \{\mathcal{B}_1, \ldots, \mathcal{B}_{i+k}\}$, the counter $\mathcal{C}_\ell$, $k \le \ell \le j+1$, can be represented homomorphically by a cascade product of factors from $\{\mathcal{B}_1, \ldots, \mathcal{B}_{i+j+1}\} \setminus \{\mathcal{N}\}$. Then, using (6) of Lemma 3.43, $\mathcal{C}_\ell$ can be represented homomorphically by a cascade product of factors from $\{\mathcal{B}_1, \ldots, \mathcal{B}_{i+j}\} \setminus \mathcal{N}$. By (5) of Lemma 3.43, then $\ell = j + 1$, that is, $\mathcal{C}_\ell = \mathcal{D}$ does not hold. But $\ell < j + 1$ contradicts (2). Therefore, (5) holds necessarily. Thus our Lemma 3.44 is true. □

Theorem 3.45. *Let* $\mathcal{C}$ *be a nontrivial counter and let* $\mathcal{K}$ *be a finite set of automata such that* $\mathcal{C}$ *can be represented homomorphically by a cascade product of factors from* $\mathcal{K}$ *and for every* $\mathcal{N} \in \mathcal{K}$ *the set* $\mathcal{K} \setminus \{\mathcal{N}\}$ *does not preserve this property. Then it can be found an enumerable set* $\mathcal{K}_1$ *of two-degree weakly nilpotent automata such that* $\mathcal{K} \cup \mathcal{K}_1$ *is a minimal homomorphically complete class under the cascade product.*

Proof. Take a set Γ of automata such that the elements of Γ are pairwise not isomorphic and for every automaton $\mathcal{A}$ there exists a $\mathcal{B} \in \Gamma$ such that $\mathcal{A}$ is isomorphic to $\mathcal{B}$. It can be easily seen that Γ is enumerable. Take an arrangement $\Gamma = \{\mathcal{A}_1, \mathcal{A}_2, \ldots\}$ of this enumerable set and construct the set $\mathcal{K}_1 = \{\mathcal{B}_1, \mathcal{B}_2, \ldots\}$ of two-degree weakly nilpotent automata and the set $\mathcal{K}_2 = \{\mathcal{C}_1, \mathcal{C}_2, \ldots\}$ of nontrivial counters such that

(1) every $\mathcal{A}_i \in \Gamma$ and $\mathcal{C}_i \in \mathcal{K}_2$, $i = 1, 2, \ldots$, can be represented homomorphically by a cascade product of factors from $\mathcal{K} \cup \mathcal{K}_1$;

(2) if $\mathcal{N} \in \mathcal{K}$, then none of elements of $\{\mathcal{C}\} \cup \mathcal{K}_2$ can be represented homomorphically by a cascade product of factors from $(\mathcal{K} \cup \mathcal{K}_1) \setminus \{\mathcal{N}\}$;

(3) if $\mathcal{B}_i \in \mathcal{K}_1$, $i = 1, 2, \ldots$, then none of elements of $\{\mathcal{C}_i, \mathcal{C}_{i+1}, \ldots\}$ can be represented homomorphically by a cascade product of factors from $(\mathcal{K} \cup \mathcal{K}_1) \setminus \{\mathcal{B}_i\}$.

Using Lemma 3.44, the existence of such sets $\mathcal{K}_1$ and $\mathcal{K}_2$ is shown directly. On the other hand, by (2) and (3) it holds that for every automaton $\mathcal{N} \in \mathcal{K} \cup \mathcal{K}_1$ there exists an automaton $\mathcal{M}$ which cannot be represented homomorphically by a cascade product of factors from $(\mathcal{K} \cup \mathcal{K}_1) \setminus \{\mathcal{N}\}$. Consequently, if $(\mathcal{K} \cup \mathcal{K}_1)$ is homomorphically complete under the cascade product, then it should be a minimal homomorphically complete class under the cascade product too. Thus, it is enough to prove that every automaton $\mathcal{A} = (A, X, \delta)$ can be represented homomorphically by a cascade product of factors from $(\mathcal{K} \cup \mathcal{K}_1)$.

Consider for an arbitrary automaton $\mathcal{A}$ the automaton $\mathcal{A}_i = (A_i, X_i, \delta_i)$ in Γ having an isomorphism onto $\mathcal{A}$. Using (1), we have that $\mathcal{A}_i$ can be represented homomorphically by a cascade product of factors from $\mathcal{K} \cup \mathcal{K}_1$. Therefore $\mathcal{A}$ also has this property. This ends the proof. □

We obtained the following direct consequence of this result.

Corollary 3.46. *There exists a minimal homomorphically complete class under the cascade product.* □

Finally, we note that, by Lemma 3.41, there exists no finite homomorphically complete class of finite automata under the cascade product.

3.5 Bibliographical Remarks

Section 3.1. Theorem 3.1 issued from K. B. Krohn and J. L. Rhodes [1962 and 1965]. A more detailed description of this result was developed by K. B. Krohn, J. L. Rhodes, and B. R. Tilson [1968]. It has a new proof in Ésik [1999]. An extension of Theorem 3.1 was given by Z. Ésik [1989a]. An attractive presentation of the Krohn–Rhodes theory was given by A. Ginzburg [1968]. The proof of Theorem 3.2 was described in Lallement [1971] and Eilenberg [1976]. Some aspects of the Krohn–Rhodes theory were studied by J. L. Rhodes and B. R. Tilson [1989], B. Austin et al. [1995], C. L. Nehaniv [1996], and Z. Ésik [2000]. Results in this section (with the notable exception of the holonomy decomposition theorem) are mostly originally due to and derived from K. B. Krohn and J. L. Rhodes [1962 and 1965] and appear in the book edited by Arbib [1968]. They have been reformulated and treated by many authors, e.g., A. Ginzburg [1968] and S. Eilenberg [1976]. Lemmas 3.4, 3.5, 3.6, and 3.7 can be derived from the results of A. Ginzburg [1968]. Theorem 3.8 is due to H. P. Zeiger [1967]. Theorem 3.10 issued from K. B. Krohn and J. L. Rhodes [1962, 1965]. The original idea of the holonomy decomposition theorem (Theorem 3.9) is due to H. P. Zeiger [1967], with a correct proof for partial transformation semigroups given by S. Eilenberg [1976]. Our proof, which makes reference only to fully defined transformation semigroups, is inspired by Eilenberg's.

Section 3.2. Proposition 3.11 and Corollary 3.12 were discovered by Z. Ésik [1991b]. Theorem 3.13 can be derived from Z. Ésik [1989a]. Proposition 3.14 was found by H. P. Zeiger [1967]. Lemma 3.15 appears in Dömösi [1984]. Lemma 3.16 is an extended form of the Lemma 1.3 of F. Gécseg's book [1986, p. 25]. Lemma 3.16 was also proved by P. Dömösi [1984]. Theorem 3.17 was elaborated by Z. Ésik and P. Dömösi [1986]. Theorem 3.18 was shown by P. Dömösi and Z. Ésik [1988a]. Proposition 3.19 was found by Z. Ésik and J. Virágh [1986]. Proposition 3.20 was developed by Z. Ésik and P. Dömösi [1986]. Lemmas 3.22, 3.23, 3.24, 3.25, and 3.26 were proved by P. Dömösi and Z. Ésik [1988a]. Lemma 3.27 is a new result. Theorem 3.28 is a strengthening of Theorem 3.17 and can be derived from Theorem 3.17 using results in Ésik and Dömösi [1986] with results from

Dömösi and Ésik [1988a]. Theorem 3.29 is from Z. Ésik and P. Dömösi [1986]. Some related connections were described by P. Dömösi and Z. Ésik [1986] and P. Dömösi and Z. Ésik [1988b, 1988c]. Lemmas 3.31, 3.32, 3.33, and 3.34 are essentially new results. Theorem 3.35 is a direct consequence of Theorem 3.17. Theorems 3.36 and 3.37 were shown by Z. Ésik [1986].

Section 3.3. Theorem 3.40 was shown by P. Dömösi [1980].

Section 3.4. A minimal homomorphically complete system under the α_0-product was presented by P. Dömösi [1976]. A nice presentation of this result is in Gécseg [1986]. The results of this section are based on Dömösi [1982].

Chapter 4

Without Letichevsky's Criterion

In Chapter 5 we will see the importance of Letichevsky's criterion in the composition of automata networks. In this chapter we consider networks of automata without Letichevsky's criterion. In particular, we describe several types of networking with very restricted structure of the permitted links. Assuming that the component automata are rather simple of particular types, the resulting networks that can be constructed are already computationally as general (with respect to homomorphic representation) as what can be constructed with unrestricted networking. We also show that the hierarchy of ν_i-products (automata networks in which there are at most i links to an automata from components of the network) is strict for this type of representation. We prove even more: The α_0-ν_i-hierarchy is strict for both homomorphic representation and homomorphic simulation. In addition, the ν_i-hierarchy also has this property.

This means that the number of permitted links may have a strong influence on the computational capacity of the network if component automata have a certain structure (i.e., satisfy the so-called semi-Letichevsky criterion).

4.1 Semi-Letichevsky Criterion

We start with the following statement.

Proposition 4.1. *Let $\mathcal{A} = (A, X, \delta)$ be an automaton having the semi-Letichevsky criterion and let $a \in A, x, y \in X, p \in X^*, \delta(a, x) \neq \delta(a, y), \delta(a, xp) = a$ (such that for every $q \in X^*$, $a \neq \delta(a, yq)$). For every automaton $\mathcal{A}'$ there exists a single-factor product $\mathcal{M}$ of $\mathcal{A}$ such that $\mathcal{A}'$ can be represented homomorphically by a diagonal product of its connected state-subautomata and $\mathcal{M}$. Moreover, we also have this property for a single-factor loop-free product $\mathcal{M}$ of $\mathcal{A}$ whenever $\delta(a, x^k) = a$ holds for some positive integer k.*

Proof. Let $\mathcal{A}' = (A', X', \delta')$ be an arbitrary automaton. Assume that $\delta(a, x^k) = a$ holds for some positive integer k and then construct $\mathcal{M} = \mathcal{A}(X', \varphi)$ such that for every $b \in A, x' \in X'$, $\varphi(b, x') = x$. In this case, $\mathcal{M}$ is a single-factor loop-free product of $\mathcal{A}$. Otherwise, put $p = x_1 \dots x_m$ with $a_1 = \delta(a, x), a_2 = \delta(a_1, x_1), \dots, a_m = \delta(a_{m-1}, x_{m-1}), a = \delta(a_m, x_m)$, and for every $b \in A,\ x' \in X'$, let

$$\varphi(b, x') = \begin{cases} x & \text{if } b = a, x' = x, \\ x_\ell & \text{if } b = a_\ell, x' = x_\ell, \ell = 1, \dots, m, \\ \text{fixed element of } X & \text{otherwise} \end{cases}$$

Then $\mathcal{M}$ is a single-factor product of $\mathcal{A}$.

In both cases we have $\delta(\delta(a,x), \varphi(\delta(a,x), q)) \neq \delta(\delta(a,y), \varphi(\delta(a,y), r)),\ q, r \in X^*$. (Otherwise, contrary of our assumptions, $\mathcal{A}$ could satisfy Letichevsky's criterion.) By Proposition 2.28, this implies the validity of our statement. □

Recall that the automaton $\mathcal{E}_2 = (\{0,1\}, \{x_1, x_2\}, \delta_{\mathcal{E}_2})$, $\delta_{\mathcal{E}_2}(0, x_1) = 0$, $\delta_{\mathcal{E}_2}(0, x_2) = \delta_{\mathcal{E}_2}(1, x_1) = \delta_{\mathcal{E}_2}(1, x_2) = 1$ is called the (two-state) elevator.

Let $\mathcal{L}$ be the class of all automata $\mathcal{A} = (\{0, \dots, n\}, X, \delta_{\mathcal{A}})$, $n = 1, 2, \dots$, $\delta_{\mathcal{A}}(0, x) = 0$, $\delta_{\mathcal{A}}(n, x) = n$, and

$$\delta_{\mathcal{A}}(j, x) \in \begin{cases} \{j, j+1\} & \text{if } 0 < j < n-1, \\ \{0, n-1, n\} & \text{if } j = n-1 \text{ and } n > 1 \end{cases}$$

for all $x \in X$. We have the following.

Lemma 4.2. *Every automaton in $\mathcal{L}$ can be represented isomorphically by an α_0-ν_2-power of the elevator.*

Proof. Let $\mathcal{A} = (\{0, \dots, n\}, X, \delta_{\mathcal{A}}) \in \mathcal{L}$. If $n = 1$, then $\mathcal{A}$ can be represented isomorphically by a quasi-direct power of the elevator having a single factor. Thus we may suppose that $n > 1$.

Consider the α_0-ν_2-power $\mathcal{E}_2^{n+1}(X, \varphi_1, \dots, \varphi_{n+1})$ of $\mathcal{E}_2$ in the following way. For arbitrary $(\ell_1, \dots, \ell_{n+1}) \in \{0,1\}^{n+1}$, $x \in X$, and $t \in \{1, \dots, n+1\}$, let

$$\varphi_t(\ell_1, \dots, \ell_{n+1}, x) = \begin{cases} x_2 & \text{if } 1 < t < n,\ \ell_{t-1} = 1, \text{ and } \delta_{\mathcal{A}}(t-1, x) = t, \\ & \text{or } t = n,\ \ell_{n-1} = 1, \text{ and } \delta_{\mathcal{A}}(n-1, x) \in \{0, n\}, \\ & \text{or } t = n+1,\ \ell_{n-1} = 1,\ \ell_n = 0, \text{ and } \delta_{\mathcal{A}}(n-1, x) = 0, \\ x_1 & \text{otherwise.} \end{cases}$$

Clearly then we can assume that $\varphi_t(\ell_1, \dots, \ell_{n+1}, x)$ really depends only on the input letter if $t = 1$; moreover, it depends only on the input letter and its $(t-1)$th variable if $1 < t \leq n$. In addition, it depends only on the input letter and its $(n-1)$th and nth variables if $t = n+1$. Thus it is true that $\mathcal{E}_2^{n+1}(X, \varphi_1, \dots, \varphi_{n+1})$ is an α_0-ν_2-power of $\mathcal{E}_2$. One can also verify by a trivial computation that, using the short notation $d_1 \dots d_{n+1}$ for $(d_1, \dots, d_{n+1}) \in \{0,1\}^{n+1}$, the mapping $\Phi : A \to \{0,1\}^{n+1}$ given by

$$\Phi(i) = \begin{cases} 1^i 0^{n+1-i} & \text{if } 1 \leq i \leq n, \\ 1^{n+1} & \text{if } i = 0 \end{cases}$$

is a state-isomorphism of $\mathcal{A}$ onto a subautomaton of $\mathcal{E}_2^{n+1}(X, \varphi_1, \dots, \varphi_{n+1})$. □

Lemma 4.3. *Every monotone automaton can be represented homomorphically by a diagonal product of automata from $\mathcal{L}$.*

Proof. Let $\mathcal{L}'$ be the class of all automata $\mathcal{A} = (\{0, \ldots, n\}, X, \delta_{\mathcal{A}}), n = 1, 2, \ldots,$ $\delta_{\mathcal{A}}(0, x) = 0, \delta_{\mathcal{A}}(n, x) = n$ and $\delta_{\mathcal{A}}(j, x) \in \{0, j, j+1\}, 0 < j < n$, for all $x \in X$. First we prove that all elements in $\mathcal{L}'$ can be represented homomorphically by a diagonal product of factors from $\mathcal{L}$. Consider an arbitrary automaton $\mathcal{A} = (\{0, \ldots, n\}, X, \delta_{\mathcal{A}}) \in \mathcal{L}'$ with $n > 1$. (We can take out of consideration the trivial case $n = 1$.) Define automata $\mathcal{A}_t = (\{0, \ldots, t\}, X, \delta_t), t = 2, \ldots, n$, having

$$\delta_t(j, x) = \begin{cases} \delta_{\mathcal{A}}(j, x) & \text{if } j = t - 1, \\ & \text{or } 1 \leq j < t - 1, \delta_{\mathcal{A}}(j, x) \in \{j, j+1\}, \\ j & \text{otherwise} \end{cases}$$

for all $x \in X$. Clearly then $\mathcal{A}_t \in \mathcal{L}$ for all $t = 2, \ldots, n$.

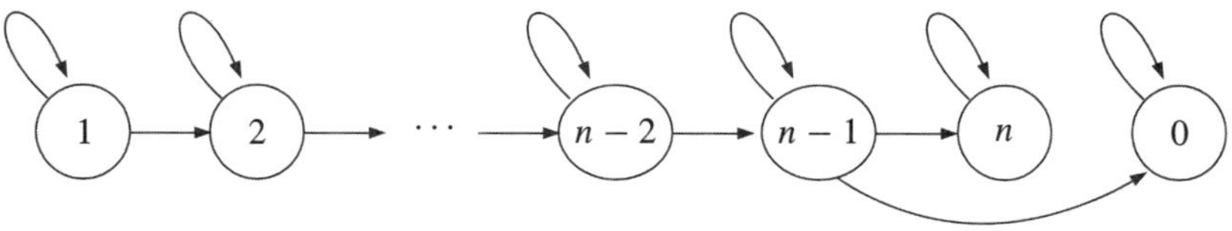

AUTOMATON $\mathcal{A}_n$

Let $B \subset A_2 \times \cdots \times A_n$ with $B = \{(1, \ldots, 1), (2, \ldots, 2), (2, 3, \ldots, 3), \ldots, (2, \ldots, \ell, \ldots, \ell), \ldots, (2, \ldots, n)\} \cup \{(\ell_1 \ldots, \ell_n) \mid 0 \in \{\ell_1, \ldots, \ell_n\}\}$. By an elementary computation we have that the mapping $\psi : A_2 \times \ldots \times A_n \to \{0, \ldots, n\}$ with

$$\psi((\ell_2, \ldots, \ell_n)) = \begin{cases} 1 & \text{if } (\ell_2, \ldots, \ell_n) = (1, \ldots, 1), \\ 2 & \text{if } (\ell_2, \ldots, \ell_n) = (2, \ldots, 2), \\ \ell & \text{if } (\ell_2, \ldots, \ell_n) = (2, 3, \ldots, \ell, \ldots, \ell), 2 < \ell \leq n \\ 0 & \text{if } 0 \in \{\ell_2, \ldots, \ell_n\} \end{cases}$$

is a state-homomorphism of an appropriate state-subautomaton of the diagonal product $\mathcal{A}_2 \Delta \cdots \Delta \mathcal{A}_n$ onto $\mathcal{A}$. Using the transitive property of the diagonal product, to establish our assertion we shall prove that every monotone automaton can be represented homomorphically by a diagonal product of factors from $\mathcal{L}'$.

Consider a monotone automaton $\mathcal{D} = (D, X, \delta_{\mathcal{D}})$ and denote $\leq$ a partial ordering on D for which $d \leq \delta_{\mathcal{D}}(d, x), d \in D, x \in X$. Take a rearrangement $d_1, \ldots, d_n$ of the elements of D such that for every $d_i, d_j \in D$, $d_i \neq d_j$, and $d_i \leq d_j$ implies $i < j$. Consider all bijective mappings of the form $f_s : \{d_1, \ldots, d_n\} \to \{1, \ldots, n\}, s = 1, \ldots, n!$. For every mapping f_s $(1 \leq s \leq n!)$, define an automaton $\mathcal{A}_s = (\{0, \ldots, n\}, X, \delta_s)$ having

$$\delta_s(k, x) = \begin{cases} k & \text{if } \delta_{\mathcal{D}}(f_s^{-1}(k), x) = f_s^{-1}(k) \text{ and } 1 \leq k \leq n, \\ k+1 & \text{if } \delta_{\mathcal{D}}(f_s^{-1}(k), x) = f_s^{-1}(k+1) \text{ and } 1 \leq k < n, \\ 0 & \text{otherwise.} \end{cases}$$

It is clear that $\mathcal{A}_s \in \mathcal{L}', 1 \leq s \leq n!$. Take the diagonal product $\mathcal{M} = \mathcal{A}_1 \Delta \cdots \Delta \mathcal{A}_n$. We shall prove that $\mathcal{M}$ homomorphically represents the automaton $\mathcal{A}$.

Define the set $B \subset \{0, \ldots, n\}^{n!}$ as follows:

(1) For every $d_k \in D$ there exists a $(b_1, \ldots, b_m) \in B$ and an $s \in \{1, \ldots, n!\}$ having $f_s(d_k) = b_s$.

(2) For every $(b_1, \ldots, b_m) \in B$ and $s, t \in \{1, \ldots, n!\}$, $0 \notin \{b_s, b_t\}$ implies $f_s^{-1}(b_s) = f_t^{-1}(b_t)$.

(3) For every pair $(b_1, \ldots, b_{n!}) \in B$, $t \in \{1, \ldots, n!\}$, $b_t \neq 0$ implies that for arbitrary $x_1, \ldots, x_r \in X$ there exists an $s \in \{1, \ldots, n!\}$ with $f_s^{-1}(b_s) = f_t^{-1}(b_t)$ and $f_s^{-1}(\delta_s(b_s, x_1 \ldots x_j)) = \delta_{\mathcal{D}}(f_t^{-1}(b_t), x_1 \ldots x_j)$, $j = 1, \ldots, r$.

Let $\psi : B \to \{d_1, \ldots, d_n\}$ be a mapping for which $\psi((b_1, \ldots, b_{n!})) = d_k$ whenever $(b_1, \ldots, b_{n!}) \in B$ has a component $b_t, t \in \{1, \ldots, n!\}$ with $f_t^{-1}(b_t) = d_k$.

By conditions (1) and (2), ψ is a well-defined surjective mapping. Using (3) and the definitions of the mappings $\delta_s, s = 1, \ldots, n!$, $\psi((\delta_1(b_1, x), \ldots, \delta_m(b_{n!}, x))) = \delta_{\mathcal{D}}(\psi((b_1, \ldots, b_{n!})), x)$, $(b_1, \ldots, b_{n!}) \in B, x \in X$. Therefore, the diagonal product $\mathcal{A}_1 \Delta \cdots \Delta \mathcal{A}_n$ has a state-subautomaton (with state set B) having a state-homomorphism ψ onto $\mathcal{A}$. □

By Lemmas 4.2 and 4.3 we can immediately derive the next statement.

Theorem 4.4. *Every monotone automaton can be represented homomorphically by an α_0-ν_2-product of the elevator.* □

Now we turn to the automata having the semi-Letichevsky criterion. Let $\mathcal{A} = (A, X, \delta)$ be an automaton satisfying the semi-Letichevsky criterion. Given a state $a \in A$, put $A'_a = A''_a = \emptyset$ if there exists no $x \in X$, $p \in X^*$ having $\delta(a, xp) = a$. Otherwise, let $p \in X^*$ be the shortest word having this property for an appropriate $x \in X$, and put

$$A'_a = \{\delta(a, xq) \mid q \text{ is a prefix of } p\},$$

$$A''_a = \{a' \in A' \mid \text{ for every } x', y' \in X : \delta(a', x') = \delta(a', y')\}.$$

Given a positive integer r, a nonnegative integer s, we say that $\mathcal{A}$ is (r, s)-weighted if there exists an $a \in A$ with $|A'_a| = r$ and $|A''_a| = s$. (Of course, it may be possible that $\mathcal{A}$ is (r, s)-weighted and, simultaneously, (r', s')-weighted such that $(r, s) \neq (r', s')$ (with $r, r' > 0$).)

Proposition 4.5. *Let $\mathcal{A}$ be an (r, s)-weighted automaton (having the semi-Letichevsky criterion). There exists a single-factor product $\mathcal{B}$ of $\mathcal{A}$ such that $\mathcal{B}$ is an $(r, r-1)$-weighted automaton.*

Proof. Let $\mathcal{A} = (A, X, \delta)$ be an (r, s)-weighted automaton (having the semi-Letichevsky criterion). There are $a \in A, x, y \in X, p \in X^*, \delta(a, x) \neq \delta(a, y), \delta(a, xp) = a$ with $|A'_a| = |xp| = r, |A''_a| = s$ (such that for every $q \in X^*$, $a \neq \delta(a, yq)$). Put $xp = x_1 \cdots x_r, x_1, \ldots, x_r \in X$. Define $\varphi : A \times X \to X$ such that for every $a' \in A$, $x' \in X$,

$$\varphi(a', x) = \begin{cases} x_t & \text{if } a' = \delta(a, x_1 \cdots x_{t-1}) \text{ and } 1 < t \leq r, \\ x & \text{otherwise.} \end{cases}$$

It is clear that the single factor product $\mathcal{A}(X, \varphi)$ of $\mathcal{A}$ is an $(r, r-1)$-weighted automaton. □

Define the automaton $\mathcal{E}'_k = (\{1, \ldots, k, *\}, \{x_1, x_2\}, \delta'_k)$ with $\delta'_k(i, x_1) = i + 1 (\text{mod } k)$, $\delta'_k(*, x_1) = *$, $\delta'_k(i, x_2) = \delta'_k(*, x_2) = *$, $1 \leq i \leq k$. We say that $\mathcal{E}'_k$ is the *semi-elevator* of length k.

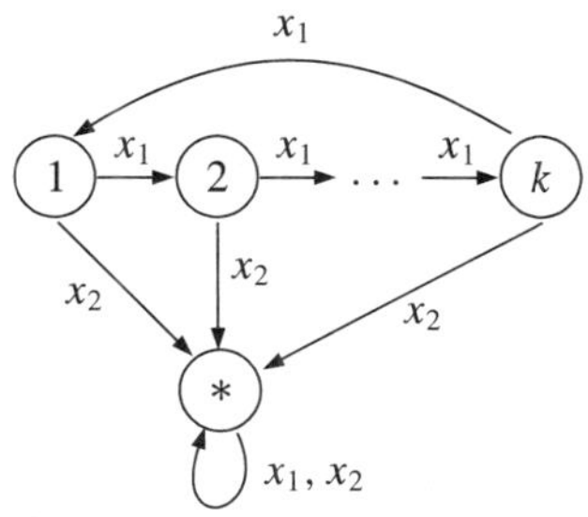

SEMI-ELEVATOR $\mathcal{E}'_k$

Lemma 4.6. *Let $\mathcal{A} = (A, X, \delta)$ be an (r, s)-weighted automaton (having the semi-Letichevsky criterion). The semi-elevator of length r can be represented homomorphically by a diagonal power $\mathcal{B}^{s+1}$, where $\mathcal{B}$ is a single-factor product of $\mathcal{A}$. In particular, if $s = r-1$, then we may assume that $\mathcal{B}$ is an input-subautomaton of $\mathcal{A}$.*

Proof. By our conditions, there are $a \in A$, $x, y \in X$, $p \in X^*$, $\delta(a, x) \neq \delta(a, y)$, $\delta(a, xp) = a$ with $|A'_a| = |xp| = r$, $|A''_a| = s$ (such that for every $q \in X^*$, $a \neq \delta(a, yq)$). For technical reasons, we put $a_1 = a$ and $xp = z_1 \cdots z_k$ such that $z_1, \ldots, z_k \in X$, $k \geq 1$. Moreover, we put $a_{i+1} = \delta(a, z_1 \cdots z_i)$, $1 \leq i < k$, and $A' = \{a_1, \ldots, a_k\}$. Let us consider the semielevator $\mathcal{E}'_k$ of length k having $\mathcal{E}'_k = (\{1, \ldots, k, *\}, \{x_1, x_2\}, \delta'_k)$ with $\delta'_k(i, x_1) = i + 1 \pmod k$, $\delta'_k(*, x_1) = *$, $\delta'_k(i, x_2) = \delta'_k(*, x_2) = *$, $1 \leq i \leq k$.

We prove that $\mathcal{E}'_k$ can be represented homomorphically by an appropriate diagonal power $\mathcal{B}^{s+1}$ such that

(1) $\mathcal{B}$ is an input-subautomaton of $\mathcal{A}$ if $s = |p|$, and
(2) $\mathcal{B}$ is a single factor product of $\mathcal{A}$ if $s < |p|$.

Let $a_{i_1}, \ldots, a_{i_s} \in A'$ be distinct states such that $\delta(a_{i_j}, y_{i_j}) \in A'$ for every $y_{i_j} \in X$, $j = 1, \ldots, s$.

If $s = |p|$, then let $\mathcal{B}$ be the input-subautomaton of $\mathcal{A}$ with input set $\{x, y\}$. Obviously, then $\delta(a_i, x) = a_{i+1 \pmod k} \in A'$ and $\delta(a_j, y) = a_{j+1 \pmod k} \in A'$ hold for every $i \in \{1, \ldots, k\}$, $j \in \{2, \ldots, k\}$. In addition, $\delta(a_1, y) \notin A'$.

If $s < |p|$, then let $\mathcal{B} = \mathcal{A}(\{x, y\}, \varphi)$ defined by

$$\varphi(b, z) = \begin{cases} z_i (= i\text{th letter of } xp) & \text{if } b = a_i, 1 \leq i \leq k, z = x, \\ y & \text{if } b = a_1, z = y, \\ \text{arbitrary fixed } y' \in X & \text{otherwise.} \end{cases}$$

Clearly, then $\mathcal{B}$ has a state-subautomaton $\mathcal{B}' = (B', \{x, y\}, \delta')$ generated by its state a_1 such that for appropriate distinct states $a_1, \ldots, a_k$ we have $\delta'(a_i, x) = a_{i+1 \pmod k}$, and simultaneously, for every $j \in \{1, \ldots, s\}, q \in X^*$, $\delta'(a_{i_j}, yq) \notin A'$. Consider the state-subautomaton $\mathcal{D} = (D, Y, \delta'')$ of the diagonal power $(\mathcal{B}')^{s+1}$ generated by the state $(a_1, \ldots, a_{s+1})$. For every positive integer m, we obtain $\delta''((a_1, \ldots, a_{s+1}), x^m) = (a_{m+1 \pmod k}, \ldots, a_{m+s+1 \pmod k})$. Simultaneously, if $r = x^u y$, then for every $q \in \{x, y\}^*$, there exists a $t \in \{1, \ldots, s+1\}$ with $\delta'(a_t, rq) \notin \{a_1, \ldots, a_k\}$. Hence $\delta''((a_1, \ldots, a_{s+1}), rq) \neq (a_{|rq|+1 \pmod k}, \ldots, a_{|rq|+s+1 \pmod k})$.

Define the mappings $\psi_1 : D \to \{1, \ldots, k, *\}$, $\psi_2 : \{x, y\} \to \{x_1, x_2\}$ with

$$\psi_1(d) = \begin{cases} m & \text{if } d = (a_{m \,(\mathrm{mod}\, k)}, \ldots, a_{m+s \,(\mathrm{mod}\, k)}),\ 1 \le m \le k, \\ * & \text{otherwise,} \end{cases}$$

$\psi_2(x) = x_1$, $\psi_2(y) = x_2$. By an elementary checking, we get that ψ is a state-homomorphism of a state-subautomaton of $\mathcal{B}^{s+1}$ onto $\mathcal{E}'_k$. The proof is complete. □

Proposition 4.7. *Let $\mathcal{A} = (A, X, \delta)$ be an (r, s)-weighted automaton (having the semi-Letichevsky criterion). Every monotone automaton can be represented homomorphically by an α_1-$\nu^{\ell}_{2(s+1)}$-product of $\mathcal{B}$, where $\mathcal{B}$ is a single-factor product of $\mathcal{A}$. In particular, if $s = r - 1$, then we may assume that $\mathcal{B}$ is an input-subautomaton of $\mathcal{A}$.*

Proof. Using Theorem 4.4, it is enough to prove that the two-state elevator can be represented homomorphically by a diagonal power $\mathcal{B}^{s+1}$ such that

(1) $\mathcal{B}$ is input-isomorphic to an input-subautomaton of $\mathcal{A}$ if $s = r - 1$, and
(2) $\mathcal{B}$ is a single factor product of $\mathcal{A}$ if $s < r - 1$.

But it is evident that every semielevator of length $k \ge 1$ can be mapped homomorphically onto the elevator. Thus, using Lemma 4.6, the proof is complete. □

Next we prove the following lemma.

Lemma 4.8. *Let $\mathcal{A} = (A, X, \delta)$ be an (r, s)-weighted automaton (having the semi-Letichevsky criterion). Then every product $\mathcal{C} \times \mathcal{D}(Y, \varphi_1, \varphi_2)$ of a counter $\mathcal{C}$ of r states and a monotone automaton $\mathcal{D}$ can be represented homomorphically by an α_1-$\nu^{\ell}_{2(s+1)}$-power of $\mathcal{A}$. In particular, if $s = r - 1$, then $\mathcal{C} \times \mathcal{D}(Y, \varphi_1, \varphi_2)$ can also be represented homomorphically by an α_0-$\nu_{2(s+1)+1}$-power of $\mathcal{A}$.*

Proof. By our conditions, there are $a \in A$, $x, y \in X$, $p \in X^*$, $\delta(a, x) \ne \delta(a, y)$, $\delta(a, xp) = a$ with $|A'_a| = |xp| = r$, $|A''_a| = s$ (such that for every $q \in X^*$, $a \ne \delta(a, yq)$). If $r = 1$, i.e., $|p| = 0$, then our statement is a direct consequence of Proposition 4.7. Thus we assume $r > 1$ (i.e., $|p| > 0$). For technical reasons, we put $a_1 = a$ and $xp = x_1 \cdots x_r$ such that $x_1, \ldots, x_r \in X, r \ge 1$. Moreover, we put $a_{i+1} = \delta(a, x_1 \cdots x_i)$, $1 \le i < k$, and $A' = \{a_1, \ldots, a_r\}$.

First we recall that all counters have singleton input sets and thus they are autonomous automata. But then an arbitrary product of a counter and any other automaton coincides with the α_0-product of the considered counter and the considered automaton. By Proposition 2.65, this single-factor product is also a monotone automaton. Thus we will assume without any restriction that $\mathcal{C} \times \mathcal{D}(Y, \varphi_1, \varphi_2)$ is an α_0-product.

If $s = r - 1$, then $\delta(a_i, x) = a_{i+1 \,(\mathrm{mod}\, r)}$ holds for every $a_i \in A'$. In this case, the subautomaton of $\mathcal{A}$ with the state set A' and the (singleton) input set $\{x\}$ forms a counter of r states. In addition, if $s = r - 1$, then by virtue of Proposition 4.7, we can also represent $\mathcal{D}$ by an α_0-$\nu_{2(s+1)}$-power of $\mathcal{A}$. Thus, $\mathcal{C} \times \mathcal{D}(Y, \varphi_1, \varphi_2)$ can be represented homomorphically by an α_0-product of two automata such that the first one is an α_0-product of $\mathcal{A}$ with a single factor and a second one is an α_0-$\nu_{2(s+1)}$-power of $\mathcal{A}$. But then, by Proposition 2.63,

$\mathcal{C} \times \mathcal{D}(Y, \varphi_1, \varphi_2)$ can be represented homomorphically by an α_0-$\nu_{2(s+1)+1}$-power of $\mathcal{A}$. Thus we are ready if $s = r - 1$.

Now let $s < r - 1$. Then, applying Proposition 4.5, we can use the above treatment for a single-factor product $\mathcal{B}$ of $\mathcal{A}$. □

We also need the following.

Lemma 4.9. *Let $c_{\mathcal{A}}$ be the least common multiple of the length of all cycles in the automaton $\mathcal{A}$ having the semi-Letichevsky criterion. Then $\mathcal{A}$ can be represented homomorphically by an α_0-product of a counter with $c_{\mathcal{A}}$ states and a monotone automaton.*

Proof. If $\mathcal{A}$ consists of a single cycle, then it does not satisfy the semi-Letichevsky criterion, a contradiction. If $\mathcal{A}$ is monotone, then our statement is trivial. Therefore, we may assume that $\mathcal{A}$ has a nontrivial cycle, and simultaneously, there exists a state of $\mathcal{A}$ which is not a member of this cycle.

In this case, we shall show the existence of automata $\mathcal{A}_0, \ldots, \mathcal{A}_s$ such that $\mathcal{A}_0 = \mathcal{A}$, the automaton $\mathcal{A}_s$ is monotone, and furthermore, for every $i \in \{0, \ldots, s-1\}$, $\mathcal{A}_i$ can be represented homomorphically by an α_0-product of $\mathcal{A}_{i+1}$ and a counter $\mathcal{C}_{c_{\mathcal{A}}}$ with $c_{\mathcal{A}}$ states. Thus, using the obvious fact that the α_0-product is associative, by consecutive applications of Proposition 2.53 we can represent homomorphically the automaton $\mathcal{A}$ by an α_0-product of automata $\mathcal{D}$ and $\mathcal{A}_s$, where $\mathcal{D}$ is an α_0-power of $\mathcal{C}_{c_{\mathcal{A}}}$ and, simultaneously, $\mathcal{A}_s$ is a monotone automaton.

If $\mathcal{A}$ is not connected, then consider a connected automaton $\mathcal{B}$ having the semi-Letichevsky criterion such that $\mathcal{A}$ is a subautomaton of $\mathcal{B}$. We can apply the above treatment substituting $\mathcal{A}$ with $\mathcal{B}$. Indeed, then $\mathcal{A}$ can be represented homomorphically by an α_0-product of $\mathcal{C}_{c_{\mathcal{A}}}$ and $\mathcal{A}_s$ whenever $\mathcal{B}$ has this property.

Now we show the existence of $\mathcal{B}$ provided that $\mathcal{A}$ is not connected. Let $\mathcal{A}$ be not connected and consider an arbitrary automaton $\mathcal{B} = (\{a_0, \ldots, a_n\}, X \cup \{x_1, \ldots, x_n\}, \delta_{\mathcal{B}})$ with $\{a_1, \ldots, a_n\} = A$, $a_0 \notin A$, $\{x_1, \ldots, x_n\} \cap X = \emptyset$ such that

$$\delta_{\mathcal{B}}(a, x) = \begin{cases} a_i & \text{if } a = a_0, x = x_i, 1 \leq i \leq n, \\ \delta(a, x) & \text{if } a \in A, x \in X, \end{cases}$$

$\delta_{\mathcal{B}}(a_0, x) \in A$ if $x \in X$, and $\delta_{\mathcal{B}}(a, x) \in \{\delta(a, x') \mid x' \in X\}$ if $a \in A, x \in \{x_1, \ldots, x_n\}$.

Clearly then $\mathcal{A}$ is a subautomaton of $\mathcal{B}$, where $\mathcal{B}$ is a connected automaton satisfying the semi-Letichevsky criterion.

It remains to prove the existence of automata $\mathcal{A}_0, \ldots, \mathcal{A}_s$ having the above properties. To this statement we shall show the existence of $\mathcal{A}_1$ such that the number of nontrivial cycles in $\mathcal{A}_1$ is fewer than the number of nontrivial cycles in $\mathcal{A}_0$. Applying this result inductively, we can reach that $\mathcal{A}_s$ does not contain any nontrivial cycle; i.e., it is a monotone automaton.

Let $a_1, \ldots, a_i$ and $b_1, \ldots, b_j$ be two cycles of $\mathcal{A}$ with $a_1 = b_1$. Observe that in this case $a_k \neq b_k$, $1 < k \leq \min(i, j)$, would imply that, contrary to our assumptions, $\mathcal{A}$ satisfies Letichevsky's criterion with $\delta(a_{\ell-1}, x) \neq \delta(a_{\ell-1}, y)$ and $\delta(a_{\ell-1}, xp) = \delta(a_{\ell-1}, yq) = a$ for some $a_{\ell-1} \in A$, $x, y \in X$, $p, q \in X^*$. We also have this consequence if $a_k = b_k$, $1 \leq k \leq \min(i, j)$ but $i \neq j$. Therefore, if two cycles have a common element, then these cycles should coincide. Let $a_{u+1}, \ldots, a_v$ be a nontrivial cycle in $\mathcal{A}$. If there are a state $a \in A \setminus \{a_1, \ldots, a_n\}$ and words $p, q \in X^*$ with $\delta(a, p) \in \{a_{u+1}, \ldots, a_v\}$ and $a \in \{\delta(a_{u+1}, q), \ldots, \delta(a_v, q)\}$, then $\mathcal{A}$ satisfies Letichevsky's criterion, a contradiction.

Therefore, for every $a \in A\backslash\{a_1, \ldots, a_n\}$ and words $p, q \in X^*$ with $\delta(a, p) \in \{a_{u+1}, \ldots, a_v\}$, it holds that $a \notin \{\delta(a_{u+1}, q), \ldots, \delta(a_v, q)\}$. Let $a_1, \ldots, a_u \in A$ denote all states for which there are words $p_1, \ldots, p_u \in X^*$ having $\delta(a_i, p_i) \in \{a_{u+1}, \ldots, a_v\}, i = 1, \ldots, u$. (We note that $\{a_1, \ldots, a_u\} = \emptyset$ may be possible.) Finally, let $A \setminus \{a_1, \ldots, a_v\} = \{a_{v+1}, \ldots, a_n\}$ for appropriate $a_{v+1}, \ldots, a_n \in A$. (Note that $\{a_{v+1}, \ldots, a_n\} = \emptyset$ may be possible but $\{a_1, \ldots, a_u, a_{v+1}, \ldots, a_n\} \neq \emptyset$ because $\mathcal{A}$ satisfies the semi-Letichevsky criterion.)

Hence, we can consider an arrangement $a_1, \ldots, a_u, a_{u+1}, \ldots, a_v, a_{v+1}, \ldots, a_n$ of the set of states in $\mathcal{A} = (A, X, \delta)$ such that $a_{u+1}, \ldots, a_v$ form a cycle of length $v - u > 1$; moreover, $\delta(a_i, x) \neq a_j$ for either $1 \leq j \leq u < i \leq v$ or $u < j \leq v < i \leq n$.

By these properties, we can define the automaton $\mathcal{A}_1 = (A, \{1, \ldots, c_{\mathcal{A}}\} \times X, \delta')$ in the following way:

For every pair $a_i \in A$, $(j, x) \in \{1, \ldots, c_{\mathcal{A}}\} \times X$ let

$$\delta'(a_i, (j, x)) = \begin{cases} \delta(a_i, x) & \text{if } a_i, \delta(a_i, x) \in \{a_1, \ldots, a_u, a_{v+1}, \ldots, a_n\}, \\ a_{k+j-u(\text{mod } v-u)+u} & \text{if } a_i \in \{a_1, \ldots, a_u\}, \delta(a_i, x) = a_k \in \{a_{u+1}, \ldots, a_v\}, \\ a_{i+j-u+1(\text{mod } v-u)+u} & \text{if } a_i \in \{a_{u+1}, \ldots, a_v\}. \end{cases}$$

Consider the counter $\mathcal{C}_{c_{\mathcal{A}}} = (\{1, \ldots, c_{\mathcal{A}}\}, \{x_{c_{\mathcal{A}}}\}, \delta_{c_{\mathcal{A}}})$ with $\delta_{c_{\mathcal{A}}}(i, x_{c_{\mathcal{A}}}) = i + 1(\text{mod } c_{\mathcal{A}}), 1 \leq i \leq c_{\mathcal{A}}$.

Let $\mathcal{M}_1 = \mathcal{C}_{c_{\mathcal{A}}} \times \mathcal{A}_1(X, \varphi_1, \varphi_2)$ be an α_0-product with $\varphi_1(i, a_j, x) = x_{c_{\mathcal{A}}}$ and $\varphi_2(i, a_j, x) = (i, x)$, whenever $(i, a_j, x) \in \{1, \ldots, c_{\mathcal{A}}\} \times \{a_1, a_{m+1}, \ldots, a_n\} \times X$.

Let $\psi : \{1, \ldots, c_{\mathcal{A}}\} \times A \to A$ be a mapping with

$$\psi((i, a_j)) = \begin{cases} a_{i+j-u(\text{mod } v-u)+u} & \text{if } j \in \{u+1, \ldots, v\}, \\ a_j & \text{otherwise.} \end{cases}$$

It is routine work to show that ψ is a state-homomorphism of $\mathcal{M}_1$ onto $\mathcal{A}$. □

Next we show the following corollary.

Corollary 4.10. *Given a finite set $\mathcal{K}$ of automata satisfying the semi-Letichevsky criterion, let $c_{\mathcal{K}}$ be the least common multiple of all positive integers which are lengths of cycles of automata in $\mathcal{K}$. Moreover, let $m_{\mathcal{K}}$ be the minimal number of cycles of automata in $\mathcal{K}$ such that every prime power divisor of $c_{\mathcal{K}}$ divides at least one of these lengths of cycles.*[18] *Consider an (r, s)-weighted automaton $\mathcal{A} = (A, X, \delta) \in \mathcal{K}$ (with semi-Letichevsky criterion, $s < r$). If an automaton $\mathcal{B}$ can be represented homomorphically by a general product of automata from $\mathcal{K}$, then it can be represented homomorphically by an α_i-ν_j^{ℓ}-product of factors in $\mathcal{K}$ with $i \leq 1$ and $j \leq m_{\mathcal{K}} + 2(s+1) + 1$. In particular, if $s = r - 1$ and for every cycle of length k in $\mathcal{B}$, the counter with k states, as well as the counter with r states, can be represented homomorphically by an α_0-product of automata in $\mathcal{K}$, then $\mathcal{B}$ can be also represented homomorphically by an α_0-ν_j^{ℓ}-product of factors from $\mathcal{K}$ with $j \leq m_{\mathcal{K}} + 2(s+1) + 1$.*

Proof. Suppose that $\mathcal{B}$ satisfies the semi-Letichevsky criterion and let r' denote the least common multiple of the lengths of all cycles in B. Then $r' \mid c_{\mathcal{K}}$. Therefore, by Lemma 4.9, $\mathcal{B}$ can be represented homomorphically by an α_0-product of a counter with r' states and a monotone automaton $\mathcal{M}$. On the other hand, by Proposition 4.7, for every (r, s)-weighted

[18]Thus $m_{\mathcal{K}} = 0$ if all cycles are trivial (having length of 1).

automaton $\mathcal{A} \in \mathcal{K}$, $\mathcal{M}$ can be represented homomorphically by an α_0-$\nu_{2(s+1)}$-power of a single-factor product $\mathcal{A}'$ of $\mathcal{A}$, where $\mathcal{A}'$ may be an input-subautomaton of $\mathcal{A}$ if $s = r - 1$. In addition, by our assumptions, $m_{\mathcal{K}}$ is the minimal number of cycles of automata in $\mathcal{K}$ such that every prime power divisor of $c_{\mathcal{K}}$ divides at least one of these lengths of cycles. If $m_{\mathcal{K}} = 0$, then $c_{\mathcal{K}} = 1$, leading to $r' = 1$. Therefore, in this case we are ready.

Consider the (r, s)-automaton $\mathcal{A} = (A, X, \delta) \in \mathcal{K}$ (having the semi-Letichevsky criterion) with $a \in A, x, y \in X, p \in X^*, \delta(a, xp) = a, \delta(a, x) \neq \delta(a, y)$, and $|A'_a| = r$, $|A''_a| = s$ (such that for every $q \in X^*, a \neq \delta(a, yq)$). Again, for technical reasons, we put $a_1 = a$ and $xp = z_1 \cdots z_r$ such that $z_1, \ldots, z_r \in X$ $(r \geq 1)$. Moreover, we put $a_{i+1} = \delta(a, z_1 \cdots z_i)$, $1 \leq i < r$, and $A' = \{a_1, \ldots, a_r\}$. Let c denote the least common multiple of r and the lengths of cycles in $\mathcal{B}$ and define the α_0-product $\mathcal{B}' = \mathcal{C}_c \times \mathcal{A}(X, \varphi_1, \varphi_2)$ of the counter $\mathcal{C}_c = (\{1, \ldots, c\}, \{x_c\}, \delta_c)$ (with c states) and $\mathcal{A}$ such that for every $x \in X$, $\varphi_1(x)$ is the (only) input letter x_c of $\mathcal{C}_c$; moreover, for every $i \in \{1, \ldots, c\}$, $x \in X$,

$$\varphi_2(i, x) = \begin{cases} x & \text{if } 1 \leq i \leq r, \\ x_t & \text{if } r < i \leq c_{\mathcal{K}}, i \equiv t (\operatorname{mod} r), \delta(a_t, x_t) = a_{t+1 (\operatorname{mod} r)}, \\ \text{fixed element of } X & \text{otherwise.} \end{cases}$$

Since $r|c$, it is obvious that for every $\ell r + j \in \{1, \ldots, c\}$, $j \in \{1, \ldots, r\}$, $x \in X$,

$$(\delta_C(\ell r + j, \varphi_1(\ell r + j, j, x)), \delta(j, \varphi_2(\ell r + j, j, x))) \neq (\ell r + j + 1 (\operatorname{mod} c), j + 1 (\operatorname{mod} r))$$

if and only if $\ell = 1$ and $\delta(j, \varphi_2(\ell r + j, j, x)) \neq j + 1 (\operatorname{mod} r)$. By this observation, $\mathcal{B}'$ is a (c, s)-automaton with $c|c_{\mathcal{K}}$.

Suppose $m_{\mathcal{K}} > 0$ and let $\mathcal{A}_1, \ldots, \mathcal{A}_m$, $m \leq m_{\mathcal{K}} \in \mathcal{K}$, denote automata for which c divides the least common multiple of $c_1, \ldots, c_m$, where c_i denotes the length of an appropriate cycle in $\mathcal{A}_i$, $i = 1, \ldots, m$. Obviously, the counter $\mathcal{C}_c$ with c states can be represented homomorphically by a direct product of counters with $c_1, \ldots, c_m$ states. On the other hand, for every $i = 1, \ldots, m$, the counter $\mathcal{C}_{c_i}$ with c_i number of states can be represented isomorphically by a single-factor product $\mathcal{B}_i$ of $\mathcal{A}_i$. Therefore, using Proposition 2.53, an appropriate α_0-ν_m-product of $\mathcal{B}_1, \ldots, \mathcal{B}_m$ and a single-factor product $\mathcal{A}'$ of $\mathcal{A}$ (which may be an input-subautomaton of $\mathcal{A}$ if $s = r - 1$) homomorphically represents the (c, s)-automaton $\mathcal{B}'$ (having the semi-Letichevsky criterion). We prove that every single-factor product $\mathcal{B}''$ of $\mathcal{B}'$ can be represented homomorphically by an appropriate α_0-ν_m-product of $\mathcal{B}_1, \ldots, \mathcal{B}_m$ and a single-factor product $\mathcal{A}'$ of $\mathcal{A}$. Recall that $\mathcal{B}'$ is an α_0-product of a counter and $\mathcal{A}$. On the other hand, all counters have singleton input sets, and thus they are autonomous automata. But then an arbitrary product of a counter and any other automaton coincides with the α_0-product of the considered counter and a single-factor product of the considered automaton. Therefore, $\mathcal{B}''$ can be represented homomorphically by an α_0-product of $\mathcal{C}_c$ and a single-factor product $\mathcal{A}'$ of $\mathcal{A}$. Hence, we can see as before that, indeed, every single-factor product $\mathcal{B}''$ of $\mathcal{B}'$ can be represented homomorphically by an appropriate α_0-ν_m-product of $\mathcal{B}_1, \ldots, \mathcal{B}_m$ and a single-factor product $\mathcal{A}'$ of $\mathcal{A}$.

Applying Lemma 4.8, an α_0-product $\mathcal{F}$ of a counter with $r'|c$ states and a monotone automaton $\mathcal{M}$ can be represented homomorphically by an α_0-$\nu_{2(s+1)+1}$-product $\mathcal{B}''$, where $\mathcal{B}''$ is a single-factor product of $\mathcal{B}'$. As we proved before, $\mathcal{B}'$ can be represented homomorphically by an appropriate α_0-ν_m-product of $\mathcal{B}_1, \ldots, \mathcal{B}_m$ and a single-factor product $\mathcal{A}'$ of $\mathcal{A}$. Thus, by Proposition 2.63, $\mathcal{F}$ can be represented homomorphically by an

α_0-$\nu_{m+2(s+1)+1}$-product of automata $\mathcal{B}_1, \ldots, \mathcal{B}_m$ and $\mathcal{A}'$ (where $\mathcal{B}_i$ is a single-factor product of $\mathcal{A}_i$, $i = 1, \ldots, m$, and $\mathcal{A}'$ is a single-factor product of $\mathcal{A}$). Hence, $\mathcal{F}$ can be represented homomorphically by an α_1-$\nu^{\ell}_{m+2(s+1)+1}$-product of $\mathcal{A}_1, \ldots, \mathcal{A}_m$ and $\mathcal{A}$. In addition, also by Lemma 4.8, if $s = r - 1$, then $\mathcal{F}$ can be represented homomorphically by an α_0-$\nu_{2(s+1)+1}$-power of $\mathcal{B}'$ too. Therefore, applying again Proposition 2.63, if for every cycle of length k in $\mathcal{B}$, the counter with k states, as well as the counter with r states, can be represented homomorphically by a single-factor α_0-product of an automaton in $\mathcal{K}$ (and $s = r - 1$), then $\mathcal{F}$ can be represented homomorphically by an α_0-$\nu_{m+2(s+1)+1}$-product of m factors in $\mathcal{K}$, and $\mathcal{A}$. But then we are ready because $m \leq m_{\mathcal{K}}$ and $\mathcal{F}$ homomorphically represents $\mathcal{B}$.

Suppose that $\mathcal{B} = (B, X_{\mathcal{B}}, \delta_{\mathcal{B}})$ is without any Letichevsky criteria and that it can be represented homomorphically by a product of factors in $\mathcal{K}$. By Propositions 4.1 and 2.54, we can restrict our investigations to the case when $\mathcal{B}$ is connected. If $\mathcal{B}$ is strongly connected, then it forms a cycle of length $|B|$ and then our statement obviously holds. Otherwise, it has a state $b_0 \in B$ which generates all states and, simultaneously, $\delta_{\mathcal{B}}(b, p) \neq p$ holds for every $p \in X^+_{\mathcal{B}}$. Then $\mathcal{B}$ can be embedded isomorphically into the automaton $\mathcal{B}' = (B, X_{\mathcal{B}} \cup \{z\}, \delta')$, where for every $b \in B$,

$$\delta'(b, x') = \begin{cases} \delta_{\mathcal{B}}(b, x') & \text{if } x' \in X_{\mathcal{B}}, \\ b_0 & \text{if } b = b_0, x' = z, \\ \text{arbitrary } \delta_{\mathcal{B}}(b, y), y \in X_{\mathcal{B}}, & \text{otherwise.} \end{cases}$$

Let m denote the least common multiple of all positive integers which are lengths of cycles in the automaton $\mathcal{B}$. Then, by the construction of $\mathcal{B}'$, m also is the least common multiple of all positive integers which are lengths of cycles in the automaton $\mathcal{B}'$. By Lemma 4.9, $\mathcal{B}'$ can be represented homomorphically by an α_0-product of a counter with m states and a monotone automaton. Because $\mathcal{B}$ can be represented homomorphically by a product of factors in $\mathcal{K}$, it is easy to see that the counter of m states also has this property. On the other hand, by Lemma 4.7, every monotone automaton can be represented homomorphically by a product of factors in $\mathcal{K}$. Thus we have that the automaton $\mathcal{B}'$ can be represented homomorphically by a product of factors in $\mathcal{K}$. Observe that $\mathcal{B}'$ satisfies the semi-Letichevsky criterion (by $\delta_{\mathcal{B}}(b_0, z) = b_0$ and $\delta_{\mathcal{B}}(b_0, x'q) \neq b_0$, $x' \in X_{\mathcal{B}}$, $q \in (X_{\mathcal{B}} \cup \{z\})^*$). Thus we can apply again Lemma 4.9, Lemma 4.8, and Proposition 2.63. The proof is complete. □

We have the following conjecture.

Conjecture 4.11. *Given a finite class $\mathcal{K}$ of automata, let $c_{\mathcal{K}}$ denote the least common multiple of all positive integers which are lengths of cycles of automata in $\mathcal{K}$. Moreover, let $m_{\mathcal{K}}$ be the minimal number of cycles of automata in $\mathcal{K}$ such that every prime power divisor of $c_{\mathcal{K}}$ divides at least one of these lengths of cycles. For every nonnegative integer s, there exist an integer $r > s$, a finite set $\mathcal{K}$ of automata, an (r, s)-weighted automaton $\mathcal{A} \in \mathcal{K}$ (having the semi-Letichevsky criterion), and an automaton $\mathcal{B}$ such that $\mathcal{B}$ can be represented by a general product of factors from $\mathcal{K}$ but $\mathcal{B}$ cannot be represented homomorphically by an α_i-ν_j-product of factors in $\mathcal{K}$ if $i \leq 1$ and $j \leq m_{\mathcal{K}} + 2(s + 1)$.*

Problem 4.12. *Prove or disprove Conjecture* 4.11.

The next observation gives a partial solution of Conjecture 4.11.

Proposition 4.13. *There exists an automaton $\mathcal{A}$ without Letichevsky criteria and a class $\mathcal{K}$ of automata having the semi-Letichevsky criterion such that $\mathcal{A}$ cannot be represented by an α_0-product of factors from $\mathcal{K}$.*

Proof. Let $\mathcal{K}$ be a singleton class having the automaton $\mathcal{B} = (\{1, 2, *\}, \{x_1, x_2\}, \delta)$ with $\delta(1, x_1) = 2, \delta(2, x_2) = 1, \delta(1, x_2) = \delta(2, x_1) = \delta(*, x_1) = \delta(*, x_2) = *$. Moreover, let $\mathcal{A} = (\{a_0, a_1, a_2\}\{y_1, y_2\}, \delta')$ be defined by $\delta'(a_0, y_1) = a_1, \delta'(a_0, y_2) = a_2$, $\delta'(a_1, y_1) = \delta'(a_1, y_2) = a_1$, $\delta'(a_2, y_1) = \delta'(a_2, y_2) = a_2$. Consider an arbitrary α_0-power $\mathcal{B}^n(X, \varphi_1, \ldots, \varphi_n)$ of $\mathcal{B}$ and prove that for every state $(b_1, \ldots, b_n)$ of $\mathcal{B}$ there exists a $p \in X^*$ such that $(\delta(b_1, \varphi_1(b_1, \ldots, b_n, p)), \ldots, (\delta(b_1, \varphi_n(b_1, \ldots, b_n, p)))) = (*, \ldots, *)$.

Indeed, if the first two letters of p are the same (assuming $|p| \geq 2$), then $\delta(b_1, \varphi_1(b_1, \ldots, b_n, p)) = *$. And then, if the third and fourth letters of p are the same (assuming $|p| \geq 4$), then $\delta(b_2, \varphi_2(b_1, \ldots, b_n, p)) = *$. Repeating this procedure, we obtain $(\delta(b_1, \varphi_1(b_1, \ldots, b_n, p)), \ldots, (\delta(b_n, \varphi_n(b_1, \ldots, b_n, p)))) = (*, \ldots, *)$ provided $p = z^{2n}$ for some $z \in X$. Suppose that, contrary to our assumptions, there exists a subautomaton $\mathcal{B}' = (B', X'\delta'')$ of this α_0-power such that $\mathcal{B}'$ can be mapped homomorphically onto $\mathcal{A}$ by a homomorphism $\psi = (\psi_1, \psi_2)$. Let $z_1, z_2 \in X'$ be given such that $\psi_2(z_1) = y_1$ and $\psi_2(z_2) = y_2$. Moreover, let $\psi_1(b) = a_0$ for some $b \in B'$. Then $\delta''(b, z_1^{2n}) = \delta''(b, z_2^{2n}) = (*, \ldots, *)$ and $a_1 = \delta'(\psi_1(b), \psi_2(z_1^{2n})) \neq \delta'(\psi_1(b), \psi_2(z_2^{2n})) = a_2$, a contradiction. Therefore, none of the α_0-powers of $\mathcal{B}$ can represent $\mathcal{A}$ homomorphically. This ends the proof. □

Problem 4.14. *Is it decidable for every positive integer k and every finite class $\mathcal{K}$ of automata having the semi-Letichevsky criterion whether or not a finite automaton can be represented homomorphically by a ν_k-product of factors in $\mathcal{K}$?*

By Corollary 4.10 we can derive the following well-known statement.

Corollary 4.15. *Given a class $\mathcal{K}$ of automata having the semi-Letichevsky criterion, let $\mathcal{A}$ be an automaton which can be represented homomorphically by a product of factors from $\mathcal{K}$. Then $\mathcal{A}$ can also be represented homomorphically by an α_1-product of factors from $\mathcal{K}$.* □

4.2 Without Any Letichevsky Criteria

Now we study automata satisfying neither Letichevsky's criterion nor the semi-Letichevsky criterion.

The following statement is obvious.

Proposition 4.16. *$\mathcal{A} = (A, X, \delta)$ is an automaton without any Letichevsky criteria if and only if for every state $a_0 \in A$, input letters $x, y \in X$, and an input word $p \in X^*$ having $\delta(a_0, xp) = a_0$, it holds that $\delta(a_0, x) = \delta(a_0, y)$.* □

Obviously, if $\mathcal{A} = (A, X, \delta)$ has the above properties, then there exists a nonnegative integer n such that for every $p \in X^*$ with $|p| \geq n$, each $\delta(a, p)$ $(a \in A)$ generates an autonomous state-subautomaton of $\mathcal{A}$. Denote by $n_{\mathcal{A}}(\leq n)$ the minimal nonnegative integer having this property.

Proposition 4.17. $n_{\mathcal{A}} \leq \max(|A| - 2, 0)$.

Proof. Take out of consideration the trivial cases. Thus we may assume $|A| > 2$. Consider $a \in A, x_1, \ldots, x_{m+2} \in X$, having $\delta(a, x_1 \cdots x_m x_{m+1}) \neq \delta(a, x_1 \cdots x_m x_{m+2})$. If $a, \delta(a, x_1), \delta(a, x_1 x_2), \ldots, \delta(a, x_1 \cdots x_m), \delta(a, x_1 \cdots x_m x_{m+1}), \delta(a, x_1 \cdots x_m x_{m+2})$ are not distinct states, then $\mathcal{A}$ satisfies either Letichevsky's criterion or the semi-Letichevsky criterion, a contradiction. Hence, $m \leq |A| - 3$. Thus $n_{\mathcal{A}} \leq |A| - 2$. □

We also note the next direct consequence of Proposition 4.16.

Proposition 4.18. *If $\mathcal{A}$ is a strongly connected automaton without any Letichevsky criteria, then $\mathcal{A}$ is autonomous.* □

By this observation, we immediately get the following.

Proposition 4.19. *Suppose that $\mathcal{A} = (A, X, \delta)$ is a strongly connected automaton without any Letichevsky criteria. There exists a $k > 0$ such that for every $a, b \in A$, $a = b$, if and only if there exists a pair $p, q \in X^*$ with $|p| \equiv |q| (\bmod\, k)$[19] and $\delta(a, p) = \delta(b, q)$.* □

Lemma 4.20. *Given an automaton $\mathcal{A} = (A, X, \delta)$ without any Letichevsky criteria, $a \in A$ is a state of a strongly connected state-subautomaton of $\mathcal{A}$ if and only if there exists a nonempty word $p \in X^*$ with $\delta(a, p) = a$.*

Proof. Let $a \in A$ be a state of a strongly connected state-subautomaton of $\mathcal{A}$. By definition, for every nonempty word $q \in X^*$, there exists a word $r \in X^*$ with $\delta(a, qr) = a$. Conversely, suppose that $\delta(a, p) = a$ for some $a \in A$ and $p \in X^*$, $p \neq \lambda$. Then for all prefixes p' of p and input letters $x, y \in X$, $\delta(a, p'x) = \delta(a, p'y)$. Therefore, for every $q \in X^*$, $\delta(a, q) = \delta(a, r)$, where r is a prefix of p with $|q| \equiv |r| (\bmod\, |p|)$. But then a generates a strongly connected state-subautomaton of $\mathcal{A}$. □

We shall use the following consequence of the above statement.

Proposition 4.21. *Let $\mathcal{A} = (A, X, \delta)$ be an automaton without any Letichevsky criteria. Moreover, suppose that $a \in A$ is not a state of any strongly connected state-subautomaton of $\mathcal{A}$. If $\delta(b, p) = a$ for some $b \in A$ and nonempty $p \in X^*$, then $\delta(a, q) \neq b$, $q \in X^*$. Conversely, if $\delta(a, r) = c$ for some $c \in A$ and nonempty $r \in X^*$, then $\delta(c, q) \neq a$, $q \in X^*$.* □

Next we prove the following proposition.

Proposition 4.22. *Given a class $\mathcal{K}$ of automata without any Letichevsky criteria, let $\mathcal{A}'$ be an automaton which can be represented homomorphically by a general product of factors from $\mathcal{K}$. There exists an automaton $\mathcal{A} \in \mathcal{K}$, a q-power $\mathcal{M}$ of $\mathcal{A}$ with a single factor such that $\mathcal{M}$ is an autonomous automaton; moreover, $\mathcal{A}'$ can be represented homomorphically by a diagonal product of its connected state-subautomata and $\mathcal{M}$.*

[19] Recall that $a \equiv b (\bmod\, n)$ means $n | a - b$. Moreover, $a \not\equiv b (\bmod\, n)$ means $n \nmid a - b$.

Proof. Consider the automaton $\mathcal{A}' = (A', X', \delta')$. We distinguish three cases.

Case 1. There exists an $\mathcal{A} = (A, X, \delta) \in \mathcal{K}$ having a nontrivial cycle. In other words, there are distinct states $a_1, \ldots, a_m \in A, m > 1$, and (not necessarily distinct) input letters $x_1, \ldots, x_m \in X$ such that $\delta(a_i, x_i) = a_{i+1 (\mathrm{mod}\, m)}$. By Proposition 4.16, this implies that for every $x \in X$, $\delta(a_i, x) = a_{i+1(\mathrm{mod}\, m)}$. Consider a fixed $x \in X$ and let $\mathcal{M} = \mathcal{A}(X, \varphi)$ be given such that $\varphi(a, x') = x$ for every $a \in A$ and $x' \in X$. Then $\mathcal{M}$ is an autonomous automaton which is a q-power of $\mathcal{A}$ with a single factor such that for every distinct $k, \ell \in \{1, \ldots, m\}$, $\delta(a_k, \varphi(a_k, p)) \neq \delta(a_\ell, \varphi(a_\ell, p))$, $p \in X^*$. Using Proposition 2.28, this ends the proof of our case.

Case 2. There exists an $\mathcal{A} = (A, X, \delta) \in \mathcal{K}$ having two distinct trivial cycles. This means that there are distinct states $a, b \in A$ and (not necessarily distinct) input letters $x_1, x_2 \in X$ having $\delta(a, x_1) = a$ and $\delta(b, x_2) = b$. But then, again applying Proposition 4.16, we obtain for every $x \in X$, $\delta(a, x) = a$ and $\delta(b, x) = b$. Thus, similar to above, we can define the q-power $\mathcal{M}$ of $\mathcal{A}$ with a single factor such that $\delta(a, \varphi(a, p)) \neq \delta(b, \varphi(b, p))$, $p \in X^*$. Then we can use Proposition 2.28 again, which completes the proof of this case.

Case 3. Neither of the two cases above apply. Using Proposition 2.23, this means that all elements of $\mathcal{K}$ are nilpotent automata. But then, using Proposition 2.64, $\mathcal{A}'$ is a nilpotent automaton whenever it can be represented homomorphically by a general product of factors in $\mathcal{K}$. Therefore, $\mathcal{A}'$ is a directable automaton. Thus, applying Proposition 2.27, it can be represented homomorphically by a diagonal product $\mathcal{M}'$ of its connected state-subautomata. Then it is obvious that $\mathcal{M}' \Delta \mathcal{M}$ also homomorphically represents $\mathcal{A}'$ whenever $\mathcal{M}$ and $\mathcal{M}'$ have the same input set and $\mathcal{M}$ is an arbitrary autonomous q-power of an automaton in $\mathcal{K}$ with a single factor. (Obviously, if X' denotes the input set of $\mathcal{M}'$ and $\mathcal{A} = (A, X, \delta)$ is an arbitrary element of $\mathcal{K}$ with a fixed input letter $x \in X$, and, moreover, $\mathcal{M} = \mathcal{A}(X, \varphi)$ is a q-product with $\varphi(a, x') = x$, $a \in A$, $x' \in X'$, then $\mathcal{M}$ has this property.) This implies the validity of our statement in this case. ❑

Lemma 4.23. *Let $\mathcal{A} = (A, X, \delta)$ be an automaton without any Letichevsky criteria. If there are $a \in A$, $q, q' \in X^*$, $|q| = |q'| \geq |A| - 1$, $\delta(a, q) \neq \delta(a, q')$, then for every pair of words $r, r' \in X^*$, $|r| = |r'|$, we have $\delta(a, qr) \neq \delta(a, q'r')$.*

Proof. Suppose that our statement does not hold, i.e., there are $a \in A$, $q, q', r, r' \in X^*$, $|q| = |q'| \geq |A| - 1$, $|r| = |r'|$ having $\delta(a, q) \neq \delta(a, q')$ and $\delta(a, qr) = \delta(a, q'r')$. Then, of course, $|r| = |r'| > 0$. We distinguish the following three cases.

Case 1. There are $q_1, r_1, q_2, r_2, q_1', r_1', q_2', r_2'$ with $q = q_1 r_1 = q_2 r_2$, $q' = q_1' r_1' = q_2' r_2'$, $|q_1| < |q_2|$, $|q_1'| < |q_2'|$ such that $\delta(a, q_1) = \delta(a, q_2)$, $\delta(a, q_1') = \delta(a, q_2')$.[20] But then, by Proposition 4.16, $\delta(a, q_1 w) = \delta(a, q_1 w')$ and $\delta(a, q_1' w) = \delta(a, q_1' w')$ for every $w, w' \in X^*$, $|w| = |w'|$. Thus, because of $\delta(a, q_1) = \delta(a, q_2)$ and $\delta(a, q_1') = \delta(a, q_2')$, we obtain that for every $w, w' \in X^*$ there are $z, z' \in X^*$ with $\delta(a, q_1 w z) = \delta(a, q_1)$ and $\delta(a, q_1' w' z') = \delta(a, q_1')$. Thus $q_1 r_1 = q$, $q_1' r_1' = q'$ imply that $\delta(a, qrz) = \delta(a, q_1)$ and $\delta(a, q'r'z') = \delta(a, q_1')$ hold for some $z, z' \in X^*$. This means that $\delta(a, qrzr_1) = \delta(a, q)$ and $\delta(a, q'r'z'r_1') = \delta(a, q')$. Put $b = \delta(a, qr) (= \delta(a, q'r'))$, $c = \delta(a, q)$, $c' = \delta(a, q')$. Then $\delta(b, zr_1) = c \neq c' = \delta(b, z'r_1')$ and $\delta(c, r) = \delta(c', r') = b$. Therefore, by Proposition 2.70, $\mathcal{A}$ satisfies Letichevsky's criterion, a contradiction.

[20]This holds automatically if $|q| = |q'| \geq |A|$.

Case 2. There are q_1, r_1, q_2, r_2 with $q = q_1r_1 = q_2r_2$, $|q_1| < |q_2|$, such that $\delta(a, q_1) = \delta(a, q_2)$, but $\delta(a, q_1') \neq \delta(a, q_2')$ holds for all distinct prefixes q_1', q_2' of q'. These assumptions imply $(|q| =)\ |q'| \leq |A| - 1$. On the other hand, $|q| = |q'| \geq |A| - 1$ is supposed. Hence, we necessarily have $|q| = |q'| = |A| - 1$. Thus we get $\delta(a, q_1') \neq \delta(a, q_2')$ for all distinct prefixes q_1', q_2' of q', where $|q'| = |A| - 1$. This implies that for every $d \in A$ there exists a prefix q_1' of q' with $\delta(a, q_1') = d$.

Then for every $d \in A$ there exists an $r_1' \in X^*$ having $\delta(d, r_1') = \delta(a, q')$. On the other hand, we may assume $\delta(a, qrzr_1) = \delta(a, q)$ as in the previous case.

Now we suppose $\delta(a, qr) = \delta(a, q'r')$ as before. Substituting d for $\delta(a, q)$ $(= \delta(a, qrzr_1))$, there exists an $r_1' \in X^*$ holding $\delta(a, qr_1') = \delta(a, q')$. Put $b = \delta(a, qr)\times$ $(= \delta(a, q'r')$, $c = \delta(a, q)$, $c' = \delta(a, q')$. Hence $\delta(b, zr_1) = c$, $\delta(b, zr_1r_1') = c'$ and $\delta(c, r) = \delta(c', r') = b$ (with $c \neq c'$). Therefore, by Proposition 2.70 we obtain again that $\mathcal{A}$ satisfies Letichevsky's criterion contrary to our assumptions.

Case 3. Let $\delta(a, q_1) \neq \delta(a, q_2)$ and $\delta(a, q_1') \neq \delta(a, q_2')$ for all distinct prefixes q_1, q_2 of q and q_1', q_2' of q', respectively. Then $|q| = |q'| \leq |A| - 1$. Recall that $|q| = |q'| \geq |A| - 1$ is also assumed. Thus $|q| = |q'| = |A| - 1$, which implies that for every $d \in A$ there are $r_1, r_1' \in X^*$ satisfying $\delta(d, r_1) = \delta(a, q)$ and $\delta(d, r_1') = \delta(a, q')$. Therefore, assuming $\delta(a, qr) = \delta(a, q'r')$ for some $r, r' \in X^*$, and substituting d for $\delta(a, qr) (= \delta(a, q'r'))$, we obtain $\delta(a, qrr_1) = \delta(a, q)$, $\delta(a, qrr_1') = \delta(a, q')$ (with $\delta(a, qr) = \delta(a, q'r')$). Put $c = \delta(a, q)$, $c' = \delta(a, q')$. Then $\delta(d, r_1) = c$, $\delta(d, r_1') = c'$, $\delta(c, r) = \delta(c', r') = d$ (with $c \neq c'$). By Proposition 2.70, this implies that $\mathcal{A}$ satisfies Letichevsky's criterion, a contradiction again. □

Lemma 4.24. *Let $\mathcal{A} = (A, X, \delta)$ be an automaton without any Letichevsky criteria. For every state $a \in A$ we have one of the following two possibilities:*

(1) *There exist $q, q' \in X^*$, $|q| = |q'| \geq |A| - 1$, such that $\delta(a, qr) \neq \delta(a, q'r')$ for every $r, r' \in X^*$, $|r| = |r'|$.*

(2) *$\delta(a, q) = \delta(a, q')$ for every $q, q' \in X^*$, $|q| = |q'| \geq |A| - 1$.*

Proof. Suppose that (1) does not hold. Then for every $q, q' \in X^*$, $|q| = |q'| \geq |A| - 1$, there exist $r, r' \in X^*$, $|r| = |r'|$, having $\delta(a, qr) = \delta(a, q'r')$. Using Lemma 4.23, $\delta(a, qr) = \delta(a, q'r')$, $|r| = |r'|$, and $|q| = |q'| \geq |A| - 1$ implies $\delta(a, q) = \delta(a, q')$. Thus (2) holds whenever (1) does not hold. □

The following statement is obvious.

Lemma 4.25. *Given a digraph $\mathcal{D} = (V, E)$, let $v \in V$, $p_1, p_2, p_2', p_3, p_4 \in V^*$ such that $p_1p_2p_3vp_4v$ and $p_1p_2'p_3vp_4v$ are walks and vp_4v is a cycle. $|p_2| \equiv |p_2'| (\mathrm{mod}\, |p_4v|)$ if and only if there are positive integers k, ℓ having $|p_1p_2p_3v(p_4v)^k| = |p_1p_2'p_3v(p_4v)^\ell|$.* □

Lemma 4.26. *Let $\mathcal{A} = (A, X, \delta)$ be an automaton without any Letichevsky criteria. Consider a state $a \in A$ and suppose that there are $q, q' \in X^*$, $|q| = |q'| \geq |A| - 1$, $\delta(a, q) \neq \delta(a, q')$. Then there are q, q' (with $|q| = |q'| \geq |A| - 1$) such that for an appropriate triplet $u, v, v' \in X^*$, $q = uv$ and $q' = uv'$, for which we have the following* *properties:*

(1) *For every prefixes z of v and z' of v' with $|z| = |z'| > 0$, $\delta(a, uz) \neq \delta(a, uz')$.*
(2) *For every prefixes wx of z and $w'x'$ of z' with $ww' \neq \lambda$ and $x, x' \in X$, $\delta(a, uw) = \delta(a, uw')$ implies $x = x'$.*
(3) *For every words $r, r' \in X^*$ with $|r| = |r'|$, $\delta(a, uvr) \neq \delta(a, uv'r')$.*

Proof. Consider $a \in A$ and suppose that our conditions hold; i.e., there are $q, q' \in X^*$ having $|q| = |q'| \geq |A| - 1$, $\delta(a, q) \neq \delta(a, q')$. In this case, by Lemma 4.24, $\delta(a, qr) \neq \delta(a, q'r')$ holds for every $r, r' \in X^*$ with $|r| = |r'| \geq 0$. Thus, it is enough to prove that we have properties (1) and (2).

Then Proposition 4.17 implies that $\delta(a, q)$ and $\delta(a, q')$ generate autonomous state-subautomata of $\mathcal{A}$. We will distinguish the following cases (omitting some of the analogous ones).

Case 1. There are $u, u', v, v', \in X^*$ such that $q = uv$, $q' = u'v'$, $\delta(a, u) = \delta(a, u')$ and for every nonempty prefixes r of v and r' of v', $\delta(a, u)(= \delta(a, u')) \neq \delta(a, u'r')$, $\delta(a, u')$ $(= \delta(a, u)) \neq \delta(a, ur)$, and $\delta(a, ur) \neq \delta(a, u'r')$.[21] Let, say, $|u| \geq |u'|$ and let v'' be a prefix of v' with $|v''| = |v|$. Submit q' with uv'' and then we will have our requirements.

Case 2. There exists a prefix u of q having $\delta(a, u) = \delta(a, q')$. We shall use the fact that, in this case, $v \neq \lambda$ for $v \in X^*$ with $q = uv$ because of $\delta(a, q) \neq \delta(a, q')$. Let $t_2 \in X^*$ be a nonempty word with minimal length having $\delta(a, q't_1t_2) = \delta(a, q't_1)$ for some word $t_1 \in X^*$ and assume that t_2 is minimal in the sense that for every nonempty $p \in X^*$, $\delta(a, q't_1p) = \delta(a, q't_1)$ implies $|t_2| \leq |p|$.[22] Then, using that $\delta(a, q')$ generates an autonomous state-subautomaton of $\mathcal{A}$, we have $q = uv$, where v is a nonempty prefix of $t_1t_2^k$ for a suitable $k \geq 0$.

To prove that in this case $|u| \equiv |q'| (\mathrm{mod}\, |t_2|)$ is impossible, assume the contrary. Recall again that $\delta(a, q')$ generates an autonomous state-subautomaton of $\mathcal{A}$. But then, applying Lemma 4.25, there are words $r, r' \in X^*$, $|r| = |r'|$, having $\delta(a, qr) = \delta(a, q'r')$. By Lemma 4.23, then $|q| = |q'| < |A| - 1$, contrary to our assumptions. Thus we have the following cases.

Case 2.1. Suppose $|u| \not\equiv |q'| (\mathrm{mod}\, |t_2|)$ such that for all prefixes u_1 of u and u_1' of q' with $u_1u_1' \neq \lambda$, $\delta(a, u_1) = \delta(a, u_1')$ implies $u_1 = u$ and $u_1' = q'$. Then we obtain our requirement again (having $q = uv$, where v is a nonempty prefix of $t_1t_2^k$ for a suitable $k \geq 0$).

Case 2.2. Assume $|u| \not\equiv |q'| (\mathrm{mod}\, |t_2|)$ and, simultaneously, let, for some prefixes u_1 of u and u_1' of q', $\delta(a, u_1) = \delta(a, u_1')$ such that $u = u_1v_1$, $q' = u_1'v_1'$; furthermore, neither $u_1 = u_1' = \lambda$ nor $v_1 = v_1' = \lambda$.[23] If $v_1 = \lambda$ and $v_1' \neq \lambda$, then $\delta(a, u_1') = \delta(a, u_1'v_1') \neq \delta(a, u_1'v_1'v)(= \delta(a, uv))$. Recall that v is a nonempty suffix of q. But then $\mathcal{A}$ has either Letichevsky's criterion or the semi-Letichevsky criterion, a contradiction. Similarly, it also leads to a contradiction if we assume $v_1 \neq \lambda$ and $v_1' = \lambda$. Thus $\lambda \notin \{v_1, v_1'\}$ can be assumed and we may also assume $\lambda \notin \{u_1, u_1'\}$ analogously.

---

[21] $u = u' = \lambda$ is possible.

[22] The finiteness of the state set of $\mathcal{A}$ implies the existence of t_1 and t_2.

[23] If $u_1 = u_1' = \lambda$ or $v_1 = v_1' = \lambda$, then we may get either the previous case or this case considering appropriate u_1, u_1', v_1, v_1' with $u_1u_1' \neq \lambda$ and $v_1v_1' \neq \lambda$.

By $|u| \not\equiv |q'| (\text{mod } |t_2|)$, either $|u_1| \not\equiv |u'_1| (\text{mod } |t_2|)$ or $|v_1| \not\equiv |v'_1| (\text{mod } |t_2|)$.

Case 2.2.1. Suppose $|u_1| \not\equiv |u'_1| (\text{mod } |t_2|)$ and let, say, $|v_1| \le |v'_1|$. Take a prefix v' of $t_1 t_2^k$ for a suitable $k \ge 0$ with $|u'_1 v_1 v'| = |q|$ and let us consider $u'_1 v_1 v'$ instead of q'.

Case 2.2.2. Suppose $|u_1| \equiv |u'_1| (\text{mod } |t_2|)$. Then $|v_1| \not\equiv |v'_1| (\text{mod } |t_2|)$. Let, say, $|u_1| \le |u'_1|$. Take a prefix v' of $t_1 t_2^k$ for a suitable $k \ge 0$ with $|u_1 v'_1 v'| = |q|$ and submit q' with $u_1 v'_1 v'$.

In both Case 2.2.1 and Case 2.2.2, we have words[24] $w, w_1, w_2, w'_1, w'_2 \in X^*$, $\lambda \notin \{w_1, w'_1\}$, $|w_1| \not\equiv |w'_1| (\text{mod } |t_2|)$, w'_2 is a prefix of w_2 (or, in the opposite case, w_2 is a prefix of w'_2), $q = ww_1w_2, q' = ww'_1w'_2$, such that $\delta(a, ww_1) = \delta(a, ww'_1)$, and $\delta(a, ww_1)$ $(= \delta(a, ww'_1))$ generates an autonomous state-subautomaton of $\mathcal{A}$. Then let $w, w_1, w_2, w'_1, w'_2 \in X^*$ be arbitrary strings with these properties for which $\min(|w_1|, |w'_1|)$ is minimal.

Suppose that for every nonempty proper prefixes z_1 of w_1 and z'_1 of w'_1 we have $\delta(a, w) \notin \{\delta(a, wz_1), \delta(a, wz'_1)\}$ and $\delta(a, wz_1) \ne \delta(a, wz'_1)$. Moreover, recall that $\delta(a, w) \ne \delta(a, ww_1) (= \delta(a, ww'_1))$ because $\delta(a, ww_1) (= \delta(a, ww'_1))$ generates an autonomous state-subautomaton of $\mathcal{A}$. By these conditions, we are done, since we have our properties for $q = ww_1w_2, q' = ww'_1w'_2$.

Now we assume $|w_1| \not\equiv |w'_1| (\text{mod } |t_2|)$ such that for some prefixes z_1 of w_1 and z'_1 of w'_1, $\delta(a, z_1) = \delta(a, z'_1)$ such that $w_1 = z_1z_2$, $w'_1 = z'_1z'_2$; furthermore, neither $z_1 = z'_1 = \lambda$ nor $z_2 = z'_2 = \lambda$.[25] We can prove $\lambda \notin \{z_1, z'_1, z_2, z'_2\}$ in a way similar to the proof of $\lambda \notin \{u_1, u'_1, v_1, v'_1\}$ in Case 2.2. Then either $|z_1| \not\equiv |z'_1| (\text{mod } |t_2|)$ or $|z_2| \not\equiv |z'_2| (\text{mod } |t_2|)$. It remains to prove that these cases are impossible.

If $|z_1| \not\equiv |z'_1| (\text{mod } |t_2|)$ and, say, $|z_2| \ge |z'_2|$, then considering the prefix w''_2 of w_2 having $|z_2 w''_2| = |z'_2 w'_2|$, we can submit w, w_1, w_2, w'_1, w'_2 with $w, z_1, z_2w_2, z'_1, z_2w''_2$, contrary to the the minimality of $\min(|w_1|, |w'_1|)$.

If $|z_1| \equiv |z'_1| (\text{mod } |t_2|)$ with $|z_2| \not\equiv |z'_2| (\text{mod } |t_2|)$ and, say, $|z_1| \ge |z'_1|$, then considering the prefix w''_2 of w'_2 having $|z_1 w''_2| = |z'_1 w'_2|$, we can submit w, w_1, w'_1, w_2, w'_2 with $wz_1, z_2, z'_2, w_2, w''_2$, contradicting the minimality of $\min(|w_1|, |w'_1|)$.

The proof is complete. □

We shall use the following definition. Let $\mathcal{A} = (A, X, \delta)$ be an automaton. For every pair $a, b \in A$, consider a fixed $x_{a,b} \in X$ with $\delta(a, x_{a,b}) = b$ if it exists and put $\rho_a(x) = x_{a,b}$ if $\delta(a, x) = b$. Moreover, put $\rho_a(\lambda) = \lambda$ and $\rho_a(x_1 \cdots x_n) = \rho_a(x_1)\rho_{\delta(a,x_1)}(x_2) \cdots \rho_{\delta(a,x_1 \cdots x_{n-1})}(x_n)$. Clearly, then $\delta(a, w) = \delta(a, \rho_a(w))$ holds for every $w \in X^*$.

Lemma 4.27. *Let $\mathcal{A} = (A, X, \delta)$ be an automaton without any Letichevsky criteria. Consider $a, a_0 \in A$, $p \in X^*$ with $\delta(a_0, p) = a$. If there are $q, q' \in X^*$, $|pq| = |pq'| \ge |A| - 1$, $\delta(a, q) \ne \delta(a, q')$, then there are an input letter $x \in X$, a word $u \in X^*$, and single-factor products $\mathcal{B} = (A, X, \delta_\mathcal{B})$, $\mathcal{C} = (A, X, \delta_\mathcal{C})$ of $\mathcal{A}$ such that for every $z, z' \in X^*$ with $|z| < |z'| \le |pu| + 1$ we obtain $(\delta_\mathcal{B}(a_0, z), \delta_\mathcal{C}(a_0, z)) \ne (\delta_\mathcal{B}(a_0, z'), \delta_\mathcal{C}(a_0, z'))$; moreover, $\{(\delta_\mathcal{B}(a_0, z), \delta_\mathcal{C}(a_0, z)) | z \in X^*, |z| \le |pu|\}$, $\{(\delta_\mathcal{B}(a_0, z), \delta_\mathcal{C}(a_0, z)) | z = z_1xz_2, z_1, z_2 \in X^*, |z_1| = |pu|\}$ and $\{(\delta_\mathcal{B}(a_0, z), \delta_\mathcal{C}(a_0, z)) | z = z_1x'z_2, x' \in X, x' \ne x, z_1, z_2 \in X^*, |z_1| = |pu|\}$ are pairwise disjoint sets.*

---

[24] In Case 2.2.1, of course, $w = \lambda$, .

[25] If $z_1 = z'_1 = \lambda$ or $z_2 = z'_2 = \lambda$, then we may get either the previously discussed case or this case considering appropriate z_1, z'_1, z_2, z'_2 with $z_1z'_1 \ne \lambda$ and $z_2z'_2 \ne \lambda$.

Proof. Let $\mathcal{A} = (A, X, \delta)$ be an automaton without any Letichevsky criteria. Consider $a, a_0 \in A$, $p \in X^*$ with $\delta(a_0, p) = a$. If there are $q, q' \in X^*, |pq| = |pq'| \geq |A| - 1, \delta(a, q) \neq \delta(a, q')$, then we may assume $q = uv$ and $q' = uv'$ for some $u, v, v' \in X^*$ such that (1)–(3) of Lemma 4.26 hold.

Put $v = x_1 z, v' = x_2 z'$ with $x_1, x_2 \in X, z, z' \in X^*, |z| = |z'|$. Thus for every prefix wx of z and $w'x'$ of z' with $ww' \neq \lambda$ and $x, x' \in X$, $x = x'$ whenever $\delta(a, ux_1w) = \delta(a, ux_2w')$. Therefore, without any contradiction, we can define the functions $\varphi : A \times X \to A$, $\varphi' : A \times X \to A$ having the following properties. Let $x', y' \in X$ be arbitrary fixed-input letters and for every $b \in A$, $y \in X$, let

$$\varphi(b, y) = \begin{cases} y' & \text{if } b \notin \{\delta(a, s) : s \text{ is a prefix of } pq \text{ or } pq'\}, \\ y'' & \text{if } \delta(a_0, p') = b, \text{ where } y'' \in X, \text{ and } p'y'' \text{ is a prefix of } pq \text{ with } p' \neq p, \\ & \text{or } \delta(a_0, p') = b, \text{ where } y'' \in X, \text{ and } p'y'' \text{ is a prefix of } pq' \text{ with } p' \neq p, \\ x_1 & \text{if } b = \delta(a, p) \text{ and } y = x, \\ x_2 & \text{if } b = \delta(a, p) \text{ and } y \neq x, \end{cases}$$

$$\varphi'(b, y) = \begin{cases} y' & \text{if } b \notin \{\delta(a, s) : s \text{ is a prefix of } pq \text{ or } pq'\}, \\ y'' & \text{if } \delta(a_0, p') = b, \text{ where } y'' \in X \text{ and } p'y'' \text{ is a prefix of } pq \text{ with } p' \neq p, \\ & \text{or } \delta(a_0, p') = b, \text{ where } y'' \in X, \text{ and } p'y'' \text{ is a prefix of } pq' \text{ with } p' \neq p, \\ x_1 & \text{if } b = \delta(a, p). \end{cases}$$

By Proposition 4.16, for every prefix p' of pu, $\delta(a_0, p') \notin \{\delta(a_0, pux_1r), \delta(a_0, pux_2r) \mid r \in X^*\}$. (We note that $\delta(a_0, p') \neq \delta(a_0, p'')$ also holds for every pair of distinct prefixes p', p'' of pu.) On the other hand, by our assumptions and Lemma 4.23, for every $r, r' \in X^*, |r| = |r'|$, it holds that $\delta(a_0, pux_1zr) \neq \delta(a_0, pux_2z'r')$. Therefore, $\delta(a_0, pux_1zr') \neq \delta(a_0, pux_2z'r')$. But then $(\delta(a_0, pux_1zr), \delta(a_0, pux_1zr)) \neq (\delta(a_0, pux_1zr'), \delta(a_0, pux_2z'r'))$ is valid for every $r, r' \in X^*$. In addition, by property (1) of Lemma 4.26, for every prefix w of z and w' of z' with $|w| = |w'|$, $(\delta(a_0, pux_1w), \delta(a_0, pux_1w)) \neq (\delta(a_0, pux_1w), \delta(a_0, pux_2w'))$. Therefore the single-factor products $\mathcal{B} = \mathcal{A}(X, \varphi)$ and $\mathcal{C} = \mathcal{A}(X, \varphi')$ satisfy the conditions of our Lemma 4.27. □

Lemma 4.28. *Let $\mathcal{A} = (A, X, \delta)$ be an automaton without any Letichevsky criteria. Consider $a, a_0 \in A$, $p \in X^*$ with $\delta(a_0, p) = a$ and suppose that $\delta(a, r) = \delta(a, r')$ holds for every $r, r' \in X^*, |pr| = |pr'| \geq |A| - 1$. Assume that $\delta(a, q) \neq \delta(a, q')$ holds for some $q, q' \in X^*, |pq| = |pq'| < |A| - 1$ and let q, q' be words of maximal length having this property. Then there are $q = uv, q' = uv'$ (having $|q| = |q'| < |A| - 1$) such that we have the following properties:*

(1) *For all prefixes r of v and r' of v' with $|r| = |r'| > 0$, we have $\delta(a, ur) \neq \delta(a, ur')$.*
(2) *For all prefixes wx of v and $w'x'$ of v' with $ww' \neq \lambda$ and $x, x' \in X$, $\delta(a, uw) = \delta(a, uw')$ implies $x = x'$.*
(3) *For all distinct prefixes p_1, p_2 of pq, $\delta(a_0, p_1) \neq \delta(a_0, p_2)$.*
(4) *For all words r, r' with $|r| = |r'| > |q| (= |q'|)$, $\delta(a, r) = \delta(a, r')$.*

Proof. Consider $a \in A$ and suppose that our conditions hold.

Suppose that for every prefix u of q and u' of q', $\delta(a, u) = \delta(a, u')$ implies $u = u' = \lambda$. Obviously then (1) and (2) hold.

Now let $q = uv$, $q' = u'v'$ such that $uu' \neq \lambda$ and $\delta(a, u) = \delta(a, u')$. Suppose that $|u|$ and $|u'|$ are minimal in the sense that for every prefixes u'' of u and u''' of u' with $u''u''' \neq \lambda$ and $\delta(a, u'') = \delta(a, u''')$, we have $\{u'', u'''\} = \{u, u'\}$.

Assume that, say, $u = \lambda$. This implies $\delta(a, u') = a$ with $u' \neq \lambda$. But then, by Proposition 4.16, a generates an autonomous state-subautomaton of $\mathcal{A}$. Therefore, $\delta(a, q) = \delta(a, q')$, a contradiction.

Now let $\lambda \notin \{u, u'\}$. Clearly, $v = v' = \lambda$ is impossible because $\delta(a, q) \neq \delta(a, q')$. Let, say, $v = \lambda$. Then, by the minimality of u and u', we have (1) and (2) again.

Next we assume $\lambda \notin \{u, u', v, v'\}$.

If $|u| = |u'|$, then we can consider, say, uv' instead of $u'v'$. Therefore, $|u| \neq |u'|$ can be supposed.

Assume that, say, $|u| > |u'|$. Then submit v with v'', where v'' is a subword of v' with $|v''| = |v|$. Because of the minimality of $|u|$ and $|u'|$ for every subword w of u and w' of u' with $|w|, |w'| > 0$, we obtain that $\delta(a, w) = \delta(a, w')$ implies $w = u$ and $w' = u'$. Submitting q, q', u, u', v, v' with uv'', q', u, u', v'', v' we get (1) and (2).

Now we prove (3), omitting some analogous cases. If there are no distinct prefixes $p_1', p_2' \in X^*$ of pq' with $\delta(a_0, p_1') = \delta(a_0, p_2')$, then change pq for pq' and pq' for pq. Therefore, in this case, we are ready. Otherwise, we may suppose $\delta(a_0, p_1') = \delta(a_0, p_2')$ for some distinct prefixes $p_1', p_2' \in X^*$ of pq'. Let, say, $p_1' = p_2'r'$ for some nonempty $r' \in X$. By Proposition 4.16 and $\delta(a_0, p_2') = \delta(a_0, p_2'r')$, this implies that $\delta(a_0, p_2')$ generates an autonomous state-subautomaton $\mathcal{B}$ of $\mathcal{A}$. Moreover, $\delta(a_0, p_1') = \delta(a_0, p_2'r') = \delta(a_0, p_2')$, $r' \neq \lambda$ implies that this autonomous state-subautomaton is strongly connected. Recall that by the maximality of $|q|(= |q'|)$, $\delta(a_0, pqx) = \delta(a_0, pq'x')$ holds for every $x, x' \in X$. Thus, $\delta(a_0, pqx)$ is also a state of the state-subautomaton $\mathcal{B}$ of $\mathcal{A}$. Then $\delta(a_0, pq) \neq \delta(a_0, pq')$ and $\delta(a_0, pqx) = \delta(a_0, pq'x')$ imply that $\delta(a_0, pq)$ is not a state of $\mathcal{B}$. Indeed, if $\delta(a_0, pq)$ and $\delta(a_0, pqx)$ are states of $\mathcal{B}$ with $\delta(a_0, pq) \neq \delta(a_0, pq')$ (and $|pq| = |pq'|$), then $\mathcal{B}$ cannot be autonomous, i.e., by Proposition 4.18, it is not strongly connected, a contradiction. Therefore, for every prefix p_1 of pq, $\delta(a_0, p_1)$ is not a state of $\mathcal{B}$.

Suppose that, contrary to our assumptions, $\delta(a_0, p_1) = \delta(a_0, p_2)$ holds for distinct prefixes p_1 and p_2 of pq and put, say, $p_1 = p_2r_1$ (where $r_1 \neq \lambda$ is assumed). In other words, $\delta(a_0, p_2r_1) = \delta(a_0, p_2)$ holds such that $\delta(a_0, p_2)$ is not a state of $\mathcal{B}$. But $\delta(a_0, pqx) = \delta(a_0, pq'x')$, $x, x' \in X$, implies that there exists an $r_2 \in X^*$ such that $\delta(a_0, p_2r_2)$ is a state of $\mathcal{B}$. Clearly, then $\mathcal{A}$ satisfies either Letichevsky's criterion or the semi-Letichevsky criterion, a contradiction.

Finally, (4) is a direct consequence of the maximality of $|q|(= |q'|)$. This completes the proof. □

Lemma 4.29. *Let $\mathcal{A} = (A, X, \delta)$ be an automaton without any Letichevsky criteria. Consider a pair $a, a_0 \in A$ of states and a word $p \in X^*$ with $\delta(a_0, p) = a$ and suppose that $\delta(a, r) = \delta(a, r')$ holds for every $r, r' \in X^*, |pr| = |pr'| \geq |A| - 1$. Assume that $\delta(a, q) \neq \delta(a, q')$ holds for some $q, q' \in X^*, |pq| = |pq'| < |A| - 1$ and let q, q' be words of maximal length having this property. Then there exist an input letter x, words $u, v, v', w \in X^*, q = uv, q' = uv', v = xw$ such that for every nonnegative integer $i \leq |pu|$, there are $\alpha_0 - \nu_1$-powers $\mathcal{B} = (B, X, \delta_{\mathcal{B}})$, $\mathcal{C} = (C, X, \delta_{\mathcal{C}})$ of $\mathcal{A}$ having states $b_0 \in B, c_0 \in C$ such that for every $z, z' \in X^*$ with $0 \leq |z| < |z'| \leq |pq| - i$ we have $(\delta_{\mathcal{B}}(b_0, z), \delta_{\mathcal{C}}(c_0, z)) \neq (\delta_{\mathcal{B}}(b_0, z'), \delta_{\mathcal{C}}(c_0, z'))$, and, moreover*

$\{(\delta_{\mathcal{B}}(b_0, z), \delta_{\mathcal{C}}(c_0, z)) | z \in X^*, |z| \leq i\}$, $\{(\delta_{\mathcal{B}}(b_0, z), \delta_{\mathcal{C}}(c_0, z)) | z = z_1 x z_2, |z_1| = i, |z_2| < |pq| - i\}$ and $\{(\delta_{\mathcal{B}}(b_0, z), \delta_{\mathcal{C}}(c_0, z)) | z \in X^*, z = z_1 x' z_2, |z_1| = i, x' \in X, x' \neq x, |z_2| < |pq| - i\}$ are pairwise disjoint sets.

Proof. Assume that our conditions hold.

Then there are $q, q', u, v, v' \in X^*$ with $q = uv$, $q' = uv'$ (having $|q| = |q'| < |A|-1$) such that we have properties (1)–(4) of Lemma 4.28.

Consider $y_1, \ldots, y_{|pq|}, y'_{|pu|+1}, \ldots, y'_{|pq|} \in X$ such that, in order, $p = y_1 \cdots y_{|p|}$, $u = y_{|p|+1} \cdots y_{|pu|}$, $v = y_{|pu|+1} \cdots y_{|pq|}$, $v' = y'_{|pu|+1} \cdots y'_{|pq|}$. Moreover, consider words p_k, v_ℓ, $v'_\ell \in X^*$, where, in order, p_k is a prefix of pu of length k, $0 \leq k \leq |pu|$, v_ℓ is a prefix of v of length ℓ, and v'_ℓ is a prefix of v' of length ℓ, $0 \leq \ell \leq |v| (= |v'|)$. Finally, let $y' \in X$ be an arbitrary fixed-input letter.

By our assumptions, without any contradiction, we can define the functions $\varphi_1 : X \to X$, $\varphi_j : A \times X \to X$, $\varphi'_1 : X \to X$, $\varphi'_j : A \times X \to X$, $j = 2, \ldots, |pu| - i + 1$, having the following properties. For every $d \in A$, $y \in Y$,

$$\varphi_1(y) = \begin{cases} y & \text{if } y = y_{|pu|+1}, \\ y'_{|pu|+1} & \text{otherwise,} \end{cases}$$

$$\varphi_j(d, y) = \begin{cases} y_{k+1} & \text{if } \delta(a_0, p_k) = d, 0 \leq k < |pu|, \\ & \text{or } \delta(a_0, pu) = d, k = |pu|, y = y_{k+1}, \\ & \text{or } \delta(a_0, puv_\ell) = d, 0 < \ell < |v|, k = |pu| + \ell, \\ y'_{k+1} & \text{if } \delta(a_0, pu) = d, k = |pu|, y \neq y_{k+1}, \\ & \text{or } \delta(a_0, puv'_\ell) = d, 0 < \ell < |v'|, k = |pu| + \ell, \\ y' & \text{otherwise,} \end{cases}$$

$$\varphi'_1(y) = y_{|pu|+1},$$

$$\varphi'_j(d, y) = \begin{cases} y_{k+1} & \text{if } \delta(a_0, p_k) = d, 0 \leq k \leq |pu|, \\ & \text{or } \delta(a_0, puv_\ell) = d, 0 < \ell < |v|, k = |pu| + \ell, \\ y' & \text{otherwise.} \end{cases}$$

Let $\mathcal{B} = (B, X, \delta_{\mathcal{B}}) = \mathcal{A}^{|pu|-i+1}(X, \varphi_1, \ldots, \varphi_{|pu|-i+1})$, $\mathcal{C} = (C, X, \delta_{\mathcal{C}}) = \mathcal{A}^{|pu|-i+1}(X, \varphi'_1, \ldots, \varphi'_{|pu|-i+1})$. Moreover, put

$$b_0 = \begin{cases} (a_0, \ldots, a_0) (\in A^{|pu|+1}) & \text{if } i = 0, \\ (\delta(a_0, p_i), \ldots, \delta(a_0, p_i)) \ (\in A^{|pu|-i+1}) & \text{if } 0 < i < |pu|, \\ \delta(a_0, p_{|pu|}) & \text{if } i = |pu| \end{cases}$$

and put $c_0 = b_0$.

By Proposition 4.16, $\delta(a_0, p') \notin \{\delta(a_0, p''), \delta(a_0, puy_{|pu|+1}r), \delta(a_0, puy'_{|pu|+1}r) \mid r \in X^*\}$ holds for every pair of distinct prefixes p', p'' of pu. But then, of course, for every $z, z' \in X^*$ with $|z| < |z'| \leq |pq| - i$, we have $\delta_{\mathcal{C}}(c_0, z) \neq \delta_{\mathcal{C}}(c_0, z')$. On the other hand, by Lemma 4.28, we may also assume $\delta(a_0, pur) \neq \delta(a_0, pur')$ for all prefixes r of v and r' of v' with $|r| = |r'| > 0$. Thus $\delta_{\mathcal{B}}(b_0, r_1 y_{|pu|+1} r_2) \neq \delta_{\mathcal{B}}(b_0, r'_1 x' r'_2)$, $x' \in X$, $x' \neq y_{|pu|+1}$, $r_1, r'_1, r_2, r'_2 \in X^*$, $|r_1| = |r'_1| = |pu| - i$, $|r_2| = |r'_2| < |v|$. Therefore, $(\delta_{\mathcal{B}}(b_0, z), \delta_{\mathcal{C}}(c_0, z)) \neq (\delta_{\mathcal{B}}(b_0, z'), \delta_{\mathcal{C}}(c_0, z'))$ if z and z' are arbitrary words with $|z| <$

$|z'| \leq |pq| - i$[26] or $z = r_1 y_{|pu|+1} r_2$, $z' = r'_1 x' r'_2$, where $x' \in X$, $x' \neq y_{|pu|+1}$, $r_1, r'_1, r_2, r'_2 \in X^*$, $|r_1| = |r'_1| = |pu| - i$, $|r_2| = |r'_2| < |v|$.[27]

Then we obtain that for every $z, z' \in X^*$ with $|z| < |z'| \leq |pq|$ we have $(\delta_{\mathcal{B}}(b_0, z), \delta_{\mathcal{C}}(c_0, z)) \neq (\delta_{\mathcal{B}}(b_0, z'), \delta_{\mathcal{C}}(c_0, z'))$, and, moreover, $\{(\delta_{\mathcal{B}}(b_0, z), \delta_{\mathcal{C}}(c_0, z)) | z \in X^*, |z| \leq i\}$, $\{(\delta_{\mathcal{B}}(b_0, z), \delta_{\mathcal{C}}(c_0, z)) | z = z_1 y_{|pu|+1} z_2, |z_1| = |pu| - i, |z_2| < |q| - |u|\}$ and $\{(\delta_{\mathcal{B}}(b_0, z), \delta_{\mathcal{C}}(c_0, z)) | z \in X^*, z = z_1 x' z_2, |z_1| = |pu| - i, |z_2| < |q| - |u|, x' \in X, x' \neq y_{|pu|+1}\}$ are pairwise disjoint sets.[28] Thus we have our statement, assuming $x = y_{|pu|+1}$. □

Lemma 4.30. *Let $\mathcal{A} = (A, X, \delta)$ be an automaton without any Letichevsky criteria. Suppose that there exist $a \in A$, $x_1, x_2 \in X$, $p, r, r' \in X^*$ having $|r| = |r'|$ such that $\delta(a, px_1 z) \neq \delta(apx_2 z')$ for every nonempty prefixes z of r and z' of r' with $|z| = |z'|$. There are a state $a' \in A$, an input letter $x \in X$, and single-factor products $\mathcal{B} = (A, X, \delta_{\mathcal{B}})$, $\mathcal{C} = (A, X, \delta_{\mathcal{C}})$ of $\mathcal{A}$ such that for every $z, z' \in X^*$ with $|z| < |z'| \leq |p| + 1$ we have $(\delta_{\mathcal{B}}(a', z), \delta_{\mathcal{C}}(a', z)) \neq (\delta_{\mathcal{B}}(a', z'), \delta_{\mathcal{C}}(a', z'))$; moreover, $\{(\delta_{\mathcal{B}}(a', z), \delta_{\mathcal{C}}(a', z)) | z \in X^*, |z| \leq |p|\}$, $\{(\delta_{\mathcal{B}}(a', z), \delta_{\mathcal{C}}(a', z)) | z = z_1 x z_2, z_1, z_2 \in X^*, |z_1| = |p|, |z_2| \leq |r|\}$ and $\{(\delta_{\mathcal{B}}(a', z), \delta_{\mathcal{C}}(a', z)) | z = z_1 x' z_2, x' \in X, x' \neq x, z_1, z_2 \in X^*, |z_1| = |p|, |z_2| \leq |r'|\}$ are pairwise disjoint sets.*

Proof. If there are words $s, s' \in X$, $|px_1 rs| = |px_2 r's'| \geq |A| - 1$ with $\delta(a, px_1 rs) \neq \delta(a, px_2 r's')$, then we have our statement applying Lemma 4.27.

Thus let $\delta(a, px_1 rs) = \delta(a, px_2 r's')$ for all words $s, s' \in X^*$, $|px_1 rs| = |px_2 r's'| \geq |A| - 1$.

Assume that, say, $\delta(a, px_1 r)$ is a state of a strongly connected state-subautomaton of $\mathcal{A}$. We will prove that in this case, $\delta(a, px_2 r')$ is not a state of any strongly connected state-subautomaton of $\mathcal{A}$. Indeed, if $\delta(a, px_2 r')$ is also a state of an appropriate strongly connected state-subautomaton of $\mathcal{A}$, then $\delta(a, px_1 r)$ and $\delta(a, px_2 r')$ are states of the same strongly connected state-subautomaton of $\mathcal{A}$ since $\delta(a, px_1 rs) = \delta(a, px_2 r's')$, $s, s' \in X^*$, $|px_1 rs| = |px_2 r's'| \geq |A| - 1$. But then, by Proposition 4.19, $\delta(a, px_1 r) = \delta(a, px_2 r')$, a contradiction.

Therefore, we may suppose that, say, $\delta(a, px_1 r)$ is not a state of any strongly connected state-subautomaton of $\mathcal{A}$.

Now we consider (not necessarily nonempty) words u, u', v, v' with $r = uv$, $r' = u'v'$, $\delta(a, u) = \delta(a, u')$ and suppose that u and u' are minimal in the sense that for every prefix u'' of u and u''' of u', $\delta(a, px_1 u'') = \delta(a, px_2 u''')$ and $u'' u''' \neq \lambda$ implies $u'' = u$ and $u''' = u'$.

Observe that by our conditions, $|u| = |u'|$ is impossible, i.e., $|u| \neq |u'|$. Let v'' be a prefix of v with $|v''| = |v'|$ if $|v| > |v'|$ and let $v'' = vz$ be for an arbitrary word $z \in X^*$ with $|vz| = |v'|$ if $|v| < |v'|$.

Now we prove that $\delta(a, p_1) \neq \delta(a, p_2)$ for all nonempty prefixes p_1 of $px_1 uv$ and p_2 of $px_2 u'v''$ with $|p_1| = |p_2| > |p|$. If $v'' = v'$, then this statement is true because of the conditions of our lemma. Thus we may assume $v'' \neq v'$ and $\max(|u|, |u'|) > 0$. If $|p_1| = |p_2| \leq \min(|px_1 u|, |px_2 u'|)$, then we have $\delta(a, p_1) \neq \delta(a, p_2)$ by our conditions.

---

[26]Then $\delta_{\mathcal{C}}(b_0, z) \neq \delta_{\mathcal{C}}(b_0, z')$.

[27]Then $\delta_{\mathcal{B}}(c_0, z) \neq \delta_{\mathcal{B}}(c_0, z')$.

[28]Clearly, $i = 0$ is possible and then $\{(\delta_{\mathcal{B}}(b_0, z), \delta_{\mathcal{C}}(c_0, z)) | z \in X^*, |z| < i\} = \emptyset$.

Suppose $|p_1| = |p_2| > \min(|px_1u|, |px_2u'|)$. If v'' is a (proper) subword of v, then $|u| < |u'|$ should hold. In other words, there exists a nonempty $q \in X^*$ with $\delta(a, p_2q) = \delta(a, p_1)$. By Proposition 4.21, $\delta(a, p_2q) = \delta(a, p_1) = \delta(a, p_2)$ would imply that $\delta(a, p_1)$ is a state of a strongly connected subautomaton of $\mathcal{A}$. But then $\delta(a, px_1r)$ also should have this property, contrary of our assumptions. Hence, we obtain $\delta(a, p_1) \neq \delta(a, p_2)$. Now we study when $v'' = vz$ for some nonempty $z \in X^*$. Then $|u| > |u'|$. We may suppose $|p_1| = |p_2| > \min(|px_1u|, |px_2u'|)$ again. But then there exists a nonempty $q \in X^*$ having $\delta(a, p_1q) = \delta(a, p_2)$. Hence, similar to before, we also receive $\delta(a, p_1) \neq \delta(a, p_2)$ as a consequence of Proposition 4.21.

Thus, we may assume for the considered words r, r' that for every $z, z_1, z_2, z', z_1', z_2'$ with $r = zz_1z_2$, $r' = z'z_1'z_2'$ we obtain $z_1 = z_1'$ whenever $\delta(a, px_1uz) = \delta(a, px_2u'z')$ with $\max(|u|, |u'|) > 0$ and $|z_1| = |z_1'|$. In other words, if p' is a subword of px_1r or px_2r' and there are $x, x' \in X$ having $\delta(a, p'x) \neq \delta(a, p'x')$, then $p' = p$ should hold. Of course, we also have this property if there are no prefixes u of r or u' of r' with $\delta(a, px_1u) = \delta(a, px_2u')$. Therefore, we can define $\varphi : A \times X \to A$ and $\varphi' : A \times X \to A$ in the following manner. Let $x', y' \in X$ be an arbitrary fixed input letter; moreover, for every pair $b \in A$, $y \in X$, let

$$\varphi(b, y) = \begin{cases} y' & \text{if } b \notin \{\delta(a, s) : s \text{ is a prefix of } pq \text{ or } pq'\}, \\ y'' & \text{if } \delta(a_0, p') = b, \text{ where } y'' \in X,\ p'y'' \text{ is a prefix of } px_1r \text{ or } px_2r, \\ & \quad \text{and } p' \neq p, \\ x_1 & \text{if } b = \delta(a, p) \text{ and } y = x, \\ x_2 & \text{if } b = \delta(a, p) \text{ and } y = x', \end{cases}$$

$$\varphi'(b, y) = \begin{cases} y' & \text{if } b \notin \{\delta(a, s) : s \text{ is a prefix of } pq \text{ or } pq'\}, \\ y'' & \text{if } \delta(a_0, p') = b, \text{ where } y'' \in X,\ p'y'' \text{ is a prefix of } px_1r \text{ or } px_2r, \\ & \quad \text{and } p' \neq p, \\ x_1 & \text{if } b = \delta(a, p). \end{cases}$$

Then, similar to the proof of Lemma 4.27, because of Proposition 4.16, $\delta(a, p') \notin \{\delta(a, p''), \delta(a, px_1z), \delta(a, px_2z') \mid z, z' \in X^*, z \text{ is a prefix of } r, z' \text{ is a prefix of } r'\}$ holds for every pair of distinct prefixes p', p'' of p. On the other hand, by our assumptions, for all prefixes z of r and z' of r' with $|z| = |z'|$, it holds that $\delta(a, px_1z) \neq \delta(a, px_2z')$. Therefore, $\delta(a, px_1z') \neq \delta(a, px_2z')$. But then $(\delta(a, px_1z), \delta(a, px_1z)) \neq (\delta(a, px_1z'), \delta(a, px_2z'))$ is valid for every prefix z of r and z' of r'. Therefore, by $a' = a$, the single-factor products $\mathcal{B} = \mathcal{A}(X, \varphi)$ and $\mathcal{C} = \mathcal{A}(X, \varphi')$ satisfy the conditions of our Lemma 4.30. □

4.3 Networks of Automata Without Any Letichevsky Criteria

In this section we show that some very restricted types of networking (in terms of the number of incoming links and feedback) already suffice for realizing the computational capacity obtainable by unrestricted networking of automata that satisfy neither the Letichevsky nor the semi-Letichevsky criterion. In particular, automata networks whose components have no more than one incoming link and receive no feedback at all (even about their own states) are all that will be required. Thus, if components are too simple (without any Letichevsky criteria), then no amount of cleverness or complexity in networking can achieve more computationally than this simple restricted type of networking.

Let $\mathcal{A} = (A, X, \delta)$ be an arbitrary automaton without any Letichevsky criteria. Consider a pair $a \in A$, $p \in X^*$ and suppose $p = x_1 \cdots x_m$ with $x_1, \ldots, x_m \in X$. Put

$p_0 = \lambda$ and $p_k = x_1 \cdots x_k, 1 \le k \le m$. By Proposition 4.16, for every $0 \le k < \ell \le m$, $\delta(a, p_k) = \delta(a, p_\ell)$ implies $\delta(a, p_k x) = \delta(a, p_\ell y), x, y \in X$. Therefore, we can define the automaton $\mathcal{A}_{a,p} = (A_{a,p}, X, \delta_{a,p})$ such that $A_{a,p} = \{a\} \cup \{\delta(a, x_1 \cdots x_i) \mid i = 1, \dots, m\} \cup \{\delta(a, x_1 \cdots x_m q) \mid q \in X^*\}$; moreover, for any $x \in X$, $\delta_{a,p}(a, x) = \delta(a, x_1)$, $\delta_{a,p}(a, x_1 \cdots x_{i-1} x) = \delta(a, x_1 \cdots x_i), i = 2, \dots, m$, and $\delta_{a,p}(a, x_1 \cdots x_m q) = \delta(a, x_1 \cdots x_m q)$. Given an automaton $\mathcal{A} = (A, X, \delta)$ without any Letichevsky criteria, let $n_{\mathcal{A}}$ denote again the minimal nonnegative integer such that $\delta(a, p)$ generates an autonomous state-subautomaton for every $a \in A$, $p \in X^*$. We shall use the following.

Lemma 4.31. *Let $\mathcal{A} = (A, X, \delta)$ be an arbitrary automaton without any Letichevsky criteria. For every pair $a \in A$, $p \in X^*$ with $|p| \ge n_{\mathcal{A}}$, we have that $\mathcal{A}_{a,p}$ is an autonomous automaton which can be represented by a single-factor product of A.*

Proof. By Proposition 4.16, for every $a \in A$, $p_1, p_2 \in X^*, x_1, x_2 \in X$, $\delta(a, p_1 x_1 p_2) = \delta(a, p_1)$ implies $\delta(a, p_1 x_1) = \delta(a, p_1 x_2)$. Hence, for every pair $y_1, y_2 \in X$, we obtain $\delta(a, q_1 y_1) = \delta(a, q_2 y_2)$ whenever $\delta(a, q_1) = \delta(a, q_2)$ such that q_1 and q_2 are distinct prefixes of the same word. Then, for every pair $a \in A$, $p \in X^*$, we can define the single-factor product $\mathcal{B} = \mathcal{A}(X, \varphi)$ such that for every pair $b \in A, x \in X$,

$$\varphi(b, x) = \begin{cases} x' & \text{if there exists a prefix } qx' \in X^* \text{ of } p, \\ & \text{such that } x' \in X \text{ and } \delta(a, q) = b, \\ x & \text{otherwise.} \end{cases}$$

By the construction of $\mathcal{B}$ and Proposition 4.17, the state $\delta(a, p)$ of the automaton $\mathcal{B}$ generates an autonomous state-subautomaton. Clearly, then the state a of the automaton $\mathcal{B}$ also generates an autonomous state-subautomaton $\mathcal{B}'$. It is easy to prove that the state-subautomaton $\mathcal{B}'$ of $\mathcal{B}$ coincides with the automaton $\mathcal{A}_{a,p}$. □

Proposition 4.32. *Let $\mathcal{B} = \mathcal{A}_1 \times \cdots \times \mathcal{A}_n(X, \varphi_1, \dots, \varphi_n)$ be a product of automata such that each of its factors is without any Letichevsky criteria and, simultaneously, the product $\mathcal{B}$ is an autonomous automaton. Then every connected state-subautomaton of $\mathcal{B}$ coincides with a (connected) state-subautomaton of a diagonal product of autonomous automata such that each of its factors is a single factor product of one of the automata $\mathcal{A}_t, t = 1, \dots, n$.*[29]

Proof. Consider the product $\mathcal{B} = \mathcal{A}_1 \times \cdots \times \mathcal{A}_n(X, \varphi_1, \dots, \varphi_n)$ of the automata $\mathcal{A}_t = (A_t, X_t, \delta_t), t = 1, \dots, n$, such that all of them are without any Letichevsky criteria and, simultaneously, $\mathcal{B}$ is an autonomous automaton. Put $m = \max(n_{\mathcal{A}_1}, \dots, n_{\mathcal{A}_n})$. If $m = 0$, then all of the factors of $\mathcal{B}$ are (connected) autonomous automata, and then our statement trivially holds. Thus we may assume $m > 0$.

Let there be given a positive integer $t \in \{1, \dots, n\}$, a state $a \in A_t$, and a nonempty word $p = z_1 \cdots z_m, z_1, \dots z_m \in X_t$. If there exists a state a' with double occurrences in the sequence $a, \delta_t(a, z_1), \dots, \delta_t(a, z_1 \cdots z_m)$, then, by Proposition 4.16, a' generates an autonomous state-subautomaton of $\mathcal{A}_t$.

Put, in order, $\mathcal{A}_{t,a,p} = (\mathcal{A}_t)_{a,p}$, $A_{t,a,p} = (A_t)_{a,p}$, $\delta_{t,a,p} = (\delta_t)_{a,p}$. Therefore $\mathcal{A}_{t,a,p} = (A_{t,a,p}, X_t, \delta_{t,a,p})$ is defined in the following manner: $A_{t,a,p} = \{a\} \cup \{\delta(a, z_1 \cdots z_i) \mid i = 1,$

[29]Recall that a diagonal product of automata with a single factor coincides with a q^ℓ-product of these single factors. (See Proposition 2.57.)

$\ldots, m\} \cup \{\delta(a, z_1 \cdots z_m q) \mid q \in X_t^*\}$, and, moreover, for any $x \in X_t$, $\delta_{t,a,p}(a, x) = \delta_t(a, z_1)$, $\delta_{t,a,p}(a, z_1 \cdots z_{i-1}x) = \delta_t(a, z_1 \cdots z_i)$, $i = 1, \ldots, m$, and $\delta_{t,a,p}(a, z_1 \cdots z_m q) = \delta_t(a, z_1 \cdots z_m q)$, $q \in X_t^*$. Then $\delta_t(a, qx) = \delta_t(a, qy)$ for every $t = 1, \ldots, n$, $a \in A_t$, $q \in X_t^*$, $|q| \geq m$, $x, y \in X_t$. Thus, for every $t \in \{1, \ldots, n\}$, $a \in A_t$, $p \in X_t^*$, $|p| = m$, $\mathcal{A}_{t,a,p}$ is an autonomous automaton; moreover, for every $q \in X_t^*$, $\delta_{t,a,p}(a, pq) = \delta_t(a, pq)$. Simultaneously, $\mathcal{A}_{t,a,p}$ preserves the property that it is without any Letichevsky criteria.

Let $\mathcal{B}' = (B', X, \delta')$ be an arbitrary connected state-subautomaton of the product $\mathcal{B}$ and let $b_0 = (b_{0,1}, \ldots, b_{0,n}) \in B'$, $b_{0,t} \in A_t$, $t = 1, \ldots, n$, denote a state for which $\mathcal{B}'$ is connected. $\mathcal{B}$ is autonomous, and thus $\mathcal{B}'$ should be also autonomous. Let $p \in X^*$ with $|p| = m$ be given and for any $t \in \{1, \ldots, n\}$, put $p_t = \varphi_t(b_{0,1}, \ldots, b_{0,n}, p)$.

Now we shall show that the diagonal product $\mathcal{M} = (B', X, \delta_{\mathcal{M}}) = \mathcal{A}_{1,b_{0,1},p_1} \Delta \cdots \Delta \mathcal{A}_{n,b_{0,n},p_n}$ homomorphically represents $\mathcal{B}'$. Of course, for every $r \in X^*$, $|r| \leq m$, $\delta_{t,b_{0,t},p_t}(b_{0,t}, r) = \delta_t(b_{0,t}, \varphi_t(b_{0,1}, \ldots, b_{0,n}, r'))$, where $\varphi_t(b_{0,1}, \ldots, b_{0,n}, r')$ is a prefix of p_t of length $|r|$. Thus, for any $r \in X^*$, $|r| \leq m$, $\delta_{\mathcal{M}}((b_{0,1}, \ldots, b_{0,n}), r) = \delta'((b_{0,1}, \ldots, b_{0,n}), r)$. Then we also have that $\delta_{\mathcal{M}}((b_{0,1}, \ldots, b_{0,n}), p) = \delta'((b_{0,1}, \ldots, b_{0,n}), p)$ is a state both of $\mathcal{M}$ and $\mathcal{B}'$. But then, for every $t \in \{1, \ldots, n\}$, the tth component of $\delta'((b_{0,1}, \ldots, b_{0,n}), p)$ generates the same autonomous state-subautomaton in $\mathcal{A}_{t,b_{0,1},p_t}$ and $\mathcal{A}_t$. Hence, by the structure of $\mathcal{M}$ and $\mathcal{B}'$, for every $p, q \in X^*$ with $|p| = m$, we can also get $\delta_{\mathcal{M}}((b_{0,1}, \ldots, b_{0,n}), pr) = \delta'((b_{0,1}, \ldots, b_{0,n}), pr)$. This completes the proof. □

By Propositions 4.32 and 4.22, we can derive the next statement.

Corollary 4.33. *Let $\mathcal{B} = \mathcal{A}_1 \times \cdots \times \mathcal{A}_n(X, \varphi_1, \ldots, \varphi_n)$ be a product of automata such that each of its factors is without any Letichevsky criteria and, simultaneously, the product $\mathcal{B}$ is an autonomous automaton. Then $\mathcal{B}$ can be represented homomorphically by a diagonal product of autonomous automata such that each of its factors is a single-factor product of one of the automata $\mathcal{A}_t$, $t = 1, \ldots, n$.* □

Lemma 4.34. *If $\mathcal{A} = (A, X, \delta) = \mathcal{A}_1 \times \cdots \times \mathcal{A}_n(X, \varphi_1, \ldots, \varphi_n)$ is a product of automata without any Letichevsky criteria, then for every $a \in A$ and $p \in X^*$, $|p| = n_{\mathcal{A}}$, the automaton $\mathcal{A}_{a,p}$ is autonomous and it can be represented homomorphically by a product of the same factors $\mathcal{A}_1, \ldots, \mathcal{A}_n$.*

Proof. It is possible that for some distinct prefixes p' and p'' of the word p there exists a $t \in \{1, \ldots, n\}$ such that the tth components of $\delta(a, p')$ and $\delta(a, p'')$ coincide. Assume that we have this situation for a pair of appropriate distinct subwords p', p'' of p and put $(a_1', \ldots, a_{t-1}', a_t', a_{t+1}', \ldots, a_n') = \delta(a, p')$, $(a_1'', \ldots, a_{t-1}'', a_t', a_{t+1}'', \ldots, a_n'') = \delta(a, p'')$. But then, by Proposition 4.16, this state component a_t' of $\mathcal{A}_t$ generates an autonomous state-subautomaton in $\mathcal{A}_t$. Thus we obtain the same product $\mathcal{A}$ if we assume that the values of $\varphi_t(a_1', \ldots, a_{t-1}', a_t', a_{t+1}', \ldots, a_n', x)$ and $\varphi_t(a_1'', \ldots, a_{t-1}'', a_t', a_{t+1}'', \ldots, a_n'', x)$ are unambiguously defined by the state a_t' and thus, for every $x', x'' \in X$, $\varphi_t(a_1', \ldots, a_{t-1}', a_t', a_{t+1}', \ldots, a_n', x') = \varphi_t(a_1'', \ldots, a_{t-1}'', a_t', a_{t+1}'', \ldots, a_n'', x'')$. On the other hand, it is clear that $n_{\mathcal{A}_t} \leq n_{\mathcal{A}}$. Thus, because of $|p| = n_{\mathcal{A}}$, the tth component of the state vector $\delta(a, p)$ generates an autonomous state-subautomaton of $\mathcal{A}_t$. Thus we get the same product automaton $\mathcal{A}$ if we suppose that for every state $(a_1', \ldots, a_n') = \delta(a, pq)$, $q \in X^*$, and $x \in X$, the value of $\varphi_t(a_1', \ldots, a_n', x)$ is unambiguously defined by the state component a_t'.

Having these assumptions, we can construct a product $\mathcal{A} = (A, X, \delta) = \mathcal{A}_1 \times \cdots \times \mathcal{A}_n(X, \varphi_1', \ldots, \varphi_n')$ of the automata $\mathcal{A}_t = (A_t, X_t, \delta_t)$ in the following way. For every $(a_1, \ldots, a_n) \in A, x \in X, t \in \{1, \ldots, n\}$,

$$\varphi_t'(a_1, \ldots, a_n, x) = \begin{cases} \varphi_t(a_1, \ldots, a_n, x') & \text{if } (a_1, \ldots, a_n) = \delta(a, p'), x' \in X, \text{ and} \\ & p'x' \text{ is a prefix of } p, \\ \varphi_t(a_1, \ldots, a_n, x) & \text{otherwise.} \end{cases}$$

But then the state a of this product generates the automaton $\mathcal{A}_{a,p}$ as a state-subautomaton. Finally, by Lemma 4.31, $\mathcal{A}_{a,p}$ should be autonomous, as we have stated. □

Consider an automaton $\mathcal{A} = (A, X, \delta)$ without any Letichevsky criteria and let $a_0 \in A$ be fixed. For an arbitrary pair k, ℓ of positive integers, put $\rho_{\mathcal{A},a_0}(x, k, \ell) = *$ if $\delta(a_0, px_1) = \delta(a_0, px_2)$ holds for every $p \in X^*, |p| = k - 1$ and $x_1, x_2 \in X$. Moreover, put $\rho_{\mathcal{A},a_0}(x, k, \ell) = *$ if for every $a = \delta(a_0, p), p \in X^*, |p| = k - 1$, property (1) of Lemma 4.24 holds, and, simultaneously, $\ell \geq \ell_k$, where ℓ_k denotes the length of the words $q, q' \in X^*, |q| = |q'|$ having maximal length such that $\delta(a, q) = \delta(a, q')$; furthermore, $\delta(a, s) \neq \delta(a, s')$ for all nonempty prefixes s of q and s' of q' with $|s| = |s'|$. Otherwise put $\rho_{\mathcal{A},a_0}(x, k, \ell) = x$.[30] In addition, put $\rho_{\mathcal{A},a_0}(x_1 \cdots x_n) = \rho_{\mathcal{A},a_0}(x_1, 1, k)\rho_{\mathcal{A},a_0}(x_2, 2, k - 1) \cdots \rho_{\mathcal{A},a_0}(x_k, k, 1)$ for every $x_1, \ldots, x_k \in X$. Finally, put $\rho_{\mathcal{A},a_0}(\lambda) = \lambda$ by definition.

We shall use the following two observations.

Proposition 4.35. *If* $\rho_{\mathcal{A},a_0}(x_1 \cdots x_n) = y_1 \cdots y_{n-1} *$ $(y_i \in \{x_i, *\}, i = 1, \ldots, n-1)$ *for some* $x_1, \ldots, x_n \in X$*, then* $\rho_{\mathcal{A},a_0}(x_1 \cdots x_n x) = y_1' \cdots y_{n-1}' **, y_i' \in \{y_i, *\}, i = 1, \ldots, n-1$ *for every* $x \in X$.

Proposition 4.36. *Let* $\mathcal{A} = (A, X, \delta)$ *be a connected automaton without any Letichevsky criteria and let* a_0 *be a state for which* $\mathcal{A}$ *is connected. For every* $p, q \in X^*$, $\rho_{\mathcal{A},a_0}(p) = \rho_{\mathcal{A},a_0}(q)$ *implies* $\delta(a_0, p) = \delta(a_0, q)$. □

Given an automaton $\mathcal{A} = (A, X, \delta)$ without any Letichevsky criteria, assume that $\mathcal{A}$ is connected for $a_0 \in A$. Define the automaton $\mathcal{D} = (D, X, \delta')$ with $D = \{\rho_{\mathcal{A},a_0}(p) \mid p \in X^*, |p| \leq |A|\}$,

$$\delta'(\rho_{\mathcal{A},a_0}(p), x) = \begin{cases} \rho_{\mathcal{A},a_0}(px) & \text{if } |p| < |A|, \\ \rho_{\mathcal{A},a_0}(p) & \text{if } |p| = |A|. \end{cases}$$

Consider a positive integer $t \in \{1, \ldots, n\}$, a state $a \in X_t$, and a nonempty word $p = z_1 \cdots z_{n_\mathcal{A}}, z_1, \ldots z_{n_\mathcal{A}} \in X_t$.

Let $p_1, \ldots, p_u$ be a complete list of all words of length of $n_\mathcal{A}$ in X^*.

Lemma 4.37. *The diagonal product* $\mathcal{D}\Delta\mathcal{A}_{a_0,p_1}\Delta \cdots \Delta\mathcal{A}_{a_0,p_u}$ *homomorphically represents* $\mathcal{A}$.

Proof. By Proposition 4.36, for every $p, q \in X^*$, $\rho_{\mathcal{A},a_0}(p) = \rho_{\mathcal{A},a_0}(q)$ implies $\delta'(a_0, p) = \delta'(a_0, q)$. Thus, we can unambiguously define the mapping $\psi : D \times A_{a_0,p_1} \times \cdots \times A_{a_0,p_u} \rightarrow$

[30] Thus we also put $\rho_{\mathcal{A},a_0}(x, k, \ell) = x$ if there are $a = \delta(a_0, p), p \in X^*, |p| = k - 1$, such that we have property (1) of Lemma 4.24.

A such that

(1) $\psi(\lambda, a_0, \ldots, a_0) = a_0$;

(2) for any $p \in X^*$ with $|p| \le n_{\mathcal{A}}$, $\psi((\delta'(\lambda, p), \delta_{a_0,p_1}(a_0, p), \ldots, \delta_{a_0,p_u}(a_0, p))) = \delta_{a_0,p_i}(a_0, p)$, where p_i is the first word in the list $p_1, \ldots, p_u$ such that for the prefix of p' of p_i of length $|p|$, $\rho_{\mathcal{A},a_0}(p') = \rho_{\mathcal{A},a_0}(p)$;

(3) for any pair $p, q \in X^*$ with $|p| = n_{\mathcal{A}}$, $\psi((\delta'(\lambda, pq), \delta_{a_0,p_1}(a_0, pq)\ldots, \delta_{a_0,p_u}(a_0, pq))) = \delta_{a_0,p_i}(a_0, pq)$, where p_i is the first word in the list $p_1, \ldots, p_u$ such that $\rho_{\mathcal{A},a_0}(p_i) = \rho_{\mathcal{A},a_0}(p)$.

Recall that for all words $r \in X^*$ of length $n_{\mathcal{A}}$, $\delta(a_0, r)$ generates an autonomous subautomaton in $\mathcal{A}$. By this observation, using Proposition 4.36, we have that ψ is a state-homomorphism of the diagonal product $\mathcal{D}\Delta\mathcal{A}_{a_0,p_1}\Delta\cdots\Delta\mathcal{A}_{a_0,p_u}$ onto $\mathcal{A}$.

By Lemma 4.34 and Proposition 4.32, each of the automata $\mathcal{A}_{a_0,p_i}$, $i = 1, \ldots, u$, can be represented homomorphically by a diagonal product of a single-factor product of the factors of $\mathcal{A}$. □

For every $i = 1, \ldots, n_{\mathcal{A}}, x \in X$, define $\mathcal{B}_{i,x} = (B_i, X, \delta_{i,x})$ in the following way.

If $\rho_{\mathcal{A},a_0}(x, i, \ell) = *$ for every positive integer ℓ, then let $B_i = \{0\}$, $\delta_{i,x}(0, z) = 0$, $z \in X$.

If $\rho_{\mathcal{A},a_0}(x, i, \ell) = x$ for every $x \in X$ and positive integer ℓ, then let $B_i = \{0, 1, \ldots, i-1, *, x\}$ with

$$\delta_{i,x}(b, z) = \begin{cases} b+1 & \text{if } 0 \le b < i-1, \\ x & \text{if } b = i-1, z = x, \\ & \text{or } b = x, \\ * & \text{otherwise.} \end{cases}$$

Otherwise, denote m_i the maximal positive integer for which $\rho_{\mathcal{A},a_0}(x, i, m_i) = x$, $x \in X$ and define $B_i = \{0, 1, \ldots, i-1, *\} \cup \{(x, \ell) \mid \ell = 1, \ldots, m_i\}$,

$$\delta_{i,x}(b, z) = \begin{cases} b+1 & \text{if } 0 \le b < i-1, \\ (x, 1) & \text{if } b = i-1, z = x, \\ (x, \ell+1) & \text{if } b = (x, \ell), 1 \le \ell < m_i, \\ * & \text{otherwise.} \end{cases}$$

Lemma 4.38. *Let $x_1, \ldots, x_s$ be an arrangement of the elements of the input set X. The diagonal product $\mathcal{B}_{1,x_1}\Delta\cdots\mathcal{B}_{1,x_s}\Delta\cdots\Delta\mathcal{B}_{n_{\mathcal{A}},x_1}\Delta\cdots\Delta\mathcal{B}_{n_{\mathcal{A}},x_s}$ homomorphically represents the above defined automaton $\mathcal{D}$.*

Proof. Let $\mathcal{D}' = (D', X, \delta'')$ denote the state-subautomaton of this diagonal product $\mathcal{B}_{1,x_1}\Delta\cdots\mathcal{B}_{1,x_s}\Delta\cdots\Delta\mathcal{B}_{n_{\mathcal{A}},x_1}\Delta\cdots\Delta\mathcal{B}_{n_{\mathcal{A}},x_s}$ generated by the state $(0, \ldots, 0)$. It is easy to prove that $\psi : B_1^{|X|}\times\cdots\times B_{n_{\mathcal{A}}}^{|X|} \to B$ with $\psi((\delta_{1,x_1}(0, p), \ldots, \delta_{1,x_s}(0, p), \ldots, \delta_{n_{\mathcal{A}},x_1}(0, p), \ldots, \delta_{n_{\mathcal{A}},x_s}(0, p))) = \delta'(\lambda, p)$ is a state homomorphism of $\mathcal{D}'$ onto $\mathcal{D}$. □

Lemma 4.39. *Given an integer $i > 0$, an automaton $\mathcal{A} = (A, X, \delta)$ without any Letichevsky criteria, states $a_0, a \in A$, words $p, q, q' \in X^*$, $|p| \ge i-1$, $|pq| = |pq'| \ge |A| - 1$, let*

$\delta(a_0, p) = a$ and $\delta(a, qr) \neq \delta(a, q'r')$ for every $r, r' \in X^*$, $|r| = |r'|$. Then $\mathcal{B}_{i,x}$, $x \in X$ can be represented homomorphically by a single factor product of $\mathcal{A}$.

Proof. By our assumptions, the state $a \in A$ of the automaton $\mathcal{A} = (A, X, \delta)$ has property (1) of Lemma 4.24. Then we can apply Lemma 4.27 provided that $|pu| \geq i - 1$. In more detail, we can define the automata $\mathcal{B}$ and $\mathcal{C}$ as in Lemma 4.27. Obviously, then for an appropriate nonnegative integer $j \geq |pu|$ and for every $u' \in X^*$, $|u'| = j$, there exists $x \in X$ such that, whenever $i > 1$, for every $x', y_1, \ldots, y_{i-1} \in X, x' \neq x$, the states $(\delta_{\mathcal{B}}(a_0, u'),\ \delta_{\mathcal{C}}(a_0, u'))$, $(\delta_{\mathcal{B}}(a_0, u'y_1),\ \delta_{\mathcal{C}}(a_0, u'y_1)), \ldots, (\delta_{\mathcal{B}}(a_0, u'y_1 \cdots y_{i-1}),\ \delta_{\mathcal{C}}(a_0, u'y_1 \cdots y_{i-1}))$, $(\delta_{\mathcal{B}}(a_0, u'y_1 \cdots y_{i-1}x), \delta_{\mathcal{C}}(a_0, u'y_1 \cdots y_{i-1}x))$ of the diagonal product $\mathcal{B}\Delta\mathcal{C}$ are distinct. Similarly, whenever $i = 1$, for every $x' \in X, x' \neq x$, the states $(\delta_{\mathcal{B}}(a_0, u'), \delta_{\mathcal{C}}(a_0, u'))$, $(\delta_{\mathcal{B}}(a_0,\ u'x), \delta_{\mathcal{C}}(a_0, u'x))$, $(\delta_{\mathcal{B}}(a_0, u'x'), \delta_{\mathcal{C}}(a_0, u'x'))$ of the diagonal product $\mathcal{B}\Delta\mathcal{C}$ are distinct. In addition, $\{(\delta(_{\mathcal{B}}(a_0, z), \delta_{\mathcal{C}}(a_0, z))|z = z_1xz_2, z_1, z_2 \in X^*, |z_1| = i + j - 1\}$ and $\{(\delta(_{\mathcal{B}}(a_0, z), \delta_{\mathcal{C}}(a_0, z))|z{=}z_1x'z_2, x' \in X, x' \neq x, z_1, z_2 \in X^*, |z_1| = i + j - 1\}$ are disjoint sets. Clearly, then the mapping ψ with $\psi((\delta_{\mathcal{B}}(a_0, u'), \delta_{\mathcal{C}}(a_0, u'))) = 0$, $\psi((\delta_{\mathcal{B}}(a_0, u'v), \delta_{\mathcal{C}}(a_0, u'v))) = m$, $v \in X^*$, $|v| = m$, $m = 1, \ldots, i - 1$, $\psi((\delta_{\mathcal{B}}(a_0, u'vxr), \delta_{\mathcal{C}}(a_0, u'vxr))) = x$, $v, r \in X^*$, $|v| = i - 1$, $\psi((\delta_{\mathcal{B}}(a_0, u'vx'r), \delta_{\mathcal{C}}(a_0, u'vx'r))) = *$, $x' \in X, x' \neq x$, $v, r \in X^*$, $|v| = i - 1$, is a state-homomorphism of an appropriate state-subautomaton of $\mathcal{B}\Delta\mathcal{C}$ onto $\mathcal{B}_{i,x}$. □

Lemma 4.40. *Given an integer $i > 0$ an automaton $\mathcal{A} = (A, X, \delta)$ without any Letichevsky criteria, let $\delta(a, q) = \delta(a, q')$ for every $a \in A$, $q, q' \in X^*$ having $|q| = |q'| \geq |A| - 1$ for which there are $a_0 \in A$, $p \in X^*$, q'', q''' with the properties $|p| = i - 1$, $|q''| = |q'''| < |A| - 1$ and $\delta(a_0, p) = a$, $\delta(a, q'') \neq \delta(a, q''')$. Suppose that $|q''| (= |q'''|)$ is maximal with this property. Then $\mathcal{B}_{i,x}$, $x \in X$ can be represented homomorphically by an α_0-ν_1-power of $\mathcal{A}$.*

Proof. By our conditions, the automaton $\mathcal{A}$ has condition (2) of Lemma 4.24. Then we can apply Lemma 4.29 assuming that $|pu| \geq i - 1$. (Then, by our assumptions, there exists a maximal positive integer m_i with $\rho_{\mathcal{A},a_0}(x, i, m_i) = x$. Moreover, $m_i = |q''| = |q'''| > 0$.) Clearly, then the mapping ψ with $\psi((\delta_{\mathcal{B}}(b_0, z), \delta_{\mathcal{C}}(c_0, z))) = |z|, z \in X^*, 0 \leq |z| < i$, $\psi((\delta_{\mathcal{B}}(b_0, pz_1xz_2), \delta_{\mathcal{C}}(c_0, z_1xz_2))) = (x, |z_2| + 1), z_1, z_2 \in X^*, |z_1| = i - 1, 0 \leq |z_2| < m_i$, $\psi((\delta_{\mathcal{B}}(b_0, z_1xz_2), \delta_{\mathcal{C}}(c_0, z_1xz_2))) = *, z_1, z_2 \in X^*, |z_1| = i - 1, |z_2| \geq m_i$, $\psi((\delta_{\mathcal{B}}(b_0, z_1x'z_2), \delta_{\mathcal{C}}(c_0, z_1x'z_2))) = *, x', x' \in X^*, x' \neq x, z_1, z_2 \in X^*, |z_1| = i - 1$, is a state-homomorphism of an appropriate state-subautomaton of $\mathcal{B}\Delta\mathcal{C}$ onto $\mathcal{B}_{i,x}$. □

For every class $\mathcal{K}$ of automata without any Letichevsky criteria, put $M_{\mathcal{K}} = \{1, \ldots, n_{\mathcal{B}}\}$ if there exists a $\mathcal{B} \in \mathcal{K}$ such that for every $\mathcal{A} \in \mathcal{K}$, $n_{\mathcal{B}} \geq n_{\mathcal{A}}$. Otherwise, let $M_{\mathcal{K}}$ be the set of all positive integers.

Lemma 4.41. *Consider a class $\mathcal{K}$ of automata without any Letichevsky criteria. Suppose that for every $\mathcal{A}' = (A', X', \delta') \in \mathcal{K}$, $a' \in A'$, $y_1, y_2 \in X'$, $p', q, q' \in X'^*$, $k \geq 0$, with $k \leq |p'|$, $|p'y_1q| = |p'y_2q'| < |A| - 1$, $\delta'(a', p'y_1q) \neq \delta'(a', p'y_2q')$, there exist $\mathcal{A} = (A, X, \delta) \in \mathcal{K}$, $a \in A$, $x_1, x_2 \in X$, $p, r, r' \in X^*$ having $|p| = k$, $|r| = |r'| = |q| (= |q'|)$ such that $\delta(a, px_1z) \neq \delta(a, px_2z')$ for all prefixes z of r and z' of r' with $|z| = |z'|$. Then for every pair $i \in M_{\mathcal{K}}$, $x \in X$, $\mathcal{B}_{i,x}$ can be represented homomorphically by a single factor product of $\mathcal{A}$.*

Proof. By our assumptions, we can apply Lemma 4.30. (Then, by our assumptions, there exists a maximal positive integer m_i such that $\rho_{\mathcal{A},a_0}(x, i, m_i) = x$. Moreover, $m_i = |q = |q'| > 0$.) Clearly, then the mapping ψ with $\psi((\delta_{\mathcal{B}}(a, z), \delta_{\mathcal{C}}(a, z))) = |z|, z \in X^*$, $0 \leq |z| < i$, $\psi((\delta_{\mathcal{B}}(a, z_1xz_2), \delta_{\mathcal{C}}(a, z_1xz_2))) = (x, |z_2|+1), z_1, z_2 \in X^*, |z_1| = i-1, 0 \leq |z_2| < m_i$, $\psi((\delta_{\mathcal{B}}(a, z_1xz_2), \delta_{\mathcal{C}}(a, z_1xz_2))) = *, z_1, z_2 \in X^*, |z_1| = i-1, |z_2| \geq m_i$, $\psi((\delta_{\mathcal{B}}(a, z_1x'z_2), \delta_{\mathcal{C}}(a, z_1x'z_2))) = *, x' \in X, x' \neq x, z_1, z_2 \in X^*, |z_1| = i-1$, is a state-homomorphism of a state-subautomaton of $\mathcal{B}\Delta\mathcal{C}$ onto $\mathcal{B}_{i,x}$. □

Lemma 4.42. *Let $\mathcal{K}$ be a class of automata without any Letichevsky criteria and assume that there exist integers $k, k' \geq 0$ such that $|p| \neq k$ whenever $\mathcal{A} = (A, X, \delta) \in \mathcal{K}, a \in A, x_1, x_2 \in X, p, r, r' \in X^*$ having $|r| = |r'| = k'$ such that $\delta(a, px_1z) \neq \delta(a, px_2z')$ for all prefixes z of r and z' of r' with $|z| = |z'|$. Then all of the q^ℓ-products of factors in $\mathcal{K}$ preserve this property.*

Proof. Consider an automaton $\mathcal{A} = (A, X, \delta) \in \mathcal{K}$. By Lemma 4.24, for every $a \in A$ we have one of the following two possibilities:

(1) There exist $q, q' \in X^*, |q| = |q'| \geq |A| - 1$ such that $\delta(a, qr) \neq \delta(a, q'r')$ for every $r, r' \in X^*, |r| = |r'|$.
(2) $\delta(a, q) = \delta(a, q')$ for every $q, q' \in X^*, |q| = |q'| \geq |A| - 1$.

By our conditions, we have $|p| \neq k$ in both cases whenever $x_1, x_2 \in X, p, r, r' \in X^*$ with $|r| = |r'| = k'$ such that $\delta(a, px_1z) \neq \delta(a, px_2z')$ for all prefixes z of r and z' of r' with $|z| = |z'|$. By the above properties (1) and (2), it is clear that all single-factor products of $\mathcal{A}$ preserve this property. On the other hand, it is also clear that a diagonal product of automata has the conditions of our statement if all of its factors have it. By Proposition 2.57, this completes the proof. □

Lemma 4.43. *Let $\mathcal{K}$ be a class of automata without any Letichevsky criteria and assume that there exist integers $k, \ell \geq 0$ such that for appropriate $p, r, r' \in X^*$ with $|p| = k, |r| = |r'| = \ell$ we have that there are an automaton $\mathcal{A} = (A, X, \delta) \in \mathcal{K}$, state $a \in A$, input letters $x_1, x_2 \in X$, such that $\delta(a, px_1z) \neq \delta(a, px_2z')$ for all prefixes z of r and z' of r' with $|z| = |z'|$. Then for every integer k' with $k' < k$ there are $p, r, r' \in X^*$ with $|p| = k', |r| = |r'| = \ell + k - k'$ such that an appropriate α_0-ν_1-power $\mathcal{A} = (A, X, \delta)$ of an automaton in $\mathcal{K}$ also will have a state $a \in A$ having the above property.*

Proof. Consider an automaton $\mathcal{A} = (A, X, \delta) \in \mathcal{K}$. By Lemma 4.24, for every $a \in A$ we have one of the following two possibilities:

(1) There exist $q, q' \in X^*, |q| = |q'| \geq |A| - 1$ such that $\delta(a, qr) \neq \delta(a, q'r')$ for every $r, r' \in X^*, |r| = |r'|$.
(2) $\delta(a, q) = \delta(a, q')$ for every $q, q' \in X^*, |q| = |q'| \geq |A| - 1$.

Suppose that (1) holds for some $\mathcal{A} = (A, X, \delta) \in \mathcal{K}, a \in A$, with a long enough q (and q') such that there are $p, r, r' \in X^*, x_1, x_2 \in X$ having $|p| \geq k, |r| = |r'| = \ell + k - k'$, px_1r is a prefix of q, and px_2r' is a prefix of q' for which $\delta(a, px_1z) \neq \delta(a, px_2z')$ for all prefixes z of r and z' of r' with $|z| = |z'|$. Then the automaton $\mathcal{A}$ with its state $a \in A$ also has the properties of our statement.

Otherwise we have to assume that (1) does not hold for any $\mathcal{A} = (A, X, \delta) \in \mathcal{K}$, $a \in A$, q, q' and $p, r, r' \in X^*$, $x_1, x_2 \in X$ having $|p| \geq k$, $|r| = |r'|$ with the above properties.

In this case we should have an automaton $\mathcal{A} = (A, X, \delta) \in \mathcal{K}$, a state $a \in A$ such that (2) holds; moreover, there are $p, r, r' \in X^*$, $x_1, x_2 \in X$ having $|p| \geq k$, $|r| = |r'| = \ell$ for which $\delta(a, px_1z) \neq \delta(a, px_2z')$ for all prefixes z of r and z' of r' with $|z| = |z'|$. Then, similar to the proof of Lemma 4.29, we can construct a diagonal product of two α_0-ν_1-powers of $\mathcal{A}$ such that it has the conditions of our statement.

Consider $y_1, \ldots, y_{|px_1z|}, y'_{|p|+1}, \ldots, y'_{|px_1z|} \in X$ such that, in order, $p = y_1 \cdots y_{|p|}$, $x_1 = y_{|p|+1}$, $x_2 = y'_{|p|+1}$, $z = y_{|p|+2} \cdots y_{|px_1z|}$, $z' = y'_{|p|+2} \cdots y'_{|px_1z|}$. Moreover, consider words $p_m, v_n, v'_n \in X^*$, where, in order, p_m is a prefix of p of length m, $1 \leq m \leq |p|$, r_n is a prefix of r of length n, and r'_n is a prefix of r' of length n, $1 \leq n \leq |r| (= |r'|)$. In addition, let $x' \in X$ be an arbitrary fixed-input letter.

By our assumptions, without any contradiction, for every nonnegative integer $i \leq |p|$ we can define the functions $\varphi_1 : X \to X$, $\varphi_j : A \times X \to X$, $\varphi'_1 : X \to X$, $\varphi'_j : A \times X \to X$, $j = 2, \ldots, |p| - i$, having the following properties:

$$\varphi_1(x) = \begin{cases} x & \text{if } x = x_1, \\ x' & \text{otherwise,} \end{cases}$$

$$\varphi_j(d, x) = \begin{cases} y_{m+1} & \text{if } \delta(a, p_m) = d, 0 \leq m < |p|, \\ & \text{or } \delta(a, p) = d, m = |p|, x = y_{m+1} (= x_1), \\ & \text{or } \delta(a, px_1) = d, m = |p| + 1, x = y_{m+1} (= x_1), \\ & \text{or } \delta(a, px_1r_n) = d, 0 < n < |r|, m = |p| + n + 1, \\ y'_{m+1} & \text{if } \delta(a, p) = d, m = |p|, x \neq y_{m+1} (= x_1), \\ & \text{or } \delta(a, px_2) = d, m = |p| + 1, \\ & \text{or } \delta(a, px_2r'_n) = d, 0 < n < |r'|, m = |p| + n + 1, \end{cases}$$

$$\varphi'_1(x) = x_1,$$

$$\varphi'_j(d, x) = \begin{cases} y_{m+1} & \text{if } \delta(a, p_m) = d, 0 \leq m < |p|, \\ & \text{or } \delta(a, px_1) = d, m = |p| + 1, \\ & \text{or } \delta(a, px_1r_n) = d, 0 < n < |r|, m = |p| + n + 1. \end{cases}$$

Let $\mathcal{B} = (B, X, \delta_{\mathcal{B}}) = \mathcal{A}^{|p|-i+1}(X, \varphi_1, \ldots, \varphi_{|p|-i+1})$ and $\mathcal{C} = (C, X, \delta_{\mathcal{C}}) = \mathcal{A}^{|p|-i+1}$ $(X, \varphi'_1, \ldots, \varphi'_{|p|-i+1})$ with $b = (\delta(a, p_{|p|-i}), \delta(a, p_{|p|-i-1}), \ldots, \delta(a, p_1), a)$, $c = b$.

By Proposition 4.16, $\delta(a, p') \notin \{\delta(a, p''), \delta(a, px_1z), \delta(a, px_2z) \mid z \in X^*\}$ holds for every pair of distinct prefixes p', p'' of p. But then, of course, for every $z, z' \in X^*$ with $|z| < |z'| \leq |px_1r|$, we have $\delta_{\mathcal{B}}(b, z) \neq \delta_{\mathcal{B}}(b, z')$. On the other hand, by Lemma 4.28, we may also assume $\delta(a, px_1w) \neq \delta(a, px_2w')$ for all prefixes w of r and w' of r' with $|w| = |w'|$. Therefore, $(\delta_{\mathcal{B}}(b, z), \delta_{\mathcal{C}}(c, z)) \neq (\delta_{\mathcal{B}}(b, z'), \delta_{\mathcal{C}}(c, z'))$ if z and z' are arbitrary words with $|z| < |z'| \leq |px_1r|$[31] or $z = z_1x_1z_2$ and $z' = z'_1xz'_2$, $|z_1| = |z'_1| = |p|$, $x \in X, x \neq x_1$, $0 < |z| = |z'|$, where z_2 is a prefix of r and z'_2 is a prefix of r'.[32]

Then we obtain that for every $z, z' \in X^*$ with $|z| < |z'| \leq |px_1r|$ we have $(\delta_{\mathcal{B}}(b, z),$ $\delta_{\mathcal{C}}(c, z)) \neq (\delta_{\mathcal{B}}(b, z'), \delta_{\mathcal{C}}(c, z'))$; moreover $\{(\delta_{\mathcal{B}}(b, z), \delta_{\mathcal{C}}(c, z)) | z \in X^*, |z| < i\}$, $\{(\delta_{\mathcal{B}}$

[31] Then $\delta_{\mathcal{C}}(b, z) \neq \delta_{\mathcal{C}}(b, z')$.

[32] Then $\delta_{\mathcal{B}}(c, z) \neq \delta_{\mathcal{B}}(c, z')$.

$(b, z), \delta_{\mathcal{C}}(c, z))|z = z_1x_1z_2, |z_1| = i-1, |z_2| < |px_1r| - i + 1\}$ and $\{(\delta_{\mathcal{B}}(b, z), \delta_{\mathcal{C}}(c, z))|z \in X^*, z = z_1xz_2, |z_1| = i-1, x' \in X, x' \neq x_1, |z_2| < |px_1r| - i + 1\}$ are pairwise disjoint sets. Then we have our statement for $k' = |p| - i$.

The proof is complete. □

Now we are ready to prove the next theorem.

Theorem 4.44. *Let $\mathcal{K}$ be a class of automata without any Letichevsky criteria. The following statements are equivalent:*

(1) *Every general product of factors from $\mathcal{K}$ can also be represented homomorphically by a q^{ℓ}-product of the factors from $\mathcal{K}$.*

(2) *For every $\mathcal{A}' = (A', X', \delta') \in \mathcal{K}$, $a' \in A'$, $y_1, y_2 \in X'$, $p', q, q' \in X'^*$, $k \geq 0$, with $k \leq |p'|$, $|p'y_1q| = |p'y_2q'| < |A| - 1$, $\delta'(a', p'y_1q) \neq \delta'(a', p'y_2q')$, there exist $\mathcal{A} = (A, X, \delta) \in \mathcal{K}$, $a \in A$, $x_1, x_2 \in X$, $p, r, r' \in X^*$ having $|p| = k$, $|r| = |r'| = |p'| + |q| - k (= |p'| + |q'| - k)$ such that $\delta(a, px_1z) \neq \delta(a, px_2z')$ for all prefixes z of r and z' of r' having $|z| = |z'|$.*

Proof. First we assume that we have condition (2) of our statement and let $\mathcal{M} = \mathcal{A}_1 \times \cdots \times \mathcal{A}_n(X, \varphi_1, \ldots, \varphi_n)$ be a product of automata $\mathcal{A}_t = (A_t, X_t, \delta_t)$, $t = 1, \ldots, n$ (such that each of $\mathcal{A}_t$, $t = 1, \ldots, n$, is without any Letichevsky criteria). Next we prove that $\mathcal{M}$ can be represented homomorphically by a diagonal product $\mathcal{N}$ of automata such that each of its factors is a single-factor product of an appropriate factor of $\mathcal{M}$. By Proposition 4.22, we can restrict ourselves to proving our statement to the connected state-subautomata of $\mathcal{M}$. On the other hand, by Proposition 2.75, $\mathcal{M}$ and thus all of its connected state-subautomata of $\mathcal{M}$ preserve the property that they are without any Letichevsky criteria.

Let $\mathcal{A} = (A, X, \delta)$ be a connected state-subautomaton of $\mathcal{M}$ without any Letichevsky criteria. If $\mathcal{A}$ is autonomous, then we are done by Corollary 4.33. Then we may assume that $\mathcal{A}$ is nonautonomous and thus $n_{\mathcal{A}} > 0$. Let us denote again by a_0 a state for which the (nonautonomous) automaton $\mathcal{A}$ is connected; moreover, let $a_0 = (a_{0,1}, \ldots, a_{0,n})$, $a_{0,t} \in A_t$, $t = 1, \ldots, n$. Consider a complete list $p_1, \ldots, p_u$ of all words of length of $n_{\mathcal{A}}$ in X^*. By Lemma 4.37, $\mathcal{D}\Delta\mathcal{A}_{a_0,p_1}\Delta \cdots \Delta\mathcal{A}_{a_0,p_u}$ homomorphically represents $\mathcal{A}$. By Lemma 4.34 and Proposition 4.32, each of the automata $\mathcal{A}_{a_0,p_i}$, $i = 1, \ldots, u$, can be represented homomorphically by a diagonal product of a single-factor product of the factors of $\mathcal{A}$. It remains to prove by (1) that $\mathcal{D}$ can be represented homomorphically by a q^{ℓ}-product of factors from $\mathcal{K}$. But it is a direct consequence of Lemmas 4.38 and 4.41.

Now we assume that (2) does not hold. This implies that there exist $\mathcal{A}' = (A', X', \delta') \in \mathcal{K}$, $a' \in A'$, $y_1, y_2 \in X'$, $p', q, q' \in X'^*$, $k \geq 0$ with $k \leq |p'|$, $|p'y_1q| = |p'y_2q'| < |A| - 1$, $\delta'(a', p'y_1q) \neq \delta'(a', p'y_2q')$ such that for every $\mathcal{A} = (A, X, \delta) \in \mathcal{K}$, $a \in A$, $x_1, x_2 \in X$, $p, r, r' \in X^*$, $|r| = |r'| = |p'| + |q| - k (= |p'| + |q'| - k)$ having $\delta(a, px_1z) \neq \delta(a, p'x_2z')$ for every prefix z of r and z' of r' with $|z| = |z'|$, we have $|p| \neq k$. By Lemma 4.42, we obtain that every q^{ℓ}-product of these automata in $\mathcal{K}$ preserves this property. But, by Lemma 4.43 it can be shown that the α_0-ν_1-product does not preserve this property. Thus we obtain this fact for the general product, too. This ends the proof. □

Next we prove the following proposition.

Proposition 4.45. *There exists an $\alpha_0 - \nu_1$-power $\mathcal{M}$ of an automaton $\mathcal{A}$ without any Letichevsky criteria such that $\mathcal{M}$ cannot be represented homomorphically by any q^{ℓ}-product of $\mathcal{A}$.*

Proof. Let $\mathcal{A} = (A, X, \delta)$ with $A = \{a_0, a_1, a_2, a_3, a_4\}$, $\{x_1, x_2\}$,

$$\delta(a_i, x_j) = \begin{cases} a_1 & \text{if } i = 0, \\ a_2 & \text{if } i = 1, j = 1, \\ a_3 & \text{if } i = 1, j = 2, \\ a_4 & \text{otherwise.} \end{cases}$$

Moreover, let $\mathcal{B} = (B, X, \delta')$ with $A = \{b_0, b_1, b_2, b_3, b_4, b_5\}$, $\{x_1, x_2\}$,

$$\delta'(b_i, x_j) = \begin{cases} b_1 & \text{if } i = 0, j = 1, \\ b_2 & \text{if } i = 0, j = 2, \\ b_3 & \text{if } i = 1, \\ b_4 & \text{if } i = 2, \\ b_5 & \text{otherwise.} \end{cases}$$

Define the functions $\varphi : A \times X \to X$, $\varphi' : A \times A \times X \to X$ in the following way:

$$\varphi(a, x) = x,$$

$$\varphi'(a_i, (a_j, x)) = \begin{cases} x_1 & \text{if } i = 2, \\ x_2 & \text{if } i = 3, \\ x & \text{otherwise.} \end{cases}$$

In addition, give $\psi : \{(a_1, a_0), (a_2, a_1), (a_3, a_1), (a_4, a_2), (a_4, a_3), (a_4, a_4)\} \to B$ as follows:

$$\psi(a_i, a_j) = \begin{cases} b_0 & \text{if } i = 1, j = 0, \\ b_1 & \text{if } i = 2, j = 1, \\ b_2 & \text{if } i = 3, j = 1, \\ b_3 & \text{if } i = 4, j = 2, \\ b_4 & \text{if } i = 4, j = 3, \\ b_5 & \text{if } i = 4, j = 4. \end{cases}$$

Obviously, ψ is a state-isomorphism of a state-subautomaton of the α_0-ν_1-power $\mathcal{A} \times \mathcal{A}(X, \varphi, \varphi')$ onto $\mathcal{B}$. On the other hand, by an elementary computation, we obtain $\delta_{\mathcal{D}}(d, y_1 y_1) = \delta_{\mathcal{D}}(d, y_2 y_1)$ for every diagonal product $\mathcal{D} = (D, Y, \delta_{\mathcal{D}})$ of arbitrary single-factor products of $\mathcal{A}$, state $d \in D$, and input letters $y_1, y_2 \in Y$. Therefore, $\mathcal{B}$ cannot be represented homomorphically by any q^{ℓ}-power of $\mathcal{A}$. Thus the α_0-ν_1-power $\mathcal{A} \times \mathcal{A}(X, \varphi, \varphi')$ also cannot be represented homomorphically by any q^{ℓ}-power of $\mathcal{A}$. □

We will also use the next statement.

Lemma 4.46. *If $\mathcal{A}$ is an automaton without any Letichevsky criteria, then every single-factor product of $\mathcal{A}$ can be represented homomorphically by an α_0-ν_1-power of $\mathcal{A}$.*

Proof. First we suppose that $\mathcal{A}$ is connected. Let us consider a single-factor product $\mathcal{A}(X, \varphi)$ of the automaton $\mathcal{A} = (A, X, \delta)$ and construct the α_0-ν_1-power $\mathcal{M} = \mathcal{A}^{n_{\mathcal{A}}}(X, \varphi_1, \dots, \varphi_{n_{\mathcal{A}}})$

of $\mathcal{A}$ in the following way. For every $(a_1, \ldots, a_{n_\mathcal{A}}) \in A^{n_\mathcal{A}}$, let $\varphi_j(a_1, \ldots, a_{n_\mathcal{A}}, x) = \varphi(a_{j-1}, x)$; moreover, let $\varphi_1(a_1, \ldots, a_{n_\mathcal{A}}, x)$ be an arbitrary fixed-element $x' \in X$. Denote by $a_0 \in A$ a state for which $\mathcal{A}$ is connected. Let $\mathcal{B} = (B, X, \delta')$ be a state-subautomaton of the α_0-ν_1-power $\mathcal{M}$ of $\mathcal{A}$ generated by its state $(a_0, \ldots, a_0)$. Then for every $z_1 \in X$, $\delta'((a_0, \ldots, a_0), z_1) = (a', a, \ldots, a)$ with $a' = \delta(a_0, x')$, $a = \delta(a_0, \varphi(a_0, z_1))$, and for every $z_1, \ldots, z_{n_\mathcal{A}} \in X, i = 1, \ldots, n_\mathcal{A}$, $\delta'((a_0, \ldots, a_0), z_1 \cdots z_i) = (a_1, \ldots, a_{i-1}, a, \ldots, a), i = 2, \ldots, n_\mathcal{A}$, for some $a_1, \ldots, a_{i-1} \in A$ such that $a = \delta(a_0, \varphi(a_0, z_1 \cdots z_i))$. On the other hand, $\delta(a_0, \varphi(a_0, p))$ generates an autonomous state-subautomaton in $\mathcal{A}(X, \varphi)$ if $|p| \geq n_\mathcal{A}$. Thus, for every $p, q, q' \in X^*$ with $|p| = n_\mathcal{A}$ and $|q| = |q'|$, $\delta(a_0, \varphi(a_0, pq)) = \delta(a_0, \varphi(a_0, pq'))$. Thus, we obtain that $\psi : B \to A$ with $\psi((a_1, \ldots, a_{n_\mathcal{A}})) = a_{n_\mathcal{A}}$ is a state-homomorphism of $\mathcal{B}$ onto $\mathcal{A}$.

If $\mathcal{A}$ is not connected, then considering Proposition 4.22, we may assume that it can be represented homomorphically by a diagonal product $\mathcal{M}'$ of its connected state-subautomata and an autonomous automaton $\mathcal{M}$ which is a q-product of $\mathcal{A}$ with a single factor. We have already proved that all connected state-subautomata of $\mathcal{A}$ can be represented homomorphically by an α_0-ν_1-power of $\mathcal{A}$. Obviously, the direct product of $\mathcal{M}$ and these α_0-ν_1-powers is an α_0-ν_1-power of $\mathcal{A}$ which homomorphically represents the diagonal product $\mathcal{M}'$. By the transitive property of homomorphic representation, this completes the proof. □

Now we are ready to prove the following result.

Theorem 4.47. *Let $\mathcal{K}$ be a class of automata without any Letichevsky criteria. Then every general product of factors from $\mathcal{K}$ can be represented homomorphically by an α_0-ν_1-product of the same factors.*

Proof. Let $\mathcal{M} = \mathcal{A}_1 \times \cdots \times \mathcal{A}_n(X, \varphi_1, \ldots, \varphi_n)$ again be a product of automata $\mathcal{A}_t = (A_t, X_t, \delta_t), t = 1, \ldots, n$, (such that each of $\mathcal{A}_t, t = 1, \ldots, n$, is without Letichevsky criteria). Using Lemma 4.46, it is enough to prove that $\mathcal{M}$ can be represented homomorphically by a diagonal product $\mathcal{N}$ of automata such that each of its factors is either a single-factor product of an appropriate factor of $\mathcal{M}$ or an α_0-ν_1-product of certain factors of $\mathcal{M}$.

By Proposition 4.22, we can restrict ourselves to proving our statement for the connected state-subautomata of $\mathcal{M}$. On the other hand, by Proposition 2.75, $\mathcal{M}$ and thus all of its connected state-subautomata of $\mathcal{M}$ preserve the property that they are without any Letichevsky criteria.

Let $\mathcal{A} = (A, X, \delta)$ be a connected state-subautomaton of $\mathcal{M}$ not having any Letichevsky criteria. If $\mathcal{A}$ is autonomous, then we are done by Corollary 4.33. Then we may assume that $\mathcal{A}$ is nonautonomous and thus $n_\mathcal{A} > 0$. Let us denote again by a_0 a state for which the (nonautonomous) automaton $\mathcal{A}$ is connected; moreover, let $a_0 = (a_{0,1}, \ldots, a_{0,n})$, $a_{0,t} \in A_t, t = 1, \ldots, n$. Consider a complete list $p_1, \ldots, p_u$ of all words of length of $n_\mathcal{A}$ in X^*. By Lemma 4.37, $\mathcal{D} \Delta \mathcal{A}_{a_0, p_1} \Delta \cdots \Delta \mathcal{A}_{a_0, p_u}$ homomorphically represents $\mathcal{A}$. By Lemma 4.34 and Proposition 4.32, each of the automata $\mathcal{A}_{a_0, p_i}, i = 1, \ldots, u$, can be represented homomorphically by a diagonal product of a single-factor product of the factors of $\mathcal{M}$. Then it is enough to prove that $\mathcal{D}$ can be represented homomorphically by a diagonal product of α_0-ν_1-products and single-factor products of factors from $\mathcal{M}$.

Let $x_1, \ldots, x_s$ be an arrangement of the elements of the input set X. By Lemma 4.38, the diagonal product $\mathcal{B}_{1,x_1} \Delta \cdots \mathcal{B}_{1,x_s} \Delta \cdots \Delta \mathcal{B}_{n_\mathcal{A},x_1} \Delta \cdots \Delta \mathcal{B}_{n_\mathcal{A},x_s}$ homomorphically represents the automaton $\mathcal{D}$.

To complete our proof we now show that each of $\mathcal{B}_{i,x}, i = 1, \ldots, n_\mathcal{A}, x \in X$, can be represented homomorphically by either a single-factor product or an α_0-ν_1-product of factors from $\{\mathcal{A}_1, \ldots, \mathcal{A}_n\}$. By Lemma 4.24, for every $\mathcal{A}_t = (A_t, X_t, \delta_t) \in \{\mathcal{A}_1, \ldots, \mathcal{A}_n\}$, there are two possibilities:

(1) There exist $a_0, a \in A_t, p, q, q' \in X_t^*, |p| \geq i - 1, |q| = |q'| \geq |A_t| - 1$ such that $\delta_t(a_0, p) = a$ and $\delta_t(a, qr) \neq \delta_t(a, q'r')$ for every $r, r' \in X^*, |r| = |r'|$.

(2) $\delta_t(a, q) = \delta_t(a, q')$ holds for every $a \in A_t, p, q, q' \in X_t^*$ having $\delta_t(a_0, p) = a, p \in X^*, |p| = i - 1, |q| = |q'| \geq |A| - 1$.

Suppose that there is an automaton $\mathcal{A}_t = (A_t, X_t, \delta_t) \in \{\mathcal{A}_1, \ldots, \mathcal{A}_n\}$ having property (1). Then, applying Lemma 4.39, we get $\mathcal{B}_{i,x}$ as a diagonal product of single-factor products of $\mathcal{A}_t$.

In case (2) we can apply Lemma 4.40, assuming that $|pu| \geq i - 1$. (Then, by our assumptions, there exists a maximal positive integer m_i with $\rho_{\mathcal{A},a_0}(x, i, m_i) = x$. Moreover, $m_i = |q - 1| = |q' - 1| > 0$.) □

This result directly implies the following two statements.

Theorem 4.48. *Let $\mathcal{K}$ be a class of automata without any Letichevsky criteria. Then every general product of factors from $\mathcal{K}$ can be represented homomorphically by an α_0-product of the same factors.* □

Theorem 4.49. *Let $\mathcal{K}$ be a class of automata without any Letichevsky criteria. Then every general product of factors from $\mathcal{K}$ can be represented homomorphically by a ν_1-product of the same factors.* □

4.4 Product Hierarchies of Automata

Theorem 4.49 shows that single links already suffice for homomorphically representing any automata network built from automata without any Letichevsky criteria. In contrast, we are going to prove that the α_0-ν_i-hierarchy becomes strict for homomorphic representation if the component automata are permitted to satisfy the semi-Letichevsky criterion as we show in this section. Theorem 4.50, the main result of this section, implies even more:

(i) The ν_i-hierarchy is strict for the homomorphic representation.

(ii) The α_0-ν_i-hierarchy is strict for the homomorphic representation.

(iii) The ν_i-hierarchy is strict for the homomorphic simulation.

(iv) The α_0-ν_i-hierarchy is strict for the homomorphic simulation.

Let $n \geq 1$ be an integer and let $\mathcal{C}_n = (C_n, \{x\}, \delta_n)$ with $C_n = \{1, \ldots, n\}$ and $\delta_n(i, x) = i + 1 \pmod{n}$ for all $i \in C_n$. Thus $\mathcal{C}_n$ is a counter with length n. Let us consider the elevator $\mathcal{E}_2 = (\{0, 1\}, \{x_1, x_2\}, \delta_{\mathcal{E}_2})$ so that $\delta_{\mathcal{E}_2}(0, x_1) = 0$ and $\delta_{\mathcal{E}_2}(0, x_2) = \delta_{\mathcal{E}_2}(1, x_1) = \delta_{\mathcal{E}_2}(1, x_2) = 1$. We set $\mathcal{K} = \{\mathcal{E}_2\} \cup \{\mathcal{C}_p \mid p > 1 \text{ is a prime}\}$ and prove the following.

Theorem 4.50. *Given a fixed positive integer i, there exists an automaton $\mathcal{M}$ which can be homomorphically represented by an α_0-ν_{i+1}-product of automata from $\mathcal{K}$ such that it cannot be simulated homomorphically by any ν_i-product of automata from $\mathcal{K}$.*

Proof. Let i be a fixed positive integer and let m be the product of the first $i+1$ prime numbers. We define $\mathcal{M} = (M, \{x, y\}, \delta')$ with $M = \{1, \dots, m+1\}$ and

$$\delta'(j, x) = \begin{cases} j + 1 \pmod m & \text{if } j \in \{1, \dots, m\}, \\ m + 1 & \text{if } j = m + 1, \end{cases}$$

$$\delta'(j, y) = \begin{cases} j + 1 \pmod m & \text{if } j \in \{2, \dots, m\}, \\ m + 1 & \text{if } j = 1 \text{ or } j = m + 1. \end{cases}$$

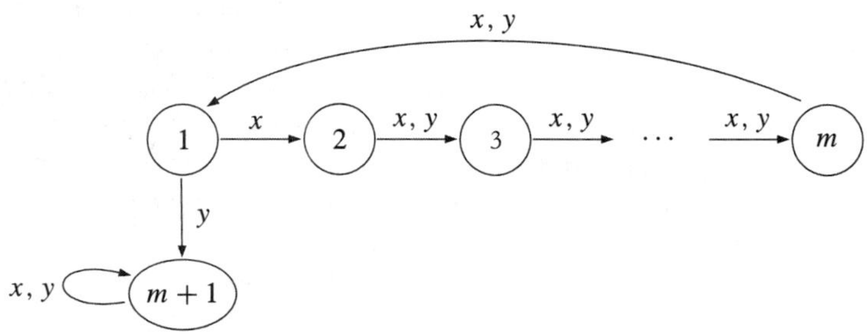

AUTOMATON $\mathcal{M}$

First we prove that $\mathcal{M}$ can be homomorphically represented by an α_0-ν_{i+1}-product of automata from $\mathcal{K}$. For each $j = 1, \dots, i+1$, let p_j denote the jth prime number. We construct an α_0-ν_{i+1}-product $\mathcal{A} = \mathcal{C}_{p_1} \times \cdots \times \mathcal{C}_{p_{i+1}} \times \mathcal{E}_2(\{x, y\}, \varphi_1, \dots, \varphi_{i+2})$ with

$$\varphi_j(k_1, \dots, k_{i+1}, k, z) = \begin{cases} x & \text{if } j \in \{1, \dots, i+1\}, \\ x_2 & \text{if } k_1 = \cdots = k_{i+1} = 1,\ j = i + 2 \text{ and } z = y, \\ x_1 & \text{otherwise.} \end{cases}$$

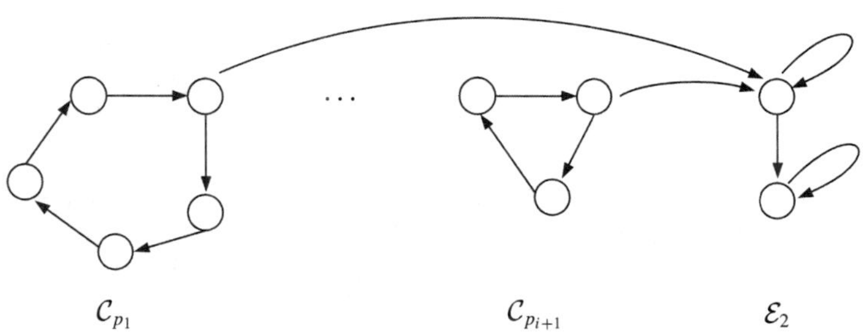

It is straightforward to show that $\mathcal{A}$ maps homomorphically onto $\mathcal{M}$.

Now we show that $\mathcal{M}$ cannot be simulated by a ν_i-product of automata from $\mathcal{K}$. Assume to the contrary that a ν_i-product $\mathcal{A} = (A, X, \delta) = \mathcal{A}_1 \times \cdots \times \mathcal{A}_n(X, \varphi_1, \dots, \varphi_n)$ of automata from $\mathcal{K}$ homomorphically simulates $\mathcal{M}$. We may suppose that n is minimal with this property, i.e., if $\mathcal{B}$ is a ν_i-product of automata from $\mathcal{K}$ which homomorphically simulates $\mathcal{M}$, then the number of factors of $\mathcal{B}$ is at least n. Let $A' \subseteq A$ and let $\tau_1 : A' \to$

M, $\tau_2 : \{x, y\} \to X^*$ be mappings such that τ_1 is onto and $\delta'(\tau_1(a), z) = \tau_1(\delta(a, \tau_2(z)))$ for all $a \in A'$ and $z \in \{x, y\}$, where it is assumed that $\delta(a, \tau_2(z)) \in A'$.

We prove that m and $|\tau_2(x)|$ are relatively prime. Suppose the contrary and let d be an arbitrary prime number which divides m and $|\tau_2(x)|$. Because of the definition of δ' there exists a positive integer k and state $a \in A'$ with $\delta(a, (\tau_2(x))^{km}) = a$ with $\tau_1(a) \in \{1, \dots, m\}$. But then, by the structure of $\mathcal{A}$, $d \mid \tau_2(x)$ implies that there exists a positive integer ℓ such that $\delta(a, (\tau_2(x))^\ell) = a$ and $d \not| \ell$ leading to $\delta'(\tau_1(a), x^\ell) = \tau_1(a)$, $\tau_1(a) \in \{1, \dots, m\}$, a contradiction. Therefore, m and $|\tau_2(x)|$ are relative primes.

We show that m and $|\tau_2(y)|$ are relatively prime. For this we suppose the contrary and let d be a prime number which divides m and $|\tau_2(y)|$. Then $d \not| |\tau_2(x)|$. Therefore, d does not divide $|\tau_2(y)(\tau_2(x))^{m-d}|$.

Because of the structure of $\mathcal{M}$ there exists a positive integer k, a state $a \in A'$ with $m \mid k$, $\tau_1(a) \in \{2, \dots, m\}$, and $\delta(a, (\tau_2(y)(\tau_2(x))^{m-d})^k) = a$. But then, by the structure of $\mathcal{A}$ and $d \mid |\tau_2(y)(\tau_2(x))^{m-d}|$, there exists a positive integer ℓ such that $d \not| \ell$ and $\delta(a, (\tau_2(y)(\tau_2(x))^{m-d})^\ell) = a$, contradicting $\tau_1(\delta(\tau_1(a), (yx^{m-d})^\ell) \neq \tau_1(a)$. (Of course, m and $m - d + 1$ are relative primes. Thus, indeed, $\tau_1(\delta(\tau_1(a), (yx^{m-d})^\ell)) \neq \tau_1(a)$ if $d \not| \ell$.)

Thus we obtain that m and $|(\tau_2(x))^{\tau_2(y)}| = |(\tau_2(y))^{\tau_2(x)}|$ are relatively prime. By this observation, $\mathcal{A}$ also homomorphically simulates $\mathcal{M}$ by the mappings $\tau_1 : A' \to M$ and $\tau_2' : \{x, y\} \to X^*$ with $\tau_2'(x) = (\tau_2(x))^{|\tau_2(y)|}$ and $\tau_2'(y) = (\tau_2(y))^{|\tau_2(x)|}$.

Then we can also assume $|\tau_2(x)| = |\tau_2(y)|$. In addition, we may choose A' such that the cardinality of A' is minimal.

Let us partition A' as $A' = A_0 \cup A_1$, where $A_0 = \tau_1^{-1}(M \setminus \{m + 1\})$ and $A_1 = \tau_1^{-1}(m + 1)$. If $a \in A_0$ and $b \in A'$, then, by the minimality of the cardinality of A, there is a word $u \in \{x, y\}^*$ with $\delta(a, \tau_2(u) = b)$. Therefore, if the jth component of a_0 is equal to 1 and $\mathcal{A}_j = \mathcal{E}_2$ for some $j = 1, \dots, n$ and $a_0 \in A_0$, then the jth component of a is equal to 1 for all $a \in A'$. But then we can get rid of the jth component obtaining a ν_i-product of $n - 1$ factors that homomorphically simulates $\mathcal{M}$. Since this contradicts the minimality of n we have that the jth component of a is necessarily 0 for all $a \in A_0$ and $j \in \{1, \dots, n\}$ with $\mathcal{A}_j = \mathcal{E}_2$.

Now we show that for every $a \in A_1$ there exists $j \in \{1, \dots, n\}$ such that the jth component of a is equal to 1 and $\mathcal{A}_j = \mathcal{E}_2$. Indeed, by the minimality of the cardinality of A', for every $a \in A_1$ $(= \tau^{-1}(m + 1))$ there exists an $a_0 \in A_0$ such that for a suitable $u \in \{x, y\}^*$, $\delta(a_0, \tau_2(u)) = a$. Because of $|\tau_2(x)| = |\tau_2(y)|$, $\delta(a_0, \tau_2(x^{|u|})) = a$ also holds whenever for every $j \in \{1, \dots, n\}$, $\mathcal{A}_j = \mathcal{E}_2$ implies that the jth component of a is 0. But $\delta(a_0, \tau_2(x^{|u|})) \in A_0$, a contradiction.

Now let $a \in \tau_1^{-1}(1)$ be a fixed state. We have $\delta(a, \tau_2(y)) \in A_1$, so that the jth component of the state $\delta(a, \tau_2(y))$ is equal to 1 and $\mathcal{A}_j = \mathcal{E}_2$ for some $j \in \{1, \dots, n\}$. Suppose that φ_j really depends on its state variables having indices $j_1, \dots, j_t$ with $t \le i$. For $s = 1, \dots, t$, define $r_s = p$ if $\mathcal{A}_j = \mathcal{C}_p$ and $r_s = 1$ if $\mathcal{A}_j = \mathcal{E}_2$. Let r be the product of the integers $r_1, \dots, r_n$. It is clear that $m \not| r$. Thus, $\delta(a, (\tau_2(x))^r) \in \tau_1^{-1}(q)$ with $q \in \{2, \dots, m\}$.

But then the j_sth component of a and $\delta(a, (\tau_2(x))^r)$ are equal for all $s = 1, \dots, t$. Therefore, the jth components of $\delta(a, \tau_2(y))$ and $\delta(a, \tau_2(x)\tau_2(y))$ are equal to 1, which contradicts $\delta(a, \tau_2(x)\tau_2(y)) \in A_0$.

This ends the proof. □

Let us summarize some of the results of this chapter for homomorphic representation by automata networks with component automata from a class $\mathcal{K}$ of non-Letichevsky automata. The α_i-product hierarchy collapses at $i = 0$ if $\mathcal{K}$ is without any Letichevsky criteria (Theorem 4.48) and at $i = 1$ if $\mathcal{K}$ satisfies the semi-Letichevsky criterion (Corollary 4.15). That is, in these two cases, no feedback at all, or only feedback of a component's own state to itself, respectively, suffices to achieve computationally what can be achieved with unrestricted feedback. Meanwhile, the ν_i-product hierarchy (bounding the number of incoming links to components by i) collapses at $i = 1$ for automata without any Letichevsky criteria (Theorem 4.49); moreover, even α_0-ν_1-products suffice (Theorem 4.47). Nevertheless, in the semi-Letichevsky case, if no feedback is allowed, then the number of permitted links can determine what can be homomorphically represented or homomorphically simulated.

Taking into consideration that the concept of homomorphic simulation is more general than the concept of homomorphic representation, and, moreover, that the concept of ν_i-product is more general than the concept of α_0-ν_i-product, by Theorem 4.50, we can derive that[33]

(1) the α_0-ν_i-hierarchy is strict for both homomorphic representation and homomorphic simulation, and
(2) the ν_i-hierarchy also has this property.

The product hierarchies for the Letichevsky case are treated in detail in the next two chapters of this monograph (in particular, Theorems 5.27 and 6.15).

4.5 Bibliographical Remarks

Section 4.1. Lemmas 4.2 and 4.3 and Theorem 4.4 are essentially contained in Dömösi and Gécseg [1992]. Corollary 4.15 is a consequence of the Ésik–Horváth characterization theorem due to Z. Ésik and Gy. Horváth [1983]. The other parts of this chapter are essentially new but they can also be derived from the results of Z. Ésik and Gy. Horváth [1983]. Some related results are in Ésik [1983].

Section 4.2. These technical observations on basic properties of automata without any Letichevsky criteria are new but more or less trivial. They are necessary to the description of some results in Section 4.3.

Section 4.3. Theorem 4.47 was obtained by F. Gécseg and H. Jürgensen [1991]. Theorem 4.48 issued from Z. Ésik and Gy. Horváth [1983]. Theorem 4.49 is given in Gécseg and Imreh [1987]. The other results are new.

Section 4.4. This section is based entirely on Dömösi and Ésik [1990] and Dömösi, Ésik, and Imreh [1989]. These results generalize the statements of P. Dömösi and Z. Ésik [1988d]. Some related characterizations are given in Imreh [1977], Gécseg [1986], Dömösi and Ésik [1987], Gécseg and Imreh [1987a, 1987b], Imreh [1988], Dömösi and Gécseg [1989], Dömösi [1990], and Dömösi and Gécseg [1992].

[33] See also the introductory part of this section.

Chapter 5
Letichevsky's Criterion

The application of two-state elements is very often used in engineering technology. We will show that this conventional solution is not the only possibility. We may build digital and electronic circuits using completely different structures. In particular, automata having Letichevsky's criterion and certain types of transformers (which realize feedback functions) can be also considered. The conventional methods make up just one special case where the applied automata are two-state flip-flop automata and the feedback functions are elementary logical units. This fact more or less is well known. But using our results, many other solutions can be derived. A further challenge of research is to determine which ones are important from the point of view of future technologies.

5.1 Homomorphic Simulation and the ν_2-Product

We start with a simple observation.

Lemma 5.1. *Let $\mathcal{K}$ denote a class of automata. Every ν_1-product of factors in $\mathcal{K}$ is an α_0-product of loop products of factors in $\mathcal{K}$.* □

Let $\mathcal{B} = (B, \{x', y'\}, \delta_{\mathcal{B}})$ be the automaton where $B = \{0, 1, 1', 2\}$ and

$$\delta_{\mathcal{B}}(a, x) = \begin{cases} 1 & \text{if } a = 0, x = x', \\ 1' & \text{if } a = 0, x = y', \\ 0 & \text{if } a = 1, x = x' \text{ or } a = 1', x = y', \\ 2 & \text{otherwise.} \end{cases}$$

Moreover, let $\mathcal{B}_0 = (B_0, \{x', y'\}, \delta_{\mathcal{B}_0})$ be the automaton with $B_0 = \{0, 1, 1'\}$ and

$$\delta_{\mathcal{B}_0}(a, x) = \begin{cases} 1 & \text{if } a = 0, x = x', \\ 1' & \text{if } a = 0, x = y', \\ 0 & \text{otherwise.} \end{cases}$$

The following statement is obvious.

Lemma 5.2. *Consider a (general) power* $\mathcal{M} = (B^n, X, \delta)$ *of* $\mathcal{B}$. *If* $a, b \in B^n$ *and* $u \in X^*$ *with* $\delta(a, u) = b$, *and if no component of* b *is* 2, *then no component of* a *is* 2. *Suppose that no component of* b *is* 2. *If* $|u|$ *is even, then for each* $i \in [n]$, *the* i*th component of* a *is* 0 *if and only if the* i*th component of* b *is* 0. *If* $|u|$ *is odd, then for each* $i \in [n]$, *the* i*th component of* a *is* 0 *if and only if the* i*th component of* b *is in* $\{1, 1'\}$. □

Next we prove the following lemma.

Lemma 5.3. *If a strongly connected automaton* $\mathcal{A}$ *is homomorphically simulated by a* ν_2*-power* $\mathcal{B}^n(X, \varphi_1, \ldots, \varphi_n)$, *then* $\mathcal{A}$ *is homomorphically simulated by some* ν_1^ℓ*-power* $\mathcal{B}^n(X, \varphi_1', \ldots, \varphi_n')$.

Proof. Suppose that a ν_2-power $\mathcal{M} = (B^n, X, \delta_{\mathcal{M}}) = \mathcal{B}^n(X, \varphi_1, \ldots, \varphi_n)$ homomorphically simulates the strongly connected automaton $\mathcal{A} = (A, Y, \delta)$ by the mappings $\tau_1 : A' \to A$ $(A' \subseteq B^n)$, $\tau_2 : Y \to X^*$. Let $\mathcal{D} = (V, E)$ denote an appropriate graph for $\mathcal{M}$ such that the in-degree of each vertex is at most 2. Since $\mathcal{A}$ is strongly connected, we may assume that A' is strongly connected in the following sense: For any $a, b \in A'$ there exists a word $u \in Y^*$ with $\delta_{\mathcal{M}}(a, \tau_2(u)) = b$. Define

$$H = \{\delta_{\mathcal{M}}(a, p) \mid a \in A', p \in X^*, \text{ there exists } y \in Y, q \in X^* : \tau_2(y) = pq\}.$$

Then, since A' is strongly connected, the set H is also strongly connected, i.e., for every pair $a, b \in H$ there exists a word $u \in X^*$ with $\delta_{\mathcal{M}}(a, u) = b$. Thus, by Lemma 5.2, if the ith component of some state in H is 2, for some $i \in [n]$, then the ith component of each state in H is 2. Moreover, if for some $i, j \in [n]$ there exists $(a_1, \ldots, a_n) \in H$ such that $a_i = a_j = 0$ or $a_i, a_j \in \{1, 1'\}$, then for all $(b_1, \ldots, b_n) \in H$ it holds that $b_i = b_j = 0$ or $b_i, b_j \in \{1, 1'\}$.

Let i denote any fixed integer with $i \in [n]$. Let $(j, i), (k, i) \in E$ denote the two edges of $\mathcal{D}$ with target i. (Without loss of generality we may assume that there are two such edges.) We show how to define the function φ_i'. We will distinguish several cases. In each case, it will hold that if both $(a_1, \ldots, a_n)$ and $\delta_{\mathcal{M}}((a_1, \ldots, a_n), x)$ are in H, then

$$\delta_{\mathcal{M}}(a_i, \varphi_i'(a_1, \ldots, a_n, x)) = \delta_{\mathcal{M}}(a_i, \varphi_i(a_1, \ldots, a_n, x)).$$

In fact, except for Case 1, we shall even have that the letters $\varphi_i'(a_1, \ldots, a_n, x)$ and $\varphi_i(a_1, \ldots, a_n, x)$ are equal. Moreover, it will be clear that the functions $\varphi_i'(a_1, \ldots, a_n, x)$ depend at most on a_i and one of a_j and a_k. If $i \in \{j, k\}$, then we may take $\varphi_i' = \varphi_i$, so from now on we assume that $i \notin \{j, k\}$.

Case 1. The ith component of some state in H is 2. Then the ith component of each state in H is 2. We define $\varphi_i'(a_1, \ldots, a_n) = x'$ for all $(a_1, \ldots, a_n) \in B^n$.

Case 2. The ith component of no state in H is 2 but there is a state in H whose jth or kth component is 2. Then the jth component of each state in H is 2, or the kth component of each state in H is 2. For each $(a_1, \ldots, a_n) \in B^n$, define $\varphi_i'(a_1, \ldots, a_n, x) = \varphi_i(b_1, \ldots, b_n, x)$, where $(b_1, \ldots, b_n)$ agrees with $(a_1, \ldots, a_n)$ except that $b_j = 2$ if the jth component of each state in H is 2 and $b_k = 2$ is the kth component of each state in H is 2.

Case 3. No state in H has its ith, jth, or kth component equal to 2. We divide this case into three subcases.

Case 3.1. For all $(a_1, \ldots, a_n) \in H$, either $a_i = a_j = 0$ and $a_k \in \{1, 1'\}$ or $a_i, a_j \in \{1, 1'\}$ and $a_k = 0$. Then for $(a_1, \ldots, a_n) \in B^n$, define

$$\varphi_i'(a_1, \ldots, a_n, x) = \begin{cases} x' & \text{if } a_i = 1, \\ y' & \text{if } a_i = 1', \\ \varphi_i(b_1, \ldots, b_n, x) & \text{otherwise,} \end{cases}$$

where $(b_1, \ldots, b_n)$ is obtained from $(a_1, \ldots, a_n)$ by setting its jth component to 0.

Case 3.2. For all $(a_1, \ldots, a_n) \in H$, either $a_i = a_k = 0$ and $a_j \in \{1, 1'\}$ or $a_i, a_k \in \{1, 1'\}$ and $a_j = 0$. This subcase is symmetrical to the previous one. For $(a_1, \ldots, a_n) \in B^n$, define

$$\varphi_i'(a_1, \ldots, a_n, x) = \begin{cases} x' & \text{if } a_i = 1, \\ y' & \text{if } a_i = 1', \\ \varphi_i(b_1, \ldots, b_n, x) & \text{otherwise,} \end{cases}$$

where $(b_1, \ldots, b_n)$ is obtained from $(a_1, \ldots, a_n)$ by setting its kth component to 0.

Case 3.3. For all $(a_1, \ldots, a_n) \in H$, either $a_i = a_j = a_k = 0$ or $a_i, a_j, a_k \in \{1, 1'\}$. For $(a_1, \ldots, a_n) \in B^n$, define

$$\varphi_i'(a_1, \ldots, a_n, x) = \begin{cases} x' & \text{if } a_i = 1, \\ y' & \text{if } a_i = 1', \\ \varphi_i(b_1, \ldots, b_n, x) & \text{otherwise,} \end{cases}$$

where $(b_1, \ldots, b_n)$ is obtained from $(a_1, \ldots, a_n)$ by setting its jth and kth components to 0.

Since we have $\delta_{\mathcal{M}}(a_i, \varphi_i(a_1, \ldots, a_n, x)) = \delta_{\mathcal{M}}(a_i, \varphi_i'(a_1, \ldots, a_n, x))$, for all states $(a_1, \ldots, a_n) \in H$ and $x \in X$ such that $\delta_{\mathcal{M}}((a_1, \ldots, a_n), x) \in H$, it follows that $\mathcal{A}$ is homomorphically simulated by $\mathcal{B}^n(X, \varphi_1', \ldots, \varphi_n')$ using the same mappings $\tau_1 : A' \to A$ $(A' \subseteq B^n)$, $\tau_2 : Y \to X^*$. Moreover, since each function $\varphi_i'(a_1, \ldots, a_n, x)$ depends at most on a_i and one of a_j and a_k, they define a ν_1^ℓ-power of $\mathcal{B}$. □

Lemma 5.4. *If a strongly connected automaton $\mathcal{A}$ is homomorphically simulated by a ν_1^ℓ-power $\mathcal{B}^n(X, \varphi_1, \ldots, \varphi_n)$, then $\mathcal{A}$ is homomorphically simulated by some ν_1-power $\mathcal{B}_0^m(X, \varphi_1', \ldots, \varphi_m')$ with $m \leq n$.*

Proof. Suppose that a ν_1-power $\mathcal{M} = (B^n, X, \delta_{\mathcal{M}}) = \mathcal{B}^n(X, \varphi_1, \ldots, \varphi_n)$ homomorphically simulates a strongly connected automaton $\mathcal{A} = (A, Y, \delta)$ by the mappings $\tau_1 : A' \to A$ $(A' \subseteq B^n)$, $\tau_2 : Y \to X^*$. We need to show that a ν_1-power $\mathcal{B}_0^m(X, \varphi_1', \ldots, \varphi_n')$ for some $m \leq n$ homomorphically simulates $\mathcal{A}$. When $\mathcal{A}$ has a single state, this is clear. Moreover, we may assume that there is no $n' < n$ such that a ν_1^ℓ-power of $\mathcal{B}$ homomorphically simulates $\mathcal{A}$.

Let $H = \{b \in B^n \mid \text{there exists } b' \in A', y \in Y, p, q \in X^* : \tau_2(y) = pq, \delta_{\mathcal{M}}(b', p) = b\}$. Suppose that $(a_1, \ldots, a_n) \in H$. If $a_i = 2$, for some $i \in [n]$, then since $\mathcal{A}$ is strongly connected, $b_i = 2$ for all $(b_1, \ldots, b_n) \in H$. But then $n > 1$ and $\mathcal{A}$ is homomorphically simulated by some ν_1^ℓ-power $\mathcal{B}^{n-1}(X, \varphi_1'', \ldots, \varphi_{n-1}'')$, contrary to our assumptions. Thus, no component of any state in H is 2. For each $i \in [n]$, $(a_1, \ldots, a_n) \in B_0^n$, and $x \in X$, define $\varphi_i'(a_1, \ldots, a_n, x) = \varphi_i(b_1, \ldots, b_n, x)$, where $(b_1, \ldots, b_n)$ is obtained from $(a_1, \ldots, a_n)$ by setting its ith component to 0. Thus, when $a_i = 0$, then $\varphi_i'(a_1, \ldots, a_n, x) = \varphi_i(a_1, \ldots, a_n, x)$, and if $a_i \in \{1, 1'\}$, then $\delta_{\mathcal{B}}(a_i, \varphi_i(a_1, \ldots, a_n, x)) = \delta_{\mathcal{B}_0}(a_i, \varphi_i'(a_1, \ldots, a_n, x))$ whenever $\delta_{\mathcal{B}}(a_i, \varphi_i(a_1, \ldots, a_n, x)) \neq 2$. Since this holds for all $i \in [n]$, it follows that $\mathcal{A}$ is also homomorphically simulated by $\mathcal{B}_0^n(X, \varphi_1', \ldots, \varphi_n')$, completing the proof. □

Lemma 5.5. *Suppose that $\mathcal{M} = \mathcal{B}_0^n(X, \varphi_1, \ldots, \varphi_n)$ is a loop power, so that its underlying graph is a cycle. Suppose that $(a_1, \ldots, a_n), (b_1, \ldots, b_n) \in B_0^n$ such that $a_i = 0$ if and only if $b_i = 0, i \in \{1, \ldots, n\}$. Then for every word $u \in X^*$ whose length is a multiple of n it holds that $a_i = b_i$ implies $\delta_{\mathcal{B}}(a_i, \varphi_i(a_1, \ldots, a_n, u)) = \delta_{\mathcal{B}}(b_i, \varphi_i(b_1, \ldots, b_n, u))$.*

Proof. Let us consider a loop power $\mathcal{M} = (B_0^n, X, \delta_{\mathcal{M}}) = \mathcal{B}_0^n(X, \varphi_1, \ldots, \varphi_n)$ with the underlying graph $\mathcal{D} = (V, \{(i, i+1 \pmod n) \mid i \in V\})$ and a pair $(a_1, \ldots, a_n), (b_1, \ldots, b_n) \in B_0^n$ having our conditions. Suppose that $a_i = b_i$ for some $i \in \{1, \ldots, n\}$. Then $\varphi_{i+1 \pmod n}(a_1, \ldots, a_n, x) = \varphi_{i+1 \pmod n}(b_1, \ldots, b_n, x)$ for every $x \in X$. Therefore, if $a_{i+1 \pmod n} = b_{i+1 \pmod n} = 0$, then the $i + 1 \pmod n$th components of $\delta_{\mathcal{M}}((a_1, \ldots, a_n), x)$ and $\delta_{\mathcal{M}}((b_1, \ldots, b_n), x)$ coincide. On the other hand, by the definition of $\mathcal{B}_0$, if $a_{i+1 \pmod n}, b_{i+1 \pmod n} \in \{1, 1'\}$, then the $i + 1 \pmod n$th components of $\delta_{\mathcal{M}}((a_1, \ldots, a_n), x)$ and $\delta_{\mathcal{M}}((b_1, \ldots, b_n), x)$ are equal to 0. By these observations, it is easy that for every word $u \in X^*$, the $i + |u| \pmod n$th components of $\delta_{\mathcal{M}}((a_1, \ldots, a_n), u)$ and $\delta_{\mathcal{M}}((b_1, \ldots, b_n), u)$ are the same. □

Lemma 5.6. *Let $\mathcal{M} = (M, X, \delta_{\mathcal{M}})$ be a loop power of $\mathcal{B}_0$ having n factors. Consider $(a_1, \ldots, a_n) \in B_0^n$ with either $a_i = a_{i+1 \pmod n} = 0$ or $a_i, a_{i+1 \pmod n} \in \{1, 1'\}$ for some $i \in \{1, \ldots, n\}$. Consider a word $w \in X^*$ of even length. Then $\delta_{\mathcal{M}}((a_1, \ldots, a_n), wz) = \delta_{\mathcal{M}}((a_1, \ldots, a_n), z)$ for every $z \in X^*$ of length n.*

Proof. Consider a loop power $\mathcal{M} = (M, X, \delta_{\mathcal{M}})$ of $\mathcal{B}_0$ having n factors and a state $(a_1, \ldots, a_n) \in B_0^n$ with either $a_i = a_{i+1 \pmod n} = 0$ or $a_i, a_{i+1 \pmod n} \in \{1, 1'\}$ for some $i \in \{1, \ldots, n\}$. To prove our statement we may assume $i = 1$ without any restriction. Let $w \in X^*$ be a word of even length; moreover, let $x_1, \ldots, x_n \in X$. Put $(a_{t,1}, \ldots, a_{t,n}) = \delta_{\mathcal{M}}((a_1, \ldots, a_n), wx_1 \ldots x_t), t = 1, \ldots, n$, and $(a_{0,1}, \ldots, a_{0,n}) = (a_1, \ldots, a_n)$. It is clear that $a_{t-1,1} \in \{1, 1'\}$ implies $a_{t,1} = 0$. In this case $a_{t,1} = \delta(a_1, \varphi_1(a_1, \ldots, a_n, x_{2k+1} \ldots x_t))$ for every $0 \leq k < t/2$. Moreover, $a_{t-1,1} = 0$ with $a_{t-1,n} = 0$ also implies that $a_{t,1} = \delta(a_1, \varphi_1(a_1, \ldots, a_n, x_{2k+1} \ldots x_t))$ if $0 \leq k < t/2$. It is clear that $a_{t-1,2} \in \{1, 1'\}$ implies $a_{t,2} = 0$. This means that $a_{t,2} = \delta(a_2, \varphi_2(a_1, \ldots, a_n, x_{2k+1} \ldots x_t))$ for every $0 \leq k < (t-1)/2$. In addition, we have for $t > 1$ and $0 \leq k < (t-1)/2$ that $a_{t-1,1}$ coincides with $\delta(a_1, \varphi_1(a_1, \ldots, a_n, x_{2k+1} \ldots x_{t-1}))$. Thus $a_{t-1,2} = 0$ also implies that $a_{t,2}$ coincides with $\delta(a_2, \varphi_2(a_1, \ldots, a_n, x_{2k+1} \ldots x_{t-1}x_t))$ for all $0 \leq k < (t-1)/2$.

Then we obtain by an induction that $a_{t,j} = \delta(a_j, \varphi_j(a_1, \ldots, a_n, x_{2k+1} \ldots x_{t-j+1} \ldots x_t)$ coincide for every $1 \leq j \leq t$ and $0 \leq k < (t - j + 1)/2$. This ends the proof. □

Lemma 5.7. *Let $\mathcal{M} = (M, X, \delta_{\mathcal{M}})$ be a loop power of $\mathcal{B}_0$ having n factors. If $\delta_{\mathcal{M}}(a, u^m) = a$ holds for some $a \in B_0^n$, $u \in X^*$, and $m > 0$ such that the length of u is a multiple of n, then $\delta_{\mathcal{M}}(a, u^2) = a$.*

Proof. Let $a = (a_1, \ldots, a_n) \in B_0^n$ and first we suppose that $|u|$ is odd (such that $|u|$ is a multiple of n). Then $\delta_{\mathcal{M}}(a, u^m) = a$ implies that m is even. On the other hand, n divides $|u|$. Therefore, in this case, n is also odd. Thus there is an $i \in \{1, \ldots, n\}$ such that either $a_i = a_{i+1\,(\mathrm{mod}\,n)} = 0$ or $\{a_i, a_{i+1\,(\mathrm{mod}\,n)}\} \in \{1, 1'\}$. Then we can apply Lemma 5.6 with $wz = u^{m-1}$ and $|z| = n$. Therefore, $\delta_{\mathcal{M}}(a, u^{m-1}) = \delta_{\mathcal{M}}(a, z)$. On the other hand, we can apply again Lemma 5.6 with $wz = u$ such that $\delta_{\mathcal{M}}(a, u) = \delta_{\mathcal{M}}(a, z)$. Then, using $\delta_{\mathcal{M}}(a, u^{m-1}) = \delta_{\mathcal{M}}(a, u)$ and $\delta_{\mathcal{M}}(a, u^m) = a$, we obtain $\delta_{\mathcal{M}}(a, u^m) = \delta_{\mathcal{M}}(a, u^2) = a$ if $|u|$ is odd.

Now consider the case when $|u|$ is even. Put $(b_1, \ldots, b_n) = \delta_{\mathcal{M}}(a, u)$ and $(c_1, \ldots, c_n) = \delta_{\mathcal{M}}(a, u^2)$. Of course, for all $i \in \{1, \ldots, n\}$, $b_i = c_i = 0$ whenever $a_i = 0$. Therefore, we are done if $a_i = c_i$ whenever $a_i \in \{1, 1'\}$. Thus let $a_i \neq c_i$ with $a_i \in \{1, 1'\}$ for some $i \in \{1, \ldots, n\}$. But then, by Lemma 5.5, $b_i = c_i$ implies $\delta(b_i, \varphi_i(b_1, \ldots, b_n, u^k)) = \delta(c_i, \varphi_i(c_1, \ldots, c_n, u^k)) = b_i$ for every $k \geq 1$ and thus $\delta_{\mathcal{M}}(a, u^m) \neq a$, a contradiction.

On the other hand, $a_i \neq c_i$ with $a_i \in \{1, 1'\}$ and $b_i \neq c_i$ implies $a_i = b_i$ and then, by Lemma 5.5, $\delta(a_i, \varphi_i(a_1, \ldots, a_n, u^k)) = \delta(b_i, \varphi_i(b_1, \ldots, b_n, u^k))$ for every $k \geq 1$ resulting in $a_i = \delta(a_i, \varphi_i(a_1, \ldots, a_n, u)) = \delta(b_i, \varphi_i(b_1, \ldots, b_n, u)) = c_i$, a contradiction. The proof is complete. □

Now we are ready to show the following statement.

Theorem 5.8. *There exists a singleton class $\mathcal{K}$ of automata which is complete with respect to the homomorphic representation under the general product but not complete with respect to the homomorphic simulation under the ν_2-product.*

Proof. Let $\mathcal{K}$ consist of the single automaton $\mathcal{B}$, defined above. Thus, by Theorem 2.69 we have that $\mathcal{K}$ is complete with respect to the homomorphic representation under the general product.

Consider an arbitrary finite simple group G other than a cyclic group of order 2. Let $\mathcal{A}_G$ denote a strongly connected automaton such that $G < S(\mathcal{A}_G)$ (e.g., $\mathcal{A}_G$ is the automaton (G, G, δ_G) with $\delta_G(g, h) = gh$ for all $g, h \in G$). Suppose that $\mathcal{A}_G$ can be homomorphically simulated by a ν_2-power $\mathcal{B}^n(X, (\varphi_1, \ldots, \varphi_n))$ of the automaton $\mathcal{B}$. Since $\mathcal{A}_G$ is strongly connected, by Lemmas 5.3 and 5.4, $\mathcal{A}_G$ can be homomorphically simulated by a ν_1-power of the automaton $\mathcal{B}_0$ defined above. But by Lemma 5.1, this ν_1-power is an α_0-product of loop powers of $\mathcal{B}_0$. Thus, by Theorem 3.1, there exists a loop power

$$\mathcal{M} = \mathcal{B}_0^k(X', \varphi_1', \ldots, \varphi_k')$$

of $\mathcal{B}_0$ with $G < \mathcal{M}$. But then, by Proposition 2.49, $G || \mathcal{M}$. Therefore, there exist a positive integer m and a subgroup H of $S(\mathcal{M})$ such that G is a homomorphic image of H and each element of H can be induced by a word over X' of length m. Since H is a group, it follows that each element of H can be induced by a word whose length is any multiple of m. In particular, each element of H can be induced by a word of length km. But then, by Lemma 5.7, it follows that the order of each element of H is 1 or 2, contradicting the assumption that G is a simple group of order > 2. □

In Theorem 6.15, we will prove that the above result is sharp. Finally, we note that we can derive the following result as a consequence of Theorem 5.8.

Theorem 5.9. *There exists a singleton class $\mathcal{K}$ of automata which is complete with respect to the homomorphic representation under the general product but not complete with respect to the homomorphic representation under the ν_2-product.* □

5.2 Automata with Control Words

We will create "control words" for any automaton that satisfies Letichevsky's criterion. These will serve as logical signals in nearly all our further constructions.

Let $\mathbf{a} = a_0 \dots a_m$ and $\mathbf{b} = b_0 b_1 \dots b_n$ denote nonempty words over an alphabet A having the following properties:

(1) $a_0 = b_0$, the letters of $\mathbf{a}$ are pairwise distinct, the letters of $\mathbf{b}$ are pairwise distinct, and b_1 does not occur in $\mathbf{a}$.

(2) If $\mathbf{a} = wxy$ and $\mathbf{b} = w'xy'$ for any factorizations with x a letter and w, w' nonempty, then $y = y'$ $(w, w' \in A^+, x \in A, y \in A^*)$.

(3) $m \leq n$ (and $n > 0$). Equivalently, $|\mathbf{a}| \leq |\mathbf{b}|$ (and $|\mathbf{b}| \geq 2$).

Given $\mathbf{a}$ and $\mathbf{b}$ as above, define *control words*, $\mathbf{u} = u_1 \dots u_s$ and $\mathbf{v} = v_1 \dots v_s$:

$$(4)\ u_1 \dots u_s = \begin{cases} {a_0}^{n+1} & \text{if } m = 0, \\ (a_1 \dots a_m a_0)^k & \text{if } m+1 \mid n+1, m \neq 0, n+1 = k(m+1), \\ a_1 \dots a_m a_0 b_1 \dots b_n a_0 & \text{if } m+1 \not\mid n+1. \end{cases}$$

$$(5)\ v_1 \dots v_s = \begin{cases} b_1 \dots b_n a_0 & \text{if } m+1 \mid n+1 \text{ (including the case } m = 0\text{)}, \\ b_1 \dots b_n a_0 a_1 \dots a_m a_0 & \text{if } m+1 \not\mid n+1. \end{cases}$$

The following lemma is obvious from (1) and (2).

Lemma 5.10. *Given control words* $\mathbf{u}$, $\mathbf{v}$, *for all* $1 \leq i, j \leq s-1$ *we have*

(1a) $u_i = u_j \neq a_0$ *implies* $u_{i+1} = u_{j+1}$,

(2a) $v_i = v_j \neq a_0$ *implies* $v_{i+1} = v_{j+1}$,

(3a) $u_i = v_j \neq a_0$ *implies* $u_{i+1} = v_{j+1}$. □

We next show the following lemma.

Lemma 5.11. *Let* $\mathcal{A} = (A, X, \delta)$ *be an automaton satisfying Letichevsky's criterion. There are states* $u_1, \dots, u_s, v_1, \dots, v_s (\in A)$ *and input letters* $x'_1, \dots, x'_s, x''_1, \dots, x''_s (\in X)$ *such that* $\delta(u_t, x'_t) = u_{t+1}, \delta(v_t, x''_t) = v_{t+1} (t = 1, \dots, s-1), \delta(u_s, x'_s) = u_1, \delta(v_s, x''_s) = v_1$. *Moreover,* $\mathbf{u} = u_1 \dots u_s$ *and* $\mathbf{v} = v_1 \dots v_s$ *are control words.*

Proof. Consider an automaton $\mathcal{A} = (A, X, \delta)$ satisfying Letichevsky's criterion; i.e., there are a state $a_0 \in A$, two input letters $x_1, y_1 \in X$, and two input words $p = x_2 \dots x_{m+1}, q = y_2 \dots y_{n+1} \in X^*$ $(x_2, \dots, x_{m+1}, y_2, \dots, y_{n+1} \in X)$, under which $\delta(a_0, x_1) \neq \delta(a_0, y_1)$ and $\delta(a_0, x_1 p) = \delta(a_0, y_1 q) = a_0$. Suppose that p and q have minimal length; i.e., $\delta(a_0, x_1 p_1) = \delta(a_0, y_1 q_1) = a_0$ $(p_1, q_1 \in X^*)$ implies $|p| \leq |p_1|$ and $|q| \leq |q_1|$. Introduce the notation $a_u = \delta(a_0, x_1 \ldots x_u)$ $(u = 1, \ldots, m)$ and $b_v = \delta(a_0, y_1 \ldots y_v)$ $(v = 1, \ldots, n)$. Moreover, we set $b_0 = a_0$, $\mathbf{a} = a_0 \ldots a_m$, and $\mathbf{b} = b_0 \ldots b_n$.

Without loss of generality, we may assume $|p| = m \leq n = |q|$. If p is the empty word $(m = 0)$, then $\delta(a_0, x_1) = a_0$ and q cannot be empty lest $\delta(a_0, y_1) = \delta(a_0, x_1)$. In any case, $n > 0$. This yields condition (3).

If for, every pair i $(= 1, \ldots, m)$, k $(= 1, \ldots, n)$ we have $a_i \neq b_k$, then we get condition (2). (And, of course, b_1 does not occur in $\mathbf{a}$.) By minimality, each of the state words $\mathbf{a} = a_0 \ldots a_m$ and $\mathbf{b} = b_0 \ldots b_n$ then has no repeated states letters. In other words, $a_i \neq a_j$ if $i \neq j$ $(i, j = 0, \ldots, m)$, and $b_k \neq b_\ell$ if $k \neq \ell$ $(k, \ell = 0, \ldots, n)$.

Otherwise, $a_i = b_k$ for some i $(= 1, \ldots, m)$ and k $(= 1, \ldots, n)$. We will take i to be the least such i and k to be least such k for this i. (Observe that $k = 1$ is not possible, for otherwise $|b_0 b_1 a_{i+1} \ldots a_m| = n + 1$ (by minimality), and then $n + 1 = m - i + 2$, whence $m - i + 1 \geq m$, implying $i \leq 1$, but then we would have $a_1 = b_1$, which is not the case.) So none of the states $a_1, \ldots, a_{i-1}$ is the same as any of the states $b_1, \ldots, b_{k-1}$. By minimality, $|x_{i+1} \ldots x_{m+1}| = |y_{k+1} \ldots y_{n+1}|$ since either of these words results in transition from $a_i = b_k$ back to a_0. Thus, we may replace $y_{k+1} \ldots y_{n+1}$ by $x_{i+1} \ldots x_{m+1}$ (or vice versa) to obtain condition (2). Under this replacement, $\mathbf{a}$ and $\mathbf{b}$ are of unchanged minimal length and so of course cannot contain repeated letters. We know $b_1 \notin \{a_0, \ldots a_{i-1}\}$ and $b_1 \notin \{b_k, \ldots b_n\} = \{a_i, \ldots, a_m\}$. Thus, b_1 does not occur in $\mathbf{a}$. Thus, conditions (1) and (2) are established in every case.

Finally, we can define $\mathbf{u} = u_1 \ldots u_s$ and $\mathbf{v} = v_1 \ldots v_s$ as in (4) and (5). The proof is complete. □

Using Lemma 5.10, we now prove the following technical lemma useful in establishing well-definedness of and performance of logical operations with control words and inputs.

Lemma 5.12. For any alphabet X, control words $\mathbf{u}$, $\mathbf{v}$ over an alphabet A, and any mapping $f : \{u_1, v_1\}^2 \times X \to \{u_1, v_1\}$ with $f(u_1, u_1, x) = u_1$ and $f(v_1, v_1, x) = v_1$ $(x \in X)$ there exists a mapping $g : A^3 \times X \to A$ satisfying

$$g(a, w_1, w_2, x) = \begin{cases} f(w_1, w_2, x) & \text{if } a \in \{u_s, v_s\} \ (= \{a_0\}),\ w_1, w_2 \in \{u_1, v_1\}, \\ u_{j+1} & \text{if } a = u_j,\ w_1, w_2 \in \{u_{j+1}, v_{j+1}\},\ j = 1, \ldots, s-1, \\ v_{j+1} & \text{if } a = v_j,\ w_1, w_2 \in \{u_{j+1}, v_{j+1}\},\ j = 1, \ldots, s-1 \end{cases}$$

$((a, w_1, w_2, x) \in A^3 \times X)$.

Proof. Let $g(a, w_1, w_2, x)$ be any fixed element of A whenever $a \in \{u_i, v_i\} \setminus \{a_0\}$ with $\{w_1, w_2\} \not\subseteq \{u_{i+1}, v_{i+1}\}$ $(i = 1, \ldots, s-1)$ or $a \notin \{u_1, \ldots, u_s, v_1, \ldots, v_s\}$. Furthermore, in the case that $w_1, w_2 \in \{u_{j+1}, v_{j+1}\}$ and $a \neq a_0$, set $g(a, w_1, w_2, x) = u_{j+1}$ if $a = u_j$, and $g(a, w_1, w_2, x) = v_{j+1}$ if $a = v_j$. Taking into consideration Lemma 5.10, g is unambiguously determined on $(A \setminus \{a_0\}) \times A^2 \times X$ and has the values given in statement of this lemma.

We still must extend g in a well-defined way to $\{a_0\} \times A^2 \times X$. That is, $a_0 \in \{u_i, v_i\} \cap \{u_j, v_j\}$, $\{w_1, w_2\} \subseteq \{u_{i+1 (\text{mod}\, s)}, v_{i+1 (\text{mod}\, s)}\}$, $\{w_1', w_2'\} \subseteq \{u_{j+1 (\text{mod}\, s)}, v_{j+1 (\text{mod}\, s)}\}$ and $(w_1, w_2) = (w_1', w_2')$ imply $g(a_0, w_1, w_2, x) = g(a_0, w_1', w_2', x)$ $(i, j = 1, \ldots, s, x \in X)$.

We distinguish the following three cases.

Case 1. $m = 0$. We put

$$g(a_0, w_1, w_2, x) = \begin{cases} b_1 & \text{if } \{w_1, w_2\} \subseteq \{u_1, v_1\} (= \{a_0, b_1\}) \text{ and } f(w_1, w_2, x) = b_1, \\ a_0 & \text{otherwise} \end{cases}$$

$(w_1, w_2 \in A, x \in X)$. Then we obtain $a_0 = u_1 = \cdots = u_s$ and $a_0 \notin \{v_1, \ldots, v_{s-1}\}$. On the other hand, $f(u_1, u_1, x) = u_1 (= a_0)$ and $f(v_1, v_1, x) = v_1 (= b_1)$ are supposed. Hence, our assertions hold whenever $a = a_0$ and $\{w_1, w_2\} \neq \{u_1, v_1\} (= \{a_0, b_1\})$. Now we suppose that $\{w_1, w_2\} = \{u_1, v_1\}$ $(= \{a_0, b_1\})$. Then $g(a_0, w_1, w_2, x) = f(w_1, w_2, x)$; moreover, for every $j = 2, \ldots, s$ we have $\{w_1, w_2\} \nsubseteq \{u_j, v_j\}$. Therefore, we have our conditions. This ends the proof of Case 1.

Case 2. $m \neq 0, m + 1 \mid n + 1$, i.e., $n + 1 = k(m + 1)$ for some positive integer k. We set

$$g(a_0, w_1, w_2, x) = \begin{cases} b_1 & \text{if } \{w_1, w_2\} \subseteq \{u_1, v_1\} \text{ and } f(w_1, w_2, x) = b_1, \\ a_1 & \text{otherwise} \end{cases}$$

$(w_1, w_2 \in A, x \in X)$. Then $v_i \neq a_0$ if $i \in \{1, \ldots, s - 1\}$. Moreover, $u_i = a_0$ implies $u_{i+1} = a_1$ for any $i \in \{1, \ldots, s - 1\}$. Therefore, similar to Case 1, we have our assertion if $i, j \in \{1, \ldots, s - 1\}$. If $\{w_1, w_2\} \subseteq \{u_1, v_1\}$ with $(w_1, w_2) \neq (u_1, u_1)$, then $b_1 \in \{w_1, w_2\}$. Hence, in this case, $\{w_1, w_2\} \nsubseteq \{u_z, v_z\}$ if $z > 1$. It remains to study the case of $(w_1, w_2) = (u_1, u_1)$. Then we supposed $f(w_1, w_2, x) = u_1$ $(= a_1)$, corresponding to $g(a_0, a_1, a_1, x) = a_1 (x \in X)$. On the other hand, by $\{w_1', w_2'\} \subseteq \{u_z, v_z\}, z > 1$ and $((a_1, a_1) =)(w_1, w_2) = (w_1', w_2')$, we have $g(a_0, w_1', w_2', x) = u_z$ $(x \in X)$ with $u_z = a_1$, whenever $a_0 \in \{u_{z-1}, v_{z-1}\}$ (or more precisely, whenever $a_0 = u_{z-1}$). This completes the proof of Case 2.

Case 3. $m + 1 \nmid n + 1$. Define

$$g(a_0, w_1, w_2, x) = \begin{cases} a_1 & \text{if } \{w_1, w_2\} \subseteq \{u_{n+2}, v_{n+2}\}, \\ b_1 & \text{if } \{w_1, w_2\} \subseteq \{u_{m+2}, v_{m+2}\}, \\ f(w_1, w_2, x) & \text{if } \{w_1, w_2\} \subseteq \{u_1, v_1\} \end{cases}$$

$(w_1, w_2 \in A, x \in X)$. Then $u_1 = a_1, v_1 = b_1, u_{n+2} = b_{n-m+1}, v_{n+2} = a_1, u_{m+2} = b_1$; furthermore, $v_{m+2} = a_0$ or $v_{m+2} = b_{m+2}$, depending on whether $m + 1 = n$ or $m + 1 < n$.

By property (1) of $a_0 a_1 \ldots a_m$ and $b_0 b_1 \ldots b_m$ (see their definition), we have, in order, $a_0 \notin \{a_1, b_1, b_{n-m+1}\}, a_1 \neq b_1$, and, if $m + 1 < n$, then $b_1 \neq b_{m+2}$. On the other hand, $m + 1 \nmid n + 1$ implies $n \neq 2m + 1$, leading to $b_{m+2} \neq a_1$ (provided $m + 1 < n$) by property (2) of $a_0 \ldots a_m$ and $b_0 b_1 \ldots b_n$ (again, see their definitions). Furthermore, $b_i = b_j$ $(i, j = 0, \ldots, n)$ implies $i = j$ by (1). Therefore, by $m + 1 < n$, $n \neq 2m + 1$ implies $b_{m+2} \neq b_{n-m+1}$, too. Similarly, since $m \leq n$ and $m + 1 \nmid n + 1$, then $m < n$ which, in addition, shows $b_{n-m+1} \neq b_1$. But then $\{a_1, b_1\}, \{a_1, b_{n-m+1}\}, \{a_0, b_1\}$ by $m + 1 = n$ or $\{a_1, b_1\}, \{a_1, b_{n-m+1}\}, \{b_1, b_{m+2}\}$ by $m + 1 < n$ are pairwise different sets. Therefore, if $w_1 \neq w_2$ and $\{w_1, w_2\} \in \{\{u_1, v_1\}, \{u_{m+2}, v_{m+2}\}, \{u_{n+2}, v_{n+2}\}\}$, then our statement is valid, where the appropriate values of $g(a_0, w_1, w_2, x)$ $(x \in X)$ are, in order, $f(w_1, w_2, x), b_1, a_1$. (By the way, $a_1 = b_{n-m+1}$ is possible. In this case, we may leave the set $\{u_{n+2}, v_{n+2}\}$ $= \{a_1, b_{n-m+1}\}$ out of consideration whenever $w_1 \neq w_2$ is assumed.) Finally, if $w_1 = w_2$, then $f(u_1, u_1, x) = u_1$ and $f(v_1, v_1, x) = v_1$ $(x \in X)$ lead to $g(a_0, a_1, a_1, x) =$

$g(a_0, b_{n-m+1}, b_{n-m+1}, x) = a_1$ $(x \in X)$ and $g(a_0, b_1, b_1, x) = g(a_0, a_0, a_0, x) = b_1$ or $g(a_0, b_1, b_1, x) = g(a_0, b_{m+2}, b_{m+2}, x) = b_1$ $(x \in X)$, depending on whether $m+1 = n$ or $m + 1 < n$. In other words, $g(a_0, u_1, u_1, x) = g(a_0, u_{n+2}, u_{n+2}, x) = g(a_0, v_{n+2}, v_{n+2}, x) = u_1 = v_{n+2}$ and $g(a_0, v_1, v_1, x) = g(a_0, u_{m+2}, u_{m+2}, x) = g(a_0, v_{m+2}, v_{m+2}, x) = v_1 = u_{m+2}$. This completes the proof. □

Considering X a singleton, we have the following consequence of Lemma 5.12.

Lemma 5.13. For any mapping $f : \{u_1, v_1\}^2 \to \{u_1, v_1\}$ with $f(u_1, u_1) = u_1$ and $f(v_1, v_1) = v_1$ there exists a mapping $g : A^3 \to A$ satisfying

$$g(a, w_1, w_2) = \begin{cases} f(w_1, w_2) & \text{if } a \in \{u_s, v_s\}\ (= \{a_0\}),\ w_1, w_2 \in \{u_1, v_1\}, \\ u_{j+1} & \text{if } a = u_j,\ w_1, w_2 \in \{u_{j+1}, v_{j+1}\},\ j = 1, \dots, s-1, \\ v_{j+1} & \text{if } a = v_j,\ w_1, w_2 \in \{u_{j+1}, v_{j+1}\},\ j = 1, \dots, s-1 \end{cases}$$

$((a, w_1, w_2) \in A^3)$. □

Lemma 5.13 leads to the following statement.

Lemma 5.14. *There exists a mapping $g : A^2 \to A$ satisfying*

$$g(a, b) = \begin{cases} b & \text{if } a \in \{u_s, v_s\}\ (= \{a_0\}),\ b \in \{u_1, v_1\}, \\ u_{j+1} & \text{if } a = u_j,\ b \in \{u_{j+1}, v_{j+1}\},\ j = 1, \dots, s-1, \\ v_{j+1} & \text{if } a = v_j,\ b \in \{u_{j+1}, v_{j+1}\},\ j = 1, \dots, s-1 \end{cases}$$

$((a, b) \in A^2)$. □

Using Lemma 5.12 and its consequences (Lemmas 5.13 and 5.14) we can conclude the following.

Corollary 5.15. *Given a positive integer n, a pair of alphabets X, A, and control words $\mathbf{u}, \mathbf{v}$ over A, let $f : \{u_1, v_1\}^n \times X \to \{u_1, v_1\}$ be a mapping with $f(u_1, \dots, u_1, x) = u_1$ and $f(v_1, \dots, v_1, x) = v_1$. There exists a mapping $g : A^n \times X \to A$ satisfying*

$$g(a, w_1, \dots, w_n, x) = \begin{cases} f(w_1, \dots w_n, x) & \text{if } a \in \{u_s, v_s\}\ (= \{a_0\}),\ w_1, \dots, w_n \in \\ & \quad \{u_1, v_1\}, \\ u_{j+1} & \text{if } a = u_j,\ w_1, \dots, w_n \in \{u_{j+1}, v_{j+1}\}, \\ & \quad j = 1, \dots, s-1, \\ v_{j+1} & \text{if } a = v_j,\ w_1, \dots, w_n \in \{u_{j+1}, v_{j+1}\}, \\ & \quad j = 1, \dots, s-1 \end{cases}$$

$((a, w_1, \dots, w_n, x) \in A^n \times X)$. □

We shall use the following lemma.

Lemma 5.16. *Let $\mathcal{A} = (A, X, \delta)$ be an automaton satisfying Letichevsky's criterion. There are states $u_1, \dots, u_s, v_1, \dots, v_s$ $(\in A)$ and input letters $x'_1, \dots, x'_s, x''_1, \dots, x''_s (\in X)$ such* *that* $\delta(u_t, x'_t) = u_{t+1}, \delta(v_t, x''_t) = v_{t+1} (t = 1, \ldots, s-1), \delta(u_s, x'_s) = u_1, \delta(v_s, x''_s) = v_1$. *Moreover,* $\mathbf{u} = u_1 \ldots u_s$ *and* $\mathbf{v} = v_1 \ldots v_s$ *are control words.*

We close this section with the following definitions. Let $\mathcal{A} = (A, X, \delta)$ be an automaton satisfying Letichevsky's criterion. Moreover, let $\mathbf{u} = u_1 \ldots u_s, \mathbf{v} = v_1 \ldots v_s (\in A^*)$ be control words as constructed in Lemma 5.11 such that for appropriate input letters $x'_1, \ldots, x'_s, x''_1, \ldots, x''_s (\in X)$ we have $\delta(u_t, x'_t) = u_{t+1}, \delta(v_t, x''_t) = v_{t+1}$ $(t = 1, \ldots, s-1), \delta(u_s, x'_s) = u_1, \delta(v_s, x''_s) = v_1$. For any $a, a', a'' \in A$ and fixed pair $u_1 \ldots u_s, v_1 \ldots v_s$ of control words we shall use the following operations on the alphabet A:

$$x[a, a']$$

$$= \begin{cases} \text{an arbitrary fixed } x \in X \text{ with } \delta(a, x) = a' & \text{if } (a, a') \in \{(u_s, u_1), (u_s, v_1)\}, \\ \text{an arbitrary fixed } x \in X \text{ with } \delta(a, x) = u_{i+1} & \text{if } (a, a') \in \{(u_i, u_{i+1}), (u_i, v_{i+1})\}, \\ & \quad i = 1, \ldots, s-1, \\ \text{an arbitrary fixed } x \in X \text{ with } \delta(a, x) = v_{i+1} & \text{if } (a, a') \in \{(v_i, u_{i+1}), (v_i, v_{i+1})\}, \\ & \quad i = 1, \ldots, s-1, \\ \text{any fixed element of } X & \text{otherwise,} \end{cases}$$

$$x[a, a' \vee a''] = \begin{cases} x[a, a'] & \text{if } (a, a') = (u_s, v_1), \\ x[a, a''] & \text{otherwise,} \end{cases}$$

$$x[a, a' \wedge a''] = \begin{cases} x[a, a'] & \text{if } (a, a') = (u_s, u_1), \\ x[a, a''] & \text{otherwise.} \end{cases}$$

(We remark that in consequence of Lemma 5.14, $x[a, a']$, $x[a, a' \vee a'']$, and $x[a, a' \wedge a'']$ are unambiguously defined.)

Proposition 5.17. *For every automaton $\mathcal{A}$ satisfying Letichevsky's criterion, we can give a single-factor product of $\mathcal{A}$ which is an m-automaton for some positive integer m.*

Proof. Let $A = (A, X, \delta)$ be an automaton satisfying Letichevsky's criterion; moreover, let $\mathbf{u} = u_1 \ldots u_s$ and $\mathbf{v} = v_1 \ldots v_s$ be appropriate control words having the properties given in Lemma 5.16. Define the single-factor product $\mathcal{M} = \mathcal{A}(A, \varphi)$ with $\varphi(a, w) = x[a, g(a, w)]$ $(a, w \in A)$, where $g : A^2 \to A$ is given as in Lemma 5.14. It is clear that $\mathcal{M}$ is an $(s, |A^2|)$-automaton with $X_i = \{u_i, v_i\}$, $i \in \{1, \ldots, s\}$. Thus $\mathcal{M}$ is an m-automaton for $m = s$. The proof is complete. □

Recall that a loop product is a $\mathcal{D}$-product, where $\mathcal{D}$ is a cycle. We now prove the following.

Proposition 5.18. *For every automaton $\mathcal{A}$ satisfying Letichevsky's criterion, we can give a single-factor product $\mathcal{M}$ of $\mathcal{A}$ such that every counter can be represented homomorphically by a loop power of $\mathcal{M}$.*

Proof. Given a counter $\mathcal{C}_\ell$ with ℓ states, let $\mathcal{A} = (A, X, \delta)$ be an automaton satisfying Letichevsky's criterion. Construct the single-factor product $\mathcal{M}$ of $\mathcal{A}$ as in the previous proposition. To prove our statement, we give a loop power of $\mathcal{M}$ which isomorphically represents a counter having $s\ell$ states. Consider the loop power $\mathcal{L} = \mathcal{M}^{s\ell}(\{x\}, \varphi_1, \ldots, \varphi_{s\ell})$ of $\mathcal{M}$ with $\varphi_i(a_1, \ldots, a_{s\ell}, a_{s\ell+1}) = a_{i+1\,(\text{mod}\, s\ell)}$ $((a_1, \ldots, a_{s\ell}) \in A^{s\ell}, a_{s\ell+1} \in A)$ and correspond the state $C(\mathbf{vu}^{\ell}, k)$ $(k \in \{1, \ldots, s\ell\})$ to the integer k. Clearly, then $\mathcal{L}$ has a subautomaton $\mathcal{B}$ with state set $C(\mathbf{vu}^{\ell}, k)$ $(k \in \{1, \ldots, s\ell\})$ such that this correspondence is an isomorphism of $\mathcal{B}$ onto the $s\ell$-state counter. This ends the proof. □

5.3 The Beauty of Letichevsky's Criterion

Recall that an automaton $\mathcal{A}$ *satisfies Letichevsky's criterion* if there are a state $a_0 \in A$, two input letters $x, y \in X$, and two input words $p, q \in X^*$ under which $\delta(a_0, x) \neq \delta(a_0, y)$ and $\delta(a_0, xp) = \delta(a_0, yq) = a_0$. If the class $\mathcal{K}$ of automata contains an automaton satisfying Letichevsky's criterion, then we also say that $\mathcal{K}$ *satisfies Letichevsky's criterion.* Otherwise we say that $\mathcal{K}$ does not satisfy it. This well-known criterion can be used not only for characterization of complete classes with respect to homomorphic representations under the general product but also for description of complete classes with respect to isomorphic and homomorphic simulations.

Under the generalized product, homomorphic representation is quivalent to both homomorphic and isomorphic simulations. The Gluškov product behaves quite differently. In this section we show that, contrary to this fact, a class of automata is complete with respect to homomorphic representations under the Gluškov product if and only if it is complete with respect to both homomorphic and isomorphic simulations.

Therefore, in this sense the Gluškov product behaves similarly to its generalized form.

For every digraph $\mathcal{D} = (V, E)$ with $V = \{1, \ldots, n\}$ and positive integer s, we define the digraph $\mathcal{D}^{[s]} = (V_s, E_s)$ having $V_s = \{1, \ldots, ns\}$, $E_s = \{(i, i-1\,(\text{mod}\, ns)) \mid i \in V_s\} \cup \{((k-1)s+1, \ell s) \mid (k, \ell) \in E\} \cup \{(j, j) \mid j \in V_s\}$.

Lemma 5.19. *Let $\mathcal{A} = (A, X, \delta)$ be an automaton having Letichevsky's criterion with s length control words. Consider an automaton $\mathcal{A}' = (A', X', \delta')$, with $A' = \{1, \ldots, |A'|\}$, and its digraph $D(\mathcal{A}') = (A', E)$ (having $E = \{(i, j) \in A' \times A' \mid$ there exists $x \in X$: $\delta'(i, x) = j\}$). Then $\mathcal{A}'$ can be simulated isomorphically by a $(D(\mathcal{A}'))^{[s]}$-power of $\mathcal{A}$.*

Proof. Let $\mathcal{A} = (A, X, \delta)$ satisfy Letichevsky's criterion; that is, there are a state $a_0 \in A$, two input letters $x, y \in X$, and two input words $p, q \in X^*$ under which $\delta(a_0, x) \neq \delta(a_0, y)$ and $\delta(a_0, xp) = \delta(a_0, yq) = a_0$. Introduce the notation, $x_1 \ldots x_s = xpyq$, $a_k = \delta(a_0, x_1 \ldots x_k)$, and $y_1 \ldots y_s = yqxp$, $b_k = \delta(a_0, y_1 \ldots y_k)$, where $k = 1, \ldots, s$ and $x_1, \ldots, x_s, y_1, \ldots, y_s \in X$. We may assume that $a_i = b_i$ implies $a_{i+1} = b_{i+1}$ and $x_{i+1} = y_{i+1}$ for any $i = 1, \ldots, s-1$. Otherwise we could exchange a_{i+1} with b_{i+1} and x_{i+1} with y_{i+1}. Clearly, then $a_s = b_s = a_0$.

Let $\mathcal{A}' = (A', X', \delta')$ be an automaton with state set $A' = \{1, \ldots, r\}$ for some positive integer r. Moreover, let $\mathcal{D} = (A', E)$ having $E = \{(i, j) \in A' \times A' \mid$ there exists $x \in X$: $\delta'(i, x) = j\}$. We shall prove that $\mathcal{A}$ has a $\mathcal{D}^{[s]}$-power which isomorphically simulates $\mathcal{A}'$. Let $Z_1, \ldots, Z_s$ be distinct finite nonempty sets for which $Z_1 = X'$. Define the power $\mathcal{A}^{rs}$ with respect to $\bigcup_{i=1}^{s} Z_i$ and $(\varphi_1, \ldots, \varphi_{rs})$ such that for any $(d_1, \ldots, d_{rs}) \in A^{rs}$,

$z \in \bigcup_{i=1}^{s} Z_i,\ t = 1, \dots, rs,$

$$\varphi_t(d_1, \dots, d_{rs}, z) = \begin{cases} \text{fixed element of } & \{x \in X \mid \delta(d_t, x) = d_{t+1}\} \\ & \quad \text{if } t \nmid s \text{ and } \{x \in X \mid \delta(d_t, x) = d_{t+1}\} \neq \emptyset, \\ x_1 & \quad \text{if } t = js,\ j \in \{1, \dots, r\}, z \in Z_1,\ d_t = a_0 \text{ and} \\ & \quad \{i \in A' \mid d_{(i-1)s+1} = b_1,\ \delta'(i, z) = j\} = \emptyset, \\ y_1 & \quad \text{if } t = js,\ j \in \{1, \dots, r\}, z \in Z_1,\ d_t = a_0 \text{ and} \\ & \quad \{i \in A' \mid d_{(i-1)s+1} = b_1,\ \delta'(i, z) = j\} \neq \emptyset, \\ x_{\ell+1} & \quad \text{if } t = js,\ j \in \{1, \dots, r\},\ z \in Z_{\ell+1},\ \ell \in \{1, \dots, s-1\}, \\ & \quad d_t = a_\ell, \\ y_{\ell+1} & \quad \text{if } t = js,\ j \in \{1, \dots, r\},\ z \in Z_{\ell+1},\ \ell \in \{1, \dots, s-1\}, \\ & \quad d_t = b_\ell, \\ \text{fixed element of } & X \text{ otherwise.} \end{cases}$$

By the above definition of $\varphi_t, t = 1, \dots, n$, it is easy to show that $\mathcal{A}^{rs}$ forms a $(D(\mathcal{A}'))^{[s]}$-power of $\mathcal{A}$. We shall now show that $\mathcal{A}^{rs}$ isomorphically simulates $\mathcal{A}'$ under suitable mappings $\tau_1 : B \to A'$ and $\tau_2 : X' \to Z_1 Z_2 \cdots Z_s$. Consider a subset B of the state set of $\mathcal{A}^{rs}$ with

$$B = \{((a_1, \dots, a_s)^k, b_1, \dots, b_s, (a_1, \dots, a_s)^{r-k-1}) \mid k = 0, \dots, r-1\}$$

and let δ'' be the transition function of $\mathcal{A}^{rs}$. Let for any $u \in A'$ and $z_1 \in Z_1,\ \delta'(u, z_1) = v$. To avoid technical difficulties, first we suppose that $u > v$. Then, taking into consideration $a_s = b_s = a_0$, for any $i = 1, \dots, r,\ \ell = 1, \dots, s-1$, we have the following equalities:

$$\delta(a_s, \varphi_{is}((a_1, \dots, a_s)^{u-1}, b_1, \dots, b_s,\ (a_1, \dots, a_s)^{r-u}, z_1)) = \begin{cases} b_1 & \text{if } i = v, \\ a_1 & \text{if } i \neq v, \end{cases}$$

$$\delta(a_\ell, \varphi_{(i-1)s+\ell}((a_1, \dots, a_s)^{u-1},\ b_1, \dots, b_s,\ (a_1, \dots, a_s)^{r-u}, z_1)) = a_{\ell+1} \text{ if } i \neq u;$$

moreover,

$$\delta(b_\ell, \varphi_{(u-1)s+\ell}((a_1, \dots, a_s)^{u-1},\ b_1, \dots, b_s,\ (a_1, \dots, a_s)^{r-u}, z_1)) = b_{\ell+1}.$$

Therefore, using the fact that $a_s = b_s = a_0$ and $u > v$,

$$\delta''(((a_1, \dots, a_s)^{u-1},\ b_1, \dots, b_s,\ (a_1, \dots, a_s)^{r-u}), z_1) = ((a_2, \dots, a_s, a_1)^{v-1},$$
$$a_2, \dots, a_s, b_1, (a_2, \dots, a_s, a_1)^{u-v-1},\ b_2, \dots, b_s, a_1, (a_2, \dots, a_s, a_1)^{r-u}).$$

On the other hand, for any $(d_1, \dots, d_{rs}) \in A^{rs},\ t = s \cdot t_1 + t_2,\ t_1 = 0, \dots, r-1,$ $t_2 = 1, \dots, s-1$, and $z_k \in Z_k,\ k = 2, \dots, s$,

$$\delta(d_t, \varphi_t(d_1, \dots, d_{rs}, z_k)) = \begin{cases} a_k & \text{if } d_t = a_{k-1}, d_{t+1} = a_k, \\ b_k & \text{if } d_t = b_{k-1}, d_{t+1} = b_k, \end{cases}$$

and similarly, for any $i = 1, \dots, r$,

$$\delta(d_{is}, \varphi_{is}(d_1, \dots, d_{rs}, z_k)) = \begin{cases} a_k & \text{if } d_{is} = a_{k-1}, \\ b_k & \text{if } d_{is} = b_{k-1}. \end{cases}$$

Thus we obtain

$$\delta''(((a_\ell, \ldots, a_s, a_1, \ldots, a_{\ell-1})^{v-1}, a_\ell, \ldots, a_s, b_1, \ldots, b_{\ell-1}, (a_\ell, \ldots, a_s, a_1, \ldots$$

$$\ldots, a_{\ell-1})^{u-v-1}, b_\ell, \ldots, b_s, a_1, \ldots, a_{\ell-1}, (a_\ell, \ldots, a_s, a_1, \ldots, a_{\ell-1})^{r-u}), z_\ell)$$

$$= ((a_{\ell+1}, \ldots, a_s, a_1, \ldots, a_\ell)^{v-1}, a_{\ell+1}, \ldots, a_s, b_1, \ldots, b_\ell, (a_{\ell+1}, \ldots, a_s, a_1, \ldots$$

$$\ldots, a_\ell)^{u-v-1}, b_{\ell+1}, \ldots, b_s, a_1, \ldots, a_\ell, \ (a_{\ell+1}, \ldots, a_s, a_1, \ldots, a_\ell)^{r-u}).$$

Then we get

$$\delta''(((a_1, \ldots, a_s)^{u-1}, b_1, \ldots, b_s, (a_1, \ldots, a_s)^{r-u}), z_1 \ldots z_s)$$

$$= ((a_1, \ldots, a_s)^{v-1}, b_1, \ldots, b_s, (a_1, \ldots, a_s)^{r-v}).$$

It is easy to check that we also have this equality if $u \leq v$. Consequently, $\mathcal{A}^{rs}$ isomorphically simulates $\mathcal{A}'$ under $\tau_1 : B \to A'$ and $\tau_2 : X' \to Z_1 Z_2 \cdots Z_s$ with $\tau_1(((a_1, \ldots, a_s)^{i-1}, b_1, \ldots, b_s, (a_1, \ldots, a_s)^{r-i})) = i \ (i = 1, \ldots, r)$, $\tau_2(x) = x z_2 \ldots z_s \quad (x \in X')$, where $z_2 \ldots z_s$ is an arbitrary fixed element of $Z_2 \cdots Z_s$. The proof is complete. □

We shall use the following direct consequence of this result.

Lemma 5.20. *Every automaton can be simulated isomorphically by a general power of an automaton satisfying Letichevsky's criterion.* □

Next we prove the following theorem.

Theorem 5.21. *A class $\mathcal{K}$ of automata is complete with respect to isomorphic or homomorphic simulations under the general product if and only if it satisfies Letichevsky's criterion.*

Proof. Suppose that $\mathcal{K}$ does not satisfy Letichevsky's criterion. Then, by Proposition 2.71, for every product $\mathcal{M}$ of automata in $\mathcal{K}$, we also get that $\mathcal{M}$ does not satisfy Letichevsky's criterion.

Moreover, by Proposition 2.76, if this product $\mathcal{M}$ isomorphically or homomorphically simulates an automaton $\mathcal{N}$, then $\mathcal{N}$ should not be noncommutative and strongly connected. In this case, $\mathcal{K}$ is not complete with respect to isomorphic or homomorphic simulations, a contradiction. This completes the proof of the necessity.

Conversely, assume that $\mathcal{K}$ contains an automaton $\mathcal{A}$ satisfying Letichevsky's criterion. Then, using Lemma 5.20, every automaton can be simulated isomorphically by a general product of factors from $\mathcal{K}$. This completes the proof. □

Given again an automaton $\mathcal{A}$ satisfying Letichevsky's criterion, its single-factor product $\mathcal{M}$ constructed in the proof of Proposition 5.17, let $\mathbf{u}$ and $\mathbf{v}$ be control words as given in Lemma 5.16. Construct the power $\mathcal{M}^s(A, \varphi_1, \ldots, \varphi_s)$ of $\mathcal{M}$ holding $\varphi_i(a_1, \ldots, a_{s+1}) = a_{i+1}\ i \in \{1, \ldots, s\}$. Clearly, then considering a state $(a_1, \ldots, a_s)$ of $\mathcal{M}^s$ such that $a_1 \ldots a_s \in \{\mathbf{u}, \mathbf{v}\}$, getting an input letter $z_1 \in \{u_1, v_1\}$, the product goes to the state $(a_2, \ldots, a_s, z_1)$. Similarly, if $a_1 \ldots a_s = w_{k+1} \ldots w_s z_1 \ldots z_k$ for some $w_{k+1} \ldots w_s \in \{u_{k+1} \ldots u_s, v_{k+1} \ldots v_s\}$, $z_1 \ldots z_k \in \{u_1 \ldots u_k, v_1, \ldots, v_k\}$ $(u_1 \ldots u_s = \mathbf{u},\ v_1 \ldots v_s = \mathbf{v})$, and $k = 1, \ldots, s-1$, then getting an input letter $z_{k+1} \in \{u_{k+1}, v_{k+1}\}$, the product goes to the state $(a_2, \ldots, a_s, z_{k+1})$. Therefore, this product simulates isomorphically the two-state reset automaton, where **u** and **v** as the states are corresponded to the states 0 and 1 (or inversely), and, moreover, the effect of the input letter 0 is simulated by the input word **u** and the effect of the input letter 1 is simulated by the input word **v**.

Given a digraph $\mathcal{D} = (V, E)$ with $V = \{1, \ldots, n\}$, let $\mathcal{P} = \mathcal{R}^m(\{0,1\}^r, \varphi_1, \ldots, \varphi_m)$ be a $\mathcal{D}$-power of the two-state reset automaton $\mathcal{R} = (\{0,1\}, \{0,1\}, \delta_{\mathcal{R}})$ such that $\varphi_i(x_1, \ldots, x_{m+r}) = x$ if $x_1 = \cdots = x_{m+r} = x$. (This special assumption is necessary to avoid difficulties when we directly apply Corollary 5.15.) Put, for example, **u** instead of 0 and **v** instead of 1 to all of the possible states of the component automata, and similarly, do it for every component of the input vectors $x \in \{0,1\}^r$. Of course, we have got a product which is isomorphic to $\mathcal{P}$, whenever we consider **u** and **v** as letters of the alphabet $\{\mathbf{u}, \mathbf{v}\}$. But we can consider the derived product as a power of the appropriate sth power of the automaton $\mathcal{A}$ and the input vectors in $\{\mathbf{u}, \mathbf{v}\}^r$ as s-length words simulating the effects of input letters of $\{0,1\}^r$ in the automaton $\mathcal{P}$. Formally, for every digraph $\mathcal{D} = (V, E)$ with $V = \{1, \ldots, n\}$ and positive integer s, we define the digraph $\mathcal{D}^{[s]} = (V_s, E_s)$ as before (having $V_s = \{1, \ldots, ns\}$, $E_s = \{(i, i-1 \,(\mathrm{mod}\, ns)) \mid i \in V_s\} \cup \{((k-1)s+1, \ell s) \mid (k, \ell) \in E\} \cup \{(j,j) \mid j \in V_s\}$, and considering a $\mathcal{D}$-power of the two-state reset automaton, we obtained as follows.

Proposition 5.22. *Given a digraph $\mathcal{D} = (V, E)$, let $\mathcal{P} = \mathcal{R}^m(\{0,1\}^r, \varphi_1, \ldots, \varphi_m)$ be a $\mathcal{D}$-power of the two-state reset automaton $\mathcal{R} = (\{0,1\}, \{0,1\}, \delta_{\mathcal{R}})$ such that $\varphi_i(0, \ldots, 0) = 0$ and $\varphi_i(1, \ldots, 1) = 1$ $(i \in \{1, \ldots, m\})$. Moreover, let $\mathcal{A} = (A, X, \delta)$ be an automaton satisfying Letichevsky's criterion, and let* **u** *and* **v** *be control words as given in Lemma* 5.16. *$\mathcal{P}$ can be simulated isomorphically by a $\mathcal{D}^{[s]}$-power of $\mathcal{A}$ with control words in $\{\mathbf{u}, \mathbf{v}\}^r$ under the mappings $\tau_1(w_{1,1}, \ldots, w_{1,m}, w_{s,1}, \ldots, w_{s,m}) = (w_1, \ldots, w_r)$, $\tau_2(z_1, \ldots, z_r) = ((z_{1,1}, \ldots, z_{1,r}), \ldots, (z_{s,1}, \ldots, z_{s,r}))$, where $w_{1,i} \ldots w_{s,i}, z_{1,j} \ldots z_{s,j} \in \{\mathbf{u}, \mathbf{v}\}$,*

$$w_i = \begin{cases} 0 & \text{if } w_{1,i} \ldots w_{s,i} = \mathbf{u}, \\ 1 & \text{if } w_{1,i} \ldots w_{s,i} = \mathbf{v}, \end{cases}$$

$$z_j = \begin{cases} 0 & \text{if } z_{1,j} \ldots z_{s,j} = \mathbf{u}, \\ 1 & \text{if } z_{1,j} \ldots z_{s,j} = \mathbf{v} \end{cases}$$

$(i \in \{1, \ldots, m\},\ j \in \{1, \ldots, r\})$. □

Obviously, we have the following consequence of this proposition.

Corollary 5.23. *Given a digraph $\mathcal{D}$, let A be an automaton having Letichevsky's criterion with s-length control words. Every $\mathcal{D}$-product of the two-state reset automaton can be simulated isomorphically by a $\mathcal{D}^{[s]}$-power of $\mathcal{A}$ using s-length words for the simulation.* □

It is an obvious consequence of Gluškov's theorem (see Theorem 2.68) that every automaton can be embedded isomorphically into a (general) product of two-state reset automata having $\lceil \log_2 n \rceil$ factors, where $\lceil x \rceil$ denotes the minimal integer k_x with $k_x \geq x$. Therefore, using Proposition 7.18, we can also derive the following statement from the above corollary.

Theorem 5.24. *Let n be a positive integer; moreover, let A be an automaton having Letichevsky's criterion with s-length control words. Consider the digraph $\mathcal{D} = (V, E)$ with $V = \{1, \ldots, ks\}$, $E = \{(i, i-1 \pmod{ks}) \mid i \in V\} \cup \{(i, i) \mid i \in V\} \cup \{(1, s), (s, ks)\}$, where $k = \lceil \log_2(n+1) \rceil$. Every n-state automaton can be simulated isomorphically by a $\mathcal{D}$-power of $\mathcal{A}$ using s-length words for the simulation.* □

We observe that the $\mathcal{D}$-power of $\mathcal{A}$ considered in the previous result is an α_2-ν_2^ℓ-power of $\mathcal{A}$. Thus we receive the next theorem.

Theorem 5.25. *A class $\mathcal{K}$ of automata is complete with respect to homomorphic or isomorphic simulations under the α_2-ν_2^ℓ-product of automata if and only if it is complete with respect to homomorphic representation under the general product.* □

We will show later a generalization of this statement. Now we give a proof of the following classical result.

Theorem 5.26. *A class $\mathcal{K}$ of automata is complete with respect to homomorphic representations under the α_2-product if (and only if) it is complete with respect to homomorphic representations under the general product. Therefore, $\mathcal{K}$ has these properties if and only if it is satisfies Letichevsky's criterion.*

Proof. It is evident that every α_2-product is a general product. Thus the necessity of the first part of our statement is trivial.

Proof of the necessity of the second part. Assume that $\mathcal{K}$ is a complete class of automata with respect to homomorphic representations under the general product. Then we can also assume that every automaton $\mathcal{A}$ can be simulated homomorphically by a general product $\mathcal{B}$ of factors in $\mathcal{K}$. Let $\mathcal{A}$ be a noncommutative strongly connected automaton. Then, by Proposition 2.76, $\mathcal{B}$ should satisfy Letichevsky's criterion. Using Proposition 2.71, $\mathcal{B}$ will have a factor in $\mathcal{K}$ having Letichevsky's criterion. (We note that an automaton satisfies the Letichevsky criterion if either one of its homomorphic images or one of its subautomata has this property. By this fact, we can also derive the second part of the necessity of our statement using Proposition 2.71.)

Proof of the sufficiency of both parts. Let us consider the following facts.

By Proposition 5.17, there exists an m-automaton as a single-factor product of $\mathcal{A}$.

Proposition 5.18 implies that for every counter $\mathcal{C}$, there exists a loop power of a single-factor product of $\mathcal{A}$ which homomorphically represents $\mathcal{C}$.

Having Corollary 5.23, the two-state reset automaton can be simulated isomorphically by an α_1-power of $\mathcal{A}$ (having s factors).

Using Theorem 5.24, every automaton can be simulated isomorphically by an α_2-power of $\mathcal{A}$.

Therefore, the class $\mathcal{M}$ of α_2-powers of $\mathcal{A}$ has all of the properties of $\mathcal{K}$ in Theorem 3.28.

By Proposition 2.51, $\mathcal{M}$ is a complete class of automata with respect to homomorphic representations under the cascade product. The proof is complete.

We note that, by Theorem 3.35, we can derive a simpler proof of the sufficiency. In addition, observe that reset automata have the properties of the automaton $\mathcal{A}$ given in

Theorem 2.68. Therefore, we can also derive the proof of the sufficiency of both parts of our theorem by Theorem 2.68, Lemma 3.34, and Proposition 5.18. □

By Theorem 5.26 it is proved that Letichevsky's criterion can be used to describe those classes which are complete with respect to homomorphic representations under the α_2-product. On the basis of this result, the next statement shows that for $i = 2$, and thus for every $i \geq 2$, the α_i-product is homomorphically as general as the Gluškov product.

Theorem 5.27 (Ésik–Horváth characterization theorem). *For every automaton $\mathcal{A}$ and class $\mathcal{K}$ of automata, $\mathcal{A}$ can be represented homomorphically by an α_2-product of automata from $\mathcal{K}$ if (and only if) $\mathcal{A}$ can be represented homomorphically by a Gluškov product of automata from $\mathcal{K}$.*

Proof. If $\mathcal{K}$ satisfies Letichevsky's criterion, then we apply Theorem 5.26. If $\mathcal{K}$ satisfies the semi-Letichevsky criterion, then we consider Corollary 4.15. It remains to study the case when $\mathcal{K}$ does not have Letichevsky's criterion. Then we consider Theorem 4.48. The proof is complete. □

Of course, the Letichevsky decomposition theorem (Theorem 2.69) can be derived from the above result. We remark it is now easy to see that a direct proof of the Letichevsky decomposition theorem can be generated in the following way.

Proof of Letichevsky decomposition theorem. The necessity of Letichevsky's criterion directly comes from Proposition 2.71. As to sufficiency, we observe that reset automata have the properties of the automaton $\mathcal{A}$ given in Gluškov's theorem (Theorem 2.68). Therefore, we can derive the direct proof of the sufficiency by Theorem 2.68, Lemma 3.34, and Proposition 5.18. □

5.4 Bibliographical Remarks

Section 5.1. Lemmas 5.5 and 5.7 and Theorem 5.9 are given in Dömösi and Ésik [2002]. All other results in this section were developed in Dömösi and Ésik [2001].

Section 5.2. The results of this section are presented in Dömösi and Nehaniv [2000].

Section 5.3. Lemmas 5.19 and 5.20 are new. Theorem 5.21 was proved by P. Dömösi [1994]. Proposition 5.22, Corollary 5.23, and Theorem 5.24 are new observations. Theorem 5.25 is a strengthened version of the main result in Dömösi [1996]. Theorem 5.26 is a well-known result of Z. Ésik [1985]. It highly improves the main result of P. Dömösi [1983]. The Ésik–Horváth theorem (Theorem 5.27), i.e., the fact that the α_2-product is homomorphically equivalent to the general product, was proved by Ésik and Gy. Horváth [1983]. A nice explanation of this statement and Theorem 5.26 is given by Gécseg [1986].

Chapter 6
Primitive Products and Temporal Products

In this section, one of our fundamental concepts is that of the primitive product. *Why is it important? A primitive product is a composition of a finite sequence of finite automata such that feedback is limited to no further than the previous factor. Furthermore, the input to each factor depends only on the global input to the system and the states of at most three factors (including the factor itself). Conversely, the state of a factor may directly influence only at most three factors (including the factor itself). Thus, the number of the possible local links is also strongly restricted. We show that the primitive product is one of the simplest type of products that preserve the completeness properties of the general product. The primitive product is general in the sense that exactly those classes are complete with respect to homomorphic representation under the primitive product which are complete with respect to homomorphic representation under the general product (i.e., with unrestricted networking). On the other hand, we will see that the primitive product is a special type of the* α_2-ν_2^ℓ*-product such that it has a strong restriction on the permitted number of local links. By our results in Chapters* 3, 4, *and* 5, *we can establish that an* α_i-ν_j*-product or an* α_i-ν_k^ℓ*-product with* $i < 2$ *or* $j \leq 2$ *or* $k < 2$ *cannot preserve the generality in the considered sense. Therefore, we would lose the generality of the primitive product if we tried to give further restrictions on the structure of permitted links. Additional conditions guarantee a strong planarity property (outerplanarity), which is desirable in the engineering of sequential circuits.*

Also studied in this chapter is the temporal product. *This is a model for multichannel automata networks, where the network may cyclically change its internal structure during its work on each channel. We will see that this concept may be much stronger with respect to homomorphic or isomorphic representation than the general product. Therefore, the study of automata networks which can modify their inner structure during their work may prove very important from the point of view of many applications.*

6.1 Primitive Products

Take the above considered general product $\mathcal{A} = \mathcal{A}_1 \times \cdots \times \mathcal{A}_n(X, \varphi_1, \ldots, \varphi_n)$ and its underlying graph $\mathcal{D} = (V, E)$. For any $t \in V$, denote by $i(t)$ and $o(t)$ the sets of incoming and outgoing edges of t, respectively, and assume that

(1) for any $t \in V$ there exist $j, k \in \{1, \ldots, t-1, t+1\}$ and $r \in \{t-1, t+1, \ldots, n\}$ such that one of the following conditions is satisfied:

(1a) $i(t) \subseteq \{(j, t), (t, t)\}$ and $o(t) \subseteq \{(t, t-1), (t, t), (t, r)\}$, or

(1b) $i(t) \subseteq \{(j, t), (k, t), (t, t)\}$ and $o(t) \subseteq \{(t, t), (t, r)\}$;

(2) if $(a, b), (c, d) \in E$ and $\{a, b\} \cap \{c, d\} = \emptyset$, then $\min\{c, d\} < a < \max\{c, d\}$ if and only if $\min\{c, d\} < b < \max\{c, d\}$.

Then we say that $\mathcal{A}$ is a *primitive product.*

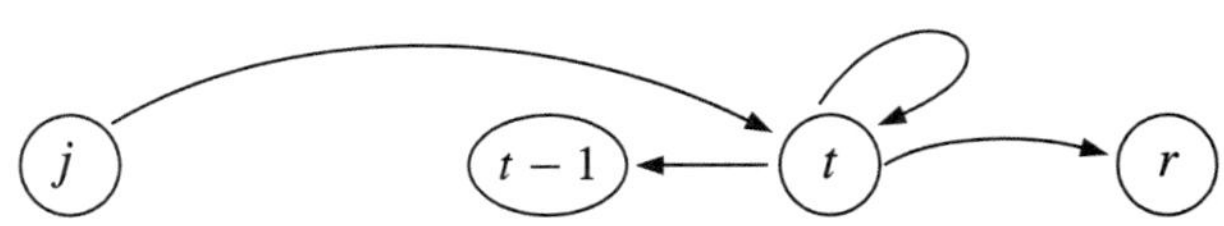

CONDITION (1a)

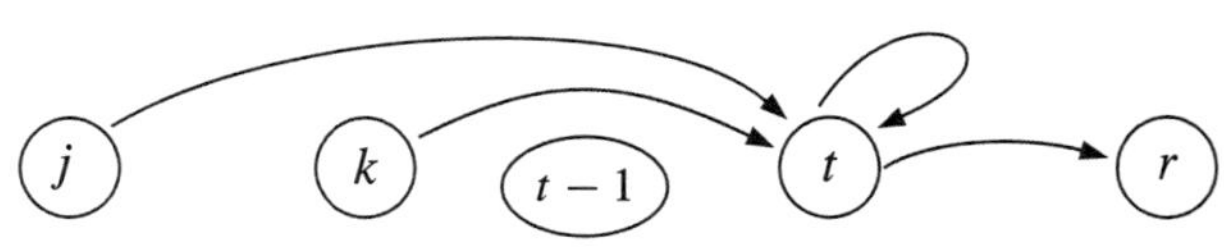

CONDITION (1b)

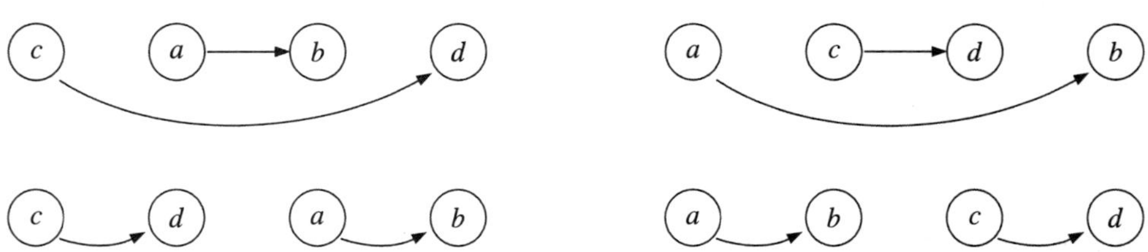

NONCROSSING CASES

The following statement is obvious.

Proposition 6.1. *Every primitive product is an* α_2-ν_2^ℓ*-product.*

On the other hand, by Corollary 2.67, we obtain the next proposition.

Proposition 6.2. *Every* α_1-ν_2^ℓ*-product of automata can be isomorphically represented by a primitive product of copies of the same automata.* □

For any class $\mathcal{K}$ of automata, let us consider the class $P(\mathcal{K})$ of primitive products having factors from $\mathcal{K}$. It is easy to see that $P(P(\mathcal{K})) \subseteq P(\mathcal{K})$ does not hold in general. However, we have the following.

Proposition 6.3. *Let $\mathcal{M} = \mathcal{M}_1 \times \cdots \times \mathcal{M}_{n+1}(X, \psi_1, \ldots, \psi_{n+1})$, $n \geq 1$, be a product of primitive products $\mathcal{M}_i = \mathcal{M}_{i,1} \times \cdots \times \mathcal{M}_{i,j_i}(X_i, \psi_{i,1}, \ldots, \psi_{i,j_i})$, $j_i \geq 2$, $i = 1, \ldots, n+1$, having the following properties. $\psi_1, \ldots, \psi_n$ may really depend only on their input variables. Moreover, $\psi_{i,1}, \ldots, \psi_{i,j_i-1}$, $i = 1, \ldots, n$, really do not depend on their last (j_i)*th*-state variables, and, if some $\psi_{n+1,k}$ $(k = 1, \ldots, j_{n+1})$ really depends on its input variable, then it may additionally depend only on its* k*th state variable and at most one other state variable, and, simultaneously, there exists at most one $\psi_{n+1,k'}$ $(k' = 1, \ldots, j_{n+1})$ with $k \neq k'$ depending on its* k*th state variable. Furthermore, the input set of $\mathcal{M}_{n+1}$ is $X_{n+1} = M_{1,j_1} \times M_{2,j_2} \times \cdots \times M_{n,j_n}$, where M_{i,j_i}, $i = 1, \ldots, n$, denotes the state set of the last factor of the product $\mathcal{M}_i$, and each $\psi_{n+1,k}$ $(k = 1, \ldots, j_{n+1})$ may depend at most on one component of X_{n+1}, and, moreover $\psi_{n+1,k}$ and $\psi_{n+1,k'}$ do not depend on the same component of X_{n+1} for $k \neq k'$ $(k, k' = 1, \ldots, j_{n+1})$.*

If ψ_{n+1} has the form $\psi_{n+1}(m_1, \ldots, m_n, m_{n+1}, x) = (m_{1,j_1}, \ldots, m_{n,j_n}) \in X_{n+1}$, where m_i is the state of $\mathcal{M}_i$ and m_{i,j_i} the state of $\mathcal{M}_i$'s last factor, then $\mathcal{M}$ is isomorphic to a primitive product of $\mathcal{M}_{i,j}$, $i = 1, \ldots, n+1$, $j = 1, \ldots, j_i$.

Proof. Let P be an arbitrary permutation over $\{1, \ldots, n\}$.[34] Considering the short notation $\mathcal{N}_\ell = \mathcal{M}_{\ell,1} \times \cdots \times \mathcal{M}_{\ell,j_\ell}$ $(\ell = 1, \ldots, n+1)$, by Proposition 2.50, we can construct the product

$$\mathcal{M}' = \mathcal{N}_{P(1)} \times \cdots \times \mathcal{N}_{P(n)} \times \mathcal{N}_{n+1}(X, \varphi'_1, \ldots, \varphi'_u)$$

with $u = j_1 + \cdots + j_{n+1}$ such that $\mathcal{M}'$ is isomorphic to $\mathcal{M}$. Denote $\psi_{n+1,s_1}, \ldots, \psi_{n+1,s_r}$ with $s_1 < \cdots < s_r$ to be all feedback functions of the product $\mathcal{M}_{n+1}$ depending on at least one component of the input set $X_{n+1} = M_{1,j_1} \times M_{2,j_2} \times \cdots \times M_{n,j_n}$. From our assumptions it follows that $r < n$. Suppose that for every $\ell \in \{1, \ldots, r\}$, $P(n - \ell + 1) = t$, whenever ψ_{n+1,s_ℓ} depends on the tth component of X_{n+1}. Clearly, then $\mathcal{M}'$ forms a primitive product of the $\mathcal{M}_{i,j}$, $i = 1, \ldots, n+1$, $j = 1, \ldots, j_i$. □

Lemma 6.4. *Let $\mathcal{D} = (V, E)$ be the underlying graph of a primitive product of automata. Then $\mathcal{D}$ has the ordered cycle property.*

Proof. The nodes of $\mathcal{D}$ are already integers, so we consider $\mathcal{D}$ under its natural labeling.
Claim. Take any pair of undirected paths $i_1 \ldots i_m$, $j_1 \ldots j_n$ consisting of nodes in $\mathcal{D}$, with $j_1 < i_1 < j_n$ and suppose either $i_m < j_1$ or $j_n < i_m$. Then the paths contain a common point.

Proof of Claim. Assume that the claim is false; then there is a minimal counterexample, with all nodes *distinct* and $n + m$ least.

Consider i_{m-1}: if $i_{m-1} < j_1$ or $j_n < i_{m-1}$, then the path $i_1 \ldots i_{m-1}$ would yield smaller counterexample unless $m = 2$. If on the other hand $j_1 < i_{m-1} < j_n$, then $i_{m-1}i_m$

[34] In other words, let P be a bijective mapping of $\{1, \ldots, n\}$ onto itself.

yields a shorter counterexample unless $m = 2$. So $m = 2$ for a minimal counterexample. Now, consider the path $j_1 \cdots j_n$. If $i_2 < j_1$, then $i_2 < j_1 < i_1 < j_n$. In this case, by (2), $i_2 < j_2 < i_1$ must hold, and thus, $j_2 \cdots j_n$ yields a shorter counterexample until $n = 2$. If $j_n < i_2$, then $j_1 < i_1 < j_n < i_2$. Then, by (2), $i_1 < j_{n-1} < i_2$ must hold, but in this case $j_1 \ldots j_{n-1}$ yields a shorter counterexample.

We have established that $n = m = 2$ in any least counterexample. Thus, $i_2 < j_1 < i_1 < j_2$ or $j_1 < i_1 < j_2 < i_2$: now by condition (2) of the definition of primitive product, since $j_1 = \min\{j_1, j_2\} < i_1 < \max\{j_1, j_2\} = j_2$, we have $j_1 < i_2 < j_2$, a contradiction. Therefore, no least counterexample can exist. This establishes the claim.

Now let c_1 denote the least node in the real cycle. It is connected by edges in the real cycle to two other nodes. Now these two nodes and c_1 are pairwise distinct. Let c_2 denote the lesser of the two and let c_k denote the greater. We have $c_1 < c_2 < c_k$. Proceeding around the real cycle in the direction from c_1 to c_2 denote the nodes c_3, c_4, etc., until we reach c_k. We assert that c_k is the greatest node in the real cycle; if not, let c_i be the node with least i such that $c_i > c_k$. Note that $i \geq 3$. By leastness of i, $c_{i-1} < c_k$, and so it must be that $c_i > c_k > c_{i-1} > c_1$, but then the path $c_k c_1$ and the path $c_i c_{i-1}$ would comprise a counterexample to the claim. Hence, c_k must indeed be the greatest node.

Furthermore, it must be true for each $i = 1, \ldots, k-1$ that $c_{i+1} > c_i$: If not, take an i such that $c_{i+1} < c_i$. Then we have $i \notin \{1, k-1\}$, and so $c_1 < c_{i+1} < c_i < c_k$. But then $c_{i+1} \ldots c_k$ is a path disjoint from the path $c_1 \ldots c_i$, and we would have contradiction to the claim.

We have established that $c_1 < c_2 < \cdots < c_k$ for the nodes $c_1, c_2, \ldots, c_k$ met in sequence traced as we go around the cycle starting in the direction from c_1 to c_2. □

Corollary 6.5. *The underlying graph of any primitive product is an outerplanar graph.*

Proof. A graph is outerplanar if and only if it contains no subdivision of K_4, the complete graph on four nodes, and no subdivision of the complete bipartite graph $K_{2,3}$. However, such a subdivision cannot have the ordered cycle property established in the lemma, since if it did, then by restriction the property would hold also for K_4 or $K_{2,3}$. But it is easy to check that K_4 and $K_{2,3}$ do not have this property. □

Remark 1. *As we see from the proofs of Lemma* 6.4 *and Corollary* 6.5, *every product of automata whose underlying graph satisfies condition* (2) *in the definition of primitive product has the ordered cycle property and an outerplanar underlying graph.*

Remark 2. *From the engineering point of view of circuit wiring, outerplanarity is an extremely desirable property, since a circuit whose components and wires comprise the nodes and edges of an outerplanar graph may be realized on a flat surface. Moreover, new wires can be run from a point outside the circuit to any or all nodes of the circuit without crossing each other or any of the existing wires.*

6.2 Primitive Products and Letichevsky's Criterion

We constructively show that if $\mathcal{A}$ is a finite automaton satisfying Letichevsky's criterion, then any finite automaton can be homomorphically represented by (i.e., is a homomorphic image of a subautomaton of or, equivalently, is a letter-to-letter (length-preserving) divisor of) a primitive product of copies of $\mathcal{A}$.

Take two alphabets X and Y. Let $n = \ell s$ ($\ell > 1$) be a fixed integer and consider a mapping $\tau : X^n \to Y^n$ having the property $\{\tau(p) \mid p \in X^n\} \subseteq \{\mathbf{w} \mid \mathbf{w} \in \{\mathbf{u}, \mathbf{v}\}^\ell\}$ for some fixed words $\mathbf{u}, \mathbf{v} \in Y^s$. We shall denote the reverse of $\tau(p)$ by $\bar{\tau}(p)$.

Set $H \subseteq \{p \in X^+ \mid |p| = n\}$, $H \neq \emptyset$. Define $\mathcal{R}_{\tau,H,d} = (R_{\tau,H,d}, X, \delta_{\tau,H,d})$ to be the automaton, where d is a positive integer, $R_{\tau,H,d} = \{(k, p, q) \in \{1, \ldots, n\} \times X^* \times Y^+ \mid k + |q| = n + d,\ |p| \in \{0, k\},\ p$ is a prefix of a word in H ($pp' \in H$ for some $p' \in X^*$), and, furthermore, $q = q'q''$, where q' is a suffix of $\mathbf{u}$ or $\mathbf{v}$ and $q'' \in \{\mathbf{u}, \mathbf{v}\}^*\}$, and, for arbitrary $(k, p, yq) \in R_{\tau,H,d}$ ($y \in \{u_t, v_t \mid t = 1, \ldots, s\}$) and $x \in X$,

$$\delta_{\tau,H,d}((k, p, yq), x) = \begin{cases} (k+1, px, q) & \text{if } k < n,\ p \neq \lambda, \text{ and } pxp' \in H \text{ for some } p' \in X^*, \\ (k+1, \lambda, q) & \text{if } k < n,\ p = \lambda, \\ & \text{or } k < n,\ p \neq \lambda, \text{ and for all } p' \in X^*,\ pxp' \notin H, \\ (1, x, q\tau(p)) & \text{if } k = n,\ p \in H, \text{ and } xp' \in H \text{ for some } p' \in X^*, \\ (1, \lambda, q\tau(p)) & \text{if } k = n,\ p \in H, \text{ and for any } p' \in X^*,\ xp' \notin H, \\ (1, x, q\mathbf{u}^\ell) & \text{if } k = n,\ p = \lambda, \text{ and } xp' \in H \text{ for some } p' \in X^*, \\ (1, \lambda, q\mathbf{u}^\ell) & \text{if } k = n,\ p = \lambda, \text{ and for any } p' \in X^*,\ xp' \notin H. \end{cases}$$

To simplify the proof of the next result, we introduce some auxiliary notions. Let $\mathcal{A} = (A, X_\mathcal{A}, \delta_\mathcal{A})$ be an automaton satisfying Letichevsky's criterion, and let $u_1 \ldots u_s$, $v_1 \ldots v_s$ be any pair of its control words. For any $w_1 \ldots w_s$ with $w_t \in \{u_t, v_t\}$ ($t = 1, \ldots, s$) we shall use the short notation $\mathbf{w}$. Consider a word $a_1 \ldots a_n \in A^+$ and an integer k ($= 1, \ldots, n$). We will denote by $c(a_1 \ldots a_n, k)$ the $(k+1)$th cyclic permutation of $a_1 \ldots a_n$. In more detail, let

$$c(a_1 \ldots a_n, k) = \begin{cases} a_{k+2} \ldots a_n a_1 \ldots a_{k+1} & \text{if } k < n-1, \\ a_1 \ldots a_n & \text{if } k = n-1, \\ a_2 \ldots a_n a_1 & \text{if } k = n. \end{cases}$$

In addition, for any pair t, k with $t = \pm 1, \pm 2, \ldots$, $k = 1, \ldots, n$, let $c(a_1 \ldots a_n, nt + k) = c(a_1 \ldots a_n, k)$ and for any integer r, denote by $\bar{c}(a_1 \ldots a_n, r)$ the reverse of $c(a_1 \ldots a_n, r)$.

Let $\mathcal{M} = (W \times Z, X_\mathcal{M}, \delta_\mathcal{M})$ be an automaton with $Y \subseteq Z$; moreover, let $\mathcal{B} = (B, X, \delta_\mathcal{B})$ be a subautomaton of $\mathcal{M}$ having a homomorphism $\psi : B \to R_{\tau,H,d}$ onto $\mathcal{R}_{\tau,H,d}$ such that $\psi((w, z)) = (k, p, yq)$ ($(w, z) \in B$, $(k, p, yq) \in R_{\tau,H,d}$, $y \in Y$) implies $z = y$. Then we say that $\mathcal{M}$ *y-represents* $\mathcal{R}_{\tau,H,d}$ (with respect to ψ).

We have the following.

Lemma 6.6. *Let $\mathcal{A} = (A, X_\mathcal{A}, \delta_\mathcal{A})$ be an automaton satisfying Letichevsky's criterion and let $u_1 \ldots u_s$, $v_1 \ldots v_s$ be any pair of its control words. Consider an alphabet X, a multiple n of s with $n = \ell s$, $\ell > 1$, a word $r \in X^n$, and a mapping $\tau : X^n \to A^n$ having the property $\tau(p) \in \{\mathbf{u}, \mathbf{v}\}^\ell$ for each $p \in X^n$. Then there exists a primitive power $\mathcal{M}$ of $\mathcal{A}$ such that $\mathcal{R}_{\tau,\{r\},1}$ is y-represented by $\mathcal{M}$. In addition, apart from the feedback functions for the last factor, the feedback functions of the factors of $\mathcal{M}$ really do not depend on their last state variable.*

Proof. For the proof of our statement, first we are going to define a product $\mathcal{N} = \mathcal{A}^{3n+1}(X, \varphi_1, \ldots, \varphi_{3n+1})$ having the following structure:

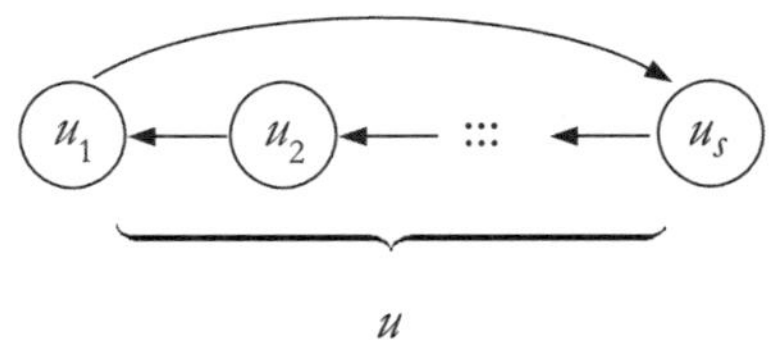

SMALL CLOCK

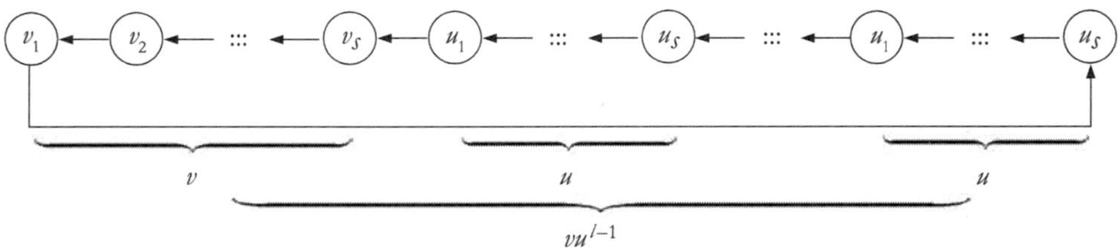

BIG CLOCK

The first s factors provide a "small clock" in which $\mathbf{u} = u_1 \ldots u_s$ cycles. The next n factors $(s+1, \ldots, s+n)$ make up a "big clock" in which $\mathbf{vu}^{\ell-1}$ cycles. The next n factors $(s+n+1, \ldots, s+2n)$ make up a buffer into which values flow from the big clock, starting with v_1. At the kth position of the buffer, if the input letter x matches the kth letter of r when the signal (headed by v_1) is about to reach this position, then the signal is permitted to continue; otherwise instead of switching to state v_1 we switch to u_1, indicating rejection of the input.[35] Finally, if the word has not been rejected by the nth input step, the acceptance signal reaches the end of the buffer, and then the buffer contains n letters which are the reverse of $\mathbf{vu}^{\ell-1}$ with $a_{2n+s} = v_1$ (and of $\mathbf{u}^{\ell}$ otherwise with $a_{2n+s} = u_1$).

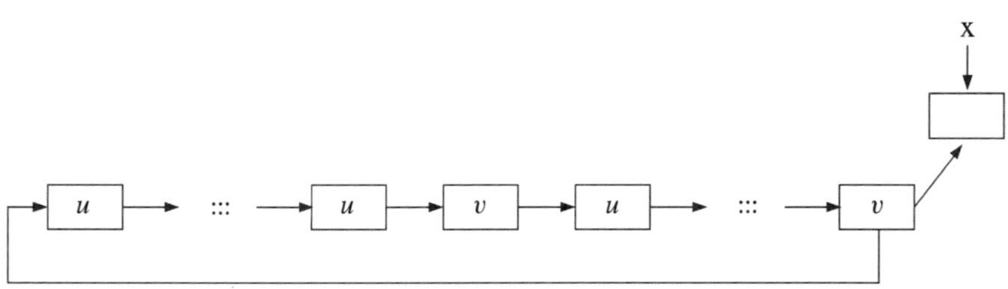

INPUT REJECTION (u ARISES)

[35] Lemma 5.12 guarantees that for each factor t of the buffer, $(a_{t-1}, a_{1-t(\mathrm{mod}\, s)}) = (v_1, u_1)$ only when $k = t - s - 1(\mathrm{mod}\, n)$, especially for the first factor of the buffer, i.e., for $t = n + s + 1$, $(a_{s+n}, a_s) = (v_1, u_1)$ if and only if $k = n$.

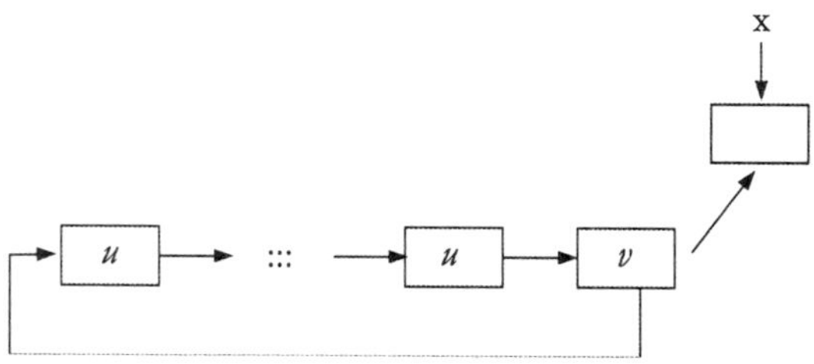

INPUT ACCEPTANCE (v ARISES)

In the next step, the buffer cycle starts again, while in the last $n - s + 1$ factors $(2n + s + 1, \ldots, 3n + 1)$, the coded word $\tau(r)$ begins to take form if the signal has arrived. Now $\tau(r) = \mathbf{w}_1 \ldots \mathbf{w}_\ell$, where each $\mathbf{w}_j \in \{\mathbf{u}, \mathbf{v}\}$. For each $j = 1, \ldots, \ell$ with $\mathbf{w}_{\ell-j+1} = \mathbf{v}$ in this step v_1 simultaneously enters factor $2n + js + 1$, while for the j with $\mathbf{w}_{\ell-j+1} = \mathbf{u}$ and u_1 enters this factor. It is important to observe that $\tau(r)$ can be fully recovered from the states of these ℓ nodes at this time, as follows from $u_1 \neq v_1$ and the form of $\tau(r) \in \{\mathbf{u}, \mathbf{v}\}^\ell$. In the next $s - 1$ steps, the letters in these factors shift to the next highest factor and the respective letters of $\mathbf{u}$ and $\mathbf{v}$ flow in. Thus, this last part will contain $\bar{\tau}(r)$ except for its first $s - 1$ letters, as $a_{2n+s+1}, \ldots, a_{3n+1}$ after s steps. Observe that the letters of $\tau(r)$ appear as n successive states a_{3n+1} of $\mathcal{A}_{3n+1}$, which is the last factor.

If the signal did not arrive, the above transition rules imply that $\mathbf{u}^\ell$ will be in the buffer after n input letters and will then flow through and out of the next part.

We will use the fact that, except for the last $(3n + 1)$th factor, the feedback function φ_t of the tth does not depend on its last state factor a_{3n+1}.

As to the mapping onto $\mathcal{R}_{\tau,\{r\},d}$, take a triplet $(k, p, yq) \in \mathcal{R}_{\tau,\{r\},1}$ ($y \in \{u_t, v_t \mid t = 1, \ldots, s\}$). We represent this triplet (k, p, yq) by an appropriate state $b = c(\mathbf{u}, k)c(\mathbf{vu}^{\ell-1}, k)\bar{c}(\mathbf{zu}^{\ell-1}, k - 1)e_1 \ldots e_{n-s+1}$ of $\mathcal{N}$. The number k is represented by the value $c(\mathbf{vu}^{\ell-1}, k)$ and $\bar{c}(\mathbf{zu}^{\ell-1}, k - 1)$, $\mathbf{z} = z_1 \ldots z_s$ represents p with $z_i \in \{u_i, v_i\}$, $i = 1, \ldots, s$. Namely, if $z_1 = u_1$, then $p = \lambda$ is assumed, and if $z_1 = v_1$, then p is understood as the k-length prefix of r. In other words, $z_1 = v_1$ means $r = pp'$ for some $p' \in X^*$ (with $|p| = k$). And $z_1 = u_1$ means $p = \lambda$. Setting $y_1 \ldots y_n \in \{\mathbf{u}, \mathbf{v}\}^\ell$, assume

$$e_1 \ldots e_{n-s+1} = \begin{cases} y_{s+k} \ldots y_{s+1} u_s \ldots u_{k+1} y_k & \text{if } 1 \leq k < s,\ \ell = 2, \\ y_{n-s+k} \ldots y_{n-s+1} u_s \ldots u_{k+1} y_{n-2s+k} \ldots y_{n-2s+1} & \\ u_s \ldots u_{k+1} \ldots y_{s+k} \ldots y_{s+1} u_s \ldots u_{k+1} y_k & \text{if } 1 \leq k < s, \ell > 2, \\ y_n \ldots y_k & \text{if } k = s, \\ u_j \ldots u_1 (u_s \ldots u_1)^{i-1} y_n \ldots y_k & \text{if } k = is + j, k < n, \\ & \quad 1 \leq j < s, \\ (u_s \ldots u_1)^{i-1} y_n \ldots y_k & \text{if } k = is \leq n. \end{cases}$$

Then k and the mirror image of $e_1 \ldots e_{n-s+1}$ represents $y_k \ldots y_n$. If $k \geq s$, then this is obvious considering the structure of $e_1 \ldots e_{n-s+1}$. (Recall that $u_s = v_s$.) If $1 \leq k < s$ and $\ell > 2$, then the mirror image of $e_1 \ldots e_{n-s+1}$ is $y_k u_{k+1} \ldots u_s y_{s+1} \ldots y_{s+k} \ldots u_{k+1} \ldots u_s$ $y_{n-2s+1} \ldots y_{n-2s+k}$ $u_{k+1} \ldots u_s y_{n-s+1} \ldots y_{n-s+k}$ representing $y_k \ldots y_n$. (Observe that k and $y_k \in \{u_k, v_k\}$ unambiguously determine $y_k \ldots y_s$; moreover, for any $i = 1, \ldots, \ell$, $y_{n-is+1} \in \{u_1, v_1\}$ unambiguously determines $y_{n-is+2} \ldots y_{n-(i-1)s}$.) We have similar consequences for $1 \leq k < s$ and $\ell = 2$. The motivation for this representation should be clear from the explanation of the buffer cycle discussed above.

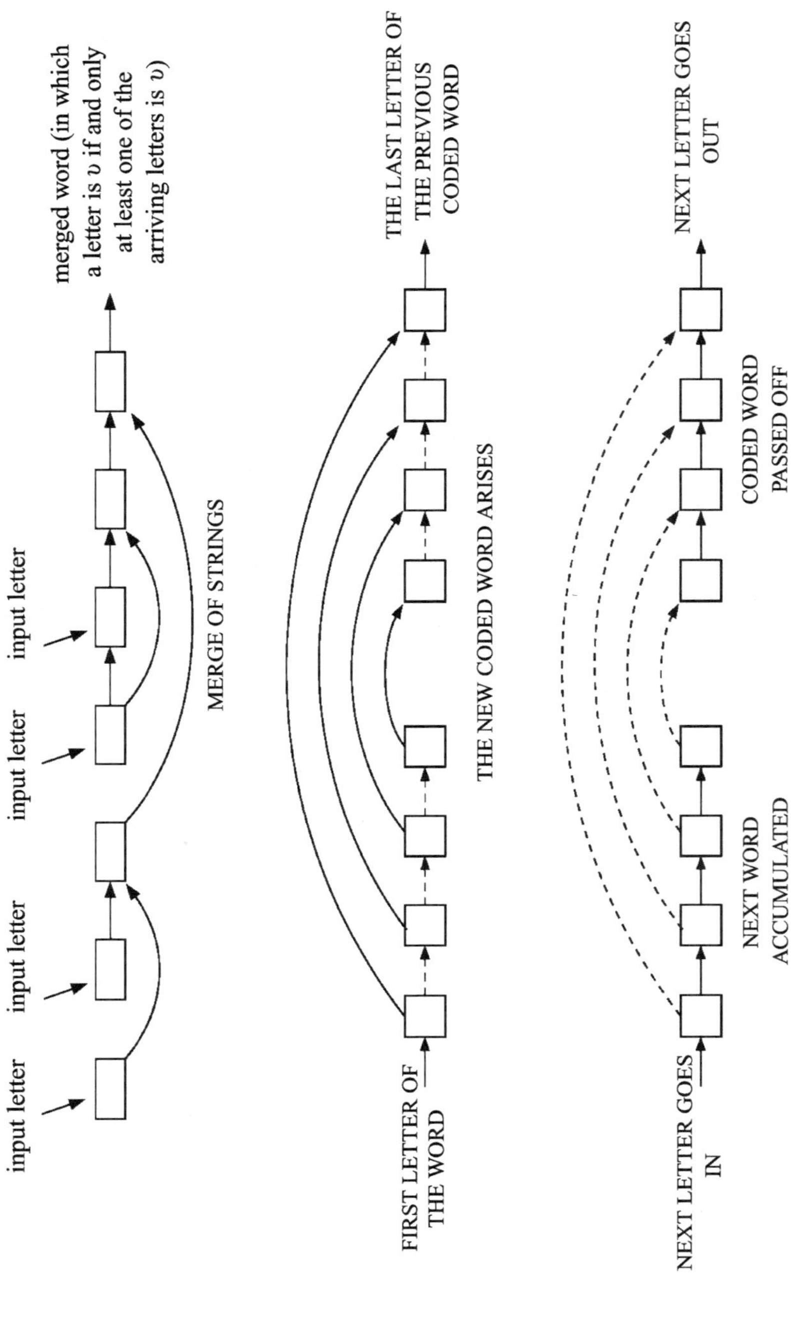
input letter
input letter
input letter
input letter
merged word (in which a letter is v if and only at least one of the arriving letters is v)
MERGE OF STRINGS
FIRST LETTER OF THE WORD
THE LAST LETTER OF THE PREVIOUS CODED WORD
THE NEW CODED WORD ARISES
NEXT LETTER GOES IN
NEXT LETTER GOES OUT
NEXT WORD ACCUMULATED
CODED WORD PASSED OFF

Formally, we define the product $\mathcal{N} = \mathcal{A}^{3n+1}(X, \varphi_1, \ldots, \varphi_{3n+1})$ such that for any $(a_1, \ldots, a_{3n+1}) \in A^{3n+1}$, $x \in X$, and $t \in \{1, \ldots, 3n+1\}$, we have

$$\varphi_t(a_1, \ldots, a_{3n+1}, x)$$

$$= \begin{cases} x[a_t, a_{t+1}] & \text{if } t = 1, \ldots, s-1, \\ & \text{or } t = s+1, \ldots, s+n-1, \\ x[a_t, a_1] & \text{if } t = s, \\ x[a_t, a_{s+1}] & \text{if } t = s+n, \\ x[a_t, u_1] & \text{if } t = s+n+1, \ldots, s+2n,\ a_t = u_s (= v_s),\ (a_{t-1}, \\ & \quad a_{1-t (\mathrm{mod}\, s)}) = (u_1, u_1), \\ & \text{or } t = s+n+1, \ldots, s+2n,\ a_t = u_s (= v_s),\ (a_{t-1}, \\ & \quad a_{1-t (\mathrm{mod}\, s)}) = (v_1, u_1), \text{ and } x \text{ is not the } t-(s+n)\text{th} \\ & \quad \text{letter of } r, \\ x[a_t, v_1] & \text{if } t = s+n+1, \ldots, s+2n,\ a_t = u_s (= v_s),\ (a_{t-1}, \\ & \quad a_{1-t (\mathrm{mod}\, s)}) = (v_1, u_1), \text{ and } x \text{ is the } t-(s+n)\text{th} \\ & \quad \text{letter of } r, \\ x[a_t, a_{t-1}] & \text{if } t = s+n+1, \ldots, s+2n,\ a_t \in \{u_i, v_i\},\ a_{t-1}, a_{1-t (\mathrm{mod}\, s)} \\ & \quad \in \{u_{i+1}, v_{i+1}\},\ i = 1, \ldots, s-1, \\ x[a_t, a_s] & \text{if } t = 2n+s+1, \text{ and the } s\text{th letter of } \bar{\tau}(r) \text{ is } u_1, \\ x[a_t, a_{s+2n} \vee a_s] & \text{if } t = 2n+s+1, \text{ and the } s\text{th letter of } \bar{\tau}(r) \text{ is } v_1, \\ x[a_t, a_{t-1}] & \text{if } t = 2n+js+1,\ j = 2, \ldots, \ell, \text{ and the } js\text{th} \\ & \quad \text{letter of } \bar{\tau}(r) \text{ is } u_1, \\ & \text{or } t = 2n+js+i,\ j = 1, \ldots, \ell-1,\ i = 2, \ldots, s, \\ x[a_t, a_{s+2n} \vee a_{t-1}] & \text{if } t = 2n+js+1,\ j = 2 \ldots, \ell, \text{ and the } js\text{th} \\ & \quad \text{letter of } \bar{\tau}(r) \text{ is } v_1. \end{cases}$$

Now we give the formal definition of $\mathcal{B}'$ and that of a mapping $\psi : B' \to R_{\tau,\{r\},1}$ under which $\mathcal{R}_{\tau,\{r\},1}$ is an (y-represented) homomorphic image of $\mathcal{B}'$.

Let B' consists of all $b \in A^{3n+1}$ for which there are $(k, p, q) \in R_{\tau,\{r\},1}$ such that $b = c(\mathbf{u}, k)c(\mathbf{v}\mathbf{u}^{\ell-1}, k)\bar{c}(\mathbf{z}\mathbf{u}^{\ell-1}, k-1)e_1 \ldots e_{n-s+1}$, $\mathbf{z} = z_1 \ldots z_s$, with $z_i \in \{u_i, v_i\}$, $i = 1, \ldots, s$, where

$$z_1 = \begin{cases} v_1 & \text{if } p \text{ is the } k\text{-length prefix of } r, \\ u_1 & \text{otherwise,} \end{cases}$$

$e_1 \ldots e_{n-s+1}$ is defined as above, and q is represented by k and $e_1 \ldots e_{n-s+1}$ as we explained. (Recall that by the structure of $R_{\tau,\{r\},1}$, $y_{n-s+k} \ldots y_{n-s+1}u_s \ldots u_{k+1}\ y_{n-2s+k} \ldots y_{n-2s+1}u_s \ldots u_{k+1} \ldots y_{s+k} \ldots y_{s+1}u_s \ldots u_{k+1}y_k$ unambiguously determines $q = y_k \ldots y_n$ whenever $k < s$ and $\ell > 2$. Similarly, by the structure of $R_{\tau,\{r\},1}$, $y_{s+k} \ldots y_{s+1}u_s \ldots u_{k+1}y_k$ unambiguously determines $q = y_k \ldots y_n$, whenever $k < s$ and $\ell = 2$. Moreover, $q = e_{n-s+1} \ldots e_{k-s+1}$ is assumed if $k \geq s$.) Furthermore, let $\psi(b) = (k, p, q)$. It is routine work to show that $\mathcal{N}$ has a subautomaton $\mathcal{B}'$ with state set B' which can be mapped homomorphically by ψ onto $\mathcal{R}_{\tau,\{r\},1}$. Finally, by $\psi(b) = (k, p, q)$, the last letter of b is the same as the first letter of q. Therefore, $\mathcal{N}$ y-represents $\mathcal{R}_{\tau,\{r\},1}$.

Applying Proposition 2.66 to the product $\mathcal{N}$, it is clear that we will get a product $\mathcal{N}'$, which also y-represents $\mathcal{R}_{\tau,\{r\},1}$; moreover, similar to $\mathcal{N}$, apart from the last factor, the feedback functions of the factors of $\mathcal{N}'$ really do not depend on their last state variable. Thus it is enough to observe that by an inductive application of Proposition 2.66, we can derive from the product $\mathcal{N}$ a primitive product $\mathcal{M}$.

In particular, every vertex of the underlying graph of $\mathcal{N}$ has not more than two incoming and two outgoing edges in the resulting product. Moreover, if there is a vertex with two outgoing edges, then it is an element of a cycle with one edge going into another element of the same cycle, and all the other cycle elements have one outgoing edge connecting them with other elements of the cycle.

In addition, cycle elements have only one incoming edge, coming from another element of the cycle.[36] Using Theorem 2.1, we may assume that $\mathcal{N}$ is a primitive product, for otherwise we could relabel its components by an appropriate permutation of their indices.

This ends the proof of Lemma 6.6. □

We next prove the following lemma.

Lemma 6.7. *Let $\mathcal{A}$ be an automaton satisfying Letichevsky's criterion, and let $\mathcal{A}^k(X, \varphi'_1, \ldots, \varphi'_k)$, $\mathcal{A}^\ell(X, \varphi''_1, \ldots, \varphi''_\ell)$ be primitive powers of $\mathcal{A}$ such that, apart from the last factors, the feedback functions of the factors really do not depend on their last state variable. Suppose that they y-represent, in order, $\mathcal{R}_{\tau,H_1,d}$ and $\mathcal{R}_{\tau,H_2,d}$ for some $\tau : X^n \to A^n$, $H_1, H_2 \subseteq \{p \in X^+ \mid |p| = n\}$ (H_1, H_2 are not necessarily disjoint sets), and $d \geq 1$, where n is a multiple of s as before. There exists a primitive power $\mathcal{M} = \mathcal{A}^{k+\ell+1}(X, \varphi_1, \ldots, \varphi_{k+\ell+1})$, which y-represents $\mathcal{R}_{\tau,H_1\cup H_2,d+1}$. Moreover, apart from the last factor, the feedback functions of the factors of $\mathcal{M}$ really do not depend on their last state variable.*

Proof. Define the power $\mathcal{A}^{k+\ell+1}(X, \varphi_1, \ldots, \varphi_{k+\ell+1})$ in the following way.

For any $(a_1, \ldots, a_{k+\ell+1}) \in A^{k+\ell+1}$, $x \in X$, $t = 1, \ldots, n + \ell + 1$,

$$\varphi_t(a_1, \ldots, a_{k+\ell+1}, x) = \begin{cases} \varphi'_t(a_1, \ldots, a_k, x) & \text{if } t \leq k, \\ \varphi''_{t-k}(a_{k+1}, \ldots, a_{k+\ell}, x) & \text{if } k < t \leq k + \ell, \\ x[a_{k+\ell+1}, a_k \vee a_{k+\ell}] & \text{otherwise.} \end{cases}$$

Clearly, this power of $\mathcal{A}$ is primitive.

Now we consider, in order, appropriate homomorphisms ψ' and ψ'' such that $\mathcal{A}^k(X, \varphi'_1, \ldots, \varphi'_k)$ y-represents $\mathcal{R}_{\tau,H_1,d}$ with respect to ψ', and, moreover, $\mathcal{A}^\ell(X, \varphi''_1, \ldots, \varphi''_\ell)$ y-represents $\mathcal{R}_{\tau,H_2,d}$ with respect to ψ''. It is clear that φ_t does not depend on its last state variable if $t \neq k + \ell + 1$. Therefore, it is a routine work to show that the power $\mathcal{M}$ y-represents $\mathcal{R}_{\tau,H_1\cup H_2,d+1}$ with respect to the homomorphism ψ having the following properties:

$$\psi(a_1, \ldots, a_{k+\ell+1}) = (c, p, a_{k+\ell+1}y_1 \ldots y_{d-c+n})$$

[36]The cycles may be wired in such a way that their first element is connected to the last one and all the others are connected to the previous ones. Then the cycles can represent clocks so that, for instance, if $d_1 \ldots d_{ms}$ is a state of a cycle (with ms length) representing the kth state of an arbitrary clock, then $d_2 \ldots d_{ms} d_1$ will represent its $k + 1 \pmod{ms}$th state.

whenever $\psi'(a_1, \ldots, a_k) = (c, p', y'_1 \ldots y'_{d-c+n})$, $\psi''(a_{k+1}, \ldots, a_{k+\ell}) = (c, p'', y''_1 \ldots y''_{d-c+n})$, $\{p', p''\} = \{p, \lambda\}$ (with $|p| \in \{0, c\}$ including the possibility of $p = \lambda$),

$$y_{d-c+n} \;(= y'_{d-c+n} = y''_{d-c+n}) = u_s \;(= v_s),$$

$$y_{j-1} = \begin{cases} u_s \;(= v_s) & \text{if } y'_{j-1} = y''_{j-1} = u_s \;(= v_s) \text{ and } y_j \in \{u_1, v_1\}, \\ u_i & \text{if } y'_{j-1} = y''_{j-1} = u_i \text{ and } y_j = u_{i+1},\ i \in \{1, \ldots, s-1\}, \\ v_i & \text{if } v_i \in \{y'_{j-1}, y''_{j-1}\} \text{ and } y_j = v_{i+1},\ i \in \{1, \ldots, s-1\}, \end{cases}$$

$j = 2, \ldots, d-c+n$, provided

$$a_{k+\ell+1} = \begin{cases} u_s \;(= v_s) & \text{if } c-d+1 \pmod s = 1, \\ u_{c-d \pmod s} & \text{if } y_1 = u_{c-d+1 \pmod s}, \\ & \quad c-d+1 \pmod s \in \{2, \ldots, s-1\}, \\ v_{c-d \pmod s} & \text{if } y_1 = v_{c-d+1 \pmod s}, \\ & \quad c-d+1 \pmod s \in \{2, \ldots, s-1\}, \\ \text{arbitrary element of } \{u_{s-1}, v_{s-1}\} & \text{if } c-d+1 \pmod s = s. \end{cases}$$

Using the definition of $\mathcal{R}_{\tau,H,d}$, by Lemma 3.5 it is obvious that $y_1 \ldots y_{d-c+n}$ and ψ are well defined. □

We shall use the following concept as well. Define the subautomaton $\mathcal{R}_{\tau,d}$ of $\mathcal{R}_{\tau,X^n,d}$ to have state set $R_{\tau,d} = R_{\tau,X^n,d} \setminus \{(k, \lambda, q) \mid (k, \lambda, q) \in R_{\tau,X^n,d}\}$. This is a subautomaton since px is a prefix of a word of X^n for every $p \in X^*$ with $|p| < n, x \in X$. For $(k, p, q) \in R_{\tau,d}$, we have $|p| = k$ always, so we will use the short notation (p, q) for $(k, p, q) \in R_{\tau,d}$. We have, for $(p, yq) \in R_{\tau,d}$ $(y \in \{u_t, v_t \mid t = 1, \ldots, s\})$ and $x \in X$,

$$\delta_{\tau,d}((p, yq), x) = \begin{cases} (px, q) & \text{if } |p| < n, \\ (x, q\tau(p)) & \text{if } |p| = n. \end{cases}$$

Let $\mathcal{M} = (W \times Z, X_{\mathcal{M}}, \delta_{\mathcal{M}})$ be an automaton with $Y \subseteq Z$; moreover, let $\mathcal{B} = (B, X, \delta_{\mathcal{B}})$ be a subautomaton of $\mathcal{M}$ having a homomorphism $\psi : B \to R_{\tau,d}$ onto $\mathcal{R}_{\tau,d}$ such that $\psi((w, z)) = (p, yq)$ $((w, z) \in B, (p, yq) \in R_{\tau,d},\ y \in Y)$ implies $z = y$. Then we also say that $\mathcal{M}$ *y-represents* $\mathcal{R}_{\tau,d}$ (with respect to ψ).

The following statement is obvious.

Proposition 6.8. *Let $\tau_i : X^n \to Y^n, i = 1, \ldots, m$, be a system of mappings; moreover, let d be a positive integer. For any $i = 1, \ldots, m$, let $\mathcal{M}_i = (W_i \times Z_i, X, \delta_i)$ be an automaton which y-represents $\mathcal{R}_{\tau_i,d}$. Consider an automaton $\mathcal{M}_{m+1} = (M_{m+1}, X_{m+1}, \delta_{m+1})$ with $X_{m+1} = Y^m$, a product $\mathcal{U} = \mathcal{R}_{\tau_1,d} \times \cdots \times \mathcal{R}_{\tau_m,d} \times \mathcal{M}_{m+1}(X, \varphi_1, \ldots, \varphi_{m+1})$ with*

$$\varphi_t((p_1, y_1q_1), \ldots, (p_m, y_mq_m), a, x) = \begin{cases} x & \text{if } t \le m, \\ (y_1, \ldots, y_m) & \text{if } t = m+1 \end{cases}$$

$((p_i, y_iq_i) \in R_{\tau_i,d}, y_i \in Y, i = 1, \ldots, m, a \in M_{m+1}, x \in X)$.

Define the product $\mathcal{M} = \mathcal{M}_1 \times \cdots \times \mathcal{M}_m \times \mathcal{M}_{m+1}(X, \psi_1, \ldots, \psi_{m+1})$ with

$$\psi_t((w_1, z_1), \ldots, (w_m, z_m), a, x) = \begin{cases} x & \text{if } t \le m, \\ (z_1, \ldots, z_m) & \text{if } t = m+1. \end{cases}$$

Then $\mathcal{M}$ homomorphically represents $\mathcal{U}$.

Lemma 6.9. *Let $\mathcal{A} = (A, X, \delta_{\mathcal{A}})$ be an automaton satisfying Letichevsky's criterion and let $\mathbf{u} = u_1 \dots u_s, \mathbf{v} = v_1 \dots v_s$, be any pair of its control words. Consider an alphabet X, a multiple n of s with $n = \ell s, \ell > 1$, and a mapping $\tau : X^n \to A^n$ having the property $\tau(p) \in \{\mathbf{u}, \mathbf{v}\}^\ell$ for each $p \in X^n$. For any integer $d \geq |X^n|$, there exists a primitive power $\mathcal{M}$ of $\mathcal{A}$ such that $\mathcal{R}_{\tau,d}$ is y-represented by $\mathcal{M}$. Moreover, apart from the last factor, the feedback functions of the factors of $\mathcal{M}$ really do not depend on their last state variable.*

Proof. By Lemma 6.6 and by an inductive application of Lemma 6.7, we can prove that $\mathcal{R}_{\tau,X^n,|X^n|}$ is y-represented by an appropriate primitive power $\mathcal{M}' = \mathcal{A}^k(X, \varphi'_1, \dots, \varphi'_k)$ of $\mathcal{A}$. Since $\mathcal{R}_{\tau,|X^n|}$ is a subautomaton of $\mathcal{R}_{\tau,X^n,|X^n|}$, this primitive power $\mathcal{M}'$ also y-represents $\mathcal{R}_{\tau,|X^n|}$. If $d = |X^n|$, then $\mathcal{M}'$ has the required conditions. Otherwise, let $\mathcal{M} = \mathcal{A}^{k+\ell}(X, \varphi_1, \dots, \varphi_{k+\ell})$ with $\ell = d - |X^n|$ such that for any $(a_1, \dots, a_{k+\ell}) \in A^{k+\ell}$, $x \in X, t = 1, \dots, k + \ell$,

$$\varphi_t(a_1, \dots, a_{k+\ell}, x) = \begin{cases} \varphi'_t(a_1, \dots, a_k, x) & \text{if } t \leq k, \\ x[a_t, a_{t-1}] & \text{otherwise.} \end{cases}$$

This power of $\mathcal{A}$ is primitive and y-represents $\mathcal{R}_{\tau,d}$. □

Lemma 6.10. *Let $\mathcal{D} = (D, X, \delta)$ and $\mathcal{B} = (B, Y, \delta')$ be automata with $D \subseteq B$. Moreover, let $\tau : X^n \to Y^n$ $(n > 0)$ be a mapping and assume that for a suitable integer $d > 0$ the following two conditions are satisfied:*

(1) *For all $a \in B$, $(p, q) \in R_{\tau,d}$, $p \in X^n$: $\delta'(a, q) \in D$ implies $\delta(\delta'(a, q), p) = \delta'(a, q\tau(p))$ $(\in D)$.*
(2) *$\{\delta(\delta'(a, q), p) \mid a \in D, (p, q) \in R_{\tau,d}, \delta'(a, q) \in D\} = D$.*

Then there exists an α_0-product $\mathcal{R}_{\tau,d} \times \mathcal{B}(X, \varphi_1, \varphi_2)$ which homomorphically represents $\mathcal{D}$ such that $\varphi_2((p, yq), x)$ $((p, yq) \in R_{\tau,d}, x \in X, y \in Y)$ really depends only on y.

Proof. Form the α_0-product $\mathcal{C} = (C, X, \delta'') = \mathcal{R}_{\tau,d} \times \mathcal{B}(X, \varphi_1, \varphi_2)$, where for arbitrary $(p, yq) \in R_{\tau,d}$ $(y \in Y)$, $b \in B$ and $x \in X$, $\varphi_1(x) = x$ and $\varphi_2((p, yq), x) = y$. Define the subautomaton $\mathcal{C}'$ of $\mathcal{C}$ with states $C' = \{((p, yq), a) \in C \mid \delta'(a, yq) \in D\}$ and input set X. We map the state $c = ((p, yq), a)$ of $\mathcal{C}'$ to the state $\delta(\delta'(a, yq), p)$ of $\mathcal{D}$.

Assume that $\mathcal{C}'$ receives an input letter x in this state c. If $|p| < n$, then $\delta''(c, x) = ((px, q), \delta'(a, y))$, which maps to the state $\delta(\delta'(\delta'(a, y), q), px) = \delta(\delta(\delta'(a, yq), p), x)$ of $\mathcal{D}$, as required.

If, on the other hand, $|p| = n$, then $\delta''(c, x) = ((x, q\tau(p)), \delta'(a, y))$. This maps to $\delta(\delta'(\delta'(a, y), q\tau(p)), x)$ in $\mathcal{D}$, that is, to $\delta(\delta'(\delta'(a, yq), \tau(p)), x) = \delta(\delta(\delta'(a, yq), p), x)$, by (1) since $\delta'(a, yq) \in D$. (Observe in the second case, $|q| = d - 1$.)

Thus, the mapping $\psi(((p, q), a)) = \delta(\delta'(a, q), p)$ $(((p, q), a) \in C', a \in D \subseteq B)$ is a homomorphism of a subautomaton of $\mathcal{C}$ into $\mathcal{D}$. By (2), ψ is a mapping onto D. Finally, as $\varphi_2((p, yq), x) = y$ $((p, yq) \in R_{\tau,d}, x \in X)$, we obtain that φ_2 really depends only on y. This ends the proof. □

We shall use the following natural extension of this result.

Lemma 6.11. *Let $\mathcal{D} = (D, X, \delta)$ be an automaton. Consider a product $\mathcal{N} = (B_1 \times \cdots \times B_m, Z^\ell, \delta') = \mathcal{B}_1 \times \cdots \times \mathcal{B}_m(Z^\ell, \varphi_1, \ldots, \varphi_m)$ of automata $\mathcal{B}_t$, $1 \le t \le m$, with $D \subseteq B_1 \times \cdots \times B_m$. Let $\tau_i : X^n \to Z^n$ $(n > 0)$, $1 \le i \le \ell$, be mappings; moreover let $\tau : X^n \to (Z^\ell)^n$ with $\tau_i(p) = z_{1,i} \ldots z_{n,i}$, $i = 1, \ldots, \ell$, whenever $\tau(p) = (z_{1,1}, \ldots, z_{1,\ell}) \ldots (z_{n,1}, \ldots, z_{n,\ell})$ such that the following two conditions hold:*

(1) *For every $a \in B_1 \times \cdots \times B_m$, $(p, q) \in R_{\tau,d}$, $p \in X^n$: $\delta'(a, q) \in D$ implies $\delta(\delta'(a, q), p) = \delta'(a, q\tau(p))$ $(\in D)$.*
(2) *$D = \{\delta(\delta'(a, q), p) \mid a \in B_1 \times \cdots \times B_m,\ (p, q) \in R_{\tau,d}, \delta'(a, q) \in D\}$.*

Then the product $\mathcal{V} = \mathcal{R}_{\tau_1,d} \times \cdots \times \mathcal{R}_{\tau_\ell,d} \times \mathcal{N}(X, \varphi'_1, \ldots, \varphi'_\ell, \varphi''_{\ell+1}) = \mathcal{R}_{\tau_1,d} \times \cdots \times \mathcal{R}_{\tau_\ell,d} \times \mathcal{B}_1 \times \cdots \times \mathcal{B}_m(X, \varphi'_1, \ldots, \varphi'_{\ell+m})$ homomorphically represents $\mathcal{D}$, where for each $(1 \le i \le \ell + m)$, we have

$$\varphi'_i((p_1, y_1q_1), \ldots, (p_\ell, y_\ell q_\ell), b_1, \ldots, b_m, x)$$
$$= \begin{cases} x & \text{if } 1 \le i \le \ell, \\ \varphi_{i-\ell}(b_1, \ldots, b_m, (y_1, \ldots, y_\ell)) & \text{otherwise} \end{cases}$$

$((p_i, y_iq_i) \in R_{\tau_i,d}$, $1 \le i \le \ell$, $x \in X$, $(y_1, \ldots, y_\ell) \in Z^\ell$, and $\varphi''_{\ell+1} = (\varphi'_{\ell+1}, \ldots, \varphi'_{\ell+m})$.

Proof. First we apply Lemma 6.10, taking $\mathcal{N}$ in the role of $\mathcal{B}$. Consider the α_0-product $\mathcal{U} = \mathcal{R}_{\tau,d} \times \mathcal{N}(X, \chi_1, \chi_2)$ given by Lemma 6.10 and the product $\mathcal{V} = \mathcal{R}_{\tau_1,d} \times \cdots \times \mathcal{R}_{\tau_\ell,d} \times \mathcal{B}_1 \times \cdots \times \mathcal{B}_m(X, \varphi'_1, \ldots, \varphi'_{\ell+m})$ just defined.

For a state $(p, yq, b_1, \ldots, b_m)$ of $\mathcal{U}$, where $(p, yq) \in R_{\tau,d}$ with $p \in X^*$, $y = y_1 \ldots y_\ell \in Z^\ell$, $q \in (Z^\ell)^*$, $b_i \in B_i$ $(1 \le i \le m)$, given $q = (x_{1,1}, \ldots, x_{1,\ell}) \ldots (x_{h,1}, \ldots, x_{h,\ell})$, we put

$$q_j = \begin{pmatrix} x_{1,j} \\ \vdots \\ x_{h,j} \end{pmatrix} \in Z^h$$

(for some $h \ge 0$, for each $j = 1, \ldots, \ell$). We write this state as

$$\left(p, \begin{pmatrix} y_1 & \cdots & y_\ell \\ x_{1,1} & \cdots & x_{1,\ell} \\ \vdots & \ddots & \vdots \\ x_{h,1} & \cdots & x_{h,\ell} \end{pmatrix}, b_1, \ldots, b_m \right).$$

Define an injective mapping $\varrho : R_{\tau,d} \times B_1 \times \cdots \times B_m \to R_{\tau_1,d} \times \cdots \times R_{\tau_\ell,d} \times B_1 \times B_m$ by

$$\varrho \left(p, \begin{pmatrix} y_1 & \cdots & y_\ell \\ x_{1,1} & \cdots & x_{1,\ell} \\ \vdots & \ddots & \vdots \\ x_{h,1} & \cdots & x_{h,\ell} \end{pmatrix}, b_1, \ldots, b_m \right) = \left(\begin{pmatrix} (p, y_1x_{1,1} \ldots x_{h,1}) \\ \vdots \\ (p, y_\ell x_{1,\ell} \ldots x_{h,\ell}) \end{pmatrix}, b_1, \ldots, b_m \right).$$

Denote by $\delta_{\mathcal{U}}$ (resp., $\delta_{\mathcal{V}}$) the transition functions of $\mathcal{U}$ (resp., $\mathcal{V}$).

If $|p| < n$, in $\delta_{\mathcal{U}}(((p, yq), b_1, \ldots, b_m), x)$, the only changes are that the row of y's is lost, p is replaced by px, and b_i is replaced by b_i', which is b_i acted on by $\varphi_{i-\ell}(b_1, \ldots, b_m, (y_1, \ldots, y_\ell))$ in $\mathcal{B}_i$ for $i = 1, \ldots, m$, whereas in ϱ of this state, the column of y's is lost and p is replaced by px, while the b's change in the same way.

If $|p| = n$, then $\varrho(\delta_{\mathcal{U}}(((p, yq), b_1, \ldots, b_m), x)$ is

$$\varrho\left(\left(x, \begin{pmatrix} x_{1,1} & \cdots & x_{1,\ell} \\ \vdots & \ddots & \vdots \\ x_{h,1} & \cdots & x_{h,\ell} \\ z_{1,1} & \cdots & z_{1,\ell} \\ \vdots & \ddots & \vdots \\ z_{n,1} & \cdots & z_{n,\ell} \end{pmatrix} \right), b_1', \ldots, b_m' \right) = \left(\begin{pmatrix} (x, x_{1,1} \ldots x_{h,1} z_{1,1} \ldots z_{n,1}) \\ \vdots \\ (x, x_{1,\ell} \ldots x_{h,\ell} z_{1,\ell} \ldots z_{n,\ell}) \end{pmatrix}, b_1', \ldots, b_m' \right),$$

$$\text{where } \tau(p) = \begin{pmatrix} z_{1,1} & \cdots & z_{1,\ell} \\ \vdots & \ddots & \vdots \\ z_{n,1} & \cdots & z_{n,\ell} \end{pmatrix}, \tau_i(p) = \begin{pmatrix} z_{1,i} \\ \vdots \\ z_{n,i} \end{pmatrix}, i = 1, \ldots, \ell.$$

This shows that for any $(p, yq) \in R_{\tau,d}$, $(b_1, \ldots, b_m) \in B_1 \times \cdots \times B_m$, $x \in X$, we have

$$\varrho(\delta_{\mathcal{U}}(((p, q), b_1, \ldots, b_m), x)) = \delta_{\mathcal{V}}(\varrho(((p, q), b_1, \ldots, b_m)), x).$$

Therefore, the product $\mathcal{U}$ can be embedded isomorphically into the product $\mathcal{V}$. But by Lemma 6.10, $\mathcal{U}$ homomorphically represents $\mathcal{D}$. Thus, $\mathcal{V}$ also has this property. □

6.3 Homomorphic Completeness Under the Primitive Product

In this section, we will establish that a primitive product of Letichevsky automata can homomorphically represent any finite automaton $\mathcal{E}$. To avoid trivialities, we note that it is enough to restrict ourselves to cases in which $\mathcal{E}$ has at least three states.

Consider an automaton $\mathcal{A}$ satisfying Letichevsky's criterion and let $\mathbf{u} = u_1 \ldots u_s$, $\mathbf{v} = v_1 \ldots v_s$ denote a pair of control words for $\mathcal{A}$ as before. We put $\mathbf{u} \wedge \mathbf{u} = \mathbf{u} \vee \mathbf{u} = \mathbf{u}$, $\mathbf{v} \wedge \mathbf{v} = \mathbf{v} \vee \mathbf{v} = \mathbf{v}$, $\mathbf{v} \wedge \mathbf{u} = \mathbf{u} \wedge \mathbf{v} = \mathbf{u}$, and $\mathbf{v} \vee \mathbf{u} = \mathbf{u} \vee \mathbf{v} = \mathbf{v}$ so that $\wedge$ and $\vee$ are, respectively, logical AND and logical OR on the set $\{\mathbf{u}, \mathbf{v}\}$.

First we show the following technical result.

Lemma 6.12. *Define the automata* $\mathcal{B} = (\{\mathbf{u}, \mathbf{v}\}^9, \{\mathbf{u}, \mathbf{v}\}^4, \delta_{\mathcal{B}})$, $\mathcal{C} = (\{\mathbf{u}, \mathbf{v}\}^{18n}, \{\mathbf{u}, \mathbf{v}\}^{6n}, \delta_{\mathcal{C}})$, $n \geq 3$, *with*

$$\begin{aligned}
\delta_{\mathcal{B}}((\mathbf{a}_1, \ldots, \mathbf{a}_9), (\mathbf{x}_1, \mathbf{x}_2, \mathbf{x}_3, \mathbf{x}_4)) &= (\mathbf{a}_2, \mathbf{a}_3 \wedge \mathbf{x}_1, \mathbf{a}_4, \mathbf{a}_5 \wedge \mathbf{x}_2, \mathbf{a}_6 \vee \mathbf{a}_3, \mathbf{a}_7 \wedge \mathbf{x}_3, \\
&\qquad \mathbf{a}_8, \mathbf{a}_9, \mathbf{x}_4), \\
\delta_{\mathcal{C}}((\mathbf{a}_1, \ldots, \mathbf{a}_{18n}), (\mathbf{x}_1, \ldots, \mathbf{x}_{6n})) &= (\mathbf{a}_1', \ldots, \mathbf{a}_{18n}'), \textit{ where for } i = 1, \ldots, n, \\
(\mathbf{a}_{18(i-1)+1}', \ldots, \mathbf{a}_{18(i-1)+9}') &= \delta_{\mathcal{B}}((\mathbf{a}_{18(i-1)+1}, \ldots, \mathbf{a}_{18(i-1)+9}), \\
&\qquad (\mathbf{x}_{6(i-1)+1}, \mathbf{x}_{6(i-1)+2}, \mathbf{x}_{6(i-1)+3}, \mathbf{a}_{18(i-1)+10})),
\end{aligned}$$

$$(\mathbf{a}'_{18(i-1)+10}, \ldots, \mathbf{a}'_{18(i-1)+18}) = \delta_{\mathcal{B}}((\mathbf{a}_{18(i-1)+10}, \ldots, \mathbf{a}_{18(i-1)+18}), (\mathbf{x}_{6(i-1)+4}, \mathbf{x}_{6(i-1)+5}, \mathbf{x}_{6(i-1)+6}, \mathbf{a}_{18(i-1)+1} \vee \mathbf{a}_{18(i-1)+19\,(\mathrm{mod}\,18n)})).$$

There exist a positive integer m and input words $\hat{\gamma}_0, \hat{\gamma}_1, \hat{\gamma}_2, \hat{\gamma}_3$ of $\mathcal{C}$ having the following properties. Given an appropriate subset $\{b_1, \ldots, b_n\}$ of the state set of $\mathcal{B}$, for every transformation γ of $\{b_1, \ldots, b_n\}$, there exists a word $\hat{\gamma} \in \{\hat{\gamma}_0, \hat{\gamma}_1, \hat{\gamma}_2, \hat{\gamma}_3\}^+$ inducing γ (i.e., $\delta_{\mathcal{B}}(b_i, \hat{\gamma}) = \gamma(b_i), i = 1, \ldots, n$) such that $|\hat{\gamma}| = m$.

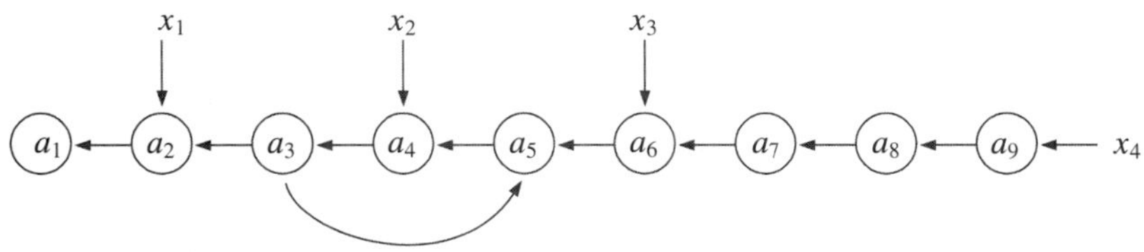

AUTOMATON $\mathcal{B}$

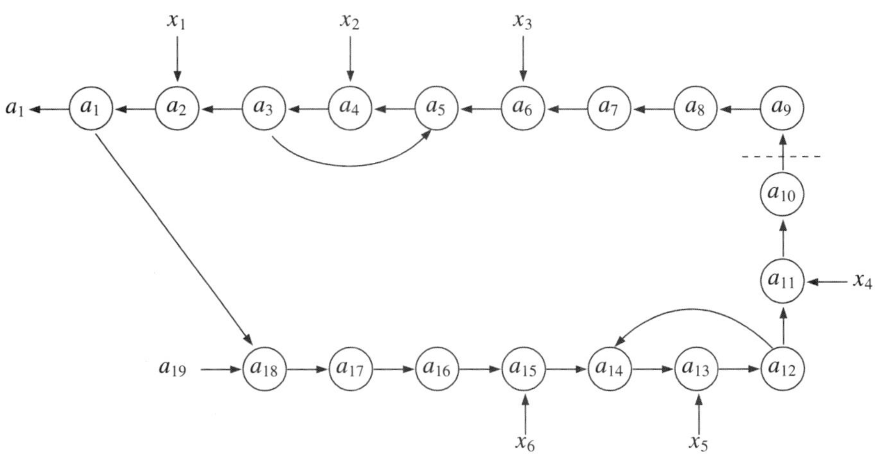

STRUCTURE OF ONE UNIT IN $\mathcal{C}$

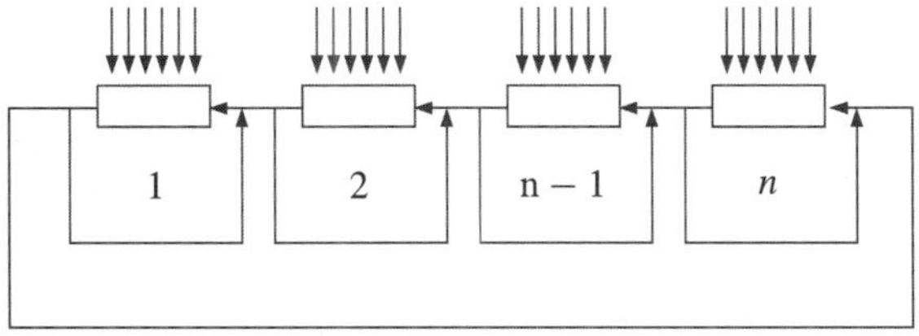

ROUGH STRUCTURE OF $\mathcal{C}$

Proof. Consider states of $\mathcal{C}$ having the form $(\mathbf{u}^4\mathbf{a}_1\mathbf{u}^{13} \ldots \mathbf{u}^4\mathbf{a}_n\mathbf{u}^{13})$, $\mathbf{a}_1, \ldots, \mathbf{a}_n \in \{\mathbf{u}, \mathbf{v}\}$, and use the short notation $(\mathbf{d}, \mathbf{e}) = \mathbf{u}^4\mathbf{d}\mathbf{u}^8\mathbf{e}\mathbf{u}^4$, $\mathbf{d}, \mathbf{e} \in \{\mathbf{u}, \mathbf{v}\}$. We represent b_i, $i \in \{1, \ldots, n\}$, by the state $\mathbf{u}^{18(i-1)}\mathbf{u}^4\mathbf{v}\mathbf{u}^{13}\mathbf{u}^{18(n-i)}$ of $\mathcal{C}$, which, using the short notation, is $(\mathbf{u}, \mathbf{u})^{i-1}(\mathbf{v}, \mathbf{u})(\mathbf{u}, \mathbf{u})^{n-i}$. First we show that we have words $q_{0,0}, q_{i,j} \in \{\mathbf{u}, \mathbf{v}\}^+$,

$i = 1, 2, 3, 4,\ j = 1, \ldots, n$, all having the same length such that

$$\delta_{\mathcal{C}}((\mathbf{d}_1, \mathbf{e}_1, \ldots, \mathbf{d}_n, \mathbf{e}_n), q_{i,j}) = (\mathbf{d}'_1, \mathbf{e}'_1, \ldots, \mathbf{d}'_n, \mathbf{e}'_n),$$

where $(\mathbf{d}'_\ell, \mathbf{e}'_\ell) = (\mathbf{d}_\ell, \mathbf{e}_\ell)$ if $\ell \neq j$ and, otherwise,

$$(\mathbf{d}'_j, \mathbf{e}'_j) = \begin{cases} (\mathbf{d}_j, \mathbf{e}_j) & \text{if } (i, j) = (0, 0), \\ (\mathbf{d}_j, \mathbf{d}_{j+1 (\mathrm{mod}\, n)}) & \text{if } i = 1,\ j = 1, \ldots, n, \\ (\mathbf{e}_j, \mathbf{d}_j) & \text{if } i = 2,\ j = 1, \ldots, n, \\ (\mathbf{u}, \mathbf{e}_j) & \text{if } i = 3,\ j = 1, \ldots, n, \\ (\mathbf{d}_j \vee \mathbf{e}_j, \mathbf{e}_j) & \text{if } i = 4,\ j = 1, \ldots, n. \end{cases}$$

Using the symmetry of the structure of $\mathcal{C}$ to show the existence of the q's, it is enough to prove the existence of $q_{0,0}, q_{i,1} \in \{\mathbf{u}, \mathbf{v}\}^+, i = 1, 2, 3, 4$. Define the following input *letters* (not words!) of $\mathcal{C}$:

$$x_0 = (\mathbf{uvu})^{2n},\ x_1 = \mathbf{uvuuuuvvu}(\mathbf{uvu})^{2n-3},\ x_2 = \mathbf{uvuuuv}(\mathbf{uvu})^{2n-2},$$
$$x_3 = \mathbf{vvuvvu}(\mathbf{uvu})^{2n-2},\ x_4 = \mathbf{uuvuuv}(\mathbf{uvu})^{2n-2},\ x_5 = \mathbf{uuu}(\mathbf{uvu})^{2n-1},$$
$$x_6 = \mathbf{uvuvvu}(\mathbf{uvu})^{2n-2},\ x_7 = \mathbf{uvv}(\mathbf{uvu})^{2n-1}.$$

It can be proved by an elementary computation that the words $q_{0,0} = {x_0}^9$, $q_{1,1} = {x_1}^5{x_2}^4, q_{2,1} = {x_3}^5{x_4}^4, q_{3,1} = {x_5}^9, q_{4,1} = {x_6}^5{x_7}^4$ satisfy our requirements. (See the detailed elementary computation below.)

Put $\hat{\gamma}_0 = (q_{0,0})^{3n^2+1}$, $\hat{\gamma}_1 = q_{1,1} \cdots q_{1,n} q_{3,1} \cdots q_{3,n} q_{2,1} \cdots q_{2,n} (q_{0,0})^{3n^2-3n+1}$, $\hat{\gamma}_2 = q_{1,1}q_{2,1}\ q_{1,2} \cdots q_{1,n}\ q_{3,1} \cdots q_{3,n} q_{2,1} \cdots q_{2,n} (q_{1,1} \cdots q_{1,n} q_{3,1} \cdots q_{3,n}\ q_{2,1} \cdots q_{2,n})^{n-1}$, $\hat{\gamma}_3 = q_{1,1}q_{3,2}q_{4,1}\ q_{2,1}q_{3,1}q_{2,1}(q_{0,0})^{3n^2-5}$, and use the short notation $b_i = \mathbf{u}^{18(i-1)}\mathbf{u}^4\mathbf{vu}^{13}\mathbf{u}^{18(n-i)}$, $i = 1, \ldots, n$. By an elementary computation we get that the $\hat{\gamma}_j$ $(j = 0, 1, 2, 3)$, which all have the same length, induce the following transformations γ_j of $\{b_1, \ldots, b_n\}$:

$$\gamma_0 = \begin{pmatrix} b_1 & b_2 & b_3 & \ldots & b_n \\ b_1 & b_2 & b_3 & \ldots & b_n \end{pmatrix},\quad \gamma_1 = \begin{pmatrix} b_1 & b_2 & \ldots & b_{n-1} & b_n \\ b_2 & b_3 & \ldots & b_n & b_1 \end{pmatrix},$$

$$\gamma_2 = \begin{pmatrix} b_1 & b_2 & b_3 & \ldots & b_n \\ b_2 & b_1 & b_3 & \ldots & b_n \end{pmatrix},\quad \gamma_3 = \begin{pmatrix} b_1 & b_2 & b_3 & \ldots & b_n \\ b_1 & b_1 & b_3 & \ldots & b_n \end{pmatrix}.$$

(The detailed computation is given below.)

Using the well-known fact that $\gamma_1, \gamma_2, \gamma_3$ generate all transformations on the n element set $\{b_1, \ldots, b_n\}$ (see also Proposition 1.5) and that γ_0 is the identity, we obtain our technical result. □

Proof of the properties of $q_{0,0}$ and $q_{i,1}, i = 1, 2, 3, 4$. The following detailed elementary computation shows that the words $q_{0,0} = {x_0}^9$, $q_{1,1} = {x_1}^5{x_2}^4, q_{2,1} = {x_3}^5{x_4}^4, q_{3,1} = {x_5}^9, q_{4,1} = {x_6}^5{x_7}^4$ satisfy our requirements in the proof of Lemma 6.12:

$$\begin{aligned}
(0)\ & \delta_{\mathcal{C}}(\mathbf{u}^4\mathbf{d}_1\mathbf{u}^8\mathbf{e}_1\mathbf{u}^8\mathbf{d}_2\mathbf{u}^8\mathbf{e}_2\mathbf{u}^8\mathbf{d}_3\mathbf{u}^8\mathbf{e}_3 \ldots \mathbf{u}^8\mathbf{d}_n\mathbf{u}^8\mathbf{e}_n\mathbf{u}^4, {x_0}^9) \\
&= \delta_{\mathcal{C}}(\mathbf{u}^3\mathbf{d}_1\mathbf{u}^8\mathbf{e}_1\mathbf{u}^8\mathbf{d}_2\mathbf{u}^8\mathbf{e}_2\mathbf{u}^8\mathbf{d}_3\mathbf{u}^8\mathbf{e}_3 \ldots \mathbf{u}^8\mathbf{d}_n\mathbf{u}^8\mathbf{e}_n\mathbf{u}^5, {x_0}^8) \\
&= \delta_{\mathcal{C}}(\mathbf{u}^2\mathbf{d}_1\mathbf{u}^8\mathbf{e}_1\mathbf{u}^8\mathbf{d}_2\mathbf{u}^8\mathbf{e}_2\mathbf{u}^8\mathbf{d}_3\mathbf{u}^8\mathbf{e}_3 \ldots \mathbf{u}^8\mathbf{d}_n\mathbf{u}^8\mathbf{e}_n\mathbf{u}^6, {x_0}^7) \\
&= \delta_{\mathcal{C}}(\mathbf{u}^4\mathbf{d}_1\mathbf{u}^8\mathbf{e}_1\mathbf{u}^8\mathbf{d}_2\mathbf{u}^8\mathbf{e}_2\mathbf{u}^8\mathbf{d}_3\mathbf{u}^8\mathbf{e}_3 \ldots \mathbf{u}^8\mathbf{d}_n\mathbf{u}^8\mathbf{e}_n\mathbf{u}^4, {x_0}^6) \\
&= \delta_{\mathcal{C}}(\mathbf{u}^3\mathbf{d}_1\mathbf{u}^8\mathbf{e}_1\mathbf{u}^8\mathbf{d}_2\mathbf{u}^8\mathbf{e}_2\mathbf{u}^8\mathbf{d}_3\mathbf{u}^8\mathbf{e}_3 \ldots \mathbf{u}^8\mathbf{d}_n\mathbf{u}^8\mathbf{e}_n\mathbf{u}^5, {x_0}^5)
\end{aligned}$$

$$= \delta_C(\mathbf{u}^2\mathbf{d}_1\mathbf{u}^8\mathbf{e}_1\mathbf{u}^8\mathbf{d}_2\mathbf{u}^8\mathbf{e}_2\mathbf{u}^8\mathbf{d}_3\mathbf{u}^8\mathbf{e}_3\ldots\mathbf{u}^8\mathbf{d}_n\mathbf{u}^8\mathbf{e}_n\mathbf{u}^6, x_0{}^4)$$
$$= \delta_C(\mathbf{u}^4\mathbf{d}_1\mathbf{u}^8\mathbf{e}_1\mathbf{u}^8\mathbf{d}_2\mathbf{u}^8\mathbf{e}_2\mathbf{u}^8\mathbf{d}_3\mathbf{u}^8\mathbf{e}_3\ldots\mathbf{u}^8\mathbf{d}_n\mathbf{u}^8\mathbf{e}_n\mathbf{u}^4, x_0{}^3)$$
$$= \delta_C(\mathbf{u}^3\mathbf{d}_1\mathbf{u}^8\mathbf{e}_1\mathbf{u}^8\mathbf{d}_2\mathbf{u}^8\mathbf{e}_2\mathbf{u}^8\mathbf{d}_3\mathbf{u}^8\mathbf{e}_3\ldots\mathbf{u}^8\mathbf{d}_n\mathbf{u}^8\mathbf{e}_n\mathbf{u}^5, x_0{}^2)$$
$$= \delta_C(\mathbf{u}^2\mathbf{d}_1\mathbf{u}^8\mathbf{e}_1\mathbf{u}^8\mathbf{d}_2\mathbf{u}^8\mathbf{e}_2\mathbf{u}^8\mathbf{d}_3\mathbf{u}^8\mathbf{e}_3\ldots\mathbf{u}^8\mathbf{d}_n\mathbf{u}^8\mathbf{e}_n\mathbf{u}^6, x_0)$$
$$= (\mathbf{u}^4\mathbf{d}_1\mathbf{u}^8\mathbf{e}_1\mathbf{u}^8\mathbf{d}_2\mathbf{u}^8\mathbf{e}_2\mathbf{u}^8\mathbf{d}_3\mathbf{u}^8\mathbf{e}_3\ldots\mathbf{u}^8\mathbf{d}_n\mathbf{u}^8\mathbf{e}_n\mathbf{u}^4).$$

(1) $\delta_C(\mathbf{u}^4\mathbf{d}_1\mathbf{u}^8\mathbf{e}_1\mathbf{u}^8\mathbf{d}_2\mathbf{u}^8\mathbf{e}_2\mathbf{u}^8\mathbf{d}_3\mathbf{u}^8\mathbf{e}_3\ldots\mathbf{u}^8\mathbf{d}_n\mathbf{u}^8\mathbf{e}_n\mathbf{u}^4, x_1{}^5x_2{}^4)$
$$= \delta_C(\mathbf{u}^3\mathbf{d}_1\mathbf{u}^{17}\mathbf{d}_2\mathbf{u}^8\mathbf{e}_2\mathbf{u}^8\mathbf{d}_3\mathbf{u}^8\mathbf{e}_3\ldots\mathbf{u}^8\mathbf{d}_n\mathbf{u}^8\mathbf{e}_n\mathbf{u}^5, x_1{}^4x_2{}^4)$$
$$= \delta_C(\mathbf{u}^2\mathbf{d}_1\mathbf{u}^{17}\mathbf{d}_2\mathbf{u}^8\mathbf{e}_2\mathbf{u}^8\mathbf{d}_3\mathbf{u}^8\mathbf{e}_3\ldots\mathbf{u}^8\mathbf{d}_n\mathbf{u}^8\mathbf{e}_n\mathbf{u}^6, x_1{}^3x_2{}^4)$$
$$= \delta_C(\mathbf{u}^4\mathbf{d}_1\mathbf{u}^{14}\mathbf{d}_2\mathbf{u}^2\mathbf{d}_2\mathbf{u}^8\mathbf{e}_2\mathbf{u}^8\mathbf{d}_3\mathbf{u}^8\mathbf{e}_3\ldots\mathbf{u}^8\mathbf{d}_n\mathbf{u}^8\mathbf{e}_n\mathbf{u}^4, x_1{}^2x_2{}^4)$$
$$= \delta_C(\mathbf{u}^3\mathbf{d}_1\mathbf{u}^{14}\mathbf{d}_2\mathbf{u}^2\mathbf{d}_2\mathbf{u}^8\mathbf{e}_2\mathbf{u}^8\mathbf{d}_3\mathbf{u}^8\mathbf{e}_3\ldots\mathbf{u}^8\mathbf{d}_n\mathbf{u}^8\mathbf{e}_n\mathbf{u}^5, x_1x_2{}^4)$$
$$= \delta_C(\mathbf{u}^2\mathbf{d}_1\mathbf{u}^{14}\mathbf{d}_2\mathbf{u}^2\mathbf{d}_2\mathbf{u}^8\mathbf{e}_2\mathbf{u}^5\mathbf{d}_2\mathbf{u}^2\mathbf{d}_3\mathbf{u}^8\mathbf{e}_3\ldots\mathbf{u}^8\mathbf{d}_n\mathbf{u}^8\mathbf{e}_n\mathbf{u}^6, x_2{}^4)$$
$$= \delta_C(\mathbf{u}^4\mathbf{d}_1\mathbf{u}^{11}\mathbf{d}_2\mathbf{u}^5\mathbf{d}_2\mathbf{u}^8\mathbf{e}_2\mathbf{u}^2\mathbf{d}_2\mathbf{u}^5\mathbf{d}_3\mathbf{u}^8\mathbf{e}_3\ldots\mathbf{u}^8\mathbf{d}_n\mathbf{u}^8\mathbf{e}_n\mathbf{u}^4, x_2{}^3)$$
$$= \delta_C(\mathbf{u}^3\mathbf{d}_1\mathbf{u}^{11}\mathbf{d}_2\mathbf{u}^5\mathbf{d}_2\mathbf{u}^8\mathbf{e}_2\mathbf{u}^2\mathbf{d}_2\mathbf{u}^5\mathbf{d}_3\mathbf{u}^8\mathbf{e}_3\ldots\mathbf{u}^8\mathbf{d}_n\mathbf{u}^8\mathbf{e}_n\mathbf{u}^5, x_2{}^2)$$
$$= \delta_C(\mathbf{u}^2\mathbf{d}_1\mathbf{u}^{11}\mathbf{d}_2\mathbf{u}^5\mathbf{d}_2\mathbf{u}^8\mathbf{e}_2\mathbf{u}^8\mathbf{d}_3\mathbf{u}^8\mathbf{e}_3\ldots\mathbf{u}^8\mathbf{d}_n\mathbf{u}^8\mathbf{e}_n\mathbf{u}^6, x_2)$$
$$= (\mathbf{u}^4\mathbf{d}_1\mathbf{u}^8\mathbf{d}_2\mathbf{u}^8\mathbf{d}_2\mathbf{u}^8\mathbf{e}_2\mathbf{u}^8\mathbf{d}_3\mathbf{u}^8\mathbf{e}_3\ldots\mathbf{u}^8\mathbf{d}_n\mathbf{u}^8\mathbf{e}_n\mathbf{u}^4).$$

(2) $\delta_C(\mathbf{u}^4\mathbf{d}_1\mathbf{u}^8\mathbf{e}_1\mathbf{u}^8\mathbf{d}_2\mathbf{u}^8\mathbf{e}_2\mathbf{u}^8\mathbf{d}_3\mathbf{u}^8\mathbf{e}_3\ldots\mathbf{u}^8\mathbf{d}_n\mathbf{u}^8\mathbf{e}_n\mathbf{u}^4, x_3{}^5x_4{}^4)$
$$= \delta_C(\mathbf{u}^3\mathbf{d}_1\mathbf{u}^8\mathbf{e}_1\mathbf{u}^8\mathbf{d}_2\mathbf{u}^8\mathbf{e}_2\mathbf{u}^8\mathbf{d}_3\mathbf{u}^8\mathbf{e}_3\ldots\mathbf{u}^8\mathbf{d}_n\mathbf{u}^8\mathbf{e}_n\mathbf{u}^5, x_3{}^4x_4{}^4)$$
$$= \delta_C(\mathbf{u}^2\mathbf{d}_1\mathbf{u}^8\mathbf{e}_1\mathbf{u}^8\mathbf{d}_2\mathbf{u}^8\mathbf{e}_2\mathbf{u}^8\mathbf{d}_3\mathbf{u}^8\mathbf{e}_3\ldots\mathbf{u}^8\mathbf{d}_n\mathbf{u}^8\mathbf{e}_n\mathbf{u}^6, x_3{}^3x_4{}^4)$$
$$= \delta_C(\mathbf{u}\mathbf{d}_1\mathbf{u}^2\mathbf{d}_1\mathbf{u}^5\mathbf{e}_1\mathbf{u}^2\mathbf{e}_1\mathbf{u}^8\mathbf{d}_2\mathbf{u}^8\mathbf{e}_2\mathbf{u}^8\mathbf{d}_3\mathbf{u}^8\mathbf{e}_3\ldots\mathbf{u}^8\mathbf{d}_n\mathbf{u}^8\mathbf{e}_n\mathbf{u}^4, x_3{}^2x_4{}^4)$$
$$= \delta_C(\mathbf{d}_1\mathbf{u}^2\mathbf{d}_1\mathbf{u}^5\mathbf{e}_1\mathbf{u}^2\mathbf{e}_1\mathbf{u}^8\mathbf{d}_2\mathbf{u}^8\mathbf{e}_2\mathbf{u}^8\mathbf{d}_3\mathbf{u}^8\mathbf{e}_3\ldots\mathbf{u}^8\mathbf{d}_n\mathbf{u}^8\mathbf{e}_n\mathbf{u}^5, x_3x_4{}^4)$$
$$= \delta_C(\mathbf{u}^2\mathbf{d}_1\mathbf{u}^5\mathbf{e}_1\mathbf{u}^2\mathbf{e}_1\mathbf{u}^5\mathbf{d}_1\mathbf{u}^2\mathbf{d}_2\mathbf{u}^8\mathbf{e}_2\mathbf{u}^8\mathbf{d}_3\mathbf{u}^8\mathbf{e}_3\ldots\mathbf{u}^8\mathbf{d}_n\mathbf{u}^8\mathbf{e}_n\mathbf{u}^5\mathbf{d}_1, x_4{}^4)$$
$$= \delta_C(\mathbf{u}^4\mathbf{d}_1\mathbf{u}^2\mathbf{e}_1\mathbf{u}^5\mathbf{e}_1\mathbf{u}^2\mathbf{d}_1\mathbf{u}^5\mathbf{d}_2\mathbf{u}^8\mathbf{e}_2\mathbf{u}^8\mathbf{d}_3\mathbf{u}^8\mathbf{e}_3\ldots\mathbf{u}^8\mathbf{d}_n\mathbf{u}^8\mathbf{e}_n\mathbf{u}^2\mathbf{d}_1\mathbf{u}, x_4{}^3)$$
$$= \delta_C(\mathbf{u}^6\mathbf{e}_1\mathbf{u}^8\mathbf{d}_1\mathbf{u}^5\mathbf{d}_2\mathbf{u}^8\mathbf{e}_2\mathbf{u}^8\mathbf{d}_3\mathbf{u}^8\mathbf{e}_3\ldots\mathbf{u}^8\mathbf{d}_n\mathbf{u}^8\mathbf{e}_n\mathbf{u}^2\mathbf{d}_1\mathbf{u}^2, x_4{}^2)$$
$$= \delta_C(\mathbf{u}^5\mathbf{e}_1\mathbf{u}^8\mathbf{d}_1\mathbf{u}^5\mathbf{d}_2\mathbf{u}^8\mathbf{e}_2\mathbf{u}^8\mathbf{d}_3\mathbf{u}^8\mathbf{e}_3\ldots\mathbf{u}^8\mathbf{d}_n\mathbf{u}^8\mathbf{e}_n\mathbf{u}^6, x_4)$$
$$= (\mathbf{u}^4\mathbf{e}_1\mathbf{u}^8\mathbf{d}_1\mathbf{u}^8\mathbf{d}_2\mathbf{u}^8\mathbf{e}_2\mathbf{u}^8\mathbf{d}_3\mathbf{u}^8\mathbf{e}_3\ldots\mathbf{u}^8\mathbf{d}_n\mathbf{u}^8\mathbf{e}_n\mathbf{u}^4).$$

(3) $\delta_C(\mathbf{u}^4\mathbf{d}_1\mathbf{u}^8\mathbf{e}_1\mathbf{u}^8\mathbf{d}_2\mathbf{u}^8\mathbf{e}_2\mathbf{u}^8\mathbf{d}_3\mathbf{u}^8\mathbf{e}_3\ldots\mathbf{u}^8\mathbf{d}_n\mathbf{u}^8\mathbf{e}_n\mathbf{u}^4, x_5{}^9)$
$$= \delta_C(\mathbf{u}^{12}\mathbf{e}_1\mathbf{u}^8\mathbf{d}_2\mathbf{u}^8\mathbf{e}_2\mathbf{u}^8\mathbf{d}_3\mathbf{u}^8\mathbf{e}_3\ldots\mathbf{u}^8\mathbf{d}_n\mathbf{u}^8\mathbf{e}_n\mathbf{u}^5, x_5{}^8)$$
$$= \delta_C(\mathbf{u}^{11}\mathbf{e}_1\mathbf{u}^8\mathbf{d}_2\mathbf{u}^8\mathbf{e}_2\mathbf{u}^8\mathbf{d}_3\mathbf{u}^8\mathbf{e}_3\ldots\mathbf{u}^8\mathbf{d}_n\mathbf{u}^8\mathbf{e}_n\mathbf{u}^6, x_5{}^7)$$
$$= \delta_C(\mathbf{u}^{13}\mathbf{e}_1\mathbf{u}^8\mathbf{d}_2\mathbf{u}^8\mathbf{e}_2\mathbf{u}^8\mathbf{d}_3\mathbf{u}^8\mathbf{e}_3\ldots\mathbf{u}^8\mathbf{d}_n\mathbf{u}^8\mathbf{e}_n\mathbf{u}^4, x_5{}^6)$$
$$= \delta_C(\mathbf{u}^{12}\mathbf{e}_1\mathbf{u}^8\mathbf{d}_2\mathbf{u}^8\mathbf{e}_2\mathbf{u}^8\mathbf{d}_3\mathbf{u}^8\mathbf{e}_3\ldots\mathbf{u}^8\mathbf{d}_n\mathbf{u}^8\mathbf{e}_n\mathbf{u}^5, x_5{}^5)$$
$$= \delta_C(\mathbf{u}^{11}\mathbf{e}_1\mathbf{u}^8\mathbf{d}_2\mathbf{u}^8\mathbf{e}_2\mathbf{u}^8\mathbf{d}_3\mathbf{u}^8\mathbf{e}_3\ldots\mathbf{u}^8\mathbf{d}_n\mathbf{u}^8\mathbf{e}_n\mathbf{u}^6, x_5{}^4)$$
$$= \delta_C(\mathbf{u}^{13}\mathbf{e}_1\mathbf{u}^8\mathbf{d}_2\mathbf{u}^8\mathbf{e}_2\mathbf{u}^8\mathbf{d}_3\mathbf{u}^8\mathbf{e}_3\ldots\mathbf{u}^8\mathbf{d}_n\mathbf{u}^8\mathbf{e}_n\mathbf{u}^4, x_5{}^3)$$
$$= \delta_C(\mathbf{u}^{12}\mathbf{e}_1\mathbf{u}^8\mathbf{d}_2\mathbf{u}^8\mathbf{e}_2\mathbf{u}^8\mathbf{d}_3\mathbf{u}^8\mathbf{e}_3\ldots\mathbf{u}^8\mathbf{d}_n\mathbf{u}^8\mathbf{e}_n\mathbf{u}^5, x_5{}^2)$$
$$= \delta_C(\mathbf{u}^{11}\mathbf{e}_1\mathbf{u}^8\mathbf{d}_2\mathbf{u}^8\mathbf{e}_2\mathbf{u}^8\mathbf{d}_3\mathbf{u}^8\mathbf{e}_3\ldots\mathbf{u}^8\mathbf{d}_n\mathbf{u}^8\mathbf{e}_n\mathbf{u}^6, x_5)$$
$$= (\mathbf{u}^{13}\mathbf{e}_1\mathbf{u}^8\mathbf{d}_2\mathbf{u}^8\mathbf{e}_2\mathbf{u}^8\mathbf{d}_3\mathbf{u}^8\mathbf{e}_3\ldots\mathbf{u}^8\mathbf{d}_n\mathbf{u}^8\mathbf{e}_n\mathbf{u}^4).$$

(4) $\delta_C(\mathbf{u}^4\mathbf{d}_1\mathbf{u}^8\mathbf{e}_1\mathbf{u}^8\mathbf{d}_2\mathbf{u}^8\mathbf{e}_2\mathbf{u}^8\mathbf{d}_3\mathbf{u}^8\mathbf{e}_3\ldots\mathbf{u}^8\mathbf{d}_n\mathbf{u}^8\mathbf{e}_n\mathbf{u}^4, x_6{}^5x_7{}^4)$
$$= \delta_C(\mathbf{u}^3\mathbf{d}_1\mathbf{u}^8\mathbf{e}_1\mathbf{u}^8\mathbf{d}_2\mathbf{u}^8\mathbf{e}_2\mathbf{u}^8\mathbf{d}_3\mathbf{u}^8\mathbf{e}_3\ldots\mathbf{u}^8\mathbf{d}_n\mathbf{u}^8\mathbf{e}_n\mathbf{u}^5, x_6{}^4x_7{}^4)$$
$$= \delta_C(\mathbf{u}^2\mathbf{d}_1\mathbf{u}^8\mathbf{e}_1\mathbf{u}^8\mathbf{d}_2\mathbf{u}^8\mathbf{e}_2\mathbf{u}^8\mathbf{d}_3\mathbf{u}^8\mathbf{e}_3\ldots\mathbf{u}^8\mathbf{d}_n\mathbf{u}^8\mathbf{e}_n\mathbf{u}^6, x_6{}^3x_7{}^4)$$
$$= \delta_C(\mathbf{u}^4\mathbf{d}_1\mathbf{u}^5\mathbf{e}_1\mathbf{u}^2\mathbf{e}_1\mathbf{u}^8\mathbf{d}_2\mathbf{u}^8\mathbf{e}_2\mathbf{u}^8\mathbf{d}_3\mathbf{u}^8\mathbf{e}_3\ldots\mathbf{u}^8\mathbf{d}_n\mathbf{u}^8\mathbf{e}_n\mathbf{u}^4, x_6{}^2x_7{}^4)$$
$$= \delta_C(\mathbf{u}^3\mathbf{d}_1\mathbf{u}^5\mathbf{e}_1\mathbf{u}^2\mathbf{e}_1\mathbf{u}^8\mathbf{d}_2\mathbf{u}^8\mathbf{e}_2\mathbf{u}^8\mathbf{d}_3\mathbf{u}^8\mathbf{e}_3\ldots\mathbf{u}^8\mathbf{d}_n\mathbf{u}^8\mathbf{e}_n\mathbf{u}^5, x_6x_7{}^4)$$
$$= \delta_C(\mathbf{u}^2\mathbf{d}_1\mathbf{u}^5\mathbf{e}_1\mathbf{u}^2\mathbf{e}_1\mathbf{u}^8\mathbf{d}_2\mathbf{u}^8\mathbf{e}_2\mathbf{u}^8\mathbf{d}_3\mathbf{u}^8\mathbf{e}_3\ldots\mathbf{u}^8\mathbf{d}_n\mathbf{u}^8\mathbf{e}_n\mathbf{u}^6, x_7{}^4)$$
$$= \delta_C(\mathbf{u}^4\mathbf{d}_1\mathbf{u}^2\mathbf{e}_1\mathbf{u}^5\mathbf{e}_1\mathbf{u}^8\mathbf{d}_2\mathbf{u}^8\mathbf{e}_2\mathbf{u}^8\mathbf{d}_3\mathbf{u}^8\mathbf{e}_3\ldots\mathbf{u}^8\mathbf{d}_n\mathbf{u}^8\mathbf{e}_n\mathbf{u}^4, x_7{}^3)$$
$$= \delta_C(\mathbf{u}^3\mathbf{d}_1\mathbf{u}^2\mathbf{e}_1\mathbf{u}^5\mathbf{e}_1\mathbf{u}^8\mathbf{d}_2\mathbf{u}^8\mathbf{e}_2\mathbf{u}^8\mathbf{d}_3\mathbf{u}^8\mathbf{e}_3\ldots\mathbf{u}^8\mathbf{d}_n\mathbf{u}^8\mathbf{e}_n\mathbf{u}^5, x_7{}^2)$$
$$= \delta_C(\mathbf{u}^2\mathbf{d}_1\mathbf{u}^2\mathbf{e}_1\mathbf{u}^5\mathbf{e}_1\mathbf{u}^8\mathbf{d}_2\mathbf{u}^8\mathbf{e}_2\mathbf{u}^8\mathbf{d}_3\mathbf{u}^8\mathbf{e}_3\ldots\mathbf{u}^8\mathbf{d}_n\mathbf{u}^8\mathbf{e}_n\mathbf{u}^6, x_7)$$
$$= (\mathbf{u}^4(\mathbf{d}_1 \vee \mathbf{e}_1)\mathbf{u}^8\mathbf{e}_1\mathbf{u}^8\mathbf{d}_2\mathbf{u}^8\mathbf{e}_2\mathbf{u}^8\mathbf{d}_3\mathbf{u}^8\mathbf{e}_3\ldots\mathbf{u}^8\mathbf{d}_n\mathbf{u}^8\mathbf{e}_n\mathbf{u}^4). \qquad \square$$

Proof of the properties of $\hat{\gamma}_j$ $(j = 0, 1, 2, 3)$. The next detailed elementary computation shows that the $\hat{\gamma}_j$ $(j = 0, 1, 2, 3)$, which all have the same length, induce the transformations γ_j of $\{b_1, \ldots, b_n\}$ given in the proof of Lemma 6.12.

(0′) Identity. $\delta_{\mathcal{C}}((\mathbf{d}_1, \mathbf{u}, \ldots, \mathbf{d}_n, \mathbf{u}), q_{0,0}) = (\mathbf{d}_1, \mathbf{u}, \ldots, \mathbf{d}_n, \mathbf{u})$.

(1′) n-cycle. $\delta_{\mathcal{C}}((\mathbf{d}_1, \mathbf{u}, \mathbf{d}_2, \ldots, \mathbf{u}, \mathbf{d}_{n-1}, \mathbf{u}, \mathbf{d}_n, \mathbf{u}), q_{1,1} \cdots q_{1,n} q_{3,1} \cdots q_{3,n} q_{2,1} \cdots q_{2,n})$
$= \delta_{\mathcal{C}}((\mathbf{d}_1, \mathbf{d}_2, \mathbf{d}_2, \ldots, \mathbf{d}_{n-1}, \mathbf{d}_{n-1}, \mathbf{d}_n, \mathbf{d}_n, \mathbf{d}_1), q_{3,1} \cdots q_{3,n} q_{2,1} \cdots q_{2,n})$
$= \delta_{\mathcal{C}}((\mathbf{u}, \mathbf{d}_2, \mathbf{u}, \ldots, \mathbf{d}_{n-1}, \mathbf{u}, \mathbf{d}_n, \mathbf{u}, \mathbf{d}_1), q_{2,1} \cdots q_{2,n})$
$= (\mathbf{d}_2, \mathbf{u}, \mathbf{d}_3, \ldots, \mathbf{u}, \mathbf{d}_n, \mathbf{u}, \mathbf{d}_1, \mathbf{u})$.

(2′) Transposition. $\delta_{\mathcal{C}}((\mathbf{d}_1, \mathbf{u}, \ldots, \mathbf{d}_n, \mathbf{u}), q_{1,1} q_{2,1} q_{1,2} \cdots q_{1,n} q_{3,1} \cdots q_{3,n} q_{2,1} \cdots q_{2,n}$
$(q_{1,1} \cdots q_{1,n} q_{3,1} \cdots q_{3,n} q_{2,1} \cdots q_{2,n})^{n-1})$
$= \delta_{\mathcal{C}}((\mathbf{d}_1, \mathbf{d}_2, \mathbf{d}_2, \mathbf{u}, \ldots, \mathbf{d}_n, \mathbf{u}), q_{2,1} q_{1,2} \cdots q_{1,n} q_{3,1} \cdots q_{3,n} q_{2,1} \cdots q_{2,n}$
$(q_{1,1} \cdots q_{1,n} q_{3,1} \cdots q_{3,n} q_{2,1} \cdots q_{2,n})^{n-1})$
$= \delta_{\mathcal{C}}((\mathbf{d}_2, \mathbf{d}_1, \mathbf{d}_2, \mathbf{u}, \mathbf{d}_3, \ldots, \mathbf{u}, \mathbf{d}_{n-1}, \mathbf{u}, \mathbf{d}_n, \mathbf{u}), q_{1,2} \cdots q_{1,n} q_{3,1} \cdots q_{3,n} q_{2,1} \cdots q_{2,n}$
$(q_{1,1} \cdots q_{1,n} q_{3,1} \cdots q_{3,n} q_{2,1} \cdots q_{2,n})^{n-1})$
$= \delta_{\mathcal{C}}((\mathbf{d}_2, \mathbf{d}_1, \mathbf{d}_2, \mathbf{d}_3, \mathbf{d}_3, \ldots, \mathbf{d}_{n-1}, \mathbf{d}_{n-1}, \mathbf{d}_n, \mathbf{d}_n, \mathbf{d}_2), q_{3,1} \cdots q_{3,n} q_{2,1} \cdots q_{2,n}$
$(q_{1,1} \cdots q_{1,n} q_{3,1} \cdots q_{3,n} q_{2,1} \cdots q_{2,n})^{n-1})$
$= \delta_{\mathcal{C}}((\mathbf{u}, \mathbf{d}_1, \mathbf{u}, \mathbf{d}_3, \ldots, \mathbf{d}_{n-1}, \mathbf{u}, \mathbf{d}_n, \mathbf{u}, \mathbf{d}_2), q_{2,1} \cdots q_{2,n}$
$(q_{1,1} \cdots q_{1,n} q_{3,1} \cdots q_{3,n} q_{2,1} \cdots q_{2,n})^{n-1})$
$= \delta((\mathbf{d}_1, \mathbf{u}, \mathbf{d}_3, \mathbf{u}, \ldots, \mathbf{u}, \mathbf{d}_n, \mathbf{u}, \mathbf{d}_2, \mathbf{u}), (q_{1,1} \cdots q_{1,n} q_{3,1} \cdots q_{3,n} q_{2,1} \cdots q_{2,n})^{n-1})$,

and now applying the n-cycle operation $n - 1$ times, we obtain

$$\delta((\mathbf{d}_1, \mathbf{u}, \mathbf{d}_3, \mathbf{u}, \ldots, \mathbf{u}, \mathbf{d}_n, \mathbf{u}, \mathbf{d}_2, \mathbf{u}), (q_{1,1} \cdots q_{1,n} q_{3,1} \cdots q_{3,n} q_{2,1} \cdots q_{2,n})^{n-1})$$
$$= (\mathbf{d}_2, \mathbf{u}, \mathbf{d}_1, \mathbf{u}, \mathbf{d}_3, \ldots, \mathbf{d}_{n-1}, \mathbf{u}, \mathbf{d}_n, \mathbf{u}).$$

(3′) Collapsing. $\delta_{\mathcal{C}}((\mathbf{d}_1, \mathbf{u}, \mathbf{d}_2, \mathbf{u}, \mathbf{d}_3, \ldots, \mathbf{d}_n, \mathbf{u}), q_{1,1} q_{3,2} q_{4,1} q_{2,1} q_{3,1} q_{2,1})$
$= \delta_{\mathcal{C}}((\mathbf{d}_1, \mathbf{d}_2, \mathbf{d}_2, \mathbf{u}, \mathbf{d}_3, \ldots, \mathbf{d}_n, \mathbf{u}), q_{3,2} q_{4,1} q_{2,1} q_{3,1} q_{2,1})$
$= \delta_{\mathcal{C}}((\mathbf{d}_1, \mathbf{d}_2, \mathbf{u}, \mathbf{u}, \mathbf{d}_3, \ldots, \mathbf{d}_n, \mathbf{u}), q_{4,1} q_{2,1} q_{3,1} q_{2,1})$
$= \delta_{\mathcal{C}}((\mathbf{d}_1 \vee \mathbf{d}_2, \mathbf{d}_2, \mathbf{u}, \mathbf{u}, \mathbf{d}_3, \ldots, \mathbf{d}_n, \mathbf{u}), q_{2,1} q_{3,1} q_{2,1})$
$= \delta_{\mathcal{C}}((\mathbf{d}_2, \mathbf{d}_1 \vee \mathbf{d}_2, \mathbf{u}, \mathbf{u}, \mathbf{d}_3, \ldots, \mathbf{d}_n, \mathbf{u}), q_{3,1} q_{2,1})$
$= \delta_{\mathcal{C}}((\mathbf{u}, \mathbf{d}_1 \vee \mathbf{d}_2, \mathbf{u}, \mathbf{u}, \mathbf{d}_3, \ldots, \mathbf{d}_n, \mathbf{u}), q_{2,1}) = (\mathbf{d}_1 \vee \mathbf{d}_2, \mathbf{u}, \mathbf{u}, \mathbf{u}, \mathbf{d}_3, \ldots, \mathbf{d}_n, \mathbf{u})$. □

Now we are ready to prove our key result for this section.

Theorem 6.13. *Let* $\mathcal{A} = (A, X_{\mathcal{A}}, \delta_{\mathcal{A}})$ *be an automaton satisfying Letichevsky's criterion. For any automaton* $\mathcal{E}$ *there exists a primitive power* $\mathcal{P}$ *of* $\mathcal{A}$ *such that* $\mathcal{E}$ *can be represented homomorphically by* $\mathcal{P}$.

Proof. Let $\mathbf{u} = u_1 \ldots u_s$, $\mathbf{v} = v_1 \ldots v_s$ denote a pair of control words of $\mathcal{A}$ as before. Consider an integer $n \geq 3$ and define the power $\mathcal{N} = \mathcal{A}^{18ns}(X_{\mathcal{N}}, \varphi_1, \ldots, \varphi_{18ns})$, $X_{\mathcal{N}} = A^{6n}$, of $\mathcal{A}$ in the following manner. For any state $(a_1, \ldots, a_{18ns}) \in A^{18ns}$, input letter $\zeta = (z_1, \ldots, z_{6n}) \in X_{\mathcal{N}}$, and t $(= 1, \ldots, 18ns)$; we have

$$\varphi_t(a_1, \ldots, a_{18ns}, \zeta)$$

$$= \begin{cases} x[a_t, a_{t+1}] & \text{if } s \text{ does not divide } t \text{ or} \\ & t = (18(i-1)+j)s,\ i = 1, \ldots, n,\ j = 1, 3, 7, 8, 9, \\ & \quad 10, 12, 16, 17, \\ x[a_t, a_{t+1} \wedge z_{6(i-1)+j/2}] & \text{if } t = (18(i-1)+j)s,\ i = 1, \ldots, n,\ j = 2, 4, 6, \\ x[a_t, a_{t+1} \wedge z_{6(i-1)+(j-3)/2}] & \text{if } t = (18(i-1)+j)s,\ i = 1, \ldots, n,\ j = 11, 13, 15, \\ x[a_t, a_{t-3s+1} \vee a_{t+1}] & \text{if } t = (18(i-1)+j)s,\ i = 1, \ldots, n,\ j = 5, 14, \\ x[a_t, a_{t-18s+1} \vee a_{t+1 (\mathrm{mod}\, 18ns)}] & \text{if } t = 18is,\ i = 1, \ldots, n. \end{cases}$$

It is easy to check that $\mathcal{N}$ is a primitive power of $\mathcal{A}$; moreover, whenever φ_t $(1 \le t \le 18ns)$ really depends on its input variable, then it may additionally depend only on its tth state variable and at most one other state variable. Therefore, $\mathcal{N}$ has the properties required by Proposition 6.3 for the last component of $\mathcal{M}$.

Denote by $\delta_{\mathcal{N}}$ the transition function of $\mathcal{N}$ and consider the automaton $\mathcal{C}$ given in Lemma 6.12. Observe that whenever $\mathcal{N}$ is in the state having the form $(w_{1,1}, \ldots, w_{1,s}, \ldots, w_{18n,1}, \ldots, w_{18n,s})$, $(w_{i,1} \ldots w_{i,s}) \in \{\mathbf{u}, \mathbf{v}\}$, $i = 1, \ldots, 18n$, by the effect of words having the form $(z_{1,1}, \ldots, z_{1,6n}) \ldots (z_{s,1}, \ldots, z_{s,6n})$, $z_{1,i} \ldots z_{s,i} \in \{\mathbf{u}, \mathbf{v}\}$, $i = 1, \ldots, 6n$, $\delta_{\mathcal{N}}((w_{1,1}, \ldots, w_{1,s}, \ldots, w_{18n,1}, \ldots, w_{18n,s}), (z_{1,1}, \ldots, z_{1,6n}) \ldots (z_{s,1}, \ldots, z_{s,6n})) = (w'_{1,1}, \ldots, w'_{1,s}, \ldots, w'_{18n,1}, \ldots, w'_{18n,s})$ if and only if $\delta_{\mathcal{C}}((\mathbf{w}^{(1)}, \ldots, \mathbf{w}^{(18n)}), (\mathbf{z}^{(1)}, \ldots, \mathbf{z}^{(6n)}) = (\mathbf{w}'^{(1)}, \ldots, \mathbf{w}'^{(18n)})$, $\mathbf{w}^{(i)} = w_{i,1} \ldots w_{i,s}$, $\mathbf{w}'^{(i)} = w'_{i,1} \ldots w'_{i,s}$, $\mathbf{z}^{(j)} = z_{1,j} \ldots z_{s,j}$, $i = 1, \ldots, 18n$, $j = 1, \ldots, 6n$. Therefore, using the short notation $b_k = \mathbf{u}^{18(k-1)+4}\mathbf{v}\mathbf{u}^{18(n-k)+13}$ for the state b_k of $\mathcal{N}$, by Lemma 6.12 we have that there exists a positive integer m having the following property. For every transformation γ on $\{b_1, \ldots, b_n\}$, there exists a word $\hat{\gamma} = \zeta_1 \ldots \zeta_{ms}$, $(\zeta_1, \ldots, \zeta_{ms} \in X_{\mathcal{N}})$ such that $\delta_{\mathcal{N}}(b_k, \hat{\gamma}) = \gamma(b_k)$ for all $1 \le k \le n$.

Every n-state automaton $\mathcal{E}$ is isomorphic to a subautomaton of an n-state automaton $\mathcal{D}$ with the following properties:

(1) For each transformation γ of the n states of D, there is an input letter x_γ inducing γ.
(2) There are at least as many distinct letters of $\mathcal{D}$ that induce γ as there are that induce γ in $\mathcal{E}$.

Thus, to complete the proof, it suffices to establish the result for the following n-state automaton having these properties whose states are a subset of those of $\mathcal{N}$. Let $\mathcal{D} = (D, X_{\mathcal{D}}, \delta_{\mathcal{D}})$, $D = \{b_1, \ldots, b_n\}$, where b_k, $k = 1, \ldots, n$, are the states of $\mathcal{N}$ discussed before. For each transformation γ of $\{b_1, \ldots, b_n\}$ let there be an input letter x_γ of $\mathcal{D}$ having $\delta_{\mathcal{D}}(b_k, x_\gamma) = \gamma(b_k)$, $k = 1, \ldots, n$. Furthermore, let there be at least as many letters of $\mathcal{D}$ which induce each given transformation γ as there are in $\mathcal{E}$.

We shall show that $\mathcal{D}$ can be represented homomorphically by a primitive power of $\mathcal{A}$. Clearly, $\{\delta_{\mathcal{D}}(b_k, x) \mid b_k \in D, x \in X_{\mathcal{D}}\} = D$.

To each length ms input word $p = x_1 \ldots x_{ms}$ of $\mathcal{D}$, we associate the transformation γ_p induced by this word on the set $D = \{b_1, \ldots, b_n\}$. Define, following Lemma 6.12, $\tau(p) = \widehat{\gamma_p}$. The mapping $\tau : (X_{\mathcal{D}})^{ms} \to (X_{\mathcal{N}})^{ms}$ satisfies $\delta_{\mathcal{D}}(b_k, p) = \delta_{\mathcal{N}}(b_k, \tau(p))$ $(b_k \in D, p \in (X_D)^{ms})$.

For every $d > 0$ and a a state of $\mathcal{N}$, $(p, q) \in R_{\tau,d}$, $p \in (X_{\mathcal{D}})^{ms}$, we clearly have, whenever $\delta_{\mathcal{N}}(a, q) \in D$, that $\delta_{\mathcal{D}}(\delta_{\mathcal{N}}(a, q), p) = \delta_{\mathcal{N}}(a, q\tau(p)) \in D$. Furthermore, by taking ι to be a letter of $X_{\mathcal{D}}$ inducing the identity under $\delta_{\mathcal{D}}$ (that is, $\delta_{\mathcal{D}}(b_i, \iota) = b_i$ for all $b_i \in D$) and letting q be $(\tau(\iota^{ms}))^j$ with $d \leq msj < ms+d$ and $p = \iota^{d-ms(j-1)}$ implying $(p, q) \in R_{\tau,d}$, we derive $\delta_{\mathcal{D}}(\delta_{\mathcal{N}}(b_i, q), p) = \delta_{\mathcal{D}}(b_i, p) = b_i$. Therefore, $D = \{\delta_{\mathcal{D}}(\delta_{\mathcal{N}}(a, q), p) \mid a$ a state of $\mathcal{N}$, $(p, q) \in R_{\tau,d}, \delta_{\mathcal{N}}(a, q) \in D\}$. This shows that conditions (1) and (2) of Lemma 6.11 hold.

For every i $(= 1, \ldots, 6n)$, define $\tau_i : (X_{\mathcal{D}})^{ms} \to A^{ms}$ as follows: for each $1 \leq j \leq ms$, the jth letter of $\tau_i(p)$ $(p \in (X_{\mathcal{D}})^{ms})$ is equal to the ith component of the jth letter $\zeta'_j = (z_{j,1}, \ldots, z_{j,6n})$ of $\tau(p)$. Therefore, as in Lemma 6.11 (taking ℓ and n of the lemma to be $6n$ and ms, respectively), we can construct the product $\mathcal{V} = \mathcal{R}_{\tau_1,d} \times \cdots \times \mathcal{R}_{\tau_{6n},d} \times \mathcal{N}(X_{\mathcal{D}}, \varphi'_1, \ldots, \varphi'_{6n}, \varphi''_{6n+1})$ which homomorphically represents $\mathcal{D}$.

By Lemma 6.9, given an integer $d \geq |X_{\mathcal{D}}|^{ms}$, for each $i = 1, \ldots, 6n$, we obtain a primitive power $\mathcal{M}_i$ of $\mathcal{A}$ such that apart from its last one, its feedback functions do not depend on the last state variable, and furthermore, $\mathcal{M}_i$ y-represents $\mathcal{R}_{\tau_i,d}$.

Now set $\psi_i(m_1, \ldots, m_{6n}, m_{6n+1}, x) = x$ for each $i = 1, \ldots, 6n$ and $\psi_{6n+1}(m_1, \ldots, m_{6n}, m_{6n+1}, x) = (z_1, \ldots, z_{6n})$, where $x \in X_{\mathcal{D}}$, z_i is the state of the last factor of $\mathcal{M}_i$ (which represents $\mathcal{R}_{\tau_i,d}$) for $1 \leq i \leq 6n$, and m_{6n+1} is the state of $\mathcal{N}$. By Proposition 6.8 (considering $\mathcal{N}$, $X_{\mathcal{D}}$, $6n$ to be $\mathcal{M}$, X, m of the proposition), we obtain $\mathcal{M} = \mathcal{M}_1 \times \cdots \times \mathcal{M}_{6n} \times \mathcal{N}(X_{\mathcal{D}}, \psi_1, \ldots, \psi_{6n+1})$, which homomorphically represents $\mathcal{V}$, hence $\mathcal{D}$, hence $\mathcal{E}$. On the other hand, observe that we have the conditions of Proposition 6.3 for the product $\mathcal{M}$ (taking $\mathcal{N}$, $X_{\mathcal{D}}$, $6n$ to be $\mathcal{M}_{n+1}$, X, n of the proposition). By Proposition 6.3, $\mathcal{M}$ is isomorphic to a primitive power $\mathcal{P}$ of $\mathcal{A}$. Therefore, $\mathcal{E}$ is homomorphically represented by the primitive power $\mathcal{P}$. This completes the proof. □

Corollary 6.14. *Let $\mathcal{K}$ be a class of finite automata. If $\mathcal{K}$ satisfies Letichevsky's criterion, then $\mathcal{K}$ is complete with respect to homomorphic representations under the primitive product.* □

By the Letichevsky decomposition theorem (Theorem 2.69), a class of finite automata is complete with respect to homomorphic representations under the Gluškov product if and only if it satisfies Letichevsky's criterion. Therefore, one obtains the following statement.

Theorem 6.15. *Suppose that $\mathcal{K}$ is a class of finite automata. Then the following statements are equivalent:*

(1) *$\mathcal{K}$ satisfies Letichevsky's criterion.*

(2) *$\mathcal{K}$ is complete with respect to homomorphic representations under the Gluškov product.*

(3) *$\mathcal{K}$ is complete with respect to homomorphic representations under the α_i-product for all $i \geq 2$.*

(4) *$\mathcal{K}$ is complete with respect to homomorphic representations under the α_i-product for some $i \geq 2$.*

(5) *$\mathcal{K}$ is complete with respect to homomorphic representations under the ν_j-product for all $j \geq 3$.*

(6) *$\mathcal{K}$ is complete with respect to homomorphic representations under the ν_j-product for some $j \geq 3$.*

(7) $\mathcal{K}$ *is complete with respect to homomorphic representations under the* α_i-ν_j-*product for all* $i \geq 2$, $j \geq 3$.

(8) $\mathcal{K}$ *is complete with respect to homomorphic representations under the* α_i-ν_j-*product for some* $i \geq 2$, $j \geq 3$.

(9) $\mathcal{K}$ *is complete with respect to homomorphic representations under the primitive product.* □

Remark. *Theorem* 3.36 *implies that there exists no finite class of automata which is complete with respect to homomorphic representations under the* α_1-*product. Therefore, by Theorems* 2.69 *and* 5.9, *there is a class of finite automata satisfying Letichevsky's criterion which is complete with respect to homomorphic representations for neither the* α_1-*product nor the* ν_2-*product. This shows that the above result is sharp.*

6.4 Temporal Products

Let $\mathcal{A} = (A, Y, \delta)$ be an automaton with $Y = X^k$ for some nonempty finite set X and positive integer k. We say that $\mathcal{A}$ is a *k-channel automaton* (with respect to X). If there exists a (one-channel) automaton $\mathcal{A}' = (A, X, \delta')$ such that for every pair $a \in A$, $(x_1, \ldots, x_k) \in X^k$, we have $\delta(a, (x_1, \ldots, x_k)) = \delta'(a, x_1 \ldots x_k)$; then it is said that $\mathcal{A}$ has a *one-channel analog* $\mathcal{A}'$. Thus $\mathcal{A}'$ isomorphically simulates $\mathcal{A}$ in equal lengths under (τ_1, τ_2), where $\tau_1 : A \to A$ and $\tau_2 : X^k \to X^*$ are defined by $\tau_1(a) = a$, $\tau_2((x_1, \ldots, x_k)) = x_1 \ldots x_k$. If $\mathcal{A}'$ is a product of automata, then $\mathcal{A}$ is a simple type of temporal power of this product (defined below). In this case, the components of this structure $\mathcal{A}'$ are fixed during its work. In the general case of temporal products, the internal structure of $\mathcal{A}'$ will be allowed to change cyclically as it operates.

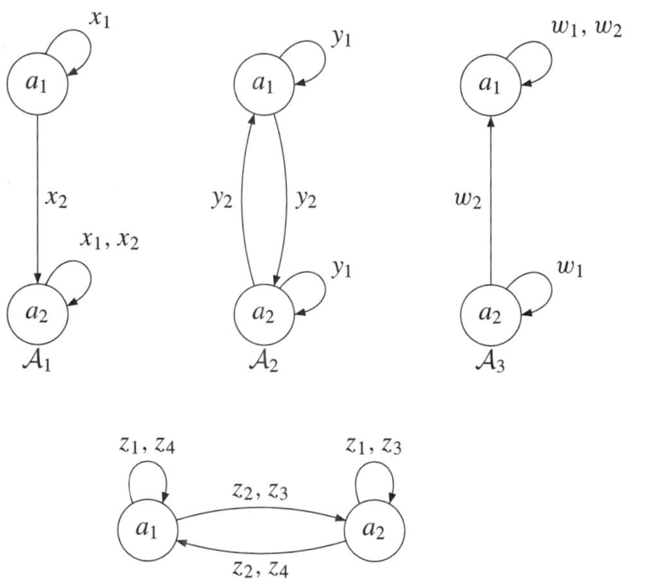

TEMPORAL PRODUCT OF AUTOMATA $\mathcal{A}_1, \mathcal{A}_2, \mathcal{A}_3$ WITH
$\varphi(z_1) = (x_1, y_1, w_1)$, $\varphi(z_2) = (x_1, y_2, w_1)$, $\varphi(z_3) = (x_2, y_1, w_1)$, $\varphi(z_4) = (x_1, y_1, w_2)$.

We will see that the temporal compositions of the general product and the cascade product can be represented by very simple models of finite automata networks which are cyclically able to modify their inner structures. We will show that, despite their simplicity, they have a very strong completeness property.

Let $\mathcal{A}_t = (A_t, X_t, \delta_t)$, $t = 1, 2$, be automata having a common state set A $(= A_1 = A_2)$. Take a finite nonvoid set X and a mapping φ of X into $X_1 \times X_2$. Then the automaton $\mathcal{A} = (A, X, \delta)$ is a *temporal product* (t-product) of $\mathcal{A}_1$ by $\mathcal{A}_2$ with respect to X and φ if for any $a \in A$ and $x \in X$, $\delta(a, x) = \delta_2(\delta_1(a, x_1), x_2)$, where $(x_1, x_2) = \varphi(x)$. The concept of the temporal product is generalized in the natural way to an arbitrary finite family of $n > 0$ automata $\mathcal{A}_t$ $(t = 1, \ldots, n)$, all with the same state set A, for any mapping $\varphi : X \to \prod_{t=1}^{n} X_t$, by defining $\delta(a, x) = \delta_n(\cdots \delta_2(\delta_1(a, x_1), x_2), \ldots, x_n)$ when $\varphi(x) = (x_1, \ldots, x_n)$. In particular, a temporal product of automata with a single factor is just a (one-to-many) relabeling by X of the input letters of some input-subautomaton of its factor or, equivalently, a quasi-direct product with just this factor. Note that the formation of the temporal product is associative.

We apply the notation β-product for each of the following concepts, respectively: g-product, q-product, q^ℓ-product, α_0-product, t-product. Given a class $\mathcal{K}$ of automata, the notions $I(\mathcal{K})$, $H(\mathcal{K})$, $S(\mathcal{K})$, $P_\beta(\mathcal{K})$ stand for the class of all isomorphic images for $\mathcal{K}$, all homomorphic images for $\mathcal{K}$, all subautomata for $\mathcal{K}$, and all β-products for $\mathcal{K}$, respectively. We define the (left) multiplication of these operators as usual in the literature. For any sequence $O_1, \ldots, O_n$ of the discussed operators we write $O_1 \ldots O_n(\mathcal{K})$ instead of $O_1(O_2(\ldots O_n(\mathcal{K}) \ldots))$. Moreover, we use the short notion T_β for SP_tIP_β. Let us first consider some simple examples:

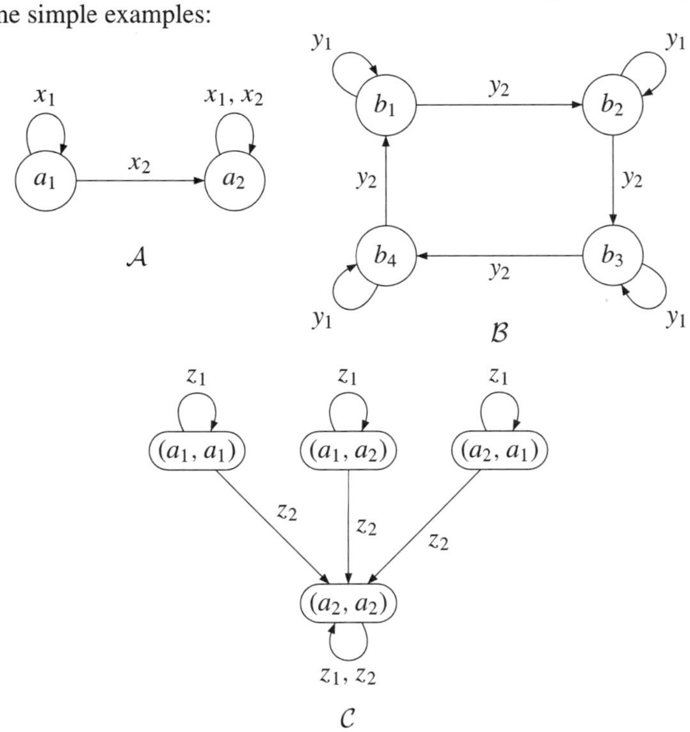

THREE AUTOMATA $\mathcal{A}$, $\mathcal{B}$, AND $\mathcal{C}$. AUTOMATON $\mathcal{C}$ IS A (SECOND) QUASI-DIRECT POWER (q-POWER) OF $\mathcal{A}$ WITH THE FEEDBACK FUNCTIONS $\varphi(a_i, a_j, z_k) = x_k$ $(i, j, k \in \{1, 2\})$.

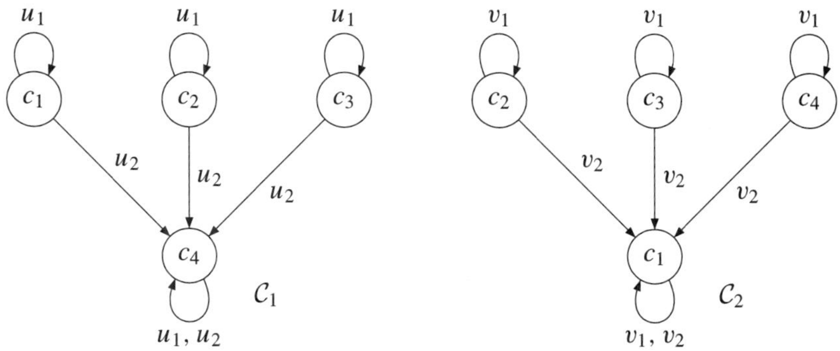

$\mathcal{C}_1, \mathcal{C}_2$ ARE ISOMORPHIC COPIES OF $\mathcal{C}$.

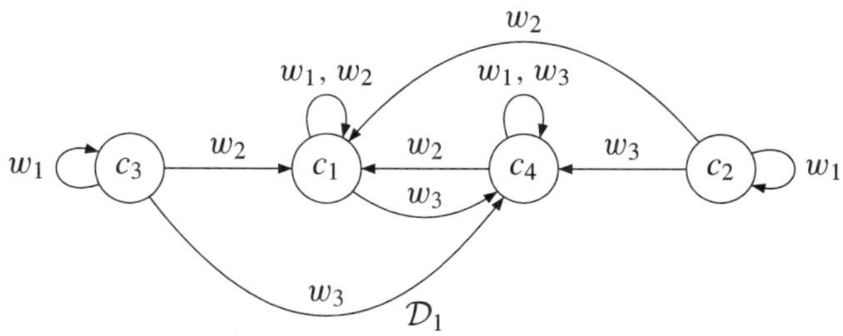

AUTOMATON $\mathcal{D}_1$ IS A t-PRODUCT OF $\mathcal{C}_1$ AND $\mathcal{C}_2$ (HENCE A T_q-POWER OF $\mathcal{A}$) WITH $\varphi(w_1) = (u_1, v_1)$, $\varphi(w_2) = (u_2, v_2)$, AND $\varphi(w_3) = (u_2, v_1)$.

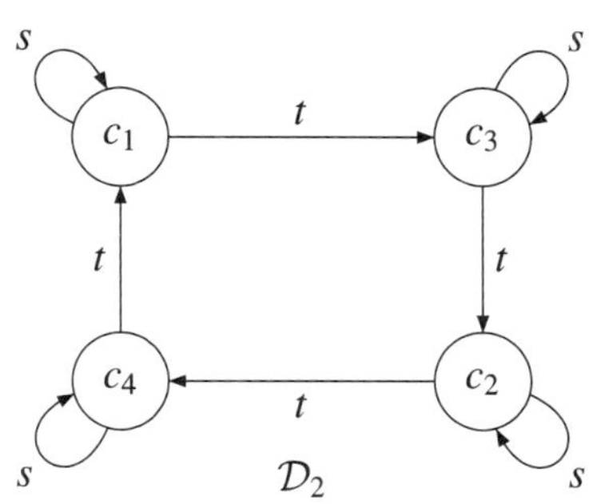

AUTOMATON $\mathcal{D}_2$ AS AN ISOMORPHIC COPY OF $\mathcal{B}$.

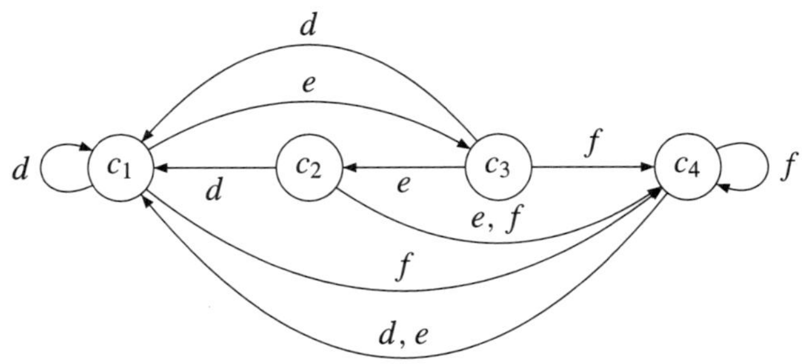

t-PRODUCT OF $\mathcal{D}_2$ BY $\mathcal{D}_1$ (HENCE A T_q-PRODUCT OF $\mathcal{A}$ AND $\mathcal{B}$)
WITH $\varphi(d) = (s, w_2)$, $\varphi(e) = (t, w_1)$, AND $\varphi(f) = (s, w_3)$.

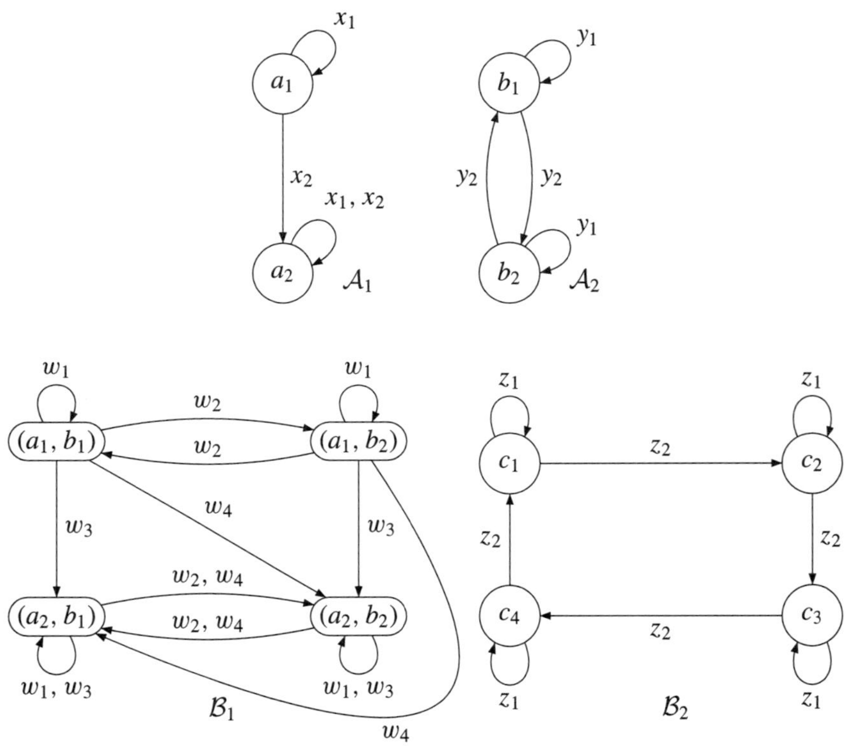

$\mathcal{B}_1$ IS A q-PRODUCT OF $\mathcal{A}_1$, $\mathcal{A}_2$ WITH THE FOLLOWING VALUES OF ITS FEEDBACK FUNCTION φ:
$\varphi(w_1) = (x_1, y_1)$, $\varphi(w_2) = (x_1, y_2)$, $\varphi(w_3) = (x_2, y_1)$, $\varphi(w_4) = (x_2, y_2)$ AND $\mathcal{B}_2$
IS AN ISOMORPHIC COPY OF $\mathcal{B}$.

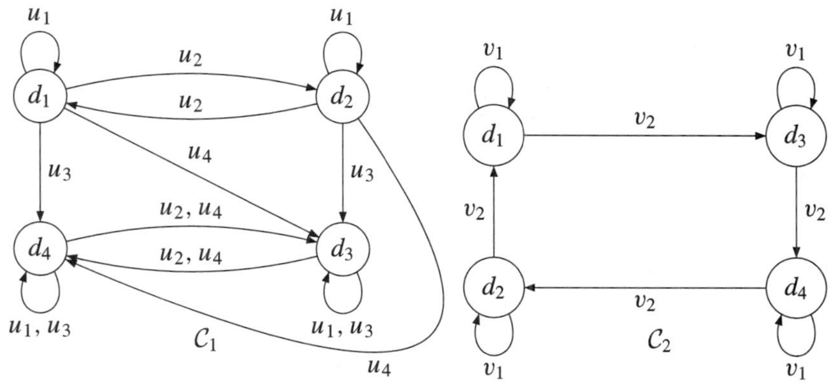

$\mathcal{C}_1$ IS AN ISOMORPHIC COPY OF $\mathcal{B}_1$ AND $\mathcal{C}_2$ IS AN ISOMORPHIC COPY OF $\mathcal{B}_2$.

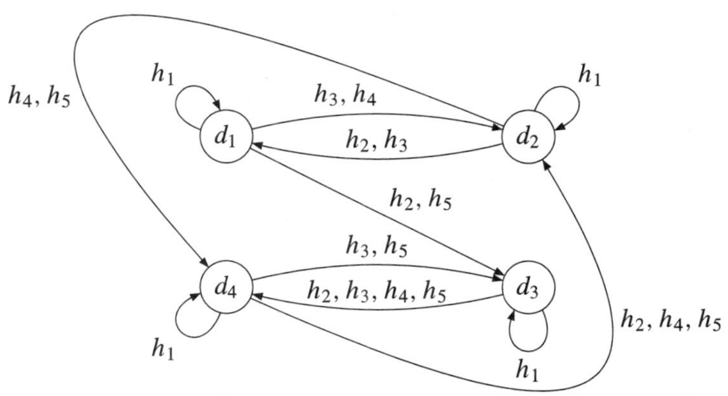

t-PRODUCT OF $\mathcal{C}_1$ BY $\mathcal{C}_2$ WITH $\varphi(h_1) = (u_1, v_1)$, $\varphi(h_2) = (u_1, v_2)$, $\varphi(h_3) = (u_2, v_1)$, $\varphi(h_4) = (u_3, v_2)$, $\varphi(h_5) = (u_4, v_1)$ IS A T_q PRODUCT OF $\mathcal{A}_1$, $\mathcal{A}_2$ AND $\mathcal{B}_2$.

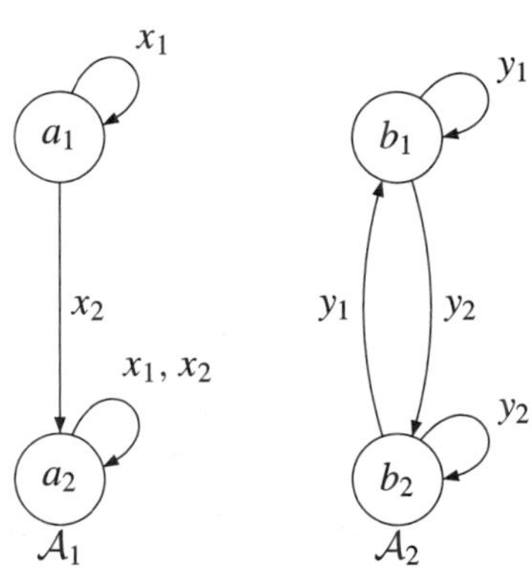

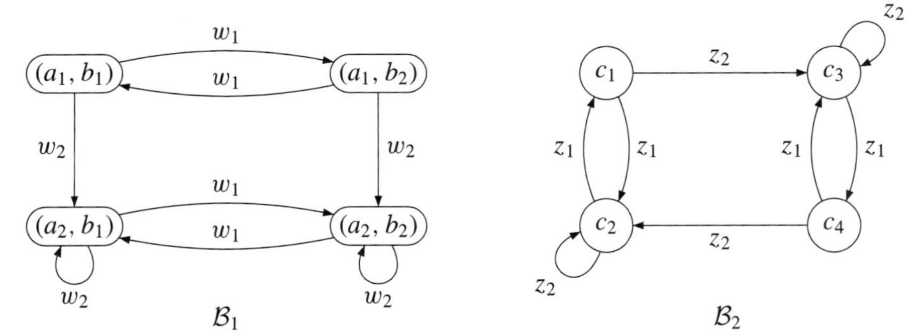

$\mathcal{B}_1$ IS A q^ℓ-PRODUCT OF $\mathcal{A}_1$ AND $\mathcal{A}_2$ HAVING THE FOLLOWING VALUES OF ITS FEEDBACK FUNCTION φ:
$\varphi(a_i, b_j, w_1) = (x_1, y_{j+1\,(\mathrm{mod}\,2)})$, $\varphi(a_i, b_j, w_2) = (x_2, y_j)$, $i, j \in \{1, 2\}$.

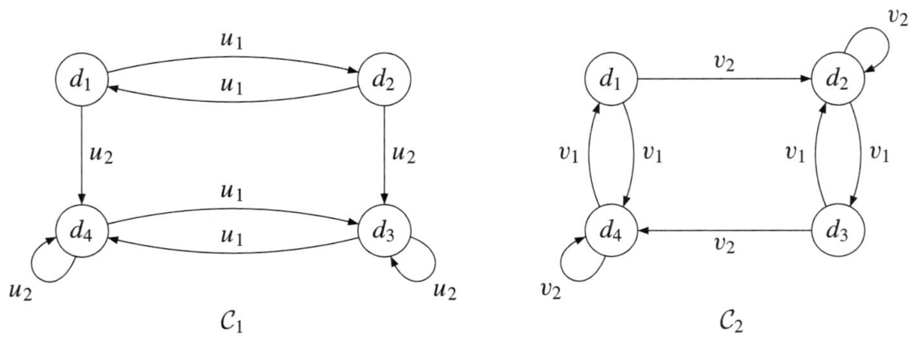

$\mathcal{C}_1$ IS AN ISOMORPHIC COPY OF $\mathcal{B}_1$ AND $\mathcal{C}_2$ IS AN ISOMORPHIC COPY OF $\mathcal{B}_2$.

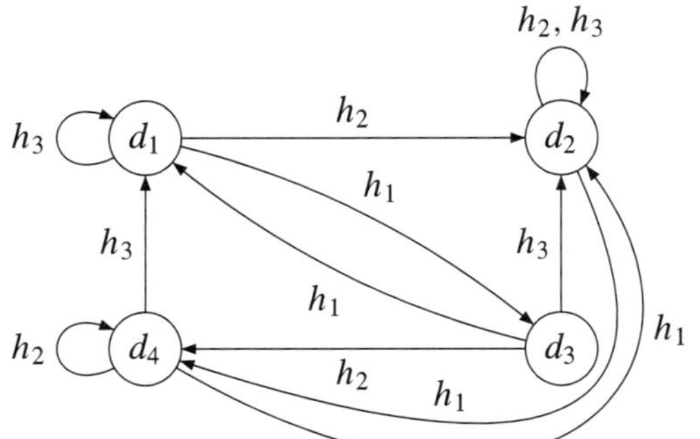

t-PRODUCT OF $\mathcal{C}_1$ BY $\mathcal{C}_2$ WITH $h_1 = (u_1, v_1)$, $h_2 = (u_1, v_2)$,
$h_3 = (u_2, v_1)$ IS A T_{q^ℓ} PRODUCT OF $\mathcal{A}_1, \mathcal{A}_2$ AND $\mathcal{B}_2$.

Let $\mathcal{K}$ again be a class of automata. We say that the automaton $\mathcal{A}$ is a T_β-product of automata from $\mathcal{K}$ if $\mathcal{A} \in T_\beta(\mathcal{K})$ $(= SP_tIP_\beta(\mathcal{K}))$.

It is said that $\mathcal{K}$ is *complete with respect to homomorphic (resp., isomorphic) representations under the T_β-product* if $HT_\beta(\mathcal{K})$ (resp., $T_\beta(\mathcal{K})$) is the class of all (finite) automata. (Note that $IST_\beta(\mathcal{K}) = T_\beta(\mathcal{K})$ and $HST_\beta(\mathcal{K}) = HT_\beta(\mathcal{K})$ hold for every class of automata.)

Now we study the T_{α_0}-product.

Theorem 6.16. *The following statements are true:*

(1) *An arbitrary T_{α_0}-product of autonomous automata is an autonomous automaton.*

(2) *An arbitrary T_{α_0}-product of permutation automata is a permutation automaton.*

(3) *An arbitrary T_q-product of reset automata is a reset automaton.* □

Let $\mathcal{A} = (A, X, \delta)$ be an automaton and define for any $x \in X$ the automaton $\mathcal{A}_x = (A, \{x_0, x\}, \delta_x)$ such that $x_0 \neq x$ is an arbitrary symbol; moreover, $\delta_x(a, x_0) = a$ and $\delta_x(a, x) = \delta(a, x)$ $(a \in A)$.

Lemma 6.17. *Let $\mathcal{K}$ be a class of automata which contains an automaton $\mathcal{B}$ with $\mathcal{A}_x \in SP_t(\{\mathcal{B}\})$, $x \in X$. Then $\mathcal{A} \in SP_t(\mathcal{K})$.*

Proof. Let $\mathcal{M}_x = (M, \{x_0, x\}, \delta'_x) \in P_t(\mathcal{B})$, $x \in X$, be automata with $\mathcal{A}_x \in S(\{\mathcal{M}_x\})$. Moreover, let $x_1, \dots, x_n$ be an arrangement of the elements of X. Take the temporal product $\mathcal{M} = \mathcal{M}_{x_1} \times \cdots \times \mathcal{M}_{x_n}(X, \varphi)$ such that for any $x_t \in X$, $\varphi(x_t) = (z_{1,x_t}, \dots, z_{n,x_t})$, where $z_{1,x_t} = \cdots = z_{t-1,x_t} = z_{t+1,x_t} \dots z_{n,x_t} = x_0$ and $z_{t,x_t} = x_t$. It is clear that $\mathcal{A}$ is a subautomaton of $\mathcal{M}$. □

Lemma 6.18. *Let $\mathcal{K}$ be a class of automata having a (not necessarily different) pair $\mathcal{A}_t = (A_t, X_t, \delta_t)$, $t = 1, 2$, of elements with the following properties:*

(1) *There is a pair $a \in A_1$, $x \in X_1$ with $a \notin \delta_1(a, x)$.*

(2) *There is a triplet $a_1, a_2 \in A_2$, $y \in X_2$ with $a_1 \neq a_2$ and $\delta_2(a_1, y) \neq \delta_2(a_2, y)$.*

Then for any positive integer m there can be found an automaton $\mathcal{N} = (N, X, \delta)$ $\in P_q(\mathcal{K})$, pairwise disjoint sets $B_1, \dots, B_{2m} \subseteq N$, and (not necessarily different) inputs y_1, $y_2, x_1, \dots, x_s \in X$, $s \geq 1$, such that $B_{m+t} = \{b_{m+t}, b_{m+t}{}^{(1)}, \dots, b_{m+t}{}^{(s)}\}$, $\delta(b^{(k)}_{m+t}, x_k)$ $= b_{m+t}$, $k = 1, \dots, s$, $b_{m+t} \notin \{b^{(1)}_{m+t}, \dots, b^{(s)}_{m+t}\}$, $t = 1, \dots, m$. Moreover,

(1a) *if $\mathcal{K}$ contains a nonautonomous automaton, then for any $t = 1, \dots, m$, $\delta(b_t, y_1)$ $\neq \delta(b_t, y_2)$ (and thus, $y_1 \neq y_2$);*

(2a) *if $\mathcal{K}$ contains a nonpermutation automaton, then $y_2 = x_1 = \cdots = x_s$, $s = m + 1$, $|B_{m+t}| = m + 2$, $t = 1, \dots, m$; and*

(3a) *if $\mathcal{K}$ is a class of permutation automata which has a nonautonomous automaton, then $s = 2$, $x_1 = y_1$, $x_2 = y_2$, $x_1 \neq x_2$, $|B_{m+t}| = 3$, $t = 1, \dots, m$.*

Proof. For a suitable system $\mathcal{D}_1, \dots, \mathcal{D}_r \in \mathcal{K}$ of automata construct a quasi-direct product $\mathcal{D}_1 \times \cdots \times \mathcal{D}_r(X, \varphi)$ and suppose that r is a sufficiently great number for the properties as we define in the following investigations.

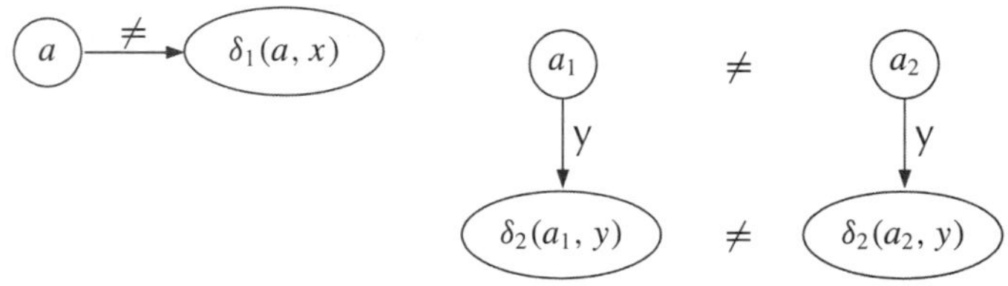

First we prove that for a suitable $\mathcal{N}$ there are pairwise disjoint sets $B_1, \ldots, B_{2m} \subseteq N$ such that $b_t \notin \{\delta(b_t, y_1), \delta(b_t, y_2)\}$, $\delta(b^{(k)}_{m+t}, x_k) = b_{m+t}$, $b_{m+t} \notin \{b^{(1)}_{m+t}, \ldots, b^{(s)}_{m+t}\}$, $t = 1, \ldots, m$, $k = 1, \ldots, s$. (Note that the special cases $y_1 = y_2$ and $s = 1$ are allowed.)

Let $\mathcal{D}_1 = \mathcal{A}_1$ and $\varphi_1(x') = x$, $x' \in X$. (If there is no danger of confusion, then we omit the arguments of any feedback function on which it really does not depend.) Then for an arbitrary $(a, d_2, \ldots, d_r) \in N$ and $x' \in X$, the first component of $\delta((a, d_2, \ldots, d_r), x')$ is $\delta_1(a, x)$. Thus, if we suppose that for every pair $b' \in \{b_u\} \cup (B_{m+u} \setminus \{b_{m+u}\})$, $u = 1, \ldots, m$, $b'' \in (B_v \setminus \{b_v\}) \cup \{b_{m+v}\}$, $v = 1, \ldots, m$, the first component of the vector b' is equal to a and the first component of the vector b'' is equal to $\delta_1(a, x)$, then we obtain $b' \neq b''$. Since $\mathcal{D}_1 = \mathcal{A}_1$ and $\varphi_1(x') = x$, $x' \in X$, this assumption does not lead to a contradiction even if $\delta(b', x') = b''$ for an $x' \in X$ (by $u = v$ and $b', b'' \in B_u$ or $b', b'' \in B_{m+u}$). Therefore, for a suitable N and $B_1, \ldots, B_{2m} \subseteq N$ we get that $\{b_u, b^{(1)}_{m+v}, \ldots, b^{(s)}_{m+v}\}$ and $\{\delta(b_u, x'), \delta(b^{(1)}_{m+v}, x'), \ldots, \delta(b^{(s)}_{m+v}, x')\}$, $x' \in X$, $u, v = 1, \ldots, m$, are pairwise disjoint sets. (In particular, if $u = v$, then we also have this fact.) Thus, if we prove that there are pairwise disjoint sets $B_1, \ldots, B_{2m} \subseteq N$ having these properties, then the first part of our lemma is shown. (Naturally, $y_1 = y_2$ and $s = 1$ are possible.) If $b' = b_{m+u}$, $b'' = b_{m+v}$, $1 \leq u < v \leq m$, then suppose the existence of an appropriate $z \in \{1, \ldots, r\}$ for which $\mathcal{D}_z = \mathcal{A}_2$, $\{d_z, d'_z\} = \{\delta_2(a_1, y), \delta_2(a_2, y)\}$, where d_z and d'_z denote the zth components of b' and b'', respectively. Then we obtain $b' \neq b''$ by (2). It remains to study the case $b' \in \{b_u\} \cup (B_{m+u} \setminus \{b_{m+u}\})$, $b'' \in \{b_v\} \cup (B_{m+v} \setminus \{b_{m+v}\})$, $1 \leq u < v \leq m$. Then assume the existence of an appropriate $z \in \{1, \ldots, r\}$ for which $\mathcal{D}_z = \mathcal{A}_2$, $\{d_z, d'_z\} = \{a_1, a_2\}$, where d_z and d'_z denote the zth components of b' and b'', respectively. Then we obtain again $b' \neq b''$ by (2). For every $x' \in X$, let $\varphi_z(x') = y$ in both of the above cases. Thus, for every pair $b' \in \{b_u\} \cup (B_{m+u} \setminus \{b_{m+u}\})$, $b'' \in \{b_v\} \cup (B_{m+v} \setminus \{b_{m+v}\})$, $1 \leq u < v \leq m$, and $x', y' \in X$, we get $\delta(b', x') \neq \delta(b'', y')$. In accordance with $\delta(b^{(k)}_{m+t}, x_k) = b_{m+t}$, $k = 1, \ldots, s$, and $b_{m+t} \notin \{b^{(1)}_{m+t}, \ldots, b^{(s)}_{m+t}\}$, $t = 1, \ldots, m$, let, for any $z \in \{1, \ldots, r\}$, the zth component of b_{m+t} be equal to $\delta_2(a_i, y)$, $i = 1, 2$, with $\mathcal{D}_z = \mathcal{A}_2$ if and only if, for every $k \in \{1, \ldots, s\}$, the zth component of $b^{(k)}_{m+t}$ is equal to a_i.

We allowed $s = 1$ (apart from cases (1a)–(3a), which we shall study soon after the general case). Thus, in the case $b' \in \{b^{(1)}_{m+t}, \ldots, b^{(s)}_{m+t}\}$ or $b'' \in \{b^{(1)}_{m+t}, \ldots, b^{(s)}_{m+t}\}$, the assumption $\{d_z, d'_z\} = \{a_1, a_2\}$ does not contradict our restrictions holding for the structure on $B_1, \ldots, B_{2m} \subseteq N$. There is no problem if $\delta(b_t, y_1) = \delta(b_t, y_2)$, too. (In this case we may get the appropriate sets by $y_1 = y_2$.)

Then, apart from cases (1a)–(3a), we obtained $b' \neq b''$ for all possible cases of $b' \in B_u$ and $b'' \in B_v$. Thus the required sets $B_1, \ldots, B_{2m}$ can be constructed.

Now suppose that $\mathcal{K}$ contains a nonautonomous automaton, i.e., for an $\mathcal{A}_3 = (A_3, X_3, \delta_3)$, in $\mathcal{K}$, we have $a \in A_3$, $x, y \in X_3$ with $\delta_3(a, x) \neq \delta_3(a, y)$. Then let $y_1, y_2 \in X$ be given with $y_1 \neq y_2$ and suppose that $\mathcal{D}_t = \mathcal{A}_3$ and $\varphi_t(y_1) = x$, $\varphi_t(y_2) = y$ hold for some $t \in \{2, \ldots, r\}$. If $a \in A_3$ is the tth component of every vector b_t, $t = 1, \ldots, m$, then we get (1a).

If $\mathcal{K}$ contains a nonpermutation automaton $\mathcal{A}_4 = (A_4, X_4, \delta_4)$, then there are $a_1, a_2 \in A_4$, $x \in X_4$ with $a_1 \neq a_2$ and $\delta(a_1, x) = \delta(a_2, x)$. Then for any pair $b_{m+t}{}^{(u)}$, $b_{m+t}{}^{(v)} \in B_{m+t}$, $t = 1, \ldots, m$, $1 \le u < v \le m+1$, let us give a $z \in \{1, \ldots, r\}$ with $\mathcal{D}_z = \mathcal{A}_4$ and $\{d_z, d'_z\} = \{a_1, a_2\}$, where d_z and d'_z denote the zth components of b^u_{m+t} and b^v_{m+t}, respectively. Thus we obtain $b^u_{m+t} \neq b^v_{m+t}$; moreover, since $\varphi_z(x') = x$ $(x' \in X)$ we do not get a contradiction assuming $\delta(b^u_{m+t}, x_u) = \delta(b^v_{m+t}, x_v) = b_{m+t}$ (by $y_2 = x_u = x_v$). Thus we have obtained (2a).

In addition, let $\mathcal{A}_5 = (A_5, X_5, \delta_5) \in \mathcal{K}$ be a nonautonomous permutation automaton, where $\mathcal{K}$ is a class of permutation automata. There exists a triplet $a \in A_5, x, y \in X_5$ with $\delta_5(a, x) \neq \delta_5(a, y)$. It is clear that for some pair $k, \ell > 1$ of integers, $\delta_5(a, x^k) = \delta_5(a, y^\ell) = a$. Thus there can be found a positive integer t such that $t < k, \ell$ and $\delta_5(a, x^{k-t}) \neq \delta_5(a, x^{\ell-t})$ and $\delta_5(a, x^{k-t+1}) = \delta_5(a, y^{\ell-t+1})$. In other words, for a suitable pair $a_1, a_2 \in A_5$ we get $a_1 \neq a_2$ and $\delta_5(a_1, x) = \delta_5(a_2, y)$.

For every pair $b^1_{m+t}, b^2_{m+t} \in B_{m+t}$, $t = 1, \ldots, m$, let us give a $z \in \{2, \ldots, r\}$ with $\mathcal{D}_z = \mathcal{A}_5$ and $\{d_z, d'_z\} = \{a_1, a_2\}$, where d_z, d'_z denote the zth components of $b_{m+t}{}^{(1)}, b_{m+t}{}^{(2)}$, in order. Then, by the choices $\varphi_z(x_1) = x$, $\varphi_z(x_2) = y$ we get $b^1_{m+t} \neq b^2_{m+t}$ and that the zth components of $\delta(b^1_{m+t}, x_1)$ and $\delta(b^2_{m+t}, x_2)$ are equal.

Now we suppose that $\mathcal{D}_w = \mathcal{A}_5$, $\varphi_w(x_1) = x$, $\varphi_w(x_2) = y$, $\{d_w, d'_w\} = \{a_1, a_2\}$ whenever the state vectors $b_{m+t}{}^{(1)}, b_{m+t}{}^{(2)}$ have different wth components d_w and d'_w for some $w \in \{1, \ldots, r\}$. Then we may assume $\delta(b^1_{m+t}, x_1) = \delta(b^2_{m+t}, x_2) = b_{m+t}$ and $x_1 = y_1, x_2 = y_2$ by (3a).

Finally, to have the conditions $x_1 = y_1$ and $x_2 = y_2$ in (3a), let $v \in \{1, \ldots, r\}$ be not necessarily different from w with $\mathcal{D}_v = \mathcal{A}_5$, $\varphi_v(y_1) = x$, $\varphi_v(y_2) = y$, respectively, and let the vth component of every b_t, $t = 1, \ldots, m$, be a suitable state $a \in A_5$ with $\delta_5(a, x) \neq \delta_5(a, y)$. Then it is clear that we can suppose $x_1 = y_1$ and $x_2 = y_2$. □

Theorem 6.19. *Any class $\mathcal{K}$ of automata is complete with respect to isomorphic representations under the T_q-product if and only if it contains (not necessarily different) automata $\mathcal{A}_t = (A_t, X_t, \delta_t)$, $t = 1, 2, 3$, with the following properties:*

(1) *There exist $a \in A_1$, $x, y \in X_1$ such that $\delta_1(a, x) \neq \delta_1(a, y)$.*

(2) *There can be found a triplet $a_1, a_2 \in A_2$, $x \in X_2$ having $a_1 \neq a_2$ and $\delta_2(a_1, x) = \delta_2(a_2, x)$.*

(3) *There are $a_1, a_2 \in A_3$, $x \in X_3$, with $a_1 \neq a_2$ and $\delta_3(a_1, x) \neq \delta_3(a_2, x)$.*

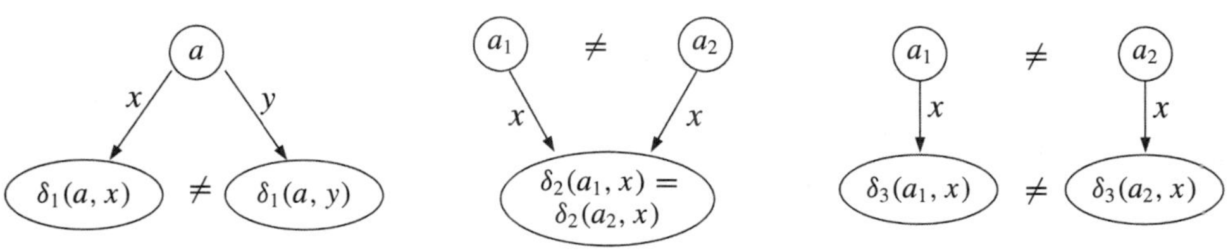

Proof. If $\mathcal{K}$ does not contain any automaton having (1), then $\mathcal{K}$ is a class of autonomous automata. If (2) does not hold for any automaton from $\mathcal{K}$, then $\mathcal{K}$ is a class of permutation automata. Finally, if $\mathcal{K}$ has no element with (3), then $\mathcal{K}$ is a class of reset automata. Thus, by Theorem 6.16, we have proved the necessity.

To show sufficiency, we show that if $\mathcal{K}$ fulfills conditions (1)–(3), then $\mathcal{K}$ is complete with respect to isomorphic representations under the T_q-product. By Lemma 6.17 it is enough to show that for any fixed nonvoid finite set A there is an automaton $\mathcal{N} \in P_q(\mathcal{K})$ having $\mathcal{A} \in SP_tI(\{\mathcal{N}\})$ for all $\mathcal{A} = (A, \{y_1, y_2\}, \delta_{\mathcal{A}})$ with $\delta_{\mathcal{A}}(a, y_1) = a, a \in A$.

By our condition (1), we get condition (1) of Lemma 6.18. Moreover, (3) is the same as condition (2) of Lemma 6.18. In addition, by (1), there is a nonautonomous automaton in $\mathcal{K}$; furthermore, by (2) the class $\mathcal{K}$ has a nonpermutation automaton. Thus conditions (1a) and (2a) of Lemma 6.18 also hold. Using these facts, consider an automaton $\mathcal{N} = (N, \{y_1, y_2\}, \delta)$ as in Lemma 6.18 and a bijective mapping $h : N \to N$ such that for every $t = 1, \dots, m, h^{-1}(b_t) = b_{m+t}, h^{-1}(\delta(b_t, y_1)) \in B_{m+t} \setminus \{b_{m+t}\}$; moreover, $h^{-1}(\delta(b_t, y_2)) \in B_{m+\ell} \setminus \{b_{m+\ell}\}$ whenever $\delta_{\mathcal{A}}(a_t, y_2) = a_\ell, \ell = 1, \dots, m$. Let $\mathcal{N}' = (N, \{y_1, y_2\}, \delta')$ be an automaton (state-isomorphic to $\mathcal{N}$) for which $\delta'(h(b), x) = h(\delta(b, x)), b \in N, x \in \{y_1, y_2\}$. Construct the temporal product $\mathcal{M} = (N, \{y_1, y_2\}, \delta_{\mathcal{M}}) = \mathcal{N} \times \mathcal{N}'(\{y_1, y_2\}, \varphi)$ such that $\varphi(y_1) = (y_1, y_2)$ and $\varphi(y_2) = (y_2, y_2)$. Suppose $\delta_{\mathcal{A}}(a_t, y_2) = a_\ell$ for a fixed $t \in \{1, \dots, m\}$. Then we get $\delta_{\mathcal{M}}(b_t, y_2) = \delta'(\delta(b_t, y_2), y_2) = \delta'(h(b'_{m+\ell}), y_2) = h(\delta(b'_{m+\ell}, y_2)) = h(b_{m+\ell}) = b_\ell$, where $b'_{m+\ell} \in B_{m+\ell} \setminus \{b_{m+\ell}\}$. Similarly, $\delta_{\mathcal{M}}(b_t, y_1) = \delta'(\delta(b_t, y_1), y_2) = \delta'(h(b'_{m+t}), y_2) = h(\delta(b'_{m+t}, y_2)) = h(b_{m+t}) = b_t$, where $b'_{m+t} \in B_{m+t} \setminus \{b_{m+t}\}$. It can be easily seen that $g : A \to N$ defined by $g(a_t) = b_t, t = 1, \dots, m$, is a state-isomorphism of $\mathcal{A}$ onto a subautomaton of $\mathcal{N}$. Therefore, $\mathcal{A} \in SP_tIP_q(\mathcal{K})$. □

By the proof of Theorem 6.19 we have also shown the following statement.

Theorem 6.20. *Let $\mathcal{K}$ be a class of automata containing a nonreset element. $\mathcal{K}$ is complete with respect to isomorphic representations under the T_{α_0}-product if and only if it is complete with respect to isomorphic representations under the T_q-product.* □

By the proof of Theorem 6.19 we have proved that if $\mathcal{K}$ does not satisfy (1)–(3) of Theorem 6.19, then $\mathcal{K}$ is either a class of autonomous automata, a class of permutation automata, or a class of reset automata, respectively. It is clear that these classes of automata are closed with respect to the homomorphism. Thus we can derive from Theorem 6.16 and Theorem 6.19 the following.

Theorem 6.21. *Any class $\mathcal{K}$ of automata is complete with respect to isomorphic representations under the T_{α_0}-product if and only if it is complete with respect to homomorphic representations under the T_{α_0}-product.* □

Now we start a complete characterization of $T_{\alpha_0}(\mathcal{K})$ (and $T_q(\mathcal{K})$) for any class $\mathcal{K}$ of automata. If a given class $\mathcal{K}$ of automata (or any member of $\mathcal{K}$) has a number of special properties, then we list them. In this way, we may omit many new definitions. (In this sense, for example, we speak about monotone reset automata, autonomous permutation automata, etc.)

Theorem 6.22. *If $\mathcal{K}$ is a class of autonomous reset automata which contains a nontrivial automaton, then $T_{\alpha_0}(\mathcal{K}) = T_q(\mathcal{K})$ is the class of all autonomous reset automata. If $\mathcal{K}$ is a class of trivial automata, then $T_{\alpha_0}(\mathcal{K}) = T_q(\mathcal{K})$ is the class of all trivial automata.*

Proof. Let $\mathcal{A}$ and $\mathcal{B}$ a pair of autonomous reset automata, where $\mathcal{A}$ is nontrivial. It can be easily seen that there exists a quasi-direct power of $\mathcal{A}$ having a subautomaton isomorphic to $\mathcal{B}$. It is also clear that any class of autonomous reset automata is closed under the general product, the α_0-product, the quasi-direct product, homomorphism, isomorphism, and the temporal product. Thus it remains to show that $T_{\alpha_0}(\mathcal{K}) = T_q(\mathcal{K})$ holds for any class of trivial automata. But this statement is obvious. □

Theorem 6.23. *If $\mathcal{K}$ is a class of autonomous permutation automata and $\mathcal{K}$ has a nondiscrete automaton with at least two states, then $T_{\alpha_0}(\mathcal{K}) = T_q(\mathcal{K})$ is the class of all autonomous permutation automata. If $\mathcal{K}$ is a class of discrete automata containing a nontrivial automaton, then $T_{\alpha_0}(\mathcal{K}) = T_q(\mathcal{K})$ is the class of all discrete automata.*

Proof. It is clear that for an arbitrary class $\mathcal{K}$ of discrete automata, $T_q(\mathcal{K})$ is a class of discrete automata with $T_{\alpha_0}(\mathcal{K}) = T_q(\mathcal{K})$. Moreover, it is easy to show that $SIP_q(\mathcal{K})$ contains all discrete automata provided that $\mathcal{K}$ is a class of discrete automata having a nontrivial automaton. Therefore, the second part of our theorem holds.

It is also obvious that $T_q(\mathcal{K})$ is a class of autonomous permutation automata with $T_{\alpha_0}(\mathcal{K}) = T_q(\mathcal{K})$, whenever $\mathcal{K}$ is a class of autonomous permutation automata. It remains to show that if this $\mathcal{K}$ has a nondiscrete automaton, then $T_q(\mathcal{K})$ is the class of all autonomous permutation automata. Observe that for any nondiscrete (autonomous) permutation automaton $\mathcal{D} \in \mathcal{K}$ we get (1) and (2) of Lemma 6.18. Therefore, for every positive integer m there can be found an $\mathcal{N} = (N, X', \delta'') \in P_q(\mathcal{K})$ and, moreover, pairwise disjoint sets $B_1, \ldots, B_{2m} \subseteq N$ and $x \in X$, such that $B_t = \{b_t, \delta(b_t, x)\}$, $B_{m+t} = \{b_{m+t}, \delta(b_{m+t}, x)\}$, $t = 1, \ldots, m$, with $|B_1| = \cdots = |B_{2m}| = 2$. ($\mathcal{N}$ is autonomous; therefore we can suppose $s = 1$ and $y_1 = y_2 = x_1 = \cdots = x_n = x$.) Let $\mathcal{A} = (\{a_1, \ldots, a_m\}, X, \delta_{\mathcal{A}})$ be any autonomous permutation automaton having m states. Give the mapping $h : N \to N$ with $h(\delta'(b_{m+t}, x)) = b_t, t \in \{1, \ldots, m\}$, and $h(b_{m+\ell}) = \delta'(b_t, x)$ whenever $\delta(a_t, x) = a_\ell, \ell \in \{1, \ldots, m\}, x \in X$. (The latter condition does not lead to a contradiction because of the fact that $\mathcal{A}$ is an autonomous permutation automaton.) Construct the automaton $\mathcal{N}' = (N, X, \delta'')$ (which is isomorphic to $\mathcal{N}$) such that for any $b' \in N$ let $\delta''(h(b'), x) = h(\delta'(b', x)), x \in X$. Take the temporal product $\mathcal{M} = (N, X, \delta_{\mathcal{M}}) = \mathcal{N} \times \mathcal{N}'(X, \varphi)$ with $\varphi(x) = (x, x), x \in X$. It can be immediately seen that by $\delta_{\mathcal{A}}(a_t, x) = a_\ell, t \in \{1, \ldots, m\}, x \in X$, we get $\delta_{\mathcal{M}}(b_t, x) = \delta''(\delta'(b_t, x), x) = \delta''(h(b_{m+\ell}), x) = h(\delta'(b_{m+\ell}, x)) = b_\ell$. Therefore, $\mathcal{A}$ can be embedded isomorphically into $\mathcal{M}$. Thus $T_q(\mathcal{K})$ contains all monotone permutation automata. The proof is complete. □

Theorem 6.24. *Let $\mathcal{K}$ be a class of autonomous automata which has (not necessarily different) automata $\mathcal{A}_t = (A_t, X_t, \delta_t)$, $t = 1, 2$, as follows:*

(1) *There is a triplet $a_1, a_2 \in A_1, x \in X_1$ with $a_1 \neq a_2$ and $\delta_1(a_1, x) = \delta_1(a_2, x)$.*
(2) *For suitable $a_1, a_2 \in A_2$ and $x \in X_2$, $a_1 \neq a_2$ and $\delta_2(a_1, x) \neq \delta_2(a_2, x)$.*

Then $T_{\alpha_0}(\mathcal{K}) = T_q(\mathcal{K})$ is the class of all autonomous automata.

Proof. Let $\mathcal{A} = (A, X, \delta)$ be any autonomous automaton. We shall show $\mathcal{A} \in T_q(\mathcal{K})$. Take an arrangement $a_1, \ldots, a_m$ of the elements of A and let $\mathcal{N} = (N, X, \delta')$ be an autonomous

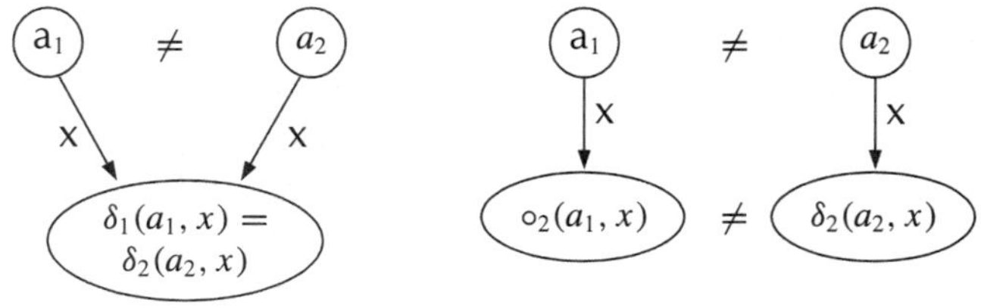

automaton having the following properties: there are pairwise disjoint sets $B_1, \ldots, B_{2m} \subseteq N$, where

(1a) $|B_t| = 2$, $|B_{m+t}| = m+1$, $t = 1, \ldots, m$;

(2a) for a suitable $b_t \in B_t$, $B_t = \{b_t, \delta'(b_t, x)\}$, $t = 1, \ldots, m$, $x \in X$;

(3a) there is a $b_{m+t} \in B_{m+t}$ such that for every $b^{(1)}_{m+t} \in B_{m+t} \setminus \{b_{m+t}\}$ it holds that $\delta'(b^{(1)}_{m+t}, x) = b_{m+t}$, $t = 1, \ldots, m$, $x \in X$.

Now let $h : N \to N$ be a bijective mapping with $h(b_{m+t}) = b_t$ and $h(b^{(1)}_{m+\ell}) = \delta'(b_t, x)$, $b^{(1)}_{m+\ell} \in B_{m+t} \setminus \{b_{m+t}\}$, $t = 1, \ldots, m$, if $\delta(a_t, x) = a_\ell$, $\ell \in \{1, \ldots, m\}$, $x \in X$.

Similar to the proofs of Theorems 6.19 and 6.23, we construct an automaton $\mathcal{N}' = (N, X, \delta'')$ (which is isomorphic to $\mathcal{N}$) such that for any $b' \in N$, $\delta''(h(b'), x) = h(\delta'(b', x))$, $x \in X$. Take the temporal product $\mathcal{M} = (N, X, \delta_{\mathcal{M}}) = \mathcal{N} \times \mathcal{N}'(X, \varphi)$ such that $\varphi(x) = (x, x)$, $x \in X$. Then for an arbitrary fixed $t \in \{1, \ldots, m\}$, $\delta_{\mathcal{A}}(a_t, x) = a_\ell$, $a_t \in A$, implies $\delta_{\mathcal{M}}(b_t, x) = \delta''(\delta'(b_t, x), x) = \delta''(h(b^{(1)}_{m+\ell}), x) = h(\delta'(b^{(1)}_{m+\ell}, x)) = h(b_{m+\ell}) = b_\ell$, where $b^{(1)}_{m+\ell} \in B_{m+\ell} \setminus \{b_{m+\ell}\}$. It can be seen directly that (h_1, h_2) with $h_1 : A \to N$, $h_1(a_t) = b_t, t = 1, \ldots, m$, and $h_2 : X \to X$, $h_2(x) = x, x \in X$, is an isomorphism of $\mathcal{A}$ onto $\mathcal{M}$. Therefore, $\mathcal{A}$ can be embedded isomorphically into $\mathcal{M}$. To prove that $T_q(\mathcal{K})$ contains all monotone permutation automata, it remains to show that $P_q(\mathcal{K})$ contains an automaton $\mathcal{N} = (N, X, \delta')$ having conditions (1a)–(3a).

Consider the automaton $\mathcal{A}_1$ in condition (1) of our theorem. By $a_1 \neq a_2$ and $\delta_1(a_1, x) \neq \delta_1(a_2, x)$, $a_1, a_2 \in A_1$, $x \in X_1$, it is clear that we have one of $\delta_1(a_1, x) \neq a_1$, $\delta_1(a_2, x) \neq a_2$. Thus we obtain conditions (1) and (2) of Lemma 6.18. (In particular, we also obtain condition (2a) of Lemma 6.18. Naturally, conditions (2a) and (3a) of Lemma 6.18 do not hold.) To obtain all required properties of $\mathcal{N}$, we consider $B_1, \ldots, B_{2m} \subseteq N$ defined in Lemma 6.18 with $s = m+1$. Thus all requirements (1a)–(3a) hold occurring in the proof of our theorem. (It is clear that in this case we get for every $t \in \{1, \ldots, m\}$ and $y_1, y_2 \in X$, $\delta(b_t, y_1) = \delta(b_t, y_2) \neq b_t$.) □

Theorems 6.22, 6.23, and 6.24 completely characterize all classes $T_{\alpha_0}(\mathcal{K})$ $(= T_q(\mathcal{K}))$ whenever $\mathcal{K}$ is a class of autonomous automata. Indeed, if $\mathcal{K}$ does not satisfy condition (1) of Theorem 6.24 and, simultaneously, it contains a nontrivial automaton, then we can apply Theorem 6.23. If either $\mathcal{K}$ is a class of trivial automata or $\mathcal{K}$ does not satisfy condition (2) of Theorem 6.24, we can apply Theorem 6.22.

Theorem 6.25. *If $\mathcal{K}$ is a class of permutation automata containing a nonautonomous permutation automaton, then $T_{\alpha_0}(\mathcal{K}) = T_q(\mathcal{K})$ is the class of all permutation automata.*

Proof. Let $\mathcal{A} = (A, X, \delta)$ be an arbitrary permutation automaton and let, for every $x \in X$, $\mathcal{M}_x = (M, \{x_0, x\}, \delta_x)$ be as before, where $x_0 \neq x$ is an arbitrary symbol, $A \subseteq M$, and moreover, for every $b \in A$, $\delta_x(b, x_0) = b$ and $\delta_x(b, x) = \delta(b, x)$. By Lemma 6.17, it is enough to show that for an appropriately chosen $\mathcal{N} \in T_q(\mathcal{K})$, we get $\mathcal{M}_x \in SP_tI(\{\mathcal{N}\})$, $x \in X$.

Consider a list $a_1, \ldots, a_m$ of the elements of A and let $\mathcal{N} = (N, X, \delta')$ be an automaton with the following properties. For any distinct pair $x, y \in X$ there exist pairwise disjoint sets $B_1, \ldots, B_{2m} \subseteq N$ having

(1a) $|B_t| = |B_{m+t}| = 3, t = 1, \ldots, m$;

(2a) for an appropriate $b_t \in B_t$, $B_t = \{b_t, \delta'(b_t, x), \delta'(b_t, y)\}, t = 1, \ldots, m$;

(3a) $B_{m+t} = \{b_{m+t}, b^1_{m+t}, b^2_{m+t}\}$, where $\delta'(b^1_{m+t}, x) = \delta'(b^2_{m+t}, y) = b_{m+t}, t = 1, \ldots, m$.

Take the bijective mapping $h : N \to N$ such that for every $t \in \{1, \ldots, m\}$, $h(b_{m+t}) = b_t$, $h(b^{(1)}_{m+t}) = \delta'(b_t, y)$ and for every $x \in X$, $h(b^{(2)}_{m+\ell}) = \delta'(b_t, x)$, whenever $\delta(a_t, x) = a_\ell, \ell \in \{1, \ldots, m\}$.

Construct again an automaton $\mathcal{N}' = (N, X, \delta'')$ which is isomorphic to $\mathcal{N}$ and $\delta''(h(b'), x) = h(\delta'(b', x)), b \in N, x \in X$.

Let x_0 be an arbitrary symbol. Construct the temporal product $\mathcal{M} = (N, \{x_0, x\}, \delta_x) = \mathcal{N} \times \mathcal{N}'(\{x_0, x\}, \varphi)$ such that $\varphi(x) = (x, y)$, $\varphi(x_0) = (y, x)$. Suppose that $\delta_{\mathcal{A}}(a_t, x) = a_\ell$, $a_t \in A$ holds for an $a_t \in A$. Thus we get $\delta_x(b_t, x) = \delta''(\delta'(b_t, x), y) = \delta''(h(b^{(2)}_{m+\ell}), y) = h(\delta'(b^{(2)}_{m+\ell}, y)) = h(b_{m+\ell}) = b_\ell$. Similarly, $\delta_x(b_t, x_0) = \delta''(\delta'(b_t, y), x) = \delta''(h(b^{(1)}_{m+t}), x) = h(\delta'(b^{(1)}_{m+t}, x)) = h(b_{m+t}) = b_t$. Therefore, (h_1, h_2) with $h_1 : A \to N$, $h_1(a_t) = b_t, t = 1, \ldots, m$, and $h_2 : X \to X$, $h_2(x_0) = x_0$, $h_2(x) = x$ is an isomorphism of $\mathcal{A}_x$ onto $\mathcal{M}_x$. Therefore, $\mathcal{A}$ can be embedded isomorphically into $\mathcal{M}$. We now show $\mathcal{N} \in T_q(\mathcal{K})$.

Let $\mathcal{D} = (D, X_{\mathcal{D}}, \delta_{\mathcal{D}})$ be any nonautonomous permutation automaton from $\mathcal{K}$. It can be seen easily that $\mathcal{D}$ has the following properties:

(1b) For a triplet $a \in A, x, y \in X_{\mathcal{D}}$, $\delta_{\mathcal{D}}(d, x) \neq \delta_{\mathcal{D}}(d, y)$.

(2b) There are distinct states $d_1, d_2 \in D$ and distinct inputs $x, y \in X_{\mathcal{D}}$ such that $\delta_{\mathcal{D}}(d_1, x) = \delta_{\mathcal{D}}(d_2, y)$.

(3b) There exists a triplet $d_1, d_2 \in D, x \in X$ having $\delta_{\mathcal{D}}(d_1, x) \neq \delta_{\mathcal{D}}(d_2, x)$.

(4b) There are $d \in D, x \in X$ with $\delta_{\mathcal{D}}(d, x) \neq d$.

By conditions (1b)–(4b) we obtain as follows. Using Lemma 6.18, for any distinct pair $x, y \in X$ there can be constructed a quasi-direct power of $\mathcal{D}$ having the conditions (1a)–(3a) given above. In detail, by Lemma 6.18 we can get $b_t \notin \{\delta'(b_t, x), \delta'(b_t, y)\}$ and $b_{m+t} \notin \{b^{(1)}_{m+t}, b^{(2)}_{m+t}\}$, with $b^{(1)}_{m+t} \neq b^{(2)}_{m+t}, t = 1, \ldots, m$. By (2b), we may obtain $\delta'(b_t, x) \neq \delta'(b_t, y)$. Finally, the disjunctivity of $B_1, \ldots, B_{2m}$ comes from Lemma 6.18. This ends the proof. □

Of course, for every nonmonotone reset automaton $\mathcal{A}$, the two-state reset automaton can be embedded isomorphically into $\mathcal{A}$. Therefore, using Proposition 3.14, every reset automaton can be embedded isomorphically into a quasi-direct power of $\mathcal{A}$ as long as $\mathcal{A}$ is a nonmonotone reset automaton. Thus, by Theorem 6.16, we have the following.

Theorem 6.26. *If $\mathcal{K}$ is a class of reset automata containing a nonmonotone reset automaton, then $T_q(\mathcal{K})$ is the class of all reset automata and $T_{\alpha_0}(\mathcal{K})$ is the class of all automata.* □

Observe that Theorems 6.22–6.26 inform us about the structure of $T_{\alpha_0}(\mathcal{K})$ and $T_q(\mathcal{K})$ whenever $\mathcal{K}$ is a class of either autonomous automata, permutation automata, or reset automata. Otherwise, we can use Theorem 6.19. This is summarized in the next statement.

Corollary 6.27. *For every class of automata, $T_{\alpha_0}T_{\alpha_0}(\mathcal{K}) = T_{\alpha_0}(\mathcal{K})$. Moreover, $T_{\alpha_0}(\mathcal{K}) = T_q(\mathcal{K})$, whenever $\mathcal{K}$ consists of autonomous automata, or $\mathcal{K}$ has a nonreset automaton. In particular, if $\mathcal{K}$ is a finite class of automata, then for an arbitrary automaton $\mathcal{A}$ it can be decided whether or not $\mathcal{A} \in T_{\alpha_0}(\mathcal{K})$ or $\mathcal{A} \in T_q(\mathcal{K})$.* □

Now we shall study the T_g-product.

Lemma 6.28. *If $\mathcal{K}$ is a class of autonomous automata, then $T_g(\mathcal{K}) = T_q(\mathcal{K})$.* □

The following statement characterizes all complete classes with respect to isomorphic representations under the T_g-product.

Theorem 6.29. *Any class $\mathcal{K}$ of automata is complete with respect to isomorphic representations under the T_g-product if and only if $\mathcal{K}$ contains a nonautonomous automaton.*

Proof. Every quasi-direct product is an α_0-product. Moreover, every α_0-product is a general product. Thus because of Lemma 6.28 and Theorem 6.16 we obtain the necessity of our condition.

To establish the sufficiency, let $\mathcal{A} = (A, X, \delta) \in \mathcal{K}$ be a nonautonomous automaton with a suitable triplet $a \in A, x, y \in X$ satisfying $\delta(a, x) \neq \delta(a, y)$. Observe that $\mathcal{K}$ satisfies condition (1) of Theorem 6.19. By Theorem 6.20, it is enough to show that for suitable single-factor products $\mathcal{A}_2, \mathcal{A}_3$ of $\mathcal{A}$, conditions (2) and (3) of Theorem 6.19 also hold. We prove this fact as follows:

(1a) There exist $a_1, a_2 \in A, x, y \in X$ with $a_1 \neq a_2$ and $\delta(a_1, x) = \delta(a_2, y)$.
(2a) There are $a_1, a_2 \in A, x \in X$ holding $a_1 \neq a_2$ and $\delta(a_1, x) \neq \delta(a_2, y)$.

If $\mathcal{A}$ is not a permutation automaton, then (by $x = y$) we have (1a), obviously. Therefore, to establish (1a) it is enough to take the case that $\mathcal{A}$ is a permutation automaton. Then we have a triplet $a \in A, x, y \in X$ with $\delta(a, x) \neq \delta(a, y)$. Thus there are a pair a_1' $(= \delta(a, x))$, a_2' $(= \delta(a, y))$ from A and positive integers k, ℓ such that $a_1' \neq a_2'$ and $\delta(a_1', x^k) = \delta(a_2', y^\ell)$ $(= a)$. By an easy computation we obtain that (1a) also holds.

Suppose that any permutation automaton $\mathcal{B}$ is trivial if $\mathcal{B}$ is a subautomaton of $\mathcal{A}$. Then by our assumptions, $\mathcal{A}$ has a nonautonomous definite subautomaton $\mathcal{A}'$. If $\mathcal{A}'$ is not a reset automaton, then we obtain (2a) by definition. If $\mathcal{A}'$ is a nonautonomous reset automaton, then there are $a, b \in A, x, y \in X$ with $a \neq b$ and $\delta(a, x) = a, \delta(b, y) = b$. Hence we get (2a).

Now let us assume that $\mathcal{A}$ has a subautomaton that is a nontrivial permutation automaton. Then for a suitable pair $b \in A, z \in X$ there is a positive integer $k > 1$ holding $\delta(b, z^k) = b$ and $\delta(b, z^t) \neq b$ if t does not divide k. Then (2a) holds whenever $a_1 = b, a_2 = \delta(b, z^{k-1})$, and $x = y = z$. Indeed, since $a_1 \neq a_2$, $\delta(a_1, x) = \delta(b, x), \delta(a_2, x) = b$, and $\delta(b, x) \neq b$, respectively, we get (2a). □

In the proof of Theorem 6.29 we also proved the following statement.

Theorem 6.30. *If any class $\mathcal{K}$ of automata is complete with respect to isomorphic representations under the T_g-product, then it is complete with respect to isomorphic representations under the T_{q^ℓ}-product.* □

It is clear that the class of all autonomous automata is closed under homomorphisms. Therefore, using Lemma 6.28, we have the following.

Theorem 6.31. *Any class $\mathcal{K}$ of automata is complete with respect to isomorphic representations under the T_g-product if and only if $\mathcal{K}$ is complete with respect to homomorphic representations under the T_{q^ℓ}-product.*

As a consequence of Lemma 6.28, moreover, via Theorems 6.29–6.31, respectively, we obtain that the T_g-product is equivalent with the T_{q^ℓ}-product from the point of view of isomorphic and homomorphic representation. Thus our statements also hold whenever for any $i > 0$ we consider either T_{α_i}-products, or T_{ν_i}-products or $T_{primitive}$-products instead of T_g-products. Thus we have obtained the next result.

Theorem 6.32. *For every positive integer i the T_g-product is equivalent to the T_{α_i}-product and also to the T_{ν_i}-product from the points of view of both homomorphic or isomorphic representations. Similarly, the T_g-product is equivalent to the $T_{primitive}$-product from the points of view of both homomorphic or isomorphic representations.*

Note that using Lemma 6.28 for any class $\mathcal{K}$ of autonomous automata Theorems 6.22–6.25 gives information about the structure of $T_g(\mathcal{K}) = T_{q^\ell}(\mathcal{K})$. Moreover, if $\mathcal{K}$ has a nonautonomous automaton, then $T_g(\mathcal{K}) = T_{q^\ell}(\mathcal{K})$ is characterized by Theorem 6.31. By these facts, we obtain the following statement.

Corollary 6.33. *For any class $\mathcal{K}$ of automata, $T_gT_g(\mathcal{K}) = T_{q^\ell}(\mathcal{K})$. In particular, if $\mathcal{K}$ is a finite class of automata, then for an arbitrary automaton $\mathcal{A}$ it is decidable whether $\mathcal{A} \in T_g(\mathcal{K})(= T_{q^\ell}(\mathcal{K}))$.* □

6.5 Bibliographical Remarks

Section 6.1. Basic properties of primitive products were described by P. Dömösi and C. L. Nehaniv [2000].

Section 6.2. Connections of primitive products and Letichevsky's criterion were also shown by P. Dömösi and C. L. Nehaniv [2000].

Section 6.3. By Letichevsky's result [1961], a class of finite automata is complete with respect to homomorphic representations under the Gluškov product if and only if it satisfies Letichevsky's criterion. Theorem 6.15 is the main result in P. Dömösi and C. L. Nehaniv [2000] showing the same fact for the primitive product.

Section 6.4. Temporal products were introduced by G. I. Ivanov [1973]. An abstract model of multichannel finite-state machines was introduced by A. Gill [1970]. Temporal compositions of α_i-products were introduced by F. Gécseg [1976a]. They were also studied by P. Dömösi [1986]. All results of this section are from P. Dömösi [1988].

Chapter 7
Finite State-Homogeneous Automata Networks and Asynchronous Automata Networks

Computer network routing, communication, and computation problems involving identical (or different) component processors connected according to a graph $\mathcal{D}$ with synchronous update are in fact just automata network problems. The interconnection digraph $\mathcal{D}$, which we call the underlying digraph of the network, is often referred to in the literature as the topology of the network.

State-homogeneous automata networks are those that have the same set Z of states at every node of $\mathcal{D}$. They are natural generalizations of the concept of cellular automata, but many receive external inputs and have different local automata.

The n-completeness of a homogeneous finite automata network means that it is able to simulate a complete homogeneous finite network in a very strong sense. The main results of Sections 7.1–7.4 *(similar to the results of Chapter* 2*) show that the homogeneous automata networks (having all loop edges) are very stable: removing many links, the network with $n \geq 1$ nodes remains n-complete as long as it remains strongly connected and has a central element. If the network has more than n nodes, then strong connectivity is enough for n-completeness (i.e., a central element is not necessary in this case). These results are in accordance with the well-known experimental results that many real-world networks are very stable against removing several links.*

We also show in this chapter a new method for the emulation of the behavior of any (synchronous) automata network by a corresponding asynchronous one which is obtained by a simple construction over very similar digraph (Section 7.5*). In particular, most results for automata networks can be carried over in a wholesale fashion to the asynchronous realm. Special cases of this result show how synchronous generalized cellular automata (automata networks with only one input symbol, a clock tick) and synchronous cellular automata (which, in addition, are state-homogeneous) can also be emulated by the corresponding type of asynchronous network. The results of this section also hold for automata networks over locally finite digraphs. So, for example, asynchronous universal cellular automata can be constructed from synchronous ones.*

7.1 State-Homogeneous Networks and Some Technical Lemmas

From now to Section 7.5, let us consider state-homogeneous automata networks, that is, automata networks in which each local component automaton has the same state set Z. We shall recover, with a new proof, an important result concerning completeness under projection for such networks, as well as an extension in Section 7.3, and in Section 7.4 further results on minimality of networks that are complete in this sense.

Here we recall the definition of automata network for the special case of state-homogeneity: Given an automaton $\mathcal{A} = (A, X, \delta)$, let $A = Z^n$ for some $|Z| \geq 1$ and $n \geq 1$ (where $|Z|$ denotes the cardinality, i.e., the number of elements in Z). Then we say that $\mathcal{A}$ is a *finite state-homogeneous automata network (of size n with respect to the basic local state set Z)*. Then the *underlying digraph* $\mathcal{D}_\mathcal{A} = (V_\mathcal{A}, E_\mathcal{A})$ of $\mathcal{A}$ is defined by $V_\mathcal{A} = \{1, \ldots, n\}$, $E_\mathcal{A} = \{(i, j) \mid$ there exists $x \in X$ such that $cp_j(\delta_x)$ really depends on its ith variable$\}$. $\mathcal{A}$ is a *$\mathcal{D}$-network* if $\mathcal{D} = (V, E)$ is a digraph with $V = V_\mathcal{A}$ and $E \supseteq E_\mathcal{A}$. In other words, $\mathcal{A}$ is a $\mathcal{D}$-network if every mapping $\delta_x : A \to A$ $(x \in X)$ is *compatible* with $\mathcal{D}$. Note that a size-n automata network may be regarded as comprising n component automata $\mathcal{A}_i = (Z, Z^n \times X, \delta_i)$, $i \in \{1, \ldots, n\}$, where the δ_i are defined by

$$\delta(z, x) = (\delta_1(z_1, (z, x)), \ldots, \delta_n(z_n, (z, x)))$$

for $z = (z_1, \ldots, z_n) \in Z^n, x \in X$. One may of course notationally suppress the components of Z^n in the inputs to $\mathcal{A}_i$ upon which δ_i does not really depend.

If $n = 1$ or $|Z| = 1$, then we say that $\mathcal{A} = (Z^n, X, \delta)$ is a *trivial* (finite) automata network. In this section we will investigate the state-homogeneous automata networks having state sets of the form Z^n for a positive integer $n > 1$ and finite set Z of cardinality at least two. Therefore, by an automata network we shall mean a nontrivial finite state-homogeneous automata network.

We shall derive a sequence of technical results on self-maps of powers G^n of a group G. In later sections, we will regard G as the local state set in an state-homogeneous automata network. We start with the following result, which can be shown by elementary computation.

Lemma 7.1. *Given a finite group G and a positive integer $n > 1$, let us define for every distinct $i, j \in \{1, \ldots, n\}$ the functions $F_{i,j}^{(t)} : G^n \to G^n, t = 1, 2, 3$, $F_j^{(4)} : G^n \to G^n$, and $U_{i,j} : G^n \to G^n$ as follows:*

$$F_{i,j}^{(1)}(g_1, \ldots, g_n) = (g_1, \ldots, g_{j-1}, g_i g_j, g_{j+1}, \ldots, g_n),$$

$$F_{i,j}^{(2)}(g_1, \ldots, g_n) = (g_1, \ldots, g_{j-1}, g_i{}^{-1} g_j, g_{j+1}, \ldots, g_n),$$

$$F_{i,j}^{(3)}(g_1, \ldots, g_n) = (g_1, \ldots, g_{j-1}, g_i, g_{j+1}, \ldots, g_n),$$

$$F_j^{(4)}(g_1, \ldots, g_n) = (g_1, \ldots, g_{j-1}, g_j{}^{-1}, g_{j+1}, \ldots, g_n),$$

$$U_{i,j}(g_1, \ldots, g_n) = (g_1, \ldots, g_{i-1}, g_j, g_{i+1}, \ldots, g_{j-1}, g_i, g_{j+1}, \ldots, g_n).$$

Then for arbitrary, pairwise distinct $i, j, k \in \{1, \ldots, n\}$ we get

$$F_{i,j}^{(1)} = F_{k,j}^{(2)} \circ F_{i,k}^{(1)} \circ F_{k,j}^{(1)} \circ F_{i,k}^{(2)},$$

$$F_{i,j}^{(2)} = F_{k,j}^{(2)} \circ F_{i,k}^{(2)} \circ F_{k,j}^{(1)} \circ F_{i,k}^{(1)},$$

$$F_{i,j}^{(3)} = F_{k,j}^{(3)} \circ F_{k,j}^{(2)} \circ F_{i,k}^{(2)} \circ F_{k,j}^{(2)} \circ F_{i,k}^{(1)} \circ F_{k,j}^{(1)},$$

$$U_{i,j} = F_{i,j}^{(1)} \circ F_{j,i}^{(2)} \circ F_{i}^{(4)} \circ F_{j}^{(4)} \circ F_{i,j}^{(1)} \circ F_{j}^{(4)}.$$

Proof. Although it is simple and elementary, we give a detailed proof of our statement to make the matter more understandable. In particular, we prove our statement by the following computations:

$$\begin{aligned}
&F_{k,j}^{(2)} \circ F_{i,k}^{(1)} \circ F_{k,j}^{(1)} \circ F_{i,k}^{(2)}(g_1, \ldots, g_n)\\
&= F_{i,k}^{(1)} \circ F_{k,j}^{(1)} \circ F_{i,k}^{(2)}(g_1, \ldots, g_{j-1}, g_k{}^{-1} g_j, g_{j+1}, \ldots, g_n)\\
&= F_{k,j}^{(1)} \circ F_{i,k}^{(2)}(g_1, \ldots, g_{j-1}, g_k{}^{-1} g_j, g_{j+1}, \ldots, g_{k-1}, g_i g_k, g_{k+1}, \ldots, g_n)\\
&= F_{i,k}^{(2)}(g_1, \ldots, g_{j-1}, g_i g_j, g_{j+1}, \ldots, g_{k-1}, g_i g_k, g_{k+1}, \ldots, g_n)\\
&= (g_1, \ldots, g_{j-1}, g_i g_j, g_{j+1}, \ldots, g_{k-1}, g_k, g_{k+1}, \ldots, g_n)\\
&= F_{i,j}^{(1)}(g_1, \ldots, g_n),
\end{aligned}$$

$$\begin{aligned}
&F_{k,j}^{(2)} \circ F_{i,k}^{(2)} \circ F_{k,j}^{(1)} \circ F_{i,k}^{(1)}(g_1, \ldots, g_n)\\
&= F_{i,k}^{(2)} \circ F_{k,j}^{(1)} \circ F_{i,k}^{(1)}(g_1, \ldots, g_{j-1}, g_k{}^{-1} g_j, g_{j+1}, \ldots, g_n)\\
&= F_{k,j}^{(1)} \circ F_{i,k}^{(1)}(g_1, \ldots, g_{j-1}, g_k{}^{-1} g_j, g_{j+1}, \ldots, g_{k-1}, g_i{}^{-1} g_k, g_{k+1}, \ldots, g_n)\\
&= F_{i,k}^{(1)}(g_1, \ldots, g_{j-1}, g_i{}^{-1} g_j, g_{j+1}, \ldots, g_{k-1}, g_i{}^{-1} g_k, g_{k+1}, \ldots, g_n)\\
&= (g_1, \ldots, g_{j-1}, g_i{}^{-1} g_j, g_{j+1}, \ldots, g_{k-1}, g_k, g_{k+1}, \ldots, g_n)\\
&= F_{i,j}^{(2)}(g_1, \ldots, g_n),
\end{aligned}$$

$$\begin{aligned}
&F_{k,j}^{(3)} \circ F_{k,j}^{(2)} \circ F_{i,k}^{(2)} \circ F_{k,j}^{(2)} \circ F_{i,k}^{(1)} \circ F_{k,j}^{(1)}(g_1, \ldots, g_n)\\
&= F_{k,j}^{(2)} \circ F_{i,k}^{(2)} \circ F_{k,j}^{(2)} \circ F_{i,k}^{(1)} \circ F_{k,j}^{(1)}(g_1, \ldots, g_{j-1}, g_k, g_{j+1}, \ldots, g_n)\\
&= F_{i,k}^{(2)} \circ F_{k,j}^{(2)} \circ F_{i,k}^{(1)} \circ F_{k,j}^{(1)}(g_1, \ldots, g_{j-1}, 1, g_{j+1}, \ldots, g_n)\\
&= F_{k,j}^{(2)} \circ F_{i,k}^{(1)} \circ F_{k,j}^{(1)}(g_1, \ldots, g_{j-1}, 1, g_{j+1}, \ldots, g_{k-1}, g_i{}^{-1} g_k, g_{k+1}, \ldots, g_n)\\
&= F_{i,k}^{(1)} \circ F_{k,j}^{(1)}(g_1, \ldots, g_{j-1}, g_k{}^{-1} g_i, g_{j+1}, \ldots, g_{k-1}, g_i{}^{-1} g_k, g_{k+1}, \ldots, g_n)\\
&= F_{k,j}^{(1)}(g_1, \ldots, g_{j-1}, g_k{}^{-1} g_i, g_{j+1}, \ldots g_{k-1}, g_k, g_{k+1}, \ldots, g_n)\\
&= (g_1, \ldots, g_{j-1}, g_i, g_{j+1}, \ldots g_{k-1}, g_k, g_{k+1}, \ldots, g_n)\\
&= F_{i,j}^{(3)}(g_1, \ldots, g_n),
\end{aligned}$$

$$
\begin{aligned}
&F_{i,j}^{(1)} \circ F_{j,i}^{(2)} \circ F_i^{(4)} \circ F_j^{(4)} \circ F_{i,j}^{(1)} \circ F_j^{(4)}(g_1, \dots, g_n)\\
&\quad = F_{j,i}^{(2)} \circ F_i^{(4)} \circ F_j^{(4)} \circ F_{i,j}^{(1)} \circ F_j^{(4)}(g_1, \dots, g_{j-1}, g_i g_j, g_{j+1}, \dots, g_n)\\
&\quad = F_i^{(4)} \circ F_j^{(4)} \circ F_{i,j}^{(1)} \circ F_j^{(4)}(g_1, \dots, g_{i-1}, g_j^{-1}, g_{i+1}, \dots, g_{j-1}, g_i g_j, g_{j+1}, \dots, g_n)\\
&\quad = F_j^{(4)} \circ F_{i,j}^{(1)} \circ F_j^{(4)}(g_1, \dots, g_{i-1}, g_j, g_{i+1}, \dots, g_{j-1}, g_i g_j, g_{j+1}, \dots, g_n)\\
&\quad = F_{i,j}^{(1)} \circ F_j^{(4)}(g_1, \dots, g_{i-1}, g_j, g_{i+1}, \dots, g_{j-1}, g_j^{-1} g_i^{-1}, g_{j+1}, \dots, g_n)\\
&\quad = F_j^{(4)}(g_1, \dots, g_{i-1}, g_j, g_{i+1}, \dots, g_{j-1}, g_i^{-1}, g_{j+1}, \dots, g_n)\\
&\quad = (g_1, \dots, g_{i-1}, g_j, g_{i+1}, \dots, g_{j-1}, g_i, g_{j+1}, \dots, g_n)\\
&\quad = U_{i,j}(g_1, \dots, g_n). \qquad \square
\end{aligned}
$$

Lemma 7.2. *Given a finite group G and a pair of relatively prime integers m, n with $1 \le m < n$, let us define for every $\ell \in \{1, \dots, n\}$, the transformations $T_i^{(0)} : G^n \to G^n$, $T_i^{(k)} : G^n \to G^n, k = 1, 2, 3, 4$, as follows:*

$$
\begin{aligned}
T^{(0)}(g_1, \dots, g_n) &= (g_n, g_1, \dots, g_{n-1}),\\
T_\ell^{(1)}(g_1, \dots, g_n) &= (g_n, g_1, \dots, g_{\ell-2}, g_{\ell-m-1\,(\mathrm{mod}\,n)} g_{\ell-1}, g_\ell, \dots, g_{n-1}),\\
T_\ell^{(2)}(g_1, \dots, g_n) &= (g_n, g_1, \dots, g_{\ell-2}, g_{\ell-m-1\,(\mathrm{mod}\,n)}^{-1} g_{\ell-1}, g_\ell, \dots, g_{n-1}),\\
T_\ell^{(3)}(g_1, \dots, g_n) &= (g_n, g_1, \dots, g_{\ell-2}, g_{\ell-m-1\,(\mathrm{mod}\,n)}, g_\ell, \dots, g_{n-1}),\\
T_\ell^{(4)}(g_1, \dots, g_n) &= (g_n, g_1, \dots, g_{\ell-2}, g_{\ell-1}^{-1}, g_\ell, \dots, g_{n-1}).
\end{aligned}
$$

Then for any fixed $\ell \in \{1, \dots, n\}$, $\mathcal{T}_{G,n} \subsetneq \langle\{T^{(0)}, T_\ell^{(k)} \mid k = 1, 2, 3, 4\}\rangle$.

Proof. For every $i \in \{1, \dots, n\},\ k \in \{1, \dots, 4\},\ T_i^{(k)} = (T^{(0)})^{n+\ell-i} \circ T_\ell^{(k)} \circ (T^{(0)})^{n+i-\ell}$. Thus we shall show only $\mathcal{T}_{G,n} \subsetneq \langle\{T^{(0)}, T_\ell^{(k)} \mid \ell \in \{1, \dots, n\}, k = 1, 2, 3, 4\rangle$. It is clear that by the simple fact that every permutation is a composite of transpositions and, moreover, transformations can be generated by permutations and elementary collapsings, using the notation in Lemma 7.1, we obtain $\mathcal{T}_{G,n} \subseteq \langle\{F_{i,j}^{(3)}, U_{i,j} \mid i, j \in \{1, \dots, n\}\rangle$. On the other hand, $\langle\{T^{(0)}, T_\ell^{(k)} \mid k = 1, 2, 3, 4, \ell = 1, \dots, n\}\rangle \setminus \mathcal{T}_{G,n} \neq \emptyset$ is clear. Thus, it is enough to prove that for every $i, j \in \{1, \dots, n\}$, $F_{i,j}^{(3)}, U_{i,j} \in \langle\{T^{(0)}, T_\ell^{(k)} \mid k = 1, 2, 3, 4, \ell = 1, \dots, n\}\rangle$. Using $F_{i+(j-1)m-1\,(\mathrm{mod}\,n), i+jm-1\,(\mathrm{mod}\,n)}^{(d)} = T_{i+jm\,(\mathrm{mod}\,n)}^{(d)} \circ (T^{(0)})^{n-1}$, $d = 1, 2, 3, i \in \{1, \dots, n\}, j = 0, 1, \dots,$ by an inductive application of Lemma 7.1, we have $F_{i-m-1\,(\mathrm{mod}\,n), i+jm-1\,(\mathrm{mod}\,n)}^{(d)} \in \langle\{T^{(0)}, T_\ell^{(k)} \mid k = 1, 2, 3, 4, \ell = 1, \dots, n\}\rangle$ ($i \in \{1, \dots, n, \}, j = 0, 1 \dots$).

Therefore, because m and n are relatively prime, we receive $F_{i,j}^{(d)} \in \langle\{T^{(0)}, T_\ell^{(k)} \mid k = 1, 2, 3, 4, \ell = 1, \dots, n\}\rangle$ ($d = 1, 2, 3, i, j \in \{1, \dots, n\}$).

Moreover, we also have $F_{i-1\,(\mathrm{mod}\,n)}^{(4)} = T_i^{(4)} \circ (T^{(0)})^{n-1}$, $i \in \{1, \dots, n\}$. Hence, applying Lemma 7.1 again, we obtain $U_{i,j} \in \langle\{T^{(0)}, T_\ell^{(k)} \mid k = 1, \dots, 4, \ell = 1, \dots, n\}\rangle$, $i, j \in \{1, \dots, n\}$, and thus, having $F_{i,j}^{(3)} \subseteq \langle\{T^{(0)}, T_\ell^{(k)} \mid k = 1, 2, 3, 4\}\rangle (i, j \in \{1, \dots, n\})$, the proof is complete. $\square$

We shall use the following lemma.

Lemma 7.3. *Given a positive integer n, let $G = \langle g \rangle$ denote a finite nontrivial cyclic group with a generator $g \in G$. There exists an arrangement $a_1, \ldots, a_m$ ($m = |G|^n$) of the elements in the* n*th direct power G^n of G such that for every $i = 1, \ldots, m-1$ there is a $j \in \{1, \ldots, n\}$ with $a_{i+1} \in \{(g_1, \ldots, g_{j-1}, g_j g^{-1}, g_{j+1}, \ldots, g_n), (g_1, \ldots, g_{j-1}, g_j g, g_{j+1}, \ldots, g_n)\}$, whenever $a_i = (g_1, \ldots, g_n)$ ($\in G^n$).*

Proof. If $n = 1$, then our statement is trivial. Now let us suppose that our statement holds for $n \geq 1$ and let $a_1, \ldots, a_m$ be a suitable arrangement of G^n. Then (g, a_1), $(g, a_2), \ldots, (g, a_m)$, (g^2, a_m), $(g^2, a_{m-1}), \ldots, (g^2, a_1)$, (g^3, a_1), $(g^3, a_2), \ldots$, $(g^3, a_m), \ldots, (g^{|G|}, a_t)$, where $t = 1$ (resp., $t = m$) if $|G|$ is even (resp., odd) is a suitable arrangement of the $(n+1)$th direct power G^{n+1} of G, where $(g^k, a_i) = (g^k, g_1, \ldots, g_n)$, whenever, $a_i = (g_1, \ldots, g_n)$ ($k = 1, \ldots, |G|, i = 1, \ldots, m$). The result now follows by induction. □

Given a nonvoid set Y, a positive integer n, let $\mathcal{T}_Y$ denote the full transformation semigroup of all functions from Y to Y. In addition, for every subset $H \subseteq \mathcal{T}_Y$, let $\langle H \rangle$ denote the subsemigroup of $\mathcal{T}_Y$ generated by H. Moreover, for any finite set X with $|X| > 1$ and positive integer $n > 1$, denote by $\mathcal{T}_{X,n}$ the subsemigroup of all transformations of $\mathcal{T}_{X^n}$ having the form $F(x_1, \ldots, x_n) = (x_{t(1)}, \ldots x_{t(n)})$, $(x_1, \ldots, x_n) \in X^n$, $t \mid \{1, \ldots, n\} \to \{1, \ldots, n\}$, and let

$$\Gamma_{X^n} = \{F : X^n \to X^n \mid F(x_1, \ldots, x_n) = (x_1, \ldots, x_{i-1}, f(x_i, x_j), x_{i+1}, \ldots, x_n),$$

where $f : X^2 \to X, i, j \in \{1, \ldots, n\}, (x_1, \ldots, x_n) \in X^n\}$. (It is understood that the case $i = j$ is allowed in the above definition of Γ_{X^n}.) Define the *elementary collapsing* $t_{j,k} : \{1, \ldots, n\} \to \{1, \ldots, n\}$ for $1 \leq j \neq k \leq n$,

$$t_{j,k}(i) = \begin{cases} j & \text{if } i = k, \\ i & \text{otherwise.} \end{cases}$$

Moreover, as usual we say that $u_{j,k} : \{1, \ldots, n\} \to \{1, \ldots, n\}$ for $1 \leq j \neq k \leq n$ is a *transposition* if

$$u_{j,k}(i) = \begin{cases} j & \text{if } i = k, \\ k & \text{if } i = j, \\ i & \text{otherwise.} \end{cases}$$

Finally, define $U_{i,j} : X^n \to X^n$ by

$$U_{i,j}(x_1, \ldots, x_n) = (x_1, \ldots, x_{i-1}, x_j, x_{i+1}, \ldots, x_{j-1}, x_i, x_{j+1}, \ldots, x_n),$$

where $(x_1, \ldots, x_n) \in X^n$ similarly as in the previous lemma.

In the following, we shall identify X in a fixed but arbitrary way with the group of residue classes of integers modulo $q = |X|$.

Now we are ready to prove the following key lemma.

Lemma 7.4. *For any fixed $\ell \in \{1, \ldots, n\}$, $\mathcal{T}_{X^n}$ is generated by the union of $\langle\{T^{(0)}, T_\ell^{(k)} \mid k = 1, \ldots, 4\}\rangle$ and the set of all functions $F : X^n \to X^n$ having the form $F(x_1, \ldots, x_n) = (x_1, \ldots, x_{\ell-1}, f(x_1, \ldots, x_n), x_{\ell+1}, \ldots, x_n)$, $f : X^n \to X$, where $x_1, \ldots, x_n \in X$.*

Proof. We can take out of consideration the trivial case $|X| = 1$. Thus we assume $|X| > 1$.

It is clear that without loss of generality we may suppose $\ell = 1$. On the other hand, using Lemma 7.2, $\{U_{i,j} \mid i, j \in \{1, \ldots, n\}\} \subsetneq \langle\{T^{(0)}, T_\ell^{(k)} \mid k = 1, \ldots, 4\rangle$. Thus it is enough to prove that the union of $\{U_{i,j} \mid i, j \in \{1, \ldots, n\}\}$ and the set of all functions $F : X^n \to X^n$ having the form $F(x_1, \ldots, x_n) = (f(x_1, \ldots, x_n), x_2, \ldots, x_n)$ generates $\mathcal{T}_{X^n}$.

For every pair $i \in \{1, \ldots, n\}$, $f : X^n \to X$, define the function $F_{i,f} : X^n \to X^n$ with $F_{i,f}(x_1, \ldots, x_n) = (x_1, \ldots, x_{i-1}, f(x_1, \ldots, x_n), x_{i+1}, \ldots, x_n)$, $(x_1, \ldots, x_n \in X)$. Thus, by letting $f' = U_{i,j} \circ f$, we have $F_{j,f} = U_{i,j} \circ F_{i,f'} \circ U_{i,j}$. So for every pair $i \in \{1, \ldots, n\}$, $f : X^n \to X$, $F_{i,f} \in \langle \mathcal{T}_{X,n} \cup \{F : X^n \to X^n \mid F(x_1, \ldots, x_n) = (f(x_1, \ldots, x_n), x_2, x_3, \ldots, x_n), f : X^n \to X, x_1, \ldots, x_n \in X\}\rangle$.

Let us identify X with a nontrivial finite cyclic group with generating element $g \in X$. Thus we also have that for any $c_1, \ldots, c_n \in X$, ${F^{(1)}}_{\epsilon,j,(c_1,\ldots,c_n)}$, ${F^{(2)}}_{\epsilon,j,(c_1,\ldots,c_n)} \in \langle \mathcal{T}_{X,n} \cup \{F : X^n \to X^n \mid F(x_1, \ldots, x_n) = (f(x_1, \ldots, x_n), x_2, x_3, \ldots, x_n), f : X^n \to X, x = (x_1, \ldots, x_n) \in X^n\}\rangle$, whenever $\epsilon \in \{1, -1\}$,

$$
\begin{aligned}
&{F^{(1)}}_{\epsilon,j,(c_1,\ldots,c_n)}(x)\\
&= \begin{cases} (c_1, \ldots, c_n) & \text{if } x = (c_1, \ldots, c_{j-1}, c_j g^\epsilon, c_{j+1}, \ldots, c_n),\\ (c_1, \ldots, c_{j-1}, c_j g^\epsilon, c_{j+1}, \ldots, c_n) & \text{if } x = (c_1, \ldots, c_n),\\ x & \text{otherwise,} \end{cases}
\end{aligned}
$$

$$
{F^{(2)}}_{\epsilon,j,(c_1,\ldots,c_n)}(x) = \begin{cases} (c_1, \ldots, c_{j-1}, c_j g^\epsilon, c_{j+1}, \ldots, c_n) & \text{if } x = (c_1, \ldots, c_n),\\ x & \text{otherwise,} \end{cases}
$$

where $x = (x_1, \ldots, x_n) \in X^n$. On the other hand, by Lemma 7.3, there exists an arrangement $a_1, \ldots, a_m$ of X^n such that for every $k = 1, \ldots, m-1$, $p_k \in \{{F^{(1)}}_{\epsilon,j,(c_1,\ldots,c_n)} \mid \epsilon \in \{-1, 1\}, j \in \{1, \ldots, n\}, c_1, \ldots, c_n \in X\}$, $t_k \in \{{F^{(2)}}_{\epsilon,j,(c_1,\ldots,c_n)} \mid \epsilon \in \{-1, 1\}, j \in \{1, \ldots, n\}, c_1, \ldots, c_n \in X\}$, where

$$
p_k(a_\ell) = \begin{cases} a_{k+1} & \text{if } \ell = k,\\ a_k & \text{if } \ell = k+1,\\ a_\ell & \text{otherwise,} \end{cases}
$$

$$
t_k(a_\ell) = \begin{cases} a_{k+1} & \text{if } \ell = k,\\ a_\ell & \text{otherwise,} \end{cases}
$$

$$
t_k(a_\ell) = \begin{cases} a_{k+1} & \text{if } \ell = k,\\ a_\ell & \text{otherwise.} \end{cases}
$$

But then $p_1, \ldots, p_{m-1}$ is a set of transpositions such that $\{p_1, \ldots, p_{m-1}\}$ generates all permutations over X^n. Furthermore, $t_1, \ldots, t_{m-1}$ is a set of elementary collapsings over X^n. Thus by the well-known fact that for every $j = 1, \ldots, m-1$, $\{p_1, \ldots, p_{m-1}, t_j\}$ generates all transformations over X^n, the proof is complete. □

Lemma 7.5. $\mathcal{T}_{X,n} \subsetneq \langle \Gamma_{X^n} \rangle$.

Proof. Using Lemma 7.1, we have that for any pair $i \neq j \in \{1, \ldots, n\}$, $U_{i,j} \in \langle \Gamma_{X^n} \rangle$. On the other hand, $\{F_{i,j}^{(3)} \mid i, j \in \{1, \ldots, n\}\} \subset \langle \Gamma_{X^n} \rangle$ holds by definition. Recall that $U_{i,j}$ transposes the elements in the ith and jth positions and that $F_{i,j}^{(3)}$ replaces the jth entry by the ith. Therefore, by the simple fact that every permutation can be composed as a product of transpositions and, moreover, transformations can be generated by permutations and elementary collapsings, we obtain the inclusion of our statement. Finally, it is trivial that $\Gamma_{X^n} \setminus \mathcal{T}_{X,n}$ is nonvoid. □

Fix arbitrary $c \neq d \in X$.

Lemma 7.6. *Given an alphabet X and a positive integer $n > 1$, let $(c_1, \ldots, c_{n-1}) \in X^{n-1}$,*

$$F_{(c_1,\ldots,c_{n-1})}(x_1, \ldots, x_n) = \begin{cases} (x_1, \ldots, x_n) & \text{if } (x_1, \ldots, x_n) = (c_1, \ldots, c_{n-1}, c), \\ (x_1, \ldots, x_{n-1}, d) & \text{if } (x_1, \ldots, x_n) \neq (c_1, \ldots, c_{n-1}, c) \end{cases}$$

$((x_1, \ldots, x_n) \in X^n)$. We have $F_{(c_1,\ldots,c_{n-1})} \in \langle \Gamma_{X^n} \rangle$.

Proof. If $n = 2$, then our statement holds by definition. Otherwise, $n > 2$ and for every $b \in X$, define

$$F_b^{(0)}(x_1, \ldots, x_n) = \begin{cases} (x_1, \ldots, x_{n-1}, c) & \text{if } x_{n-1} = b, x_n = c, \\ (x_1, \ldots, x_{n-1}, d) & \text{otherwise}, \end{cases}$$

where $(x_1, \ldots, x_n) \in X^n$.

For every $i \in \{1, \ldots, n-1\}$, $(c_i, \ldots, c_{n-1}) \in X^{n-i}$, let

$$F_{(c_i,\ldots,c_{n-1})}(x_1, \ldots, x_n) = \begin{cases} (x_1, \ldots, x_{n-1}, c) & \text{if } (x_i, \ldots, x_n) = (c_i, \ldots, c_{n-1}, c), \\ (x_1, \ldots, x_{n-1}, d) & \text{otherwise}, \end{cases}$$

where $x = (x_1, \ldots, x_n) \in X^n$). It is clear that $F_{(c_{n-1})} = F_{c_{n-1}}^{(0)}$. On the other hand, for every $i \in \{2, \ldots, n-1\}$, $F_{(c_{i-1},\ldots,c_{n-1})} = U_{i-1,n-1} \circ F_{c_{i-1}}^{(0)} \circ U_{i-1,n-1} \circ F_{(c_i,\ldots,c_{n-1})}$. Simultaneously, we have by definition that $F_{c_{i-1}}^{(0)} \in \Gamma_{X^n}$ holds for every $i \in \{2, \ldots, n-1\}$. Moreover, by Lemma 7.5 we have $U_{i,j} \in \langle \Gamma_{X^n} \rangle$. Thus we get our result by induction. □

Lemma 7.7. *Given an alphabet X and a positive integer $n > 1$, let $(c_1, \ldots, c_{n-1})$, $(d_1, \ldots, d_{n-1}) \in X^{n-1}$, $d \in X$ ($(c_1, \ldots, c_{n-1}) = (d_1, \ldots, d_{n-1})$ is allowed) and let*

$$F_{(c_1,\ldots,c_{n-1}),(d_1,\ldots,d_{n-1})}^{(1)}(x) = \begin{cases} (c_1, \ldots, c_{n-1}, d) & \text{if } (x_1, \ldots, x_{n-1}) = (d_1, \ldots, d_{n-1}), \\ (d_1, \ldots, d_{n-1}, d) & \text{if } (x_1, \ldots, x_{n-1}) = (c_1, \ldots, c_{n-1}), \\ (x_1, \ldots, x_{n-1}, d) & \text{otherwise}, \end{cases}$$

$$F_{(c_1,\ldots,c_{n-1}),(d_1,\ldots,d_{n-1})}^{(2)}(x) = \begin{cases} (c_1, \ldots, c_{n-1}, d) & \text{if } (x_1, \ldots, x_{n-1}) = (d_1, \ldots, d_{n-1}), \\ (x_1, \ldots, x_{n-1}, d) & \text{otherwise}, \end{cases}$$

where $x = (x_1, \ldots, x_n) \in X^n$.

Then we have $F_{(c_1,\ldots,c_{n-1}),(d_1,\ldots,d_{n-1})}^{(i)} \in \langle \Gamma_{X^n} \rangle$, $i = 1, 2$.

Proof. We have $c \in X$ arbitrary with $c \neq d$, and set $F_c^{(3)}(x) = (x_1, \ldots, x_{n-1}, c)$, $F_d^{(3)}(x) = (x_1, \ldots, x_{n-1}, d)$, and

$$F_{(c_1,\ldots,c_{n-1}),(d_1,\ldots,d_{n-1})}^{(4)}(x) = \begin{cases} (c_1, \ldots, c_{n-1}, c) & \text{if } (x_1, \ldots, x_n) = (d_1, \ldots, d_{n-1}, c), \\ (x_1, \ldots, x_{n-1}, d) & \text{otherwise}, \end{cases}$$

where $x = (x_1, \dots, x_n) \in X^n$; moreover, define $F_{(c_1,\dots,c_{n-1})}$ as in Lemma 7.6. In addition, let

$$F^{(5)}(x_1, \dots, x_n) = \begin{cases} (x_1, \dots, x_{n-1}, c) & \text{if } x_n = d, \\ (x_1, \dots, x_{n-1}, d) & \text{if } x_n = c, \\ (x_1, \dots, x_{n-1}, x_n) & \text{otherwise,} \end{cases}$$

and for every $a \in X$, let

$$F_a^{(6)}(x_1, \dots, x_n) = \begin{cases} (x_1, \dots, x_{n-2}, a, x_n) & \text{if } x_n = c, \\ (x_1, \dots, x_{n-1}, x_n) & \text{otherwise} \end{cases}$$

($(x_1, \dots, x_n) \in X^n$). It is clear that $F_c^{(3)}, F_d^{(3)}, F^{(5)}, F_a^{(6)} \in \Gamma_{X^n}$. Next we show that $F^{(4)}_{(c_1,\dots,c_{n-1}),(d_1,\dots,d_{n-1})} \in \langle \Gamma_{X^n} \rangle$. Indeed, by an easy computation we get $F^{(4)}_{(c_1,\dots,c_{n-1}),(d_1,\dots,d_{n-1})} = F_{(d_1,\dots,d_{n-1})} \circ F^{(6)}_{c_{n-1}} \circ U_{n-2,n-1} \circ F^{(6)}_{c_{n-2}} \circ U_{n-3,n-1} \circ \cdots \circ U_{2,n-1} \circ F^{(6)}_{c_2} \circ U_{1,n-1} \circ F^{(6)}_{c_1} \circ U_{1,n-1} \circ U_{2,n-1} \circ \cdots \circ U_{n-2,n-1}$. On the other hand, by Lemma 7.7 we have $U_{i,j} \in \langle \Gamma_{X^n} \rangle$. But then $F^{(2)}_{(c_1,\dots,c_{n-1}),(d_1,\dots,d_{n-1})} = F_c^{(3)} \circ F^{(4)}_{(c_1,\dots,c_{n-1}),(d_1,\dots,d_{n-1})} \circ F_d^{(3)}$ implies $F^{(2)}_{(c_1,\dots,c_{n-1}),(d_1,\dots,d_{n-1})} \in \langle \Gamma_{X^n} \rangle$. It remains to prove that $F^{(1)}_{(c_1,\dots,c_{n-1}),(d_1,\dots,d_{n-1})} \in \langle \Gamma_{X^n} \rangle$. This connection, completing the proof, comes from $F^{(1)}_{(c_1,\dots,c_{n-1}),(d_1,\dots,d_{n-1})} = F_c^{(3)} \circ F^{(4)}_{(c_1,\dots,c_{n-1}),(d_1,\dots,d_{n-1})} \circ F^{(5)} \circ F^{(4)}_{(d_1,\dots,d_{n-1}),(c_1,\dots,c_{n-1})} \circ F_d^{(3)}$. □

Let $\mathcal{F}_{X^{n-1}\times\{d\}}$ be the semigroup of functions $\{F \in \mathcal{T}_{X^n} \mid F(x_1, \dots, x_n) \in X^{n-1} \times \{d\}$, $x_1, \dots, x_n \in X$, where F is really independent of $x_n\}$. By the above statement we get the following result.

Lemma 7.8. $\mathcal{F}_{X^{n-1}\times\{d\}} \subsetneq \langle \Gamma_{X^n} \rangle$.

Proof. For every pair $(c_1, \dots, c_{n-1}), (d_1, \dots, d_{n-1}) \in X^{n-1}$, let us define the mappings $F^{(1)}_{(c_1,\dots,c_{n-1}),(d_1,\dots,d_{n-1})}$, $F^{(2)}_{(c_1,\dots,c_{n-1}),(d_1,\dots,d_{n-1})}$ as in Lemma 7.7. Observe that $F^{(1)}_{(c_1,\dots,c_{n-1}),(d_1,\dots,d_{n-1})}$ acts as a transposition in the permutation group over the set $X^{n-1} \times \{d\}$, while $F^{(2)}_{(c_1,\dots,c_{n-1}),(d_1,\dots,d_{n-1})}$ acts as an elementary collapsing in the transformation semigroup over the set $X^{n-1} \times \{d\}$. By Lemma 7.7 we obtain that all of these transpositions and elementary collapsings are in $\langle \Gamma_{X^n} \rangle$. It is well known that the set of all transpositions and elementary collapsings on a set generates all mappings on that set, so any map taking $X^{n-1} \times \{d\}$ to itself may be written as the restriction to $X^{n-1} \times \{d\}$ of a composite of the above functions. A moment's reflections shows that the set of all these $F^{(1)}_{(c_1,\dots,c_{n-1}),(d_1,\dots,d_{n-1})}$, $F^{(2)}_{(c_1,\dots,c_{n-1}),(d_1,\dots,d_{n-1})}$ in fact generates all of $\mathcal{F}_{X^{n-1}\times\{d\}}$, since a function in the latter is uniquely determined by its behavior on $X^{n-1} \times \{d\}$. In addition, it is clear that $\Gamma_{X^n} \setminus \mathcal{F}_{X^{n-1}\times\{d\}}$ is nonvoid. This completes the proof. □

Now we are ready to prove the key lemma.

Lemma 7.9. *$\mathcal{T}_{X^n}$ is generated by the union of Γ_{X^n} and the set of all functions $F : X^n \to X^n$ having the form $F(x_1, \dots, x_n) = (f(x_1, \dots, x_n), x_2, x_3, \dots, x_n)$, $f : X^n \to X$, where $x_1, \dots, x_n \in X$.*

Proof. We can take out of consideration the trivial case $|X| = 1$. Thus we assume $|X| > 1$. For every pair $i \in \{1, \dots, n\}$, $f : X^n \to X$, define the function $F_{i,f} : X^n \to X^n$ with

$F_{i,f}(x_1,\ldots,x_n) = (x_1,\ldots,x_{i-1}, f(x_1,\ldots,x_n), x_{i+1},\ldots,x_n)$ $(x_1,\ldots,x_n \in X)$. Thus, by letting $f' = U_{i,j} \circ f$, we have $F_{j,f} = U_{i,j} \circ F_{i,f'} \circ U_{i,j}$. So for every pair $i \in \{1,\ldots,n\}$, $f : X^n \to X$, $F_{i,f} \in \langle \Gamma_{X^n} \cup \{F : X^n \to X^n \mid F(x_1,\ldots,x_n) = (f(x_1,\ldots,x_n), x_2, x_3,\ldots,x_n),$ $f : X^n \to X, x_1,\ldots,x_n \in X\}\rangle$.

Let us identify X with a nontrivial finite cyclic group with a generating element $g \in X$. Thus we also have that for any $c_1,\ldots,c_n \in X$, $F^{(1)}{}_{\epsilon,j,(c_1,\ldots,c_n)}, F^{(2)}{}_{\epsilon,j,(c_1,\ldots,c_n)} \in \langle \Gamma_{X^n} \cup \{F : X^n \to X^n \mid F(x_1,\ldots,x_n) = (f(x_1,\ldots,x_n), x_2, x_3, \ldots, x_n), f : X^n \to X, x = (x_1,\ldots,x_n) \in X^n\}\rangle$, whenever $\epsilon \in \{1,-1\}$,

$$F^{(1)}{}_{\epsilon,j,(c_1,\ldots,c_n)}(x) = \begin{cases} (c_1,\ldots,c_n) & \text{if } x = (c_1,\ldots,c_{j-1}, c_j g^{\epsilon}, c_{j+1},\ldots,c_n), \\ (c_1,\ldots,c_{j-1}, c_j g^{\epsilon}, c_{j+1},\ldots,c_n) & \text{if } x = (c_1,\ldots,c_n), \\ x & \text{otherwise}, \end{cases}$$

$$F^{(2)}{}_{\epsilon,j,(c_1,\ldots,c_n)}(x) = \begin{cases} (c_1,\ldots,c_{n-1}, d) & \text{if } x = (c_1,\ldots,c_{j-1}, c_j g^{\epsilon}, c_{j+1},\ldots,c_n), \\ x & \text{otherwise}, \end{cases}$$

where $x = (x_1,\ldots,x_n) \in X^n$. On the other hand, by Lemma 7.3, there exists an arrangement $a_1,\ldots,a_m$ of X^n such that for every $k = 1,\ldots,m-1$, $p_k \in \{F^{(1)}{}_{\epsilon,j,(c_1,\ldots,c_r)} \mid \epsilon \in \{-1,1\}, j \in \{1,\ldots,n\}, c_1,\ldots,c_n \in X\}$, $t_k \in \{F^{(2)}{}_{\epsilon,j,(c_1,\ldots,c_n)} \mid \epsilon \in \{-1,1\}, j \in \{1,\ldots,n\}, c_1,\ldots,c_n \in X\}$, where

$$p_k(a_\ell) = \begin{cases} a_{k+1} & \text{if } \ell = k, \\ a_k & \text{if } \ell = k+1, \\ a_\ell & \text{otherwise}, \end{cases}$$

$$t_k(a_\ell) = \begin{cases} a_{k+1} & \text{if } \ell = k, \\ a_\ell & \text{otherwise}. \end{cases}$$

But then $p_1,\ldots,p_{m-1}$ is a set of transpositions such that $\{p_1,\ldots,p_{m-1}\}$ generates all permutations over X^n. At the same time, $t_1,\ldots,t_{m-1}$ is a set of elementary collapsings over X^n. Thus by the well-known fact that for every $j = 1,\ldots,m-1$, $\{p_1,\ldots,p_{m-1}, t_j,\}$ generates all transformations over X^n, the proof is complete. □

7.2 Network Completeness for Digraphs Having All Loop Edges

We start with the following.

Lemma 7.10. *Let $\mathcal{D} = (V, E)$ be a strongly connected digraph containing all loop edges. Using the notation of Lemma* 7.1, *let $F_{i,j} : X^n \to X^n$ denote any of $F^{(1)}_{i,j}$, $F^{(2)}_{i,j}$, $F^{(3)}_{i,j}$, or $U_{i,j}$, where $i, j \in V$. Then*

$$F^{(4)}_i, F_{i,j} \in \langle F^{(\epsilon)}_{\ell,k}, F^{(4)}_\ell \mid \ell = k \text{ or } (\ell,k) \in E, \epsilon \in \{1,2,3\}\rangle.$$

Thus $F^{(4)}_i$ and $F_{i,j}$ $(i, j \in V)$ are composites of functions in Γ_{X^n} that are compatible with the graph.

Proof. For $F_i^{(4)}$ this is clear. We first establish that if there is a directed walk on digraph $\mathcal{D}$ from i to j, then $F_{i,j}^{(\epsilon)}$ is a composite of functions compatible with the graph, where $\epsilon \in \{1, 2, 3\}$. Since by Lemma 7.1 $U_{i,j}$ is a composite of such, the result will then follow. We shall proceed by induction on the length L (the number of edges counting repetitions) of the walk. For $L = 0$, we have $i = j$ and by definition $F_{i,j}$ is compatible with $\mathcal{D}$. Also when $i = j$ or $L = 1$, clearly $F_{i,j}$ is compatible with $\mathcal{D}$. Now suppose that $i \neq j$, $L > 1$, and that we have a walk length $L + 1$ from vertex i to vertex j. Denote the penultimate vertex on the walk by v. If $v = j$, then $F_{i,j} = F_{i,v}$, or, if $v = i$, $F_{i,j} = F_{v,j}$. In either case, we have a shorter walk connecting the vertices so $F_{i,j}$ has the required property. Otherwise i, j, and v are pairwise distinct, so by the previous lemma $F_{i,j}$ is the composite of F's whose subscripts are among (i, v), (v, j), j and i, and which have the required property by induction hypothesis. □

Recall that

$$\Gamma_{X^n} = \{F : X^n \to X^n \mid F(x_1, \dots, x_n) = (x_1, \dots, x_{i-1}, f(x_i, x_j), x_{i+1}, \dots, x_n),$$
$$\text{where } f : X^2 \to X, i, j \in \{1, \dots, n\}, (x_1, \dots, x_n) \in X^n\}$$

(where the case $i = j$ is allowed in the above definition). We have the following direct consequence of the above lemma.

Proposition 7.11. *Consider the subsemigroup $\langle \Gamma_{X^n} \rangle$ of $\mathcal{T}_{X^n}$ generated by the elements of Γ_{X^n}. For any strongly connected graph on n vertices containing all loop edges, this subsemigroup is also generated by a subset of Γ_{X^n} consisting of some maps compatible with the graph.* □

Let $\mathcal{A} = (Z^n, X, \delta)$, $\mathcal{B} = (Z^m, Y, \delta')$ be networks (having the same basic set Z). We say that $\mathcal{B}$ *simulates* $\mathcal{A}$ *by projection* if there exists an $H \subseteq \{1, \dots, m\}$ such that every $\delta_x : Z^n \to Z^n (x \in X)$ is an H-projection of a mapping $\delta'_p : Z^m \to Z^m$ $(p \in Y^+)$. If there exists a $\mathcal{D}$-network $\mathcal{B}$ which simulates a given network $\mathcal{A}$ by projection, then it is said that $\mathcal{A}$ *can be simulated* on $\mathcal{D}$ by projection. A digraph $\mathcal{D}$ is called *n-complete with respect to simulation by projection* if every network of size n can be simulated on $\mathcal{D}$ by projection.

The following statement is obvious.

Proposition 7.12. *Given a positive integer $n > 1$, a digraph is n-complete with respect to simulation by projection if and only if it has a strongly connected subdigraph having this property.* □

Consider a digraph $\mathcal{D} = (V, E)$ and a $\mathcal{D}$-network $\mathcal{A} = (Z^n, X, \delta)$. If $(i, i) \notin V$ for some $i \in V$, then we obviously have that the ith component of δ does not depend on its ith-state variable. In this case the ith component-automaton of the network must be a *reset automaton without identity.* ($\mathcal{B} = (B, X_{\mathcal{B}}, \delta_{\mathcal{B}})$ is called reset automaton without identity if for every input letter $x \in X$, $\{\delta_{\mathcal{B}}(b, x) \mid b \in B\}$ is a singleton.) On the basis of this fact, it is clear that an important special case of the finite automata networks is when the communication link has all loop edges. In this case all components of the network may depend on their own states, too. We shall use the characterization as follows.

Theorem 7.13. *Given a positive integer $n > 1$, a digraph of order n having all loop edges is n-complete with respect to simulation by projection if and only if it is strongly connected and centralized.*

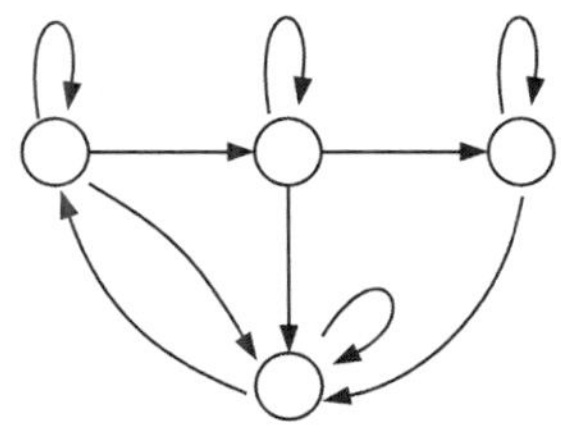

$n = 4$

Proof. Sufficiency of the condition follows directly from Proposition 7.11 and Lemma 7.9, since the members of Γ_{X^n} are composites of compatible maps and centralization (without loss of generality at vertex 1) implies that all of $\mathcal{T}_{X^n}$ is generated by maps compatible with the digraph. For necessity, obviously $\mathcal{D}$ must clearly be strongly connected. Let $q = |Z|$, and identify the elements of Z in a fixed but arbitrary way with the elements $\{0, 1, \ldots, q-1\}$ of the modulo q residue ring of integers. Suppose $F \in \mathcal{T}_{Z^n}$, such that the image of F excludes exactly one element of Z^n:

$$|\mathrm{Im}\, F| = q^n - 1, \text{ where } \mathrm{Im}\, F = \{F(z) \mid z = (z_1, \ldots, z_n) \in Z^n\}.$$

Clearly, if F is a written as a composite of functions on Z^n, then at least one of them also has $q^n - 1$ elements in its image. Thus it is enough to show that such a function cannot be a transition function of a network which is not centralized.

For such a function $F = (f_1, \ldots, f_n)$ there exist unique $a = (a_1, \ldots, a_n)$ and $b = (b_1, \ldots, b_n)$ such that $F^{-1}(a) = \emptyset$ and $|F^{-1}(b)| = 2$. Since we identify Z with the elements of the ring of integers modulo q, one has the sum $(\star)$. Notice that for a bijective $F' = (f_1', \ldots, f_n')$ one has $\sum_{x \in Z^m} f_i'(z) = \sum_{c_i \in Z} q^{n-1} c_i$; hence, it follows, since F is "almost" bijective,

$$(\star)\quad \sum_{x \in Z^m} f_i(x) = \left(\sum_{c_i \in Z} q^{n-1} c_i \right) + b_i - a_i \pmod q = b_i - a_i \pmod q \quad \text{since } n \geq 2.$$

Since $a \neq b$, this sum is nonzero for at least one component i.

Now suppose that the network is not centralized. Then for each i, the component function f_i of F depends on only j variables with $j < n$. (Of course, the exact value of j may depend on i.) Suppose, without loss of generality, these variables are $z_1, \ldots, z_j$. It follows that the cardinality of $f_i^{-1}(f_i(z_1, \ldots, z_j, 0, \ldots, 0))$ is a multiple of q^{n-j}. Thus,

$$\sum_{z \in Z^m} f_i(z) = \sum_{y \in Z} y \cdot \text{a multiple of } q^{n-j} = 0 \pmod q$$

for all components i. This contradicts $(\star)$ and $a \neq b$, so no such F may be a transition function of the network. $\square$

Using the above theorem and the preliminary results above, we recover the following characterization.

Theorem 7.14. *Given a positive integer $n > 1$, a digraph $\mathcal{D} = (V, E)$ having all loop edges is n-complete with respect to simulation by projection if and only if one of the following conditions holds:*

(1) *$\mathcal{D}$ has a strongly connected subdigraph of order $m > n$.*

(2) *$\mathcal{D}$ has a strongly connected centralized subdigraph of order n.*

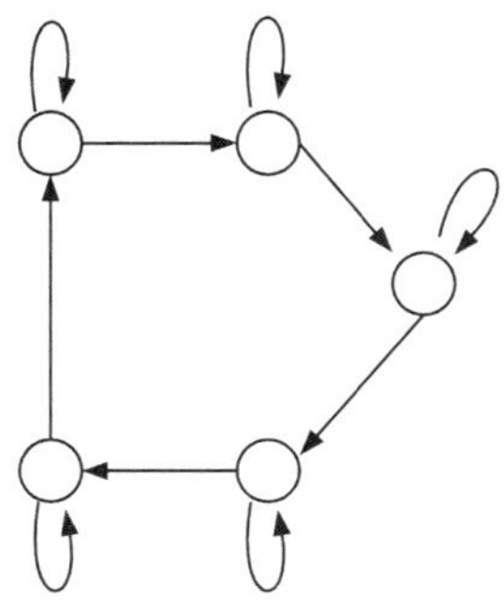

$m = 5$

$n = 4$

CONDITION (1)

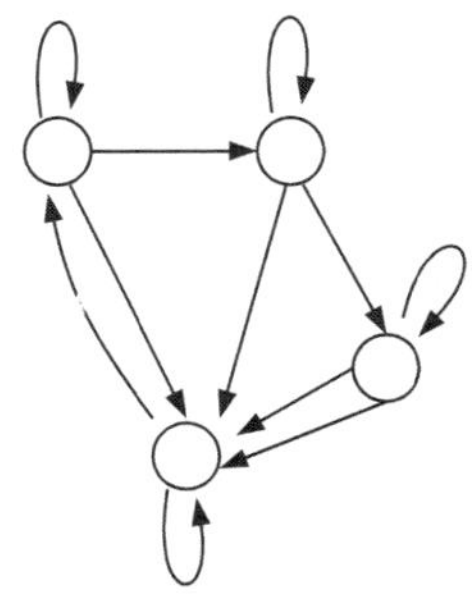

$n = 4$

CONDITION (2)

Proof. Let $\mathcal{D} = (V, E)$ be a digraph having all loop edges and consider a $\mathcal{D}$-network $\mathcal{A} = (Z^{|V|}, X, \delta)$ having the underlying graph $\mathcal{D}_{\mathcal{A}} = (V_{\mathcal{A}}, E_{\mathcal{A}})$ with $V_{\mathcal{A}} = V, E_{\mathcal{A}} = E$. Suppose that $\mathcal{A}$ is maximal in the sense that $\{\delta_x \mid x \in X\}$ is the set of all functions which are compatible with $\mathcal{D}_{\mathcal{A}}$.

Using Proposition 7.12 and Theorem 7.13, it is enough to prove that D is n-complete if it has a strongly connected subdigraph $\mathcal{D}' = (V', E')$ of order $m > n$. $pr_{V'}(\delta_x)$ exists if δ_x is really independent of all $j \notin V'$. Define the network $\mathcal{B} = (Z^m, X, \delta')$ such that for any pair $b \in Z^m, x \in X$, $\delta'(b, x) = pr_{V'}(\delta_x)(b)$ and let $V' = \{v_1, \ldots, v_m\}$ with $(1 \leq) v_1 < \cdots < v_m (\leq |V|)$. Then $\mathcal{B}$ has the underlying graph $V_{\mathcal{B}} = \{1, \ldots, m\}$, $E_{\mathcal{B}} = \{(i, j) \mid (v_i, v_j) \in E'\}$. On the other hand, by the maximality of $\mathcal{A}$, $\{\delta'_x \mid x \in X\}$ is the set of all functions which are compatible with $V_{\mathcal{B}}$. Therefore, by Lemma 7.10, for all $i, j \in V_{\mathcal{B}}$, there exists $p \in X^+$ such that $U_{i,j} = \delta'_p$, where $U_{i,j}(z_1, \ldots, z_m) = (z_1, \ldots, z_{i-1}, z_j, z_{i+1}, \ldots, z_{j-1}, z_i, z_{j+1}, \ldots, z_m)$, $((z_1, \ldots, z_m) \in Z^m)$. Therefore, by the strongly connectivity of $D_{\mathcal{B}}$ and Lemma 7.10, for any pair $i, j \in V_{\mathcal{B}}$, there exists a $p \in X^+$ such that $U_{i,j} = \delta'_p$. (Clearly, then $\delta'_p = \delta'_{x_s} \circ \cdots \circ \delta'_{x_1}$ whenever $p = x_1 \ldots x_s$ $(x_1, \ldots, x_s \in X)$.) Simultaneously, by the maximality of $\mathcal{A}$, for any $(i, j) \in E_{\mathcal{B}}$, there exists an $x \in X$ such that $F_{i,j,f} = \delta'_x$, where $F_{i,j,f}(z_1, \ldots, z_m) = (z_1, \ldots, z_{j-1}, f(z_i, z_j), z_{j+1}, \ldots, z_n)$. Hence, taking into consideration that for every pairwise distinct $i, j, k \in V_{\mathcal{B}}$, $F_{k,j,f} = U_{i,k} \circ F_{i,j,f} \circ U_{i,k}$, we obtain that for all distinct $i, j \in V_{\mathcal{B}}$ and $f : Z^2 \to Z$, there is a $p \in X^+$ having $F_{i,j,f} = \delta'_p$. Applying Lemma 7.8 (taking X, n, d to be Z, m, z, where $z \in Z$ is arbitrary), $\mathcal{F}_{Z^{m-1} \times \{z\}} \subset \{\delta'_p \mid p \in X^+\}$. Therefore, by $n < m$ we obtain that for every network having the form $\mathcal{C} = (Z^n, Y, \delta'')$ and input letter $y \in Y$, there exists a word

$p \in X^+$ such that δ''_y is a $\{1, \ldots, n\}$-projection of δ'_p. But then there is an $H \subseteq V$ such that for every $y \in Y$, there can be found a $p \in X^+$ having $\delta''_y = pr_H(\delta_p)$. This means that in this case $\mathcal{D}$ is n-complete. This ends the proof of our theorem. □

7.3 Complete Finite Automata Network Graphs with Minimal Number of Edges

In this section, we extend the investigation of $\mathcal{D}$-networks by taking into consideration digraphs not necessarily having loop edges at every node. (As we have already remarked, if we consider $\mathcal{A} = (Z^n, X, \delta)$ as a network of finite automata, then its ith component-automaton is a reset automaton without identity whenever (i, i) is not an edge of $\mathcal{D}$.)

Problem 7.15. *For every positive integer $n > 1$, give a complete characterization of n-complete digraphs with respect to simulation by projection.*

The first steps into this direction are the characterizations of n-complete networks (with respect to simulation by projection) having minimal number of edges. Recall that the n-complete digraph $\mathcal{D} = (V, E)$ has *minimal number of edges* if for every n-complete digraph $\mathcal{D}' = (V', E')$, $|V| = |V'|$ implies $|E| \leq |E'|$.

First we show the next statement.

Theorem 7.16. *Given a positive integer $n > 1$, a digraph with n vertices is an n-complete digraph with minimal number of edges if and only if it is isomorphic to the digraph $\mathcal{D} = (V, E)$ with $V = \{1, \ldots, n\}$ and $E = \{(i, j) \mid i, j \in V, j = i + 1 \pmod n \text{ or } j = 1\}$.*

Proof. It is clear that for an arbitrary $m \in \{1, \ldots, n\}$, the functions $T^{(0)}, T_\ell^{(k)} \mid k = 1, 2, 3, 4$, defined in Lemma 7.2 are compatible with $\mathcal{D}$. Suppose that m is chosen such that it is relatively prime to n. Then the sufficiency of this statement is a direct consequence of Lemma 7.4.

For necessity, we may assume the n vertices are $V = \{1, \ldots, n\}$. First we show the existence of $j \in V$ with $\{(i, j) \mid i \in V\} \subseteq E$, whenever $\mathcal{D}$ is n-complete. Then by suitable relabeling we show that the digraph is isomorphic to the one in the statement of the lemma

Let $T : X^n \to X^n$ such that $|\{T(x_1, \ldots, x_n) : x_1, \ldots, x_n \in X\}| = |X^n| - 1$. First we show that for every $F_1, \ldots, F_m \in \mathcal{T}_{X^n}$, $T = F_1 \circ \cdots \circ F_m$ implies the existence of an index i preserving the property $|\{F_i(x_1, \ldots, x_n) : x_1, \ldots, x_n \in X\}| = |X^n| - 1$. Of course, if $F_1, \ldots, F_m$ are injective, then $T = F_1 \circ \cdots \circ F_m$ should be also injective, a contradiction. On the other hand, $T = F_1 \circ \cdots \circ F_m$ implies $|\{F(x_1, \ldots, x_n) : x_1, \ldots, x_n \in X\}| \leq \min\{|\{F_i(x_1, \ldots, x_n) : x_1, \ldots, x_n \in X\}| : i = 1, \ldots, m\}$. Therefore, we obtain our assumption regarding the existence of an index i preserving the property $|\{F_i(x_1, \ldots, x_n) : x_1, \ldots, x_n \in X\}| = |X^n| - 1$.

Now we identify the elements of X in a fixed but arbitrary way with the elements of $\{1, \ldots, |X|\}$ and consider X^n as a subset of the nth direct power of integers. For every $(a_{1,1}, \ldots, a_{1,n}), \ldots, (a_{m,1}, \ldots, a_{m,n}) \in X^n$, let $\sum\{(a_{i,1}, \ldots, a_{i,n}) \mid i = 1, \ldots, m\} = (\sum_{i=1}^m a_{i,1}, \ldots, \sum_{i=1}^m a_{i,n})$. Let $a = (a_1, \ldots, a_n)$, $b = (b_1, \ldots, b_n) \in X^n$ denote distinct elements with $|F_i^{-1}(a)| = 0$ and $|F_i^{-1}(b)| = 2$. And let $j \in \{1, \ldots, n\}$ be an index with $a_j \neq b_j$.

We prove that $|X|$ does not divide $pr_j(\sum\{F_i(x_1,\dots,x_n) : x_1,\dots,x_n \in X\})$. Indeed, then $pr_j(\sum\{F_i(x_1,\dots,x_n) : x_1,\dots,x_n \in X\}) = pr_j(\sum\{(x_1,\dots,x_n) : x_1,\dots,x_n \in X\}) + b_j - a_j) = |X^{n-1}|(\sum_{k=0}^{|X|-1} k) + b_j - a_j$. Of course, by this equality we have that $|X|$ does not divide $pr_j(\sum\{F_i(x_1,\dots,x_n) : x_1,\dots,x_n \in X\})$.

Suppose that for every $j \in V$ there exists an $i \in V$ with $(i,j) \notin E$. Consider the set $\mathcal{D}_X$ of all functions of the form $X^n \to X^n$ which are compatible with $\mathcal{D}$. Now we show that for every $F \in \mathcal{D}_X$, $|X|$ divides $pr_j(\sum\{F(x_1,\dots,x_n) : x_1,\dots,x_n \in X\})$, implying $F_i \notin \mathcal{D}_X$.

By $F \in \mathcal{D}_X$ we have that for an appropriate $\ell \in \{1,\dots,n\}$, $pr_j(F(x_1,\dots,x_n)) = pr_j(F(x_1,\dots,x_{\ell-1},x'_\ell,x_{\ell+1},\dots,x_n))$ $((x_1,\dots,x_n) \in X^n, x'_\ell \in X, \ell = j$ is allowed). Therefore, for an arbitrary fixed $c \in X$, $pr_j(\sum\{F(x_1,\dots,x_n) : x_1,\dots,x_n \in X\}) = |X|pr_j(\sum\{F(x_1,\dots,x_{\ell-1},c,x_{\ell+1},\dots,x_n) : x_1,\dots,x_{\ell-1},x_{\ell+1},\dots,x_n \in X\})$. But then $|X|$ divides $pr_j(\sum\{F(x_1,\dots,x_n) : x_1,\dots,x_n \in X\})$ for every $j = 1,\dots,n$. Hence we get $F_i \notin \mathcal{D}_X$. Consequently, there exists a $T \in \mathcal{T}_{X^n}$ with $T \notin \langle\mathcal{D}_X\rangle$. This ends the proof of the existence of $j \in V$ with $\{(i,j) \mid i \in V\} \subseteq E$ whenever $\mathcal{D}$ is n-complete. Then we are done if we can prove the existence of a permutation $p : V \to V$ with $\{(p(i),p(j)) \mid i,j \in V, p(j) = p(i)+1 \pmod n\} \subseteq E$.

Consider the mapping $T^{(0)} : X^n \to X^n$ defined by $T^{(0)}(x_1,\dots,x_n) = (x_n,x_1,\dots,x_{n-1})$ $(x_1,\dots,x_n \in X)$. To complete the proof of our theorem, we will show $T^{(0)} \notin \mathcal{D}_X$ if there exists no such a permutation p.

It is also clear that an n-complete digraph $\mathcal{D}$, having n vertices, should be strongly connected. Therefore, all vertices have (nonloop) incoming edges. Thus, by the minimality of $|E|$, we get $|E \setminus \{(i,j) : i \in V\}| = n-1$. Simultaneously, the strongly connectivity of $\mathcal{D}$ implies $\{j\} \times (V \setminus \{j\}) \cap E \neq \emptyset$ (where $j \in V$ with $\{(i,j) \mid i \in V\} \subseteq E$). On the other hand, if there exists no permutation p having the above discussed property, then by the strong connectivity of $\mathcal{D}$, $V \times \{j\} \subseteq E$, and $|E \setminus \{(i,j) : i \in V\}| = n-1$, we can prove $|\{j\} \times (V \setminus \{j\}) \cap E| \geq 2$, implying the existence of two distinct vertices $i_1, i_2 \in V$ with $\{(\ell,i_r) \mid r = 1,2, \ell \in V\} \cap E = \{(j,i_1),(j,i_2)\}$.

It is enough to prove that in this case $T^{(0)} \notin \mathcal{D}_X$. Clearly, $F_1 \in \mathcal{D}_X$ implies the existence of functions $f_k : X \to X$, $k = 1,2$, with $pr_{i_k}(F_1(x_1,\dots,x_n)) = f_k(x_j)$. Therefore, the cardinality of $\{(y_1,y_2) \mid y_k = pr_{i_k}(F_1(x_1,\dots,x_n)), k = 1,2, x_1,\dots,x_n \in X\}$ is not greater than $|X|$. In a similar way, for every $F_1,\dots,F_m \in \mathcal{D}_X$, $m > 1$, there exist functions $f_k : X \to X$, $k = 1,2$, such that $pr_{i_k}(F_1 \circ \cdots \circ F_m(x_1,\dots,x_n)) = f_k(pr_j(F_1 \circ \cdots \circ F_{m-1}(x_1,\dots,x_m)))$, implying that the cardinality of $\{(y_1,y_2) \mid y_k = pr_{i_k}(F_1 \circ \cdots \circ F_m(x_1,\dots,x_n)), k = 1,2, x_1,\dots,x_n \in X\}$ is not greater than $|X|$. On the other hand, the cardinality of $\{(y_1,y_2) \mid y_k = pr_{i_k}(T^{(0)}(x_1,\dots,x_n)), k = 1,2, x_1,\dots,x_n \in X\}$ is $|X|^2$, yielding to $T^{(0)} \notin \mathcal{D}_X$. The proof is complete. □

Now we prove the following characterization.

Theorem 7.17. *Given a positive integer $n > 1$, $\mathcal{D} = (V,E)$ with $V = \{1,\dots,m\}$, $m > n$, is an n-complete digraph with minimal number of edges if and only if there exists a permutation $p : \{1,\dots,m\} \mapsto \{1,\dots,m\}$ such that $E = \{(p(i),p(j)) \mid p(i),p(j) \in \{1,\dots,n+1\}, p(j) = p(i+1 \pmod{n+1})\} \cup \{(p(1),p(r))\}$, where $r \in \{1,\dots,n+1\}$, $r \neq 2$, and, moreover, $r-2$ and $n+1$ are relatively prime.*[37]

[37]The case $r = 1$ is not excluded. Moreover, if there are more than $n+1$ vertices, then all except for $n+1$ are isolated.

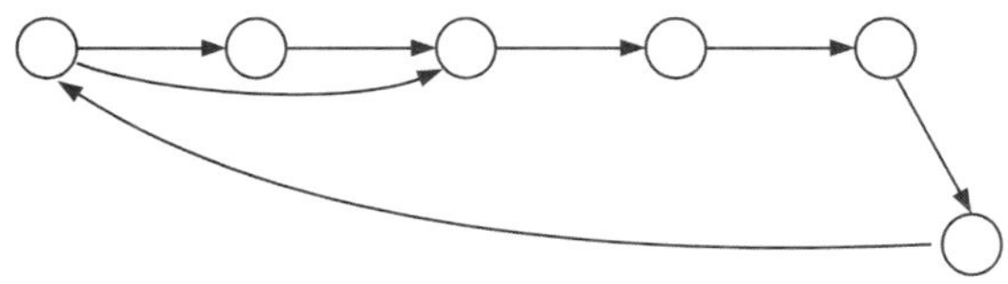

$r = 3,\ n = 5,\ m = 6$

Proof. To show sufficiency it is enough to prove for any $n > 2$ the n-completeness of $\mathcal{D} = (\{1,\ldots,n+1\}, \{(i, i+1\,(\mathrm{mod}\, n+1)) \mid i \in \{1,\ldots,n+1\}\} \cup \{(1,r)\})$, where $r \in \{1,\ldots,n+1\}, r \neq 2$, and, in addition, $r-2$ and $n+1$ are relative primes.

Consider the set $\mathcal{D}_X$ of all functions of the form $X^{n+1} \to X^{n+1}$ which are compatible with $\mathcal{D}$. By definition, we obtain $\{T^{(0)}, T_\ell^{(k)} \mid k = 1,\ldots,4,\} \subsetneq \mathcal{D}_X$, where $T^{(0)}, T_\ell^{(k)}, k = 1,\ldots,4$, are defined as in Lemma 7.2 (taking m, ℓ, n of the lemma to be $r-2$, r, and $n+1$, respectively). Identifying X with a finite group and using Lemma 7.2, we get $\mathcal{T}_{X,n+1} \subsetneq \langle \mathcal{D}_X \rangle$, too. On the other hand, we have by definition $\{F : X^{n+1} \to X^{n+1} \mid F(x_1,\ldots,x_{n+1}) = (x_{n+1}, x_1, \ldots, x_{i-2\,(\mathrm{mod}\,n+1)}, f(x_1, x_{i-1\,(\mathrm{mod}\,n+1)}), x_i, \ldots, x_r)$, $f : X^2 \to X, i \in \{1,\ldots,n+1\}$, $(x_1,\ldots,x_{n+1}) \in X^{n+1}\} \subset \mathcal{D}_X$. But then, $\{F : X^{n+1} \to X^{n+1} \mid F(x_1,\ldots,x_{n+1}) = (x_1,\ldots,x_{i-1\,(\mathrm{mod}\,n+1)}, f(x_i, x_{i+1\,(\mathrm{mod}\,n+1)}), x_{i+1\,(\mathrm{mod}\,n+1)}, \ldots, x_{n+1}), f : X^2 \to X, i \in \{1,\ldots,n+1\}$, $(x_1,\ldots,x_{n+1}) \in X^{n+1}\} \cup \mathcal{T}_{X,n+1} \subseteq \langle \mathcal{D}_X \rangle$, implying $\Gamma_{X^{n+1}} \subseteq \langle \mathcal{D}_X \rangle$. Applying Lemma 7.8, this shows the n-completeness of $\mathcal{D}$.

Using the obvious fact that n-complete digraph must have a strongly connected n-complete subdigraph, by our minimality conditions, we consider only digraphs which have a strongly connected subdigraph, and all vertices outside of this digraph are isolated. Thus, the sufficiency just established implies that by the minimality conditions, we may restrict our investigations to the strongly connected n-complete digraphs having not more than $n+2$ edges. (We can take out of consideration the isolated vertices.) If we have $n+1$ vertices and fewer than $n+1$ edges, then our digraph is not strongly connected. On the other hand, if we consider a strongly connected digraph $\mathcal{D}$ with $n+1$ vertices and $n+1$ edges, i.e., a cycle having $n+1$ length, then for every $F \in \langle \mathcal{D}_X \rangle$, there exist $k \in \{1,\ldots,n+1\}$, $f_i : X \to X, i = 1,\ldots,n+1$, with $F(x_1,\ldots,x_n) = (f_1(x_k), f_2(x_{k+1\,(\mathrm{mod}\,n+1)}), \ldots, f_{n+1}(x_{n+k\,(\mathrm{mod}\,n+1)}))$ $(x_1,\ldots,x_n \in X)$. Therefore, for any $1 \leq i_1 < i_2 < \cdots < i_m \leq n+1$, $pr_{i_1,\ldots,i_m}(F(x_1,\ldots,x_{n+1}))$ $= (f_{i_1}(x_{i_1+k\,(\mathrm{mod}\,n+1)}), \ldots, f_{i_m}(x_{i_m+k\,(\mathrm{mod}\,n+1)}))$ $(x_1,\ldots,x_{n+1} \in X)$, which obviously shows that this type of digraph cannot be n-complete.

But then it is enough to consider only strongly connected digraphs having $n+1$ vertices and $n+2$ edges.

By the strong connectivity of $\mathcal{D}$ we may suppose that $\mathcal{D} = (V, E)$, with $|V| = n+1, |E| = n+2$, has a cycle $\mathcal{C} = (C, E_C)$ with k length for some $1 \leq k \leq n+1$, where $C = \{v_1,\ldots,v_k\} (\subsetneq V)$, $E_C = \{(v_i, v_{i+1\,(\mathrm{mod}\,k)}) \mid i = 1,\ldots,k\} (\subsetneq E)$.

Using the strong connectivity of $\mathcal{D}$ again, for every $C \subsetneq V$ there are distinct (v_i, v_j), $(v_s, v_t) \in E$ with $v_i, v_t \in C$, $v_j, v_s \in V \setminus C$. Therefore, by an induction we get the structure of $\mathcal{D}$ in the following manner.

If $k < n+1$, then $V = \{v_1,\ldots,v_k,v_{k+1},\ldots,v_{n+1}\}$, $E = E_C \cup \{(v_{k+i-1}, v_{k+i}) \mid i = 1,\ldots,n-k+1\} \cup \{(v_{n+1}, v_\ell)\}$, where $\ell \in \{1,\ldots,k\}$ is arbitrarily fixed.

If $k = n + 1$, then, of course, $V = C$, and $E = E_C \cup \{(v_{n+1}, v_\ell)\}$ for some $\ell \in \{2, \dots, n + 1\}$.

To complete the case $k = n + 1$, it suffices to study digraphs having the form $\mathcal{D} = (\{v_1, \dots, v_{n+1}\}, \{(v_i, v_{i+1\,(\mathrm{mod}\, n+1)}) \mid i \in \{1, \dots, n + 1\}\} \cup \{(v_1, v_\ell)\})$, where $\ell \in \{1, \dots, n + 1\}$, $\ell \neq 2$, such that $\ell - 2\,(\mathrm{mod}\, n + 1)$ and $n + 1$ are not relative primes. Then $n + 1$ and $\ell - 2$ have a divisor $d > 1$. We claim that for each mapping $F \in \langle \mathcal{D}_X \rangle$, the following holds:

There exists an integer $c(F)$, such that if $pr_i(F)$ really depends on its kth coordinate, then $i \equiv k - c(F)(\mathrm{mod}\ d)$.

Trivially, our property holds for each compatible map $F \in \mathcal{D}_X$, as can be seen by taking $c(F) = 1$. Moreover, if G and F both have this property, one easily checks that so does $F \circ G$ with $c(F \circ G) = c(F) + c(G)$. By induction, this establishes the above property for all maps generated by composing compatible maps. Therefore, for every $F \in \langle \mathcal{D}_X \rangle$ and $i \in \{1, \dots, n + 1\}$, $pr_i(F)$ depends only on a proper divisor of $n + 1$ many variables, which is fewer than n. Therefore, digraphs having structure like this are not n-complete.

It remains to study the case $k < n + 1$. Then $V = \{v_1, \dots, v_k, v_{k+1}, \dots, v_{n+1}\}$, $E = E_C \cup \{(v_{k+i-1}, v_{k+i}) \mid i = 1, \dots, n - k + 1\} \cup \{(v_{n+1}, v_\ell)\}$, where $\ell \in \{1, \dots, k\}$ is arbitrarily fixed. Of course, if $k = 1$ or $\ell = 1$, then we have one of the cases discussed previously. Thus we assume $k, \ell \neq 1$.

Given a set X with $|X| \geq 2$, let $\mathcal{M}_X = \{F : X^n \to X^n : |X^n| - 1 \leq |\{F(x_1, \dots, x_n) : (x_1, \dots, x_n) \in X^n\}|(\leq |X^n|)\}$. Clearly, then for every $F : X^n \to X^n$, $F \in \langle \mathcal{M}_X \rangle$.

To complete our proof, by Theorem 7.13, it is enough to show that there exists a digraph $\mathcal{D}' = (V', E')$ with n vertices such that it is not centralized, and, simultaneously, for every pair $F \in \langle \mathcal{D}_X \rangle$, $H \subsetneq \{1, \dots, n + 1\}$, $|H| = n$, the existence of $pr_H(F)$ implies $pr_H(F) \in \langle \mathcal{D}'_X \rangle$ whenever $pr_H(F) \in \mathcal{M}_X$ (where $\mathcal{D}'_X$ denotes the set of all functions of the form $F : X^n \mapsto X^n$ to be compatible with $\mathcal{D}'$).

By an elementary computation it can be proved that $\mathcal{D}' = (V', E')$ has this property whenever $E' = V' \times V' \setminus \{(v_i, v_i) \mid v_i \in V'\}$ (and $|V'| = n$). (See the detailed proof below.) Therefore, there exists a $T \in \mathcal{M}_X$ with $T \notin \langle \mathcal{D}'_X \rangle$. But then for every $F \in \langle \mathcal{D}_X \rangle$, $H \subsetneq \{1, \dots, n + 1\}$, $|H| = n$, $pr_H(F) \neq T$. Therefore, $\mathcal{D}$ cannot be n-complete as we stated. □

Proof. *We prove the following: For every pair $F \in \langle \mathcal{D}_X \rangle$, $H \subsetneq \{1, \dots, n + 1\}$, $|H| = n$, the existence of $pr_H(F)$ implies $pr_H(F) \in \langle \mathcal{D}'_X \rangle$ whenever $pr_H(F) \in \mathcal{M}_X$ (where $\mathcal{D}'_X$ denotes the set of all functions of the form $F : X^n \mapsto X^n$ to be compatible with $\mathcal{D}'$).* Observe that for every $F \in \mathcal{D}_X$ there are $f_j : X \to X$, $j = 1, \dots, \ell - 1, \ell + 1, \dots, n + 1$, $f_\ell : X^2 \to X$ with $F(x_1, \dots, x_n) = (f_1(x_k), f_2(x_1), \dots, f_{\ell-1}(x_{\ell-2}), f_\ell(x_{\ell-1}, x_{n+1}), f_{\ell+1}(x_\ell), \dots, f_{n+1}(x_n))$ $((x_1, \dots, x_{n+1}) \in X^{n+1})$. Assume $H = \{1, \dots, n + 1\} \setminus \{i\}$ with $i \notin \{\ell - 1, n + 1\}$. If $i = \ell$, then $f_{\ell+1}$ really does not depend on its variable. Moreover, f_1 and f_k depend on the same variable. In addition, $pr_H(F)$ has only n variables. This implies $|\{pr_H(F)(x_1, \dots, x_{i-1}, x_{i+1}, \dots, x_{n+1}) : (x_1, \dots, x_{i-1}, x_{i+1}, \dots, x_n) \in X^n\}| \leq |X|^{n-2}$. If $i = k$, then f_1 and f_{k+1} really do not depend on their variable (and $pr_H(F)$ has only n variables), which also leads to the above result. In addition, if $i \notin \{\ell - 1, \ell, k, n + 1\}$, then f_{i+1} really does not depend on its variable (and $pr_H(F)$ has only n variables). Hence we get $|\{pr_H(F)(x_1, \dots, x_{i-1}, x_{i+1}, \dots, x_{n+1}) : (x_1, \dots, x_{i-1}, x_{i+1}, \dots, x_n) \in X^n\}| \leq |X|^{n-1}$.

Therefore, $H = \{1, \ldots, n+1\} \setminus \{i\}$, $i \in \{1, \ldots, n+1\} \setminus \{\ell - 1, n+1\}$ and $F = F_1 \circ \cdots \circ F_m$, $F_1, \ldots, F_m \in \mathcal{D}_X$ implies $|\{pr_H(F)(x_1, \ldots, x_n) : (x_1, \ldots, x_n) \in X^n\}| \le |X^{n-1}|$. Hence, in this case, $pr_H(F) \notin \mathcal{M}_X$. Thus we may assume $H = \{1, \ldots, n+1\} \setminus \{i\}$, $i \in \{\ell - 1, n+1\}$.

Let $F = F_1 \circ \cdots \circ F_m$ with $F_1, \ldots, F_m \in \mathcal{D}_X$, such that $pr_H(F)$ exists for a suitable $H = \{1, \ldots, i-1, i+1, \ldots, n+1\}$, $i \in \{\ell - 1, n+1\}$. It remains to prove that there exists a mapping $T \in D'_X$ satisfying $pr_H(F) = T$ (either $pr_H(F) \in \mathcal{M}_X$ or not).

First we study the case $m = 1$. Consider a mapping $F \in D_X$, a set $H = \{1, \ldots, i-1, i+1, \ldots, n+1\}$ with $i \in \{\ell - 1, n+1\}$ (such that the existence of $pr_H(F)$ is not supposed). First we prove that $pr_H(F)$ exists and there exists $T \in D'_X$ having $pr_H(F) = T$. Define functions $f_j : X \to X$, $j \in \{1, \ldots, \ell-1, \ell+1, \ldots, n+1\}$, $f_\ell : X^2 \to X$ with $F(x_1, \ldots, x_{n+1}) = (f_1(x_k), f_2(x_1), \ldots, f_{\ell-1}(x_{\ell-2}), f_\ell(x_{\ell-1}, x_{n+1}), f_{\ell+1}(x_\ell), \ldots, f_{n+1}(x_n))$ $((x_1, \ldots, x_{n+1}) \in X^{n+1})$. Assume $i = \ell - 1$. Clearly, then f_ℓ really may not depend on its first variable; i.e., there exists a $g : X \to X$ with $f_\ell(x_{\ell-1}, x_{n+1}) = g(x_{n+1})$ $(x_{\ell-1}, x_{n+1} \in X)$. Therefore, we can write $F(x_1, \ldots, x_{n+1}) = (f_1(x_k), f_2(x_1), \ldots, f_{\ell-1}(x_{\ell-2}), g(x_{n+1}), f_{\ell+1}(x_\ell), \ldots, f_{n+1}(x_n))$ $((x_1, \ldots, x_{n+1}) \in X^{n+1})$. Take $T : X^n \to X^n \in D'_X$ with $T(x_1, \ldots, x_{\ell-3}, x_{\ell-2}, x_\ell, \ldots, x_{n+1}) = (f_1(x_k), f_2(x_1), \ldots, f_{\ell-2}(x_{\ell-3}), g(x_{n+1}), f_{\ell+1}(x_\ell), \ldots, f_{n+1}(x_n))$ $((x_1, \ldots, x_{n+1}) \in X^{n+1})$. Obviously, then $pr_H(F)(x_1, \ldots, x_{\ell-2}, x_\ell, \ldots, x_{n+1}) = T(x_1, \ldots, x_{\ell-2}, x_\ell, \ldots, x_{n+1})$. Then we get the existence of $pr_H(F)$ and that $pr_H(F) = T$ with $T \in \mathcal{D}'_X$. We can handle the case $i = n+1$ similarly.

Now let $i = \ell - 1$ and $m > 1$. Take a pair $F_1 \in \langle D_X \rangle$ and $F_2 \in D_X$. Studying the case $m = 1$, we have already proved the existence of appropriate mappings $f_{2,j} : X \to X$, $j = 1, \ldots, \ell - 1, \ell + 1, \ldots, n+1$ and $g_2 : X \to X$ for which $F_2(x_1, \ldots, x_{n+1}) = (f_{2,1}(x_k), f_{2,2}(x_1), \ldots, f_{2,\ell-2}(x_{\ell-3}), f_{2,\ell-1}(x_{\ell-2}), g_2(x_{n+1}), f_{2,\ell+1}(x_\ell), \ldots, f_{2,n+1}(x_n))$ $((x_1, \ldots, x_{n+1}) \in X^{n+1})$. Then there exists $T_2 \in D'_X$ with $T_2(x_1, \ldots, x_{\ell-3}, x_{\ell-2}, x_\ell, \ldots, x_{n+1}) = (f_{2,1}(x_k), f_{2,2}(x_1), \ldots, f_{2,\ell-2}(x_{\ell-3}), g_2(x_{n+1}), f_{2,\ell+1}(x_\ell), \ldots, f_{2,n+1}(x_n))$ $((x_1, \ldots, x_{\ell-1}, x_\ell, \ldots, x_{n+1}) \in X^n)$. Clearly, then $pr_H(F_2) = T_2$. Let us suppose inductively that $pr_H(F_1)(x_1, \ldots, x_{\ell-3}, x_{\ell-2}, x_\ell, \ldots, x_{n+1}) = T'_1 \circ \cdots \circ T'_{m-1}(x_1, \ldots, x_{\ell-3}, x_{\ell-2}, x_\ell, \ldots, x_{n+1})$ $((x_1, \ldots, x_{\ell-2}, x_\ell, \ldots, x_{n+1}) \in X^n)$ for appropriate $T'_1, \ldots, T'_{m-1} \in D'_X$. (Note that, considering the case $m = 1$, we have the existence of $T'_1 \in D'_X$ for $m = 2$.) Put $T_1 = T'_1 \circ \cdots \circ T'_{m-1}$ and let $T_1(x_1, \ldots, x_{\ell-3}, x_{\ell-2}, x_\ell, \ldots, x_{n+1}) = (f_{1,1}(x_1, \ldots, x_{\ell-2}, x_\ell, \ldots, x_{n+1}), \ldots, f_{1,\ell-2}(x_1, \ldots, x_{\ell-2}, \ldots, x_{n+1}), f_{1,\ell}(x_1, \ldots, x_{\ell-2}, x_\ell, \ldots, x_{n+1}), f_{1,\ell+1}(x_1, \ldots, x_{\ell-2}, x_\ell, \ldots, x_{n+1}), \ldots, f_{1,n+1}(x_1, \ldots, x_{\ell-2}, x_\ell, \ldots, x_{n+1}))((x_1, \ldots, x_{\ell-2}, x_\ell, \ldots, x_{n+1}) \in X^n)$ for appropriate $f_j : X^n \to X^n$, $j = 1, \ldots, \ell - 2, \ell, \ldots, n+1$. Hence $T_1 \circ T_2(x_1, \ldots, x_{\ell-3}, x_{\ell-2}, x_\ell, \ldots, x_{n+1}) = (f_{2,1}(f_{1,k}(x_1, \ldots, x_{\ell-2}, x_\ell, \ldots, x_{n+1})), f_{2,2}(f_{1,1}(x_1, \ldots, x_{\ell-2}, x_\ell, \ldots, x_{n+1})), \ldots, f_{2,\ell-2}(f_{1,\ell-3}(x_1, \ldots, x_{\ell-2}, x_\ell, \ldots, x_{n+1})), g_2 (f_{1,n+1}(x_1, \ldots, x_{\ell-2}, x_\ell, \ldots, x_{n+1})), f_{2,\ell+1} (f_{1,\ell}(x_1, \ldots, x_{\ell-2}, x_\ell, \ldots, x_{n+1})), \ldots, f_{2,n+1} (f_{1,n}(x_1, \ldots, x_{\ell-2}, x_\ell, \ldots, x_{n+1})))$. On the other hand, the existence of $T_1 \in \langle D'_X \rangle$ with $pr_H(F_1)(x_1, \ldots, x_{\ell-3}, x_{\ell-2}, x_\ell, \ldots, x_{n+1}) = T_1(x_1, \ldots, x_{\ell-3}, x_{\ell-2}, x_\ell, \ldots, x_{n+1})$ $((x_1, \ldots, x_{\ell-2}, x_\ell, \ldots, x_{n+1}) \in X^n)$ and the formula $T_1(x_1, \ldots, x_{\ell-3}, x_{\ell-2}, x_\ell, \ldots, x_{n+1}) = (f_{1,1}(x_1, \ldots, x_{\ell-2}, x_\ell, \ldots, x_{n+1}), \ldots, f_{1,\ell-2}(x_1, \ldots, x_{\ell-2}, x_\ell, \ldots, x_{n+1}), f_{1,\ell}(x_1, \ldots, x_{\ell-2}, x_\ell, \ldots, x_{n+1}), f_{1,\ell+1}(x_1, \ldots, x_{\ell-2}, x_\ell, \ldots, x_{n+1}), \ldots, f_{1,n+1}(x_1, \ldots, x_{\ell-2}, x_\ell, \ldots, x_{n+1}))$ $((x_1, \ldots, x_{\ell-2}, x_\ell, \ldots, x_{n+1}) \in X^n)$ implies the existence of $g_1 : X^{n+1} \to X^{n+1}$ such that $F_1(x_1, \ldots, x_{n+1}) = (f_{1,1}(x_1, \ldots, x_{\ell-2}, x_\ell, \ldots, x_{n+1}), \ldots, f_{1,\ell-2}(x_1, \ldots, x_{\ell-2}, x_\ell, \ldots, x_{n+1}), f_{1,\ell-1}(x_1, \ldots, x_{\ell-2}, x_\ell, \ldots, x_{n+1}), g_1(x_1, \ldots, x_{n+1}), f_{1,\ell+1}(x_1, \ldots, x_{\ell-2}, x_\ell, \ldots, x_{n+1}), \ldots, f_{1,n+1}(x_1, \ldots, x_{\ell-2}, x_\ell, \ldots, x_{n+1}))((x_1, \ldots, x_{n+1}) \in X^{n+1})$. Obviously, then

$pr_H(F_1 \circ F_2)$ exists and coincides with $T_1 \circ T_2$. Therefore, we have that whenever $F_2 \in D_X$ and $F_1 \in \langle D_X \rangle$, the existence of $pr_H(F_1)$ with $pr_H(F_1) \in \langle D'_X \rangle$ implies the existence of $pr_H(F_1 \circ F_2)$ such that $pr_H(F_1 \circ F_2) \in \langle D'_X \rangle$. It is clear that considering the case $m > 1, i = n+1$, we obtain the same consequence. Therefore, we could prove our statement by induction. □

7.4 Completeness and Computation

The next statement shows a very simple example for n-complete digraphs. (We should note that it can be derived as a consequence of Theorem 7.17.)

Proposition 7.18. *Given a positive integer $n > 1$, $\mathcal{D} = (V, E)$ with $V = \{1, \ldots, n+1\}$ and $E = \{(i, i+1 \pmod{n+1}) \mid i \in V\} \cup \{(1,1)\}$ is an n-complete digraph.*[38]

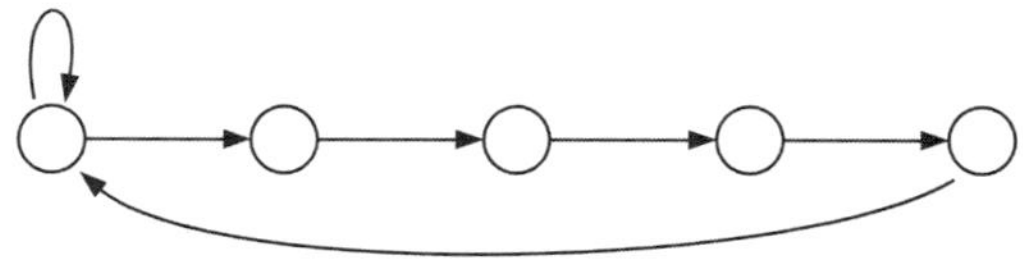

Proof. Let $\Delta_{X^{n+1}} = \{F : X^{n+1} \to X^{n+1} \mid F(x_1, \ldots, x_{n+1}) = (f(x_1, x_{n+1}), x_1, x_2, \ldots, x_n)$, $f : X^2 \to X, x_1, \ldots, x_{n+1} \in X\}$. We have by definition that all $F \in \Delta_{X^{n+1}}$ are compatible with $\mathcal{D}$. In particular, $T : X^{n+1} \to X^{n+1}$ with $T(x_1, \ldots, x_{n+1}) = (x_{n+1}, x_1, \ldots, x_n)(x_1, \ldots, x_{n+1} \in X)$ is also compatible with $\mathcal{D}$, so for every $i \in \{1, \ldots, n+1\}$ and $F \in \Delta_{X^{n+1}}$, $T^{(i-1 \,(\mathrm{mod}\, n+1))} \circ F \circ T^{n-i+1} \in \langle \Delta_{X^{n+1}} \rangle$. In other words, for all functions $F^{[i]} : X^{n+1} \to X^{n+1}$ having $F^{[i]}(x_1, \ldots, x_{n+1}) = (x_1, \ldots, x_{i-2\,(\mathrm{mod}\, n+1)}, f(x_i, x_{i-1\,(\mathrm{mod}\, n+1)}), x_i, \ldots, x_{n+1}), i \in \{1, \ldots, n+1\}, f : X^2 \to X, x_1, \ldots, x_{n+1} \in X$, we get $F^{[i]} \in \langle \Delta_{X^{n+1}} \rangle$. Considering the functions defined in Lemma 7.1, then we get $F^{(k)}_{i-1,i}, F^{(4)}_{\ell}, F^{(k)}_{n+1,1} \in \langle \Delta_{X^{n+1}} \rangle, i \in \{2, \ldots, n+1\}, k \in \{1,2,3\}, \ell \in \{1, \ldots, n+1\}$. Applying Lemma 7.1, this leads to $F^{(k)}_{i,j}, F^{(4)}_{j}, U_{i,j} \in \langle \Delta_{X^{n+1}} \rangle, i, j \in \{1, \ldots, n+1\}, k \in \{1,2,3\}$. But then, using Lemma 7.8, we receive $\mathcal{F}_{X^n \times \{d\}} \subseteq \langle \Delta_{X^{n+1}} \rangle$. In other words, for an arbitrary $F : X^{n+1} \to X^{n+1}$ with $F(x_1, \ldots, x_{n+1}) \in X^n \times \{d\}$ $(x_1, \ldots, x_{n+1} \in X$, d is fixed and F independent of $x_{n+1})$, we have $F \in \langle \Delta_{X^{n+1}} \rangle$. On the other hand, of course, for any $F' : X^n \to X^n$ there exists an $F \in \mathcal{F}_{X^n \times \{d\}}$ such that $F' = F_H$, when $H = \{1, \ldots, n\}$.

Therefore, $\mathcal{D}$ is n-complete. □

Now let for any fixed $i, j \in \{1, \ldots, n+1\}$ and $f : X^2 \to X$, $F \in \Gamma_{X^{n+1}}$ with $F(x_1, \ldots, x_{n+1}) = (x_1, \ldots, x_{i-1}, f(x_i, x_j), x_{i+1}, \ldots, x_{n+1})$ $(x-1, \ldots, x_{n+1} \in X)$, and moreover, let $F^{[i+1\,(\mathrm{mod}\, n+1)]}(x_1, \ldots, x_{n+1}) = (x_1, \ldots, x_{i-1(\mathrm{mod}\, n+1)}, f(x_{i+1\,(\mathrm{mod}\, n+1)}, x_i), x_i, \ldots, x_{n+1})$. Using $F^{[i+1\,(\mathrm{mod}\, n+1)]} \in \langle \Delta_{X^{n+1}} \rangle$ and $F = U_{i,i+1\,(\mathrm{mod}\, n+1)} \circ U_{i,j} \circ F^{[i+1\,(\mathrm{mod}\, n+1)]} \circ U_{i+1\,(\mathrm{mod}\, n+1),j}$, we get $F \in \langle \Delta_{X^{n+1}} \rangle$, which leads to $\Gamma_{X^{n+1}} \subseteq \langle \Delta_{X^{n+1}} \rangle$.

For a positive integer $n > 1$, let $K = \{i_1, \ldots, i_k\}$, $L = \{j_1, \ldots, j_\ell\}$ be subsets of $\{1, \ldots, n\}$. Moreover, let $F : X^n \to X^n$ be a transformation. The (K, L)-*projection* of F, if it exists, is defined as the function $pr_{K,L}(F) : X^k \to X^\ell$ having $pr_L(F(x_1, \ldots, x_n)) = pr_{K,L}(F)(pr_K(x_1, \ldots, x_n))$. Let $\mathcal{A} = (Z^n, X, \delta)$, $\mathcal{B} = (Z^m, Y, \delta')$

[38] As before, $k \pmod{m}$ denotes the least *positive* integer ℓ for which $m | k - \ell$.

be finite automata networks (having the same basic set Z). We say that $\mathcal{B}$ *computes* $\mathcal{A}$ if there exist subsets $K, L \subseteq \{1, \ldots, m\}$ such that every δ_x $(x \in X)$ is a (K, L)-projection of a mapping $\delta'_p : Z^m \to Z^m$ $(p \in Y^+)$. (Of course, in this case we should have $|K| = |L| = n$.) If $\mathcal{B}$ is a $\mathcal{D}$-network and $\mathcal{B}$ computes $\mathcal{A}$, then we also say that $\mathcal{D}$ *computes* $\mathcal{A}$. $\mathcal{D}$ is called *n-complete with respect to computation by projection* if every network of size n is computable on $\mathcal{D}$.

Problem 7.19. *For every positive integer $n > 1$, give a complete characterization of n-complete digraphs having all loop edges with respect to computation by projection.*

We show the following partial result regarding the above problem.

Theorem 7.20. *Let us consider the family of linear digraphs of the form*

$$\mathcal{L}_n = (\{1, \ldots, n\}, \{(1,2), (2,3), \ldots, (n-1, n)\}), \quad n = 1, 2, \ldots.$$

$\mathcal{L}_m$ is n-complete with respect to computation by projection if $m \geq 2n + 1$.

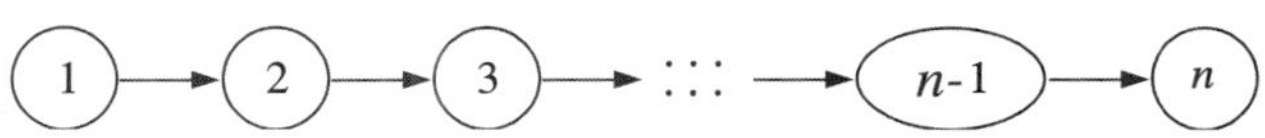

Proof. We will omit the study of the trivial case $n = 1$. Thus we assume $n \geq 2$. Let us identify X with the $|X|$-degree cyclic group with a generator $g \in X$ and identity element $e \in X$ (i.e., $g^{|X|} = e$ holds). First we prove that for every finite set X with $|X| \geq 2$, a pair of positive integers n, ℓ with $\ell \leq n$, and elements $c_1, \ldots, c_\ell \in X$, the function $F_{c_1,\ldots,c_\ell}$ is a composite of $\mathcal{L}_{2n+1}$-compatible functions whenever

$$F_{c_1,\ldots,c_\ell}(x_1, \ldots, x_{2n+1})$$

$$= \begin{cases} (x_1, \ldots, x_\ell, e, g, x_{\ell+3}, \ldots, x_{2n+1}) & \text{if } (x_1, \ldots, x_\ell) = (c_1, \ldots, c_\ell), \\ (x_1, \ldots, x_\ell, e, e, x_{\ell+3}, \ldots, x_{2n+1}) & \text{otherwise} \end{cases}$$

$(x_1, \ldots, x_{2n+1} \in X)$.

Put $T_\ell(x_1, \ldots, x_{2n+1}) = (x_1, \ldots, x_\ell, e, g, x_{\ell+3}, \ldots, x_{2n+1})$. In addition, for every $\epsilon \in \{-1, 1\}, i, j = 1, \ldots, \ell + 1, i < j$, let

$$R_{\epsilon,i,j,c}(x_1, \ldots, x_{2n+1}) = \begin{cases} (x_1, \ldots, x_{j-1}, x_i x_j, x_{j+1}, \ldots, x_{2n+1}) & \text{if } x_i = c, \\ (x_1, \ldots, x_{2n+1}) & \text{otherwise} \end{cases}$$

$(x_1, \ldots, x_{2n+1} \in X)$. Of course, T_ℓ and all $R_{\epsilon,i,i+1,c} (\epsilon \in \{-1, 1\}, i = 1, \ldots, \ell, c \in X)$ are compatible with $\mathcal{L}_{2n+1}$. On the other hand, observe that $F^{(1)}_{i,i+1}, F^{(2)}_{i,i+1}, i = 1, \ldots, \ell$, defined in Lemma 7.1 (taking n of the lemma to be $2n + 1$), are also $\mathcal{L}_{2n+1}$-compatible. In addition, using computations similar to those in Lemma 7.1, for every pairwise distinct $i, j, k \in \{1, \ldots, \ell + 1\}$ and $c \in X$, we obtain

$$R_{1,i,j,c} = F^{(2)}_{k,j} \circ R_{1,i,k,c} \circ F^{(1)}_{k,j} \circ R_{-1,i,k,c},$$

$$R_{-1,i,j,c} = F^{(2)}_{k,j} \circ R_{-1,i,k,c} \circ \circ F^{(1)}_{k,j} \circ R_{1,i,k,c}.$$

Hence, it is clear that all functions $R_{\epsilon,i,j,c}$ ($\epsilon \in \{-1, 1\}, i, j = 1, \ldots, \ell + 1, i < j, c \in X$) can be composed of $\mathcal{L}_{2n+1}$-compatible functions. Thus, applying the fact that for every $i = 1, \ldots, \ell$,

$$T_\ell \circ R_{1,i,\ell+1,c}(x_1, \ldots, x_{2n+1}) = \begin{cases} (x_1, \ldots, x_\ell, x_i, g, x_{\ell+3}, \ldots, x_{2n+1}) & \text{if } x_i = c, \\ (x_1, \ldots, x_\ell, e, g, x_{\ell+3}, \ldots, x_{2n+1}) & \text{otherwise} \end{cases}$$

($x_1, \ldots, x_{2n+1} \in X$), and using the obviously $\mathcal{L}_{2n+1}$-compatible function

$$T_{c,\ell}(x_1, \ldots, x_{2n+1}) = \begin{cases} (x_1, \ldots, x_\ell, e, x_{\ell+2}, \ldots, x_{2n+1}) & \text{if } x_{\ell+1} = c, \\ (x_1, \ldots, x_\ell, e, e, x_{\ell+3}, \ldots, x_{2n+1}) & \text{otherwise} \end{cases}$$

($x_1, \ldots, x_{2n+1} \in X$), we obtain $F_{c_1,\ldots,c_\ell} = T_\ell \circ R_{1,1,\ell+1,c_1} \circ T_{c_1,\ell} \circ \cdots \circ R_{1,\ell-1,\ell+1,c_{\ell-1}} \circ T_{c_{\ell-1},\ell} \circ R_{1,\ell,\ell+1,c_\ell} \circ T_{c_\ell,\ell}$ ($c_1, \ldots, c_\ell \in X$). This ends the proof that for every $\ell \le n$ and $c_1, \ldots, c_\ell \in X$, $F_{c_1,\ldots,c_\ell}$ is a composite of $\mathcal{L}_{2n+1}$-compatible functions.

Now we define the $\mathcal{L}_{2n+1}$-compatible functions

$$R_c(x_1, \ldots, x_{2n+1}) = \begin{cases} (x_1, \ldots, x_{n+2}, c, x_{n+4}, \ldots, x_{2n+1}) & \text{if } x_{n+2} = g, \\ (x_1, \ldots, x_{2n+1}) & \text{otherwise,} \end{cases}$$

$$T(x_1, \ldots, x_{2n+1}) = (x_1, \ldots, x_{n+3}, x_{n+3}, x_{n+4}, \ldots, x_{2n})$$

($x_1, \ldots, x_{2n+1} \in X$), and for every $f : X^n \to X$, let

$$F_f = \prod_{(c_1,\ldots,c_n) \in X^n} F_{c_1,\ldots,c_n} \circ R_{f(c_1,\ldots,c_n)}.$$

Put

$$T^{(i)}(x_1, \ldots, x_{2n+1}) = \begin{cases} (x_1, \ldots, x_{n-1}, e, x_n, \ldots, x_{2n}) & \text{if } i = 1, \\ (x_1, \ldots, x_n, g, x_{n+1}, \ldots, x_{2n}) & \text{if } i = 2, \end{cases}$$

and consider for every $c, d \in X$,

$$Z_{c,d}(x_1, \ldots, x_{2n+1})$$
$$= \begin{cases} (x_1, \ldots, x_{n+1}, c, x_{n+3}, \ldots, x_{2n+1}) & \text{if } (x_{n+1}, x_{n+2}) = (g, d), \\ (x_1, \ldots, x_{2n+1}) & \text{otherwise} \end{cases}$$

($x_1, \ldots, x_{2n+1} \in X$). For every $f : X^n \to X$, set

$$Z_f = T^{(1)} \circ T^{(2)} \circ \prod_{(c_1,\ldots,c_{n-1}) \in X^{n-1}} \left(\prod_{c \in X} F_{c_1,\ldots,c_{n-1}} \circ Z_{f(c_1,\ldots,c_{n-1},c),c} \right).$$

Then we have for every $f_i : X^n \to X, i = 1, \ldots, n, x_1, \ldots, x_{2n+1} \in X$,

$$\left(\prod_{i=1}^{n-2} F_{f_i} \circ T \right) \circ F_{f_2} \circ Z_{f_1}(x_1, \ldots, x_{2n+1}) = Z_{f_1}(x_1, \ldots, x_n, e, y, f_2(x_1, \ldots, x_n), \ldots,$$

$$f_n(x_1, \ldots, x_n)) = (x_1, \ldots, x_{n-1}, e, z, f_1(x_1, \ldots, x_n), f_2(x_1, \ldots, x_n), \ldots, f_n(x_1, \ldots, x_n))$$

(where $y, z \in \{e, g\}$). By our constructions, the above-defined function is a composite of $\mathcal{L}_{2n+1}$-compatible functions. Hence, taking into consideration that $f_i : X^n \to X, i = 1, \dots, n$, are arbitrary, this completes the proof for the case $m = 2n + 1$. It is also clear that this implies the validity of our result for every $m \geq 2n + 1$, too. □

Problem 7.21. *For every positive integer $n > 1$, give a complete characterization of n-complete digraphs with respect to computation by projection (such that we take into consideration the loop edges of the communication links as we discussed in this section).*

7.5 Asynchronous Automata Networks

In this section we derive a general result which shows how it is possible to emulate the behavior of a given synchronously updated automata network by a corresponding asynchronous one. This allows one to transfer results concerning the usual (synchronous) automata networks to the asynchronous realm, including, for example, cellular automata and their generalizations. Moreover, the result also holds for infinite automata networks over locally finite underlying graphs.

We show that any automata network $\mathcal{A}$ with global synchronous updates can be emulated by another one, $\widehat{\mathcal{A}}$, whose structure derives from that of $\mathcal{A}$ by a simple construction but whose updates are made asynchronously at its various component automata (e.g., possibly randomly or sequentially, with or without possible simultaneous updates at different nodes). By emulation, we refer to the existence of a spatial-temporal covering (local time), allowing one to project the behavior of $\widehat{\mathcal{A}}$ continuously onto that of $\mathcal{A}$. We also show the existence of a spatial-temporal section of the asynchronous automata network's behavior which completely determines the synchronous global state of $\mathcal{A}$ at every time step.

We give the construction of the asynchronous automata network, establish its freedom from deadlocks, and construct local time functions and spatial-temporal sections relating any possible behavior of $\widehat{\mathcal{A}}$ to the single corresponding behavior of $\mathcal{A}$ on a given input sequence starting from a given initial global state.

This establishes that the behavior of any synchronous automata network actually can be emulated without the restriction of synchronous update, freeing us from the need of a global clock signal. Local information is sufficient to guarantee that the synchronous behavior of $\mathcal{A}$ is completely determined by any asynchronous behavior of $\widehat{\mathcal{A}}$ starting from a corresponding global state and given the same input sequence as $\mathcal{A}$. Moreover, the relative passage of corresponding local time at any two nodes in $\widehat{\mathcal{A}}$ is bounded in a simple way by approximately one-third of the distance between them.

As corollaries, any synchronous generalized cellular automaton or synchronous cellular automaton can be emulated by an asynchronous one of the same type.

Implementation aspects of these asynchronous automata are also discussed, and open problems and research directions are indicated.

Relaxing our usual assumptions, for this section we will allow consideration of automata networks over possibly infinite digraphs. For a digraph $\mathcal{D} = (V, E)$, we say node w is a *neighbor* of v if there is an incoming edge from w to v, that is, $(w, v) \in E$. The *neighborhood of* v is the set $N(v) \subseteq V$ of all neighbors of node w. The *reflexive-symmetric closure* $\widehat{\mathcal{D}}$ of a digraph $\mathcal{D} = (V, E)$ is a digraph the same set of vertices V and edges

$$\widehat{E} = \{(w, v) \in V \times V \mid (w, v) \in E \text{ or } (v, w) \in E \text{ or } v = w\}.$$

Thus $\widehat{\mathcal{D}}$ has as edge set the symmetric and reflexive closure of the relation E. The digraph $\widehat{\mathcal{D}}$ is essentially the same as the underlying undirected graph $U(\mathcal{D})$ of $\mathcal{D}$ with all loop edges added. The digraph $\mathcal{D}$ is *locally finite* if its reflexive-symmetric closure $\widehat{\mathcal{D}}$ has no node with infinitely many neighbors. Here we will define a slightly extended concept of automata network that allows for synchronous or asynchronous updates of the states of local automata which are connected according to a locally finite digraph. As before, an *automata network* (or $\mathcal{D}$-product of automata) is defined by giving a digraph $\mathcal{D} = (V, E)$, a V-indexed set of automata $\mathcal{A}_v$, an external input alphabet X, and a V-indexed set of feedback functions that are compatible with $\mathcal{D}$.

Specifically, to each vertex $v \in V$, let an automaton $\mathcal{A}_v = (A_v, X_v, \delta_v : A_v \times X_v \to A_v)$ be associated. The set A_v is the set of local states, X_v is the set of local input letters, and δ_v is the local transition function at node v. As before, if there is no danger of confusion, we write $a_v \cdot x_v$ for the state $\delta_v(a_v, x_v)$ whenever $a_v \in A_v$ and $x_v \in X_v$. A *global state* of the automata network $\mathcal{A}$ is an element a of $A = \prod_{v \in V} A_v$. For a vertex $v \in V$, denote by $a_v \in A_v$ the v-component of a. Let $X \neq \emptyset$ be an external alphabet, and let $X^{\eth} = X \cup \{\eth\}$ where $\eth \notin X$ may be regarded as a wait symbol. For each node $v \in V$, let there be a *feedback function*

$$\varphi_v : \prod_{w \in N(v)} A_w \times X^{\eth} \to X_v.$$

This determines a local input letter $x_v \in X_v$ to $\mathcal{A}_v$ as a function of the external input letter (or wait symbol) $x \in X^{\eth}$ and the state nodes in the neighborhood of v. We may extend φ_v to $\varphi_v : A \times X^{\eth} \to X_v$ by letting $\varphi_v(a, x) = \varphi_v((a_w)_{w \in N(v)}, x)$. Here $(a_w)_{w \in N(v)} \in \prod_{w \in N(v)} A_w$ is the assignment of states to all components in the neighborhood of v according to global state a. In this way, φ_v does not really depend on its w-component unless $w \in N(v)$. Given the digraph $\mathcal{D} = (V, E)$, automata $\{\mathcal{A}_v\}_{v \in V}$, feedback functions $\{\varphi_v\}_{v \in V}$, and external alphabet X as above, the associated *synchronous automata network* $\mathcal{A}$ is an automaton with states $A = \prod_{v \in V} A_v$, inputs X, and transition function $\delta : A \times X \to A$ defined for all $a \in A$ and $x \in X$ by giving the new v-component of state $a \cdot x$ as

$$\delta_v(a_v, x) = \delta_v(a_v, \varphi_v(a, x)) = a_v \cdot \varphi_v(a, x),$$

where $a_v \in A_v$ is the v-component of global state $a \in A$. With these data $\mathcal{A}$ is the (synchronous) automata network (or a general product or Gluškov product) of local automata $\mathcal{A}_v$ over the digraph $\mathcal{D}$ according to the feedback functions φ_v. In this synchronous case, the wait symbol $\eth$ is actually superfluous and X rather than $X^{\eth}$ may be used throughout, as we shall see. It is only required in the asynchronous generalization, so in the synchronous case the definition of automata network here is essentially identical to the standard one.

For all natural numbers $n \in \mathbb{N}$, let x_n be a letter in X. If the sequence $x_1, x_2, x_3, \ldots$ is input to $\mathcal{A}$ in a synchronous network, starting from an initial global state $a(0) \in A$ with $v \mapsto a_v(0)$, then the global state $a(n)$ of $\mathcal{A}$ at time n is given inductively by

$$a_v(n) = a_v(n-1) \cdot \varphi_v(a(n-1), x_n)$$

for all $n \geq 1$. Note that we are using a discrete model of time. Thus the successive states of the local automaton $\mathcal{A}_v$ at node v, $a_v(0), a_v(1), a_v(2), \ldots$ and the successive global states $a(0), a(1), a(2), \ldots$ of the entire network $\mathcal{A}$ depend in general on the particular values of external inputs, except in the case $|X| = 1$. The function $a : \mathbb{N} \to A$ with $a(n)$ having v-component $a_v(n) \in A_v$ is called the *behavior* of the synchronous network $\mathcal{A}$ on the given input sequence $\{x_n\}_{n \in \mathbb{N}}$ and initial state $a(0)$.

Note also that in this synchronous case, a letter is read at each update, so the wait symbol is never used in place of an input letter by any feedback function φ_v. Thus we could have equivalently used X rather than $X^{\eth}$ in the definition of the φ_v. This is what the classical definition does. $X^{\eth}$ is needed for the general asynchronous case below. In the synchronous case and, as we shall see, for generalized cellular automata, this is equivalent to the classical definition.

Our concept of *asynchronous automata network* $\mathcal{A}$ requires again giving a digraph $\mathcal{D} = (V, E)$, a V-indexed set of automata $\mathcal{A}_v$, an external input alphabet X, and a V-indexed set of feedback functions $\{\varphi_v\}_{v \in V}$ that are compatible with $\mathcal{D}$. A V-indexed family of *read functions*

$$\rho_v : \prod_{w \in N(v)} A_w \to \{0, 1\},$$

which are used to determine whether the feedback function for node v receives the wait symbol $\eth$ or a letter of external input, is also required. These ingredients completely determine the asynchronous automata network $\mathcal{A}$.

We will allow local update at a node v without necessarily changing local state at any other node, and local automata will be allowed to read the global input sequence asynchronously and independently according to their update times and local state in their neighborhoods. In particular, local automata will be allowed to wait (as a function of the state of their local neighborhood) before reading the next letter of external input.

To capture the notion of asynchronous local updates, embed $\mathbb{N}$ as a model of time arbitrarily into the nonnegative real numbers $\mathbb{R}^+$ (or nonnegative rationals $\mathbb{Q}^+$, or $\mathbb{N}$),

$$\tau : \mathbb{N} \to \mathbb{R}^+ \text{ (or } \mathbb{Q}^+ \text{ or } \mathbb{N}),$$

with $\tau(0) = 0$ and $i < j$ implying $\tau(i) < \tau(j)$. At time $\tau(n)$ with n positive, a set of local updates will occur simultaneously in the asynchronous automata network. During the half-open interval $[0, \tau(1))$, the state of the automata network is an initial global state $a(0)$ as above. For each $n > 0$, during open interval $(\tau(n), \tau(n+1))$, the state of the network does not change at all. At time $\tau(n)$, let $U_{\tau(n)} \subseteq V$ denote the *nodes updated at time* $\tau(n)$. Formally, a *(local) update* is said to occur at node $v \in V$ at time $\tau(n) \in \mathbb{R}^+$ if and only if v lies in the update set $U_{\tau(n)}$. Thus subsets of the local automata of $\mathcal{A}$ will be updated instantaneously at time points $\tau(1), \tau(2), \ldots,$ with all local automata having nodes in the update set $U_{\tau(n)}$ updated simultaneously as a function of the current states of their neighbors and possibly the input letters they are currently reading. We require that each node $v \in V$ is updated an unbounded number of times, i.e., $v \in U_{\tau(n)}$ for infinitely many $n \in \mathbb{N}$.

An *update pattern* (τ, U) of an asynchronous network is an order-preserving function (as above) $\tau : \mathbb{N} \to \mathbb{R}^+$ together with a family of update sets $U_{\tau(n)} \subseteq V, n > 0$. (Sometimes we will suppress the update sets and refer to τ as an update pattern.) For $t \in \mathbb{R}^+ \setminus \tau(\mathbb{N})$

and also for $t = 0$, one may define $U_t = \emptyset$. Then for all moments in time $t \in \mathbb{R}^+$, an update occurs at node v at time t if and only if $v \in U_t$. A *run* of a network is a sequence of global states $a(t)$ of the network, and we will soon see how an update pattern together with a infinite input word $\{x_n\}_{n>0}$ (with x_n in X for all positive natural numbers n) determine a well-defined run $a : \mathbb{R}^+ \to A$, with component values $a_v(t) = (a(t))_v \in A_v$ at time $t \in \mathbb{R}^+$, called the *(continuous) behavior* of the asynchronous network $\mathcal{A}$ for this update pattern and input sequence, with initial global state $a(0)$. The restriction $a : \tau(\mathbb{N}) \to A$, of a to $\tau(\mathbb{N})$, is called the *(discrete) behavior* of $\mathcal{A}$ and clearly determines the continuous behavior a on $\mathbb{R}^+$, since nothing in the network may change at any $t \notin \tau(\mathbb{N})$. An update pattern is not a part of the specification of the automata network and need not be given in advance. An update pattern and external input sequence are, however, required to determine a behavior of the network.

Local Reading and Waiting. For each node $v \in V$, we assume a next available letter $x_{n^*(v)} \in X$—*which one will depend on how far $\mathcal{A}_v$ has read(!)*—is available at time $\tau(n)$ to be read from the sequence of global external inputs $x_1, x_2, x_3, \ldots$ which are read sequentially but not synchronously by the local automata in the asynchronous network. That is, the letters of external input $x_1, x_2, x_3, \ldots$ are read in sequence at each node v, but node v is also permitted to wait and update itself before reading (or "consuming") a letter. This is why the feedback function must handle the case when the next letter is not to be read yet, i.e., $\varphi_v : A \times X^{\eth} \to X_v$, where $\eth$, the wait symbol, is used as the second argument to φ_v when the next input letter is not read.

However, whether the next letter at node v is read may depend at most on the states of the local automata at the neighbors of node v and may not depend on external input letter itself. Thus this is determined, for each node $v \in V$, by the read function $\rho_v : \prod_{w \in N(v)} A_w \to \{0, 1\}$ for node v. Like the feedback functions φ_v, the ρ_v are given when specifying the asynchronous automata network and may be extended to all global states $\rho_v : A \to \{0, 1\}$ by defining

$$\rho_v(a) = \rho_v((a_w)_{w \in N(v)}).$$

Thus $\rho_v(a)$ does not really depend on the w-component a_w of $a \in A$ unless $w \in N(v)$. If $\rho_v(a) = 1$ when $v \in U_{\tau(n)}$, then the next external input letter is read and passed to the feedback function φ_v; but if $\rho_v(a) = 0$ when $v \in U_{\tau(n)}$, then the wait symbol $\eth$ is passed to the feedback function, and the external letter remains available. Letters of a copy of the infinite external input sequence are thus consumed in order at every node. Each local automaton gets to consume a copy of the same external input sequence, but the local automaton may wait before consuming the next letter, depending on the state of its local neighborhood and the function ρ_v.

Thus at time $t = \tau(n)$ with $n > 0$, if $v \in U_{\tau(n)}$ and the next letter locally available for reading is $x \in X$, a local update of state at node v is given by

$$a_v(\tau(n)) = \begin{cases} a_v(\tau(n-1)) \cdot \varphi_v(a(\tau(n-1)), x) \text{ if } \rho_v(a(\tau(n-1))) = 1, \\ a_v(\tau(n-1)) \cdot \varphi_v(a(\tau(n-1)), \eth) \text{ if } \rho_v(a(\tau(n-1))) = 0. \end{cases}$$

Note that the values of φ_v and ρ_v here do not depend on the state of local automata other than the neighbors of $\mathcal{A}_v$ at time $\tau(n-1)$. Since ρ_v is a function, there is no non-determinism in deciding whether the next letter is to be read or not for a given state $a \in A$.

Details of Asynchronous Behavior. To keep track, as external observers, of which letter of the external input sequence is currently available at time $\tau(n)$ at node v, we note that the *index* $n^*(v)$ *to the next letter for node* v *at time* $\tau(n)$ is given inductively by $1^*(v) = 1$ and

$$(n+1)^*(v) = \begin{cases} n^*(v) + 1 & \text{if } v \in U_{\tau(n)} \text{ and } \rho_v(a(\tau(n-1))) = 1, \\ n^*(v) & \text{otherwise.} \end{cases}$$

Thus, starting from the first letter of the global input sequence $\{x_n\}_{n \in \mathbb{N}^+}$, the index to the next input letter is advanced if and only if the input letter in position $n^*(v)$ has been read when $\mathcal{A}_v$ was last updated. Thus *the next external letter which may be read by the local automaton at node* $v \in V$ at time $\tau(n)$, with n a positive natural number, is denoted $x_{n^*(v)} \in X$.

Up to but not including time $\tau(n)$, the local automaton at v will have consumed $x_1, \ldots, x_{n^*(v)-1}$. Thus $x_{n^*(v)}$ indicates the next letter that the local automaton may consume at time $\tau(n)$. It is important to note that the local automata $\mathcal{A}_v$ do not themselves carry any information on what the next letter will be or where to find it, any more than do standard finite automata reading an input sequence. The notation $x_{n^*(v)}$ merely allows an external observer to describe which is the next letter of the external input sequence that is available to the local automaton.

The updates of state at node v for $n > 0$ are formally described by

$$a_v(\tau(n)) = \begin{cases} a_v(\tau(n-1)) \cdot \varphi_v(a(\tau(n-1)), x_{n^*(v)}) & \text{if } v \in U_{\tau(n)} \text{ and } \rho_v(a(\tau(n-1))) = 1, \\ a_v(\tau(n-1)) \cdot \varphi_v(a(\tau(n-1)), \eth) & \text{if } v \in U_{\tau(n)} \text{ and } \rho_v(a(\tau(n-1))) \neq 1, \\ a_v(\tau(n-1)) & \text{otherwise,} \end{cases}$$

where $x_{n^*(v)} \in X$ is the letter in position $n^*(v)$ of the external input sequence.

Recall that no change of state occurs at time t unless $t = \tau(n)$ for some $n > 0$. Therefore at every node $v \in V$, for all times t in the half-open interval $[\tau(n-1), \tau(n))$, where $n > 0$, we have $a_v(t) = a_v(\tau(n-1))$. Thus the state of node v during this interval, i.e., the state from time $\tau(n-1)$ up to and including any time just before time $\tau(n)$, is exactly $a_v(\tau(n-1))$. Given an initial global state $a(0)$, the above update rule determines the state $a_v(t)$ of $\mathcal{A}_v$ and hence the state $a(t)$ of the entire network for all $t \in \mathbb{R}^+$, so we have a well-defined run, the (continuous) behavior of $\mathcal{A}$.

Note that if external inputs are always read (i.e., $\rho_v(a) = 1$ for all $a \in A$, $v \in V$) and $U_{\tau(n)} = V$ for every $n > 0$, then the sequence of global states $a(\tau(0)), a(\tau(1)), a(\tau(2)), \ldots$ is exactly the behavior of a uniquely determined corresponding synchronous automata network.

To state our main result, we introduce notions of spatial-temporal covering and of asynchronous emulation (related to the notion of isomorphic simulation for automata).[39]

A *spatial-temporal covering* for digraph $\mathcal{D}$ is any function $\lambda : \mathbb{R}^+ \times V \to \mathbb{N}$ such that following conditions hold:

(1) The restriction $\lambda : \mathbb{R}^+ \times \{v\} \twoheadrightarrow \mathbb{N}$ of λ to every given vertex $v \in V$ is surjective.

[39]More general definitions of emulation allowing differing sets of nodes and alphabets for $\mathcal{A}$ and $\widehat{\mathcal{A}}$, partial definition of π, etc., are possible (in analogy to the classical notion of division for synchronous automata), but for simplicity we shall use this one, which suffices for our purposes here.

(2) λ is locally monotonically increasing, i.e., for all $t, t' \in \mathbb{R}^+$ and $v \in V$,

$$t \geq t' \text{ implies } \lambda(t, v) \geq \lambda(t', v).$$

(3) For all $t, t' \in \mathbb{R}^+$ and $v \in V$,

$$|\lambda(t, v) - \lambda(t'v)| \leq d(v, v'),$$

where d denotes the distance metric in the associated digraph $\widehat{\mathcal{D}}$ (the reflexive-symmetric closure of the relation E).

Let $\mathcal{A}$ be an synchronous automata network over a digraph $\mathcal{D} = (V, E)$ with global state set A, and let $\widehat{\mathcal{A}}$ be an asynchronous automata network with the same input alphabet X, a digraph $\mathcal{D}' = (V, E')$ with the same set of nodes, and global state set $\widehat{A}$. Let $\pi : \widehat{A} \to A$ be a function from global states of the asynchronous automata network to global states of the synchronous one, such that $\pi_v(\hat{a}) = (\pi(\hat{a}))_v$ depends only on $\hat{a}_v$ for all $\hat{a} \in \widehat{A}$. Thus we can denote $(\pi(\hat{a}))_v$ by $\pi(\hat{a}_v)$.

We then say that the behavior $\hat{a} : \mathbb{R}^+ \to \widehat{A}$ of $\widehat{\mathcal{A}}$ starting in state $\hat{a}(0)$ for update pattern (τ, U) and input sequence $x_1, x_2, \ldots$ ($x_i \in X$ for $i \in \mathbb{N}$) *emulates* the behavior $a : \mathbb{N} \to A$ of $\widehat{\mathcal{A}}$ starting in state $a(0)$ with the same input sequence under the projection π if there exists a spatial-temporal covering $\lambda : \mathbb{R}^+ \times V \to \mathbb{N}$ such that the following diagram commutes for each $v \in V$:

$$\begin{array}{rcll} \mathbb{R}^+ & \xrightarrow{\hat{a}_v} & \widehat{A}_v & \text{(asynchronous)} \\ \lambda(-, v) \downarrow & & \downarrow \pi & \\ \mathbb{N} & \xrightarrow{a_v} & A_v & \text{(synchronous)} \end{array}$$

That is, $\pi(\hat{a}_v(t)) = a_v(\lambda(t, v))$ with $a_v(n)$ = state in $\mathcal{A}$ of node v at time $n \in \mathbb{N}$ and $\hat{a}_v(t)$ = state in $\widehat{\mathcal{A}}$ of node v at time $t \in \mathbb{R}^+$.

Theorem 7.22 (asynchronous emulation of synchronous automata networks). *Let any synchronous automata network $\mathcal{A}$ over a locally finite digraph $\mathcal{D} = (V, E)$ with local automata $\mathcal{A}_v = (A_v, X_v, \delta_v)$ $(v \in V)$ and external input alphabet X be given.*

We construct an asynchronous automata network $\widehat{\mathcal{A}}$ (with the same input alphabet X) such that every possible behavior of $\widehat{\mathcal{A}}$ with input sequence $\{x_n\}_{n>0}$ emulates the (only possible) behavior of $\mathcal{A}$ with input sequence $\{x_n\}_{n>0}$, when $\widehat{\mathcal{A}}$ starts in an initial global state $\hat{a}(0)$ depending only on the initial global state $a(0)$ of $\mathcal{A}$.

Moreover, the following hold:

(1) *The underlying digraph for $\widehat{\mathcal{A}}$ is the reflexive-symmetric closure of the digraph for $\mathcal{A}$.*

(2) *For each vertex v, the local automaton $\widehat{\mathcal{A}}_v$ of $\widehat{\mathcal{A}}$ are not much more complicated than the local automaton $\mathcal{A}_v$ of $\mathcal{A}$. Moreover, $\widehat{\mathcal{A}}_v$ is a direct product of $\mathcal{A}_v$, an identity-reset automaton, and a modulo three counter (with identity).*[40] *In fact, $\mathcal{A}_v$ has state set $\widehat{A}_v = A_v \times A_v \times \{1, 2, 3\}$.*

[40]Recall that for $n \geq 1$, a modulo n counter with identity is an automaton $\mathcal{C}_n^1 = (\{1, \ldots, n\}, \{+0, +1\}, \delta_{\mathcal{C}_n^1})$ with $\delta_{\mathcal{C}_n^1}(i, +1) = i + 1 \pmod n$, and $\delta_{\mathcal{C}_n^1}(i, +0) = i$, $i = 1, \ldots, n$.

(3) *The projection* $\pi : \widehat{A} \to A$ *is given locally by* $\pi_v(a_v, b_v, r) = a_v$ *for* $(a_v, b_v, r) \in \widehat{A}_v$.

(4) *The starting state of* $\widehat{\mathcal{A}}$ *is given by* $\hat{a}_v(0) = (a_v(0), a_v(0), 3)$ *for all* $v \in V$.

(5) *Furthermore, the spatial-temporal covering of the emulation satisfies*

$$|\lambda(t, v) - \lambda(t, v')| \leq \left\lfloor \frac{d(v, v') + 2}{3} \right\rfloor .$$

We call $\lambda(t, v)$ the *local time* of the synchronous automaton $\mathcal{A}$ at vertex v for time t in the emulating asynchronous automaton $\widehat{\mathcal{A}}$. Of course, λ depends in general on the update pattern (τ, U) for the particular behavior of $\widehat{\mathcal{A}}$. Thus (5) above says that the difference in local time at two nodes in the emulating asynchronous automata network is bounded above by approximately one-third the distance between them.

Proof. We give the construction of $\widehat{\mathcal{A}}$ and show by a series of lemmas that it has the required properties. As before, let $\widehat{\mathcal{D}}$ be the reflexive and symmetric closure of $\mathcal{D}$. $\widehat{\mathcal{D}} = (V, \widehat{E})$, where

$$\widehat{E} = \{(v, v') \times V \times V : v = v' \text{ or } (v, v') \in E \text{ or } (v', v) \in E\}.$$

Let $\hat{N}(v)$ denote the neighborhood of v in $\widehat{\mathcal{D}}$. We construct $\widehat{\mathcal{A}}$ as an automata network over $\widehat{\mathcal{D}}$. The local automaton $\widehat{\mathcal{A}}_v$ at node $v \in V$ has states $\widehat{A}_v = A_v \times A_v \times \{1, 2, 3\}$, and its input alphabet is

$$\widehat{X}_v = (X_v \cup \{1\}) \times (\text{constants on } A_v \cup \{1\}) \times \{+0, +1\},$$

where 1 is a new symbol that acts as the identity on the corresponding component, and, for each $a_v \in A_v$, the middle component includes input letter *constant* a_v, which acts as a constant resetting the middle component to a_v, whereas, in the third component, $+0$ acts as the identity and $+1$ increases that component by 1 modulo 3.[41]

We write $\hat{a}_v = (a_v, b_v, r_v) \in \widehat{A}_v = A_v \times A_v \times \{1, 2, 3\}$ for the local state $\hat{a}_v$ at node $v \in V$ of $\widehat{\mathcal{A}}$. The read functions of $\widehat{\mathcal{A}}$ are $\hat{\rho}_v : \widehat{A} \to \{0, 1\}$ with

$$\hat{\rho}_v(\hat{a}) = \begin{cases} 1 & \text{if } r_w \neq r_v - 1 \pmod 3 \text{ for all } w \in \widehat{N}(v) \text{ and } r_v = 3, \\ 0 & \text{otherwise.} \end{cases}$$

The feedback functions of $\widehat{\mathcal{A}}$ are $\widehat{\varphi}_v : A \times X^{\eth} \to \widehat{X}_v$ with

$$\widehat{\varphi}_v(\hat{a}, x) = \begin{cases} (1, 1, +0) & \text{if } r_w = r_v - 1 \pmod 3 \text{ for some } w \in \widehat{N}(v), \\ (1, 1, +1) & \text{if } r_w \neq r_v - 1 \pmod 3 \text{ for all } w \in \widehat{N}(v) \\ & \quad \text{and } r_v \neq 3, \\ (\varphi_v(c, x), \text{constant } a_v, +1) & \text{if } r_w \neq r_v - 1 \pmod 3 \text{ for all } w \in \widehat{N}(v) \\ & \quad \text{and } r_v = 3, \end{cases}$$

[41] The asynchronous construction here uses a modulo n counter for the third component of local state with $n = 3$, but with obvious modifications $n \geq 3$ could be used just as well. The reader may observe that the method would not work with $n = 2$. Intuitively, the construction relies on being able to locally distinguish whether neighbors are in the *future*, *present*, or *past* relative to a given node (in the corresponding synchronous network). Thus, at least three values are required. Having these distinctions in the asynchronous realm, we can guarantee that local time never gets too far "out of sync" for nearby nodes, as we shall establish formally.

where a_v is the first component of $\hat{a}_v$ in state $\hat{a}$ and c is an arbitrary state of $\mathcal{A}$ such that for each $w \in N(v)$,

$$c_w = \begin{cases} a_w & \text{if } r_w = 3, \\ b_w & \text{if } r_w = 1. \end{cases}$$

Note also r_w must lie in $\{1, 3\}$ in the determining the c_w of the third case, as necessarily $r_v = 3$ in third case and $w \in N(v) \subseteq \widehat{N}(v)$ implies $r_w \neq 2 \pmod 3$.

Updates at node v in $\widehat{\mathcal{A}}$ are thus given by the local update function with

$$\begin{aligned} &\widehat{\delta}_v((a_v, b_v, r_v), \widehat{\varphi}_v(\hat{a}, x))(a_v, b_v, r_v) \cdot \widehat{\varphi}_v(\hat{a}, x) \\ &= \begin{cases} (a_v \cdot 1, b_v \cdot 1, r_v + 0) & \text{if } r_w = r_v - 1 \pmod 3 \\ & \quad \text{for some } w \in \widehat{N}(v), \\ (a_v \cdot 1, b_v \cdot 1, r_v + 1 \pmod 3) & \text{if } r_w \neq r_v - 1 \pmod 3 \\ & \quad \text{for all } w \in \widehat{N}(v), \text{ and } r_v \neq 3 \\ (a_v \cdot \varphi_v(c, x), b_v \cdot \text{constant } a_v, 3 + 1 \pmod 3) & \text{otherwise,} \end{cases} \end{aligned}$$

where c is as above.

That is,

$$\begin{aligned} &\widehat{\delta}_v((a_v, b_v, r_v), \widehat{\varphi}_v(\hat{a}, x)) \\ &= \begin{cases} (a_v, b_v, r_v) & \text{if } r_w = r_v - 1 \pmod 3 \text{ for some } w \in \widehat{N}(v), \\ (a_v, b_v, r_v + 1 \pmod 3) & \text{if } r_w \neq r_v - 1 \pmod 3 \text{ for all } w \in \widehat{N}(v) \text{ and } r_v \neq 3, \\ (a_v \cdot \varphi_v(c, x), a_v, 1) & \text{otherwise,} \end{cases} \end{aligned}$$

where c is as above.

Notice that the transition function of $\mathcal{A}_v$ and the feedback function φ_v from the synchronous network are used to give the input to $\mathcal{A}_v$ in the third case. Of course the value of $a_v \cdot \varphi_v(c, x)$ depends only on x, a_v and the c_w with w in $N(v)$, the neighborhood of v in the original digraph $\mathcal{D}$.

In computing $\widehat{\delta}_v$, $x \in X^{\eth}$ is the letter currently available for possible reading by the local automaton at node v if $\hat{\rho}_v(\hat{a}) = 1$ but is $x = \eth$ in case $\hat{\rho}_v(\hat{a}) = 0$ (see the discussion of local reading and waiting above). By our choice of reading functions, the letter x is the wait symbol $\eth$ in the first two cases of the local update rule and lies in X if and only if the third case applies.[42] Thus, the third case applies if and only if the next available letter is consumed.

Suppose the initial state of $\mathcal{A}$ in a synchronous run is $a(0)$ with each node $v \in V$ in state $a_v(0) \in A_v$. Let the initial state of $\widehat{\mathcal{A}}$ have the automaton at each node v in state $\hat{a}_v(0) = (a_v(0), a_v(0), 3)$.

For a given behavior of $\widehat{\mathcal{A}}$, we say there is a $+1$-*update* at vertex v whenever the transition rule $\widehat{\delta}_v$ is applied to update the local state using either its second or third case, i.e., exactly when the last component of state is incremented by $+1$ modulo 3. We say there is a *real update* at node v (corresponding to a synchronous update in $\mathcal{A}$ at that node) whenever

[42] If X is a singleton, the above definition of $\widehat{\delta}_v$ is even consistent with $\hat{\rho}_v(\hat{a}) = 1$ for all $\hat{a} \in \widehat{A}$, since the current letter is not different from the next one in an infinite input word consisting of identical letters. This observation will be used when we specialize Theorem 7.22 in Corollaries 7.29 and 7.30, respectively, to emulating generalized cellular automata and cellular automata by asynchronous automata networks which can be chosen to be asynchronous generalized cellular automata and asynchronous cellular automata, respectively.

transition rule $\widehat{\delta}_v$ is applied to update the local state using its third case, i.e., exactly when the the last component of state changes from 3 to 1. Let $p(t, v)$ be the number of $+1$-updates that have occurred at vertex v during a behavior with update pattern τ up to and including time $t \in \mathbb{R}^+$.

Lemma 7.23. *For each pair of neighboring nodes $v, v' \in V$ in $\widehat{\mathcal{D}}$ and for all $t \in \mathbb{R}^+$,*

$$|p(t, v) - p(t, v')| \leq 1.$$

Proof. It suffices to consider the values of $p(t, v)$ for $t \in \{\tau(0), \tau(1), \tau(2), \tau(3), \ldots\}$ since no applications of local transition rules occur between them. At $t = 0 = \tau(0)$, $p(t, w) = 0$ for all $w \in V$, so $p(t, v) = p(t, v')$ holds. Now by induction, we suppose that the inequality holds at time $\tau(k)$ for all $k \leq n \in \mathbb{N}$.

If $p(\tau(n), v) = p(\tau(n), v')$, then at time $n + 1$, a $+1$-update will occur at none, one, or both v and v', so, as a result, the inequality will always hold at time $\tau(n + 1)$.

Otherwise $|p(\tau(n), v) - p(\tau(n), v')| = 1$. Without loss of generality, we may assume

$$p(\tau(n), v') = p(\tau(n), v) - 1.$$

Since the last component of state at a node is increased by 1 (modulo 3) for each $+1$-update at that node and otherwise remains unchanged, obviously the last component $r_w(t)$ of state at any node w in V is congruent to $p(t, w)$ modulo 3. Therefore

$$r_{v'}(\tau(n)) = r_v(\tau(n)) - 1 \pmod 3.$$

It follows from the definition of local update $\widehat{\delta}_v$ that there will be no $+1$-update at node v at $\tau(n + 1)$, so $p(\tau(n), v) = p(\tau(n + 1), v)$. Now there are two possibilities: Either there is also no $+1$-update at node v' at this time, so $p(\tau(n), v') = p(\tau(n + 1), v')$ and the inequality is preserved; or, otherwise, there is a $+1$-update at node v', so then

$$p(\tau(n + 1), v') = p(\tau(n), v') + 1 = p(\tau(n), v) = p(\tau(n + 1), v),$$

and again the inequality holds.

It follows by induction that it holds for all n, and hence for all $t \in \mathbb{R}^+$. □

Corollary 7.24. *Let v and v' be vertices at distance d in the graph $\widehat{\mathcal{D}}$. Then for all $t \in \mathbb{R}^+$,*

$$|p(t, v) - p(t, v')| \leq d.$$

Proof. This follows immediately by considering a path of minimal length from v to v' and applying the above lemma. □

Define $\lambda(t, v)$ to be the number of real updates that have occurred at node v in $\widehat{\mathcal{A}}$ up to and including time $t \in \mathbb{R}^+$.

Corollary 7.25 (continuity of local time). *Let v and v' be vertices at distance d in the graph $\widehat{\mathcal{D}}$. Then for all $t \in \mathbb{R}^+$,*

$$|\lambda(t, v) - \lambda(t, v')| \leq \left\lfloor \frac{d + 2}{3} \right\rfloor.$$

Proof. By definition we have

$$\lambda(t, v) = \left\lceil \frac{p(t, v)}{3} \right\rceil.$$

We may assume $\lambda(t, v) \geq \lambda(t, v')$, whence

$$\begin{aligned} |\lambda(t, v) - \lambda(t, v')| &= \lambda(t, v) - \lambda(t, v') \\ &= \left\lceil \frac{p(t, v)}{3} \right\rceil - \left\lceil \frac{p(t, v')}{3} \right\rceil \\ &\leq \frac{p(t, v) + 2}{3} - \frac{p(t, v')}{3} \\ &\leq \frac{d+2}{3} \text{ (by the corollary above).} \end{aligned}$$

Since $|\lambda(t, v) - \lambda(t, v')|$ is an integer, it can be no more than $\lfloor \frac{d+2}{3} \rfloor$.

Lemma 7.26 (freedom from deadlocks). *With asynchronous automata network $\widehat{\mathcal{A}}$ in the initial configuration with $\hat{a}_v(0) = (a_v(0), a_v(0), 3)$ corresponding to an initial configuration of $\mathcal{A}$ with node $v \in V$ in $a_v(0)$, for any update pattern (τ, U) and any input sequence $\{x_n\}_{n>0}$ of letters in X, the number of real updates at each node is unbounded. That is, for each fixed $v \in V$, always*

$$\lim_{n\to\infty} \lambda(\tau(n), v) = \infty.$$

It follows that $\lambda : \mathbb{R}^+ \times \{v\} \to \mathbb{N}$ is surjective for each $v \in V$.

Proof. It suffices to show, for each fixed $v = v_0 \in V$, $p(\tau(n), v_0)$ increases without bound. Hence it is enough to show that if $p(\tau(n), v_0) = R$, then there is an $r > n$ with $p(\tau(r), v_0) \geq R$.

Since $\widehat{\mathcal{D}}$ is locally finite, there are finitely many nodes $v_0, v_1, \dots v_k$ with distance $d(v_0, v_i) \leq R$. By our hypotheses on update patterns each of these $v_i \in U_{\tau(m)}$ for infinitely many $m > n$.

Suppose, for a contradiction, that no node among these can receive a $+1$-update. That is, for all $m > n$, $p(\tau(n), v_i) = p(\tau(m), v_i)$ for all $0 \leq i \leq k$. Let $w(0) = v_0$. Inductively, starting with $i = 0$, since $w(i)$ cannot get a $+1$-update, by definition of $\hat{\delta}_v$, $w(i)$ must have a neighbor $w(i+1)$ with $r_{w(i+1)} = r_{w(i)} - 1 \pmod 3$, and hence by Lemma 7.23, $p(\tau(n), w(i+1)) = p(\tau(n), w(i)) - 1$. Thus we can find distinct nodes, $w(0), w(1), w(2), \dots, w(R)$ within distance R of node $v = w(0)$ such that $p(\tau(n), w(i)) = p(\tau(n), v) - i$ for $0 \leq i \leq R$. In particular the node $w(R)$ has $p(\tau(n), w(R)) = 0$. By Lemma 7.23, the neighboring nodes w' to $w(R)$ have $p(\tau(n), w') \in \{0, 1\}$, and $0 \leq p(\tau(m'), w') \leq 1$ for all $m' > n$ as long as $w(R)$ has not received a $+1$-update. Let $m > n$ be the least integer, such that $w(R) \in U_{\tau(m)}$. Then by the local update rule, $w(R)$ gets a $+1$-update at time $\tau(m)$ but lies with distance R of v_0, a contradiction.

Therefore, within the finitely many nodes within distance R of v, some node must indeed be $+1$-updated at some time $\tau(m)$ with $m > n$. Repeating this argument, for any time $\tau(n')$, we can always find a time $\tau(m')$ with $m' > n'$ such that some node within distance R of v_0 gets a $+1$-update. Since there are only finitely many such nodes, eventually (for some $r > n$) some node $w = v_i$ $(0 \leq i \leq k)$ among them will have $p(\tau(r), w) > 2R$.

But then by Corollary 7.24,

$$|p(\tau(r), w) - p(\tau(r), v_0)| \leq R,$$

implying that

$$p(\tau(r), v_0) > R;$$

i.e., node v_0 must get a $+1$-update as well. $\square$

The local time function λ is clearly locally monotonically increasing, so Lemma 7.26 and Corollary 7.25 (together with the fact that $\lfloor \frac{d+2}{3} \rfloor \leq d$) show that λ is a spatial-temporal covering, as required.

Proposition 7.27 (emulation using local time). *Let the initial states, inputs, and update pattern be as in Lemma 7.26. Then the first component of state at node v in the asynchronous automata network $\widehat{\mathcal{A}}$ at time t equals the state of node v in the synchronous automata network $\mathcal{A}$ at time $\lambda(t, v)$. That is,*

$$a_v(\lambda(t, v)) = \pi(\hat{a}_v(t)).$$

Proof. It suffices to prove the assertion for all $t \in \tau(\mathbb{N})$.

If $\hat{a}_v(t) = (x, y, r) \in \widehat{\mathcal{A}}_v$, then let $\pi(\hat{a}_v(t)) = x$, its first component as before; let $\pi_2(\hat{a}_v(t)) = y$, its second component, and let $r_v(t) = r$, its third component.

We proceed by induction on m to show that

(1) $a_v(\lambda(\tau(m), v)) = \pi(\hat{a}_v(\tau(m)))$,

(2) $\lambda(\tau(m), v) \geq 1$ implies that $\pi_2(\hat{a}_v(\tau(m))) = a_v(\lambda(\tau(m), v) - 1)$,

(3) $m > 0$ implies that $m^*(v) = \lambda(\tau(m-1), v) + 1$.

We first carry out the induction for (3). If $m = 1$, then by definition $1^*(v) = 1$, but since at time $\tau(0)$ at v there have been no real updates, we have $\lambda(\tau(0), v) + 1 = 1$. For $m > 1$, by definition $m^*(v) = (m-1)^*(v) + 1$ if and only if $v \in U_{\tau(m)}$ and $\rho_v(\hat{a}(\tau(m-1))) = 1$, i.e., if and only if the third case in the definition of $\hat{\delta}_v$ is applied. Thus $m^*(v)$ increases if and only if there is a real update at node v; i.e., every time $\lambda(\tau(m-1), v)$ increases by 1, so does $m^*(v)$. Thus,

$$\lambda(\tau(m), v) - \lambda(\tau(m-1), v) = (m+1)^*(v) - m^*(v).$$

This and the induction hypothesis yields (3).

For $m = 0$, we have $\tau(0) = 0$, $\lambda(\tau(0), v) = 0$ and $\hat{a}_v(0) = (a_v(0), a_v(0), 3)$, so (1) is immediate. (2) holds vacuously.

Suppose by induction hypothesis that (1) and (2) hold for all m with $0 \leq m < n$. We show that these assertions follow also for $m = n$.

Case 1. If $v \in U_\tau(n)$ but there is no real update at node v, it follows from the definition of λ that $\lambda(\tau(n), v) = \lambda(\tau(n-1), v)$ and, using the definition of $\hat{\delta}_v$ that the first and second coordinates of $\hat{a}_v$ are unchanged. Thus, $\pi(\hat{a}_v(\tau(n))) = \pi(\hat{a}_v(\tau(n-1)))$, and therefore

$$a_v(\lambda(\tau(n), v)) = a_v(\lambda(\tau(n-1), v)) = \pi(\hat{a}_v(\tau(n-1))) = \pi(\hat{a}_v(\tau(n))),$$

as required for (1). For implication (2), if $\lambda(\tau(n), v) > 1$, then

$$\begin{aligned}\pi_2(\hat{a}_v(\tau(n))) &= \pi_2(\hat{a}_v(\tau(n-1))) \text{ as second component is unchanged,}\\ &= a_v(\lambda(\tau(n-1), v) - 1) \text{ by induction hypothesis,}\\ &= a_v(\lambda(\tau(n), v) - 1) \text{ since } \lambda(\tau(n), v) = \lambda(\tau(n-1), v),\end{aligned}$$

as required. If $\lambda(\tau(n), v) = 1$, then it is clear from the definition of $\hat{\delta}_v$, since the value of the second coordinate can only be changed in Case 3, below, that $\pi_2(\hat{a}_v(\tau(n))) = a_v(0)$, which is just $a_v(\lambda(\tau(n), v) - 1)$, as required.

Case 2. If $v \notin U_{\tau(n)}$, then $\hat{a}_v$ is unchanged and everything follows as above.

Case 3. Finally, if $v \in U_{\tau(n)}$ and there is a real update at v at time $\tau(n)$, then by definition of $\hat{\delta}_v$ we have $r_v(\tau(n-1)) = 3$ and $r_w(\tau(n-1)) \neq r_v(\tau(n-1)) - 1 \pmod 3$ for all $w \in \widehat{N}(v)$, where $r_w(t)$ denotes the third component of $\hat{a}_w(t)$. So each $r_w(\tau(n-1)) \in \{1, 3\}$. It also is clear by induction that $r_w(t) \equiv p(t, w) \pmod 3$ always holds. By Lemma 7.23, $|p(t, w) - p(t, v)| \leq 1$ for the neighboring nodes w and v.

We consider the c_w's that are used in the third case of the local update rule in updating node v.

If $r_w(\tau(n-1)) = 3$, then $p(\tau(n-1), w) = p(\tau(n-1), v)$ follows and hence $\lambda(\tau(n-1), w) = \lambda(\tau(n-1), v)$. In this case, by induction hypothesis for node w, $\pi(\hat{a}_w(\tau(n-1))) = a_w(\lambda(\tau(n-1), w)) = a_w(\lambda(\tau(n-1), v))$. It follows that c_w in the third case of the local update rule equals $a_w(\lambda(\tau(n-1), v))$.

Otherwise, $r_w(\tau(n-1)) = 1$, and since $r_v(\tau(n-1)) = 3$, we have $p(\tau(n-1), w) = p(\tau(n-1), v) + 1$ and hence $\lambda(\tau(n-1), w) = \lambda(\tau(n-1), v) + 1$.[43]

Thus $\lambda(\tau(n-1), w) \geq 1$, and so then

$$\begin{aligned}c_w &= \pi_2(\hat{a}_w(\tau(n-1))) \text{ since } r_w(\tau(n-1)) = 1,\\ &= a_w(\lambda(\tau(n-1), w) - 1) \text{ by induction hypothesis at node } w,\\ &= a_w(\lambda(\tau(n-1), v)).\end{aligned}$$

It follows that for each $w \in \widehat{N}(v)$, the c_w in the third case of the local update rule applied at node v equals $a_w(\lambda(\tau(n-1), v))$.

By the induction hypothesis that $\pi(\hat{a}_v(\tau(n-1))) = a_v(\lambda(\tau(n-1), v))$ and by (3), $n^*(v) = \lambda(\tau(n-1), v) + 1$. Thus, the first component of $\hat{a}_v(\tau(n))$ is

$$\begin{aligned}\pi(\hat{a}_v(\tau(n))) &= \pi(\hat{a}_v(\tau(n-1))) \cdot \varphi_v(c, x_{n^*(v)}) \text{ by definition of } \widehat{\delta_v},\\ &= a_v(\lambda(\tau(n-1), v)) \cdot \varphi_v(c, x_{n^*(v)}) \text{ by induction hypothesis,}\\ &= a_v(\lambda(\tau(n-1), v)) \cdot \varphi_v(a(\lambda(\tau(n-1), v)), x_{n^*(v)})\\ &\qquad \text{since for all neighbors } w \in N(v), \text{ we have}\\ &\qquad\qquad c_w = a_w(\lambda(\tau(n-1), v)),\\ &= a_v(\lambda(\tau(n-1), v)) \cdot \varphi_v(a(\lambda(\tau(n-1), v)), x_{\lambda(\tau(n-1),v)+1}) \text{ by (3),}\end{aligned}$$

[43]Notice that the most recent +1-update at node w must have been a real update, since $r_w(\tau(n-1)) = 1$. Since $r_v(\tau(n-1)) = 3$ and nodes v and w differ by at most one +1-update, local time at node v is behind local time at node w by exactly 1.

$$= a_v(\lambda(\tau(n-1), v) + 1) \text{ by definition of local update in } \mathcal{A},$$
$$= a_v(\lambda(\tau(n), v)) \text{ since at time } \tau(n) \text{ there is a real update at } v.$$

This shows (1).

Also by definition of $\widehat{\delta}_v$, we have that the second component of $\hat{a}_v(\tau(n))$ equals the first component of $\hat{a}_v(\tau(n-1))$, so

$$\begin{aligned}\pi_2(\hat{a}_v(\tau(n))) &= \pi(\hat{a}_v(\tau(n-1))),\\ &= a_v(\lambda(\tau(n-1), v)) \text{ by induction hypothesis,}\\ &= a_v(\lambda(\tau(n), v) - 1) \text{ again since at time } \tau(n) \text{ there is a real}\\ &\qquad \text{update at } v.\end{aligned}$$

This shows (2) and completes the induction. The lemma is proved. □

Remark. *By Lemma* 7.26, *for each node* $v_0 \in V$, $\lambda(\tau(n), v_0)$ *takes all nonnegative integer values, so Proposition* 7.27 *shows how to recover the entire history of node* v *in* $\mathcal{A}$: *just record the first component of the state at node* v *in* $\widehat{\mathcal{A}}$ *every time there is a real update at* v. *Thus the sequence of global states* $a(0), a(1), \ldots$ *of* $\mathcal{A}$ *under a given external input sequence* $x_1, x_2, \ldots$ *can be recovered from any behavior of* $\widehat{\mathcal{A}}$ *started in state* $\hat{a}_v(0) = (a_v(0), a_v(0), 3)$ *for all* $v \in V$ *with the same input sequence.*

The results above establish all the properties asserted for $\widehat{\mathcal{A}}$ in the statement of Theorem 7.22, whose proof is now complete. □

Definition. *For a given external input sequence* $\{x_n\}_{n>0}$ *and update pattern* (τ, U), *any function* $\eta : \mathbb{N} \times V \to \mathbb{R}^+$ *satisfying, for all* $v \in V$, $n \in \mathbb{N}$,

$$a_v(n) = \pi(\hat{a}_v(\eta(n, v)))$$

is called a spatial-temporal section *of the behavior of the asynchronous automata network* $\widehat{\mathcal{A}}$ *mapping onto the behavior of the synchronous automata network* $\mathcal{A}$.

Corollary 7.28 (existence of spatial-temporal sections). *Let* $\eta_\tau : \mathbb{N} \times V \to \mathbb{R}^+$ *be defined by*

$$\eta_\tau(n, v) = \text{the least } \tau(m) \text{ with } m \in \mathbb{N} \text{ such that } \lambda(\tau(m), v) \text{ is } n.$$

Then η_τ *is a spatial-temporal section of the behavior of the asynchronous automata network* $\widehat{\mathcal{A}}$ *mapping onto the behavior of the synchronous automata network* $\mathcal{A}$.

Proof. As noted in the above remark, for every node v, local time $\lambda(\tau(m), v)$ takes all values $n \in \mathbb{N}$, so η_τ is well defined. By Proposition 7.27, this function is a spatial-temporal section. □

The function η_τ of Corollary 7.28 is called the *natural spatial-temporal section* for fixed external input sequence $\{x_n\}_{n>0}$ and update pattern (τ, U) of the behavior of $\widehat{\mathcal{A}}$.

Generalized Cellular Automata and Cellular Automata. A synchronous or asynchronous automata network is said to be a *generalized cellular automaton* if its external input alphabet

X is a singleton. In the synchronous case, the external input serves, in effect, only as a global synchronous update signal for $\mathcal{A}$, i.e., a clock tick, but does not otherwise affect the state of $\mathcal{A}$.[44] Similarly, in the asynchronous case, the external input letter serves as a local update signal for $\mathcal{A}_v$ whenever $v \in U_{\tau(n)}$ and $\rho_v(\tau(n-1)) = 1$.

A generalized cellular automaton $\mathcal{A}$ is a *cellular automaton* if it satisfies the following: (1) The edge relation E is a symmetric relation on V; (2) for every $v, w \in V$, $\mathcal{A}_w = \mathcal{A}_v$, that is, a copy of the same local automaton occurs at each node; (3) for every $v, w \in V$, there is a corresponding graph automorphism $\pi : \mathcal{D} \to \mathcal{D}$ with $\pi(v) = w$ such that for all $a \in A$,

$$\varphi_v((a_u)_{u \in N(v)}, x) = \varphi_w((a_{\pi(u)})_{\pi(u) \in N(w)}, x).$$

Conditions (2) and (3) imply that the automata network is highly homogeneous in that every vertex in the network has an isomorphic local neighborhood and, also, the component automaton at each vertex computes the same transition function of the states of component automata in its local neighborhood as computed by any other component automaton in the network. (In particular a cellular automaton is state-homogeneous.)

The isomorphic local automata at any two nodes v and v' in a cellular automaton are sometimes referred to as *cells* of $\mathcal{A}$. (For automata networks in general, the local automata $\mathcal{A}_v$ are sometimes referred to as the *cells* of $\mathcal{A}$, too.)

Corollary 7.29 (asynchronous emulation of generalized cellular automata). *If $\mathcal{A}$ is a synchronous generalized cellular automaton, then there is an emulating asynchronous automata network satisfying the same conclusions as in Theorem 7.22, but also $\widehat{\mathcal{A}}$ is an asynchronous generalized cellular automaton.*

Proof. This is of course just a special case of Theorem 7.22, with the additional observation that $\widehat{\mathcal{A}}$ is an asynchronous generalized cellular automaton since it has the same singleton alphabet as $\mathcal{A}$.

But it is worth remarking that in this case a further simplification is possible. Since the external input alphabet $X = \{x\}$, all input letters in the infinite external input sequence are identical. Thus it is possible to completely ignore the external input letter in the feedback functions φ_v, and, moreover, it is unnecessary to keep track locally of which letter in the infinite word is available for reading. Therefore the definition of $\hat{\rho}_v(\hat{a})$ can be simplified to be constant $\hat{\rho}_v(\hat{a}) = 1$ for all $v \in V$ and $\hat{a} \in \widehat{A}$ without affecting the global run. The resulting asynchronous generalized cellular automata never waits before reading a letter, node v is updated exactly when $v \in U_t$, and no wait symbol cases are needed for the feedback functions; i.e., we restrict φ_v to $\widehat{A} \times X \to X_v$ using X rather than $X^{\mathfrak{d}}$. Since X is a singleton we may as well simplify feedback functions to have form $\varphi_v : \widehat{A} \to X_v$. Thus, the local transition functions then simplify to

[44]The reader familiar with cellular automata may find it helpful to think of $\mathcal{A}$ as a cellular automaton, except that the interconnection graph $\mathcal{D}$ is not required to be regular, the local automata $\mathcal{A}_v$ are not required to be isomorphic, and neighborhoods are not required to be symmetric. ($v \in N(v')$ does not necessarily imply $v' \in N(v)$ for vertices v, v'.)

$$\widehat{\delta}_v((a_v, b_v, r_v), \widehat{\varphi}_v(\hat{a}))$$
$$= \begin{cases} (a_v, b_v, r_v) & \text{if } r_w = r_v - 1 \pmod 3 \text{ for some } w \in \widehat{N}(v), \\ (a_v, b_v, r_v + 1 \pmod 3) & \text{if } r_w \neq r_v - 1 \pmod 3 \text{ for all } w \in \widehat{N}(v) \\ & \quad \text{and } r_v \neq 3, \\ (a_v \cdot \varphi_v(c), a_v, 1) & \text{otherwise,} \end{cases}$$

where c is an arbitrary state of $\mathcal{A}$ such that for each $w \in N(v)$,

$$c_w = \begin{cases} a_w & \text{if } r_w = 3, \\ b_w & \text{if } r_w = 1. \end{cases}$$ □

Corollary 7.30 (asynchronous emulation of cellular automata). *If $\mathcal{A}$ is a synchronous cellular automaton, then there is an emulating asynchronous generalized cellular automaton satisfying the same conclusions as in Corollary* 7.29, *but also $\widehat{\mathcal{A}}$ is an asynchronous cellular automaton.*

Proof. This is clear since the digraph $\widehat{\mathcal{D}}$ of $\widehat{\mathcal{A}}$ is identical to the digraph $\mathcal{D}$ of $\mathcal{A}$ except possibly that all nodes become neighbors of themselves in $\widehat{\mathcal{D}}$, and so $\widehat{\mathcal{A}}$ inherits the conditions in the definition of cellular automata satisfied by $\mathcal{A}$. □

Remarks on Local External Inputs. *By applying the asynchronous emulation theorem for automata networks, the results of this section easily generalize to the case where local automata at each node v have access to separate input external sequences from possibly different external alphabets depending on v: One simply replaces the external alphabet X by a new external alphabet Z which is the product of the external alphabets for each node and only allows the feedback function at v to depend on the projection of an external letter of Z giving the external input for node v.*

Remark on Asynchronous Universal Computation. *One-dimensional cellular automata (cellular automata over the digraph $\mathbb{Z}$ of integers as nodes, with edges from an integer to integers differing by a most one) are computationally universal: One can encode any universal Turing machine U into this cellular space by letting state at a node keep track of* (1) *contents of the corresponding cell on the Turing machine's tape,* (2) *whether the read-write head of the Turing machine is present in the given location, and* (3) *the state of the finite-state memory of read-write head controller if it is present in the given cell. Since transitions in the Turing machine's computation (changes of contents of the cell being read, movement left or right, and changes in finite-state controller) are determined locally, these transitions can be encoded in a corresponding cellular automaton rule giving a cellular automaton $\mathcal{U}$ over $\mathbb{Z}$ whose behavior on any starting configuration is that the same as that of U. By the results on asynchronous emulation of cellular automata (Corollary* 7.30), *there is a one-dimensional asynchronous cellular automaton $\widehat{\mathcal{U}}$ over the same digraph that emulates $\mathcal{U}$, i.e., can compute any partial recursive function.*

Remarks on Implementations and State Number. (1) *The need for a global clock that is required by synchronous automata networks and (generalized) cellular automata is eliminated by the asynchronous emulation theorem. It constructively shows how an asynchronous emulation can be implemented, e.g., on parallel, distributed, and/or asynchronous* *computational devices, without global clocks, and how the synchronous behavior can be recovered.* (2) *In computational implementations of synchronous cellular automata on present-day sequential computers it is usual to keep two copies of the state space, one for current state of the entire space and one for the next state into which updated local states are written as they are computed. Before the next global time step, the two global copies are exchanged, and then the process is repeated. Thus in practice for each cell $\mathcal{A}_v$ in the space, one keeps two copies of local states. So if there are $|A_v| = n$ possible states in each cell, this corresponds to n^2 possible states for each cell in an implementation.*

For asynchronous cellular automata, our construction of $\widehat{\mathcal{A}}$ for Corollary 7.30 *(and for asynchronous automata networks more generally in Theorem* 7.22*) uses local automata that for each of the corresponding synchronous local automaton keep a copy of current local state (their first component), which is current according to local time $\lambda(t, v)$, and a copy of the previous local state (in their second component), and in addition a modulo* 3 *counter value. There are thus $3n^2 = |\widehat{A}_v| = |A_v| \times |A_v| \times 3$ possible local states. But if $v, v' \in U_t$ implies $d(v, v') \neq 1$ (e.g., if only a single random node is updated at a time), then it unnecessary to keep auxiliary copies of the entire state space (or even of the portion to be updated) in a sequential implementation. The only essential increase in memory usage is then the addition of local modulo* 3 *counters at each node.*

Remark on Local Synchronization with Modulo n Counters. *It is straightforward to modify the proof of Theorem* 7.22 *to obtain a variant result using modulo n counters, for $n \geq 3$, rather than modulo* 3 *counters for local synchronization. This of course results in corresponding variants of Corollaries* 7.29 *and* 7.30 *for asynchronous emulation in the realms of generalized cellular and cellular automata.*

Problem 7.31. *The asynchronous emulation theorem (Theorem* 7.22*) allows the update sets $U_{\tau(n)}$ for $n > 0$ to be arbitrary subject only to having $v \in U_{\tau(n)}$ for infinitely many n. Thus, for example, deterministic or nondeterministic, sequential, uniformly random, or Poisson-distributed, locally synchronous, and other choices of update patterns are permitted.*

(1) *For particular types of update patterns and network topologies, derive precise bounds on the rate of local real update. Given $v \in V$ study the relative passage of local time in the asynchronous model at node v as compared to the synchronous one; i.e., for synchronous global time $t \in \mathbb{N}$ determine bounds on the ratio*

$$\lambda(\tau(t), v)/t$$

for $t > 0$, and study its behavior as $t \to \infty$. Under what circumstances does the ratio remain bounded away from zero?

(2) *Also determine the precise (or expected) number $E(t_0)$ of asynchronous updates for local time to exceed t_0; i.e., determine $E(t_0) \in \mathbb{R}^+$ with*

$$n \geq E(t_0) \Rightarrow \lambda(v, \tau(n)) \geq t_0.$$

Under what circumstances is $E(t_0)$ independent of v?

(3) *Extend the asynchronous emulation theorem and its corollaries to networks in which the underlying graph is permitted to change over time, i.e., with addition or deletion of new edges and nodes.*

(4) *Extend the results to the case when state changes are not instantaneous, and a node may receive delayed information concerning the states of its neighbors.*

(5) *Develop methods for synchronous and asynchronous automata networks to cope with defective local automata and errors in transmission of local state to neighbors.*

(6) *Is it possible to obtain analogous results to those of this paper if sometimes letters of external input have not yet arrived at nodes reading them? This would represent a strengthening of the asynchronous emulation theorem (but not for the generalized cellular automata analogues), since it would then not need to be assumed that the next letter of global input were always available for local reading, thus allowing for delays in the external input reaching local nodes of the network.*

7.6 Bibliographical Remarks

Section 7.1. Technical results on finite state-homogeneous networks given in this section were described by P. Dömösi and C. L. Nehaniv [1997, 1998].

Section 7.2. Lemma 7.10 and Propositions 7.11 and 7.12 can also be derived from results in Dömösi and Nehaniv [1998]. Theorem 7.13 is due to M. Tchuente [1988]. Theorem 7.14 was proved in Dömösi and Kovács [1992].

Section 7.3. Complete finite network graphs with minimal number of edges were completely characterized by P. Dömösi and C. L. Nehaniv [1999].

Section 7.4. The results of this section was given in Dömösi and Nehaniv [1998].

Section 7.5. Cellular automata were introduced by J. von Neumann and S. Ulam; an important early study is by J. von Neumann [1966]. See also E. F. Codd [1968] and A. W. Burks [1970]. Studies of asynchronous automata networks with applications to computer and electrical engineering include, e.g., Varshavsky [1965 (with B. L. Osievich), 1968, 1969, 1990] and his collaborators M. Kishinevsky et al. [1994a, 1994b] and also J. A. Brzozowski and C.-J. Seger [1995], J. A. Brzozowski [2000], and many others. Theorem 7.22, its corollaries, and all other results of Section 7.5 are due to C. L. Nehaniv [2002a and in press] and follow the proofs in C. L. Nehaniv [in press]. The observation that synchronous cellular automata can perform universal computation is due to A. R. Smith III [1971].

The asynchronous emulation theorem (Theorem 7.22), its corollaries (Corollaries 7.29 and 7.30, which extend previously known constructions but incompletely proved results), and all other results shown here are due independently to C. L. Nehaniv [2002a and in press]. Formulations equivalent to Corollary 7.30 were found by K. Nakamura [1974], who sketched a proof of freedom from deadlocks, and independently by C. L. Nehaniv [2002a and in press], and also in related, weaker form but without complete proof by T. Toffoli [1978] and T. Toffoli and N. Margolus [1987].

Applications exhibiting the details of the construction in the simplified case of emulating synchronous cellular automata by asynchronous cellular automata, and the first examples of self-replication and of evolution in implemented asynchronous cellular automata, as well as remarks on universal computation in asynchronous cellular automata, were given by C. L. Nehaniv [2002b, 2002c].

Bibliography

A. Ádám [1996], *Behavior and Simplicity of Moore Automata*. Akadémiai Kiadó, Budapest, 1996.

G. P. Agibalov and N. V. Evtušenko [1979], Characterization condition of existence and other problems of cascade connections of finite automata (in Russian). *MTA SZTAKI Tanulmányok*, **99** (1979), 181–197.

G. P. Agibalov and N. V. Evtušenko [1984], Algebraic characterization of permutation automata decomposable into cascade connection of smaller components (in Russian). *Kibernetika,* **1** (1984), 9–15.

S. V. Alešin [1970a], Bases in groups of automaton permutations (in Russian). *Diskret. Analiz.,* **17** (1970), 3–8.

S. V. Alešin [1970b], On the absence of bases in certain classes of finite automata (in Russian). *Problemy Kibernet.,* **22** (1970), 67–74, 296–297.

D. S. Ananichev, P. Dömösi, and C. L. Nehaniv. Some properties of digraphs and semigroups, *Acta Cybernet.*, to appear.

M. A. Arbib (ed.) [1968], *Algebraic Theory of Machines, Languages and Semigroups, with a Major Contribution by K. Krohn and J. L. Rhodes.* Academic Press, New York, 1968.

B. Austin, K. Henckell, C. Nehaniv, and J. Rhodes [1995], Subsemigroups and complexity via the presentation lemma. *J. Pure Appl. Algebra*, **101** (1995), 245–289.

I. Babcsányi and A. Nagy [2004], Homomorphic direct product of automata, *Publ. Math.*, in press.

A.-L. Barabási [2002], *Linked, The New Science of Networks.* Perseus, Cambridge, Mass., 2002.

A.-L. Barabási and E. Bonabeau [2003], Scale-free networks. *Scientific American*, May 2003, 60–69.

W. A. Barret and J. D. Couch, *Compiler Construction: Theory and Practice.* Science Research Associates, Inc., Chicago, 1979.

M. Bartha and M. Krész [2000], Elementary decomposition of soliton automata. *Acta Cybernet.,* **14** (2000), 631–652.

R. Bartlett and M. Garzon [1993], Monomial cellular automata. *Complex Systems,* **7** (1993), 367–388.

R. Bartlett and M. Garzon [1995], Bilinear cellular automata. *Complex Systems,* **9** (1995), 455–476.

Z. Bavel [1983], *Introduction to the Theory of Automata.* Reston Publishing Company, Inc., Reston, Va., 1983.

T. L. Booth [1967], *Sequential Machines and Automata Theory.* John Wiley and Sons, Inc., New York, 1967.

J. G. Brookshear [1989], *Theory of Computation: Formal Languages, Automata, and Complexity.* Benjamin/Cummings Publishing Company, Inc., Redwood City, Calif., 1989.

A. Brüggemann, L. Priese, D. Rödding and R. Schätz [1983], Modular decomposition of automata. In *Logics and Machines: Decision Problems and Complexity*, E. Börger, G. Hasenjaeger, D. Rödding, eds., Lecture Notes in Comput. Sci. **171,** Springer-Verlag, New York, 1983, 198–236.

A. Brüggermann-Klein and R. Klein [1987], On the minimality of K, F, and D or: Why Löten is non-trivial. In *Computation Theory and Logic*, Egon Börger ed., Lecture Notes in Comput. Sci. **270,** Springer-Verlag, New York, 1987, 59–66.

J. A. Brzozowski [2000], Delay-insensitivity and ternary simulation. *Theoret. Comput. Sci.*, **245** (2000), 3–25.

J. A. Brzozowski and C.-J. Seger [1995], *Asynchronous Circuits.* Springer-Verlag, New York, 1995.

A. W. Burks (ed.) [1970], *Essays on Cellular Automata.* University of Illinois Press, Urbana, 1970.

G. Chartrand and F. Harary [1967], Planar permutation graphs. *Ann. Inst. Henri Poincaré, N. Ser., Sect. B,* **3** (1967), 433–438.

C. Choffrut (ed.) [1986], *Automata Networks.* Lecture Notes in Comput. Sci. **316,** Springer-Verlag, New York, 1986.

M. Čirič, B. Imreh, and M. Steinby [1999], Subdirectly irreducible definite, reverse definite, and generalized definite automata. *Publ. Electrotechn. Fak. Ser. Mat.,* **10** (1999), 69–79.

A. H. Clifford and G. B. Preston [1961], *The Algebraic Theory of Semigroups,* Vol. I. Math. Surveys 7.1, AMS, Providence, R.I., 1961.

A. H. Clifford and G. B. Preston [1967], *The Algebraic Theory of Semigroups,* Vol. II. Math. Surveys 7.2, AMS, Providence, R.I., 1967.

E. F. Codd [1968], *Cellular Automata.* Academic Press, New York, 1968.

D. I. A. Cohen [1977], *Introduction to Computer Theory*, 2nd ed. John Wiley and Sons, New York, 1991.

M. Conner [1990], Sequential machines realized by group representations. *Inform. and Comput.,* **85** (1990), 183–201.

M. Conner and R. Tolimieri [1986], Group representation for multi-valued sequential machines. In *Proc. Sixteenth International Symposium on Multivalued Logic*, Blacksburg, Va., 1986, 258–265.

J. H. Conway [1971], *Regular Algebra and Finite Machines.* Chapman and Hall, London, 1971.

B. Csákány and F. Gécseg [1965], On the group of automaton permutations (in Russian). *Kibernetika,* **1** (1965), 14–17; translated as *Cybernetics,* **1** (1965), 13–15.

J. Dassow [1981], *Completeness Problems in the Structural Theory of Automata.* Akademie-Verlag, Berlin, 1981.

J. Dassow and H. Jürgensen [1990], Soliton automata. *J. Comput. System Sci.,* **40** (1990), 154–181.

J. Dassow and H. Jürgensen [1991], Soliton automata with a single exterior node. *Theoret. Comput. Sci.,* **84** (1991), 281–292.

J. Dassow and H. Jürgensen [1993], Soliton automata with at most one cycle. *J. Comput. System Sci.,* **46** (1993), 155–197.

M. Demlová, J. Demel, and V. Koubek [1981], On subdirectly irreducible automata. *RAIRO Inform. Théor.,* **15** (1981), 23–46.

J. Dénes [1986], Research problem 40. *Period. Math. Hung.,* **17** (1986), 245–246.

J. Dénes and P. Hermann [1982], On the product of all elements in a finite group. *Ann. Discrete Math.,* **15** (1982), 105–109.

P. Denning, J. B. Dennis, and J. E. Qualitz [1978], *Machines, Languages, and Computation.* Prentice-Hall, Inc., Englewood Cliffs, N.J., 1978.

P. Dömösi [1972], On the semigroup of automaton mappings with finite alphabet. *Acta Cybernet.,* **1** (1972), 251–254.

P. Dömösi [1975], On superpositions of automata. *Acta Cybernet.,* **2** (1975), 335–343.

P. Dömösi [1976], On minimal R-complete systems of finite automata. *Acta Cybernet.,* **3** (1976), 37–41.

P. Dömösi [1980], On homomorphically QD-complete systems of automata. In *Papers on Automata Theory*, II, Karl Marx University of Economics, Budapest, 1980, 79–97.

P. Dömösi [1981], On homomorphically α_i-simple automata. In *Papers on Automata Theory*, III, Karl Marx University of Economics, Budapest, 1981, 93–124.

P. Dömösi [1982], On homomorphically α_0-complete systems of automata. In *Papers on Automata Theory*, IV, Karl Marx University of Economics, Budapest, 1988, 57–68.

P. Dömösi [1983], On homomorphically α_i-complete systems of finite automata. *Acta Cybernet.*, **6** (1983), 85–88.

P. Dömösi [1984], On cascade products of standard automata. In *Conference on Automata, Languages and Mathematical Systems, Salgótarján*, Karl Marx University of
Economics, Budapest, 1984, 37–45.

P. Dömösi [1986], On complete systems of automata. In *Conference on Automata, Languages and Programming Systems, Salgótarján*, Karl Marx University of Economics, Budapest, 1986, 87–95.

P. Dömösi [1988], On temporal products of automata. In *Papers on Automata and Languages, X,* Karl Marx University of Economics, Budapest, 1988, 49–62.

P. Dömösi [1990], Products of automata and homomorphic simulation. *Pure Math. Appl., Ser. A*, **1** (1990), 11–19.

P. Dömösi [1992], Complete classes of finite automata and simulation. In *Proceedings of the International Conference on Words, Languages and Combinatorics, II*, Kyoto Sangyo University, Kyoto, Japan, M. Ito and H. Jürgensen, eds., World Scientific, 1994, 98–104.

P. Dömösi [1996], On α_2-ν_3-products of automata. *Publ. Math.,* **48** (1996), 233–242.

P. Dömösi and Z. Ésik [1986], On homomorphic realization of automata with α_0-products. In *Papers on Automata and Languages*, VIII, Karl Marx University of Economics, Budapest, 1986, 63–97.

P. Dömösi and Z. Ésik [1987], On homomorphic simulation of automata by ν_1-products. In *Papers on Automata and Languages*, IX, Karl Marx University of Economics, Budapest, 1987, 91–112.

P. Dömösi and Z. Ésik [1988a], Critical classes for the α_0-product. *Theoret. Comput. Sci.,* **61** (1988), 17–24.

P. Dömösi and Z. Ésik [1988b], On homomorphic simulation of automata by α_0-products. *Acta Cybernet.,* **8** (1988), 315–323.

P. Dömösi and Z. Ésik [1988c], On homomorphic realization and homomorphic simulation of automata by α_0-products. In *Second Conference on Automata, Languages and Programming Systems, Salgótarján*, Karl Marx University of Economics, Budapest, 1988, 89–98.

P. Dömösi and Z. Ésik [1988d], On the hierarchy of ν_i-products of automata. *Acta Cybernet.,* **8** (1988), 253–258.

P. Dömösi and Z. Ésik [1990], Product hierarchies of automata and homomorphic simulation. *Acta Cybernet.,* **9** (1990), 371–373.

P. Dömösi and Z. Ésik [2001], Homomorphic simulation and Letichevsky's criterion. *J. Automata Lang. Comb.*, **6** (2001), 427–436.

P. Dömösi and Z. Ésik [2002], A note on complete classes of automata for the ν_3-products. *Publ. Math.,* **60** (2002), 539–550.

P. Dömösi, Z. Ésik, and B. Imreh [1989], On product hierarchies of automata. In *Fundamentals of Computation Theory*, Lecture Notes in Comput. Sci. **380**, Springer-Verlag, New York, 1989, 137–144.

P. Dömösi and F. Gécseg [1989], Simulation by ν_1^*-products of automata. *Publ. Math.,* **36** (1989), 51–56.

P. Dömösi and F. Gécseg [1992], Simulation and representation by ν_i^*-products of automata. *Publ. Math. Debrecen,* **40** (1992), 75–83.

P. Dömösi and B. Imreh [1989], On ν_i-products of automata. *Acta Cybernet.,* **6** (1989), 149–162.

P. Dömösi and B. Kovács [1992], Simulation on finite networks of automata. In *Words, Languages and Combinatorics*, M. Ito, ed., World Scientific, River Edge, N.J., 1992, 131–138.

P. Dömösi and C. L. Nehaniv [1997], Some results and problems on finite homogeneous automata networks. *Proc. Japanese Association of Mathematical Sciences, Annual Meeting on Languages, Computation and Algebra,* T. Imaoka, ed., Kobe University, 1997.

P. Dömösi and C. L. Nehaniv [1998], Algebraic theory of finite automata networks. *Math. Japon.,* **48** (1998), 481–509.

P. Dömösi and C. L. Nehaniv [1999], Complete finite automata network graphs with minimal number of edges. *Acta Cybernet.,* **14** (1999), 37–50.

P. Dömösi and C. L. Nehaniv [2000], On complete systems of automata. *Theoret. Comput. Sci.*, **245** (2000), 27–54.

V. Drobot [1989], *Formal Languages and Automata Theory.* Computer Science Press, Inc., Rockville, Md., 1989.

H. Ehrig, K.-D. Kiermeier, H.-J. Kreowski, and W. Kühnel [1974], *Universal Theory of Automata. A Categorical Approach.* B. G. Teubner, Stuttgart, 1974.

S. Eilenberg [1974], *Automata, Languages, and Machines*, Vol. A. Academic Press, London, New York, 1974.

S. Eilenberg [1976], *Automata, Languages, and Machines*, Vol. B. Academic Press, London, New York, 1976.

Z. Ésik [1983], On homomorphic realization of monotone automata. In *Papers on Automata Theory*, V, Karl Marx University of Economics, Budapest, 1983, 63–76.

Z. Ésik [1985], Homomorphically complete classes of automata with respect to the α_2-product. *Acta Sci. Math.*, **48** (1985), 135–141.

Z. Ésik [1986], Complete classes of automata for the α_1-product. *Found. Control Engr.,* **11** (1986), 95–107.

Z. Ésik [1987a], Loop products and loop-free products. *Acta Cybernet.,* **8** (1987), 41–43.

Z. Ésik [1987b], On isomorphic realization of automata with α_0-products. *Acta Cybernet.,* **8** (1987), 119–127.

Z. Ésik [1989a], An extension of the Krohn-Rhodes decomposition of automata. In *Machines, Languages, and Complexity*, Lecture Notes in Comput. Sci. **381**, Springer-Verlag, New York, 1989, 66–71.

Z. Ésik [1989b], On α_1^λ-products of automata, *Acta Sci. Math.,* **53** (1989), 245–253.

Z. Ésik [1991a], A note on isomorphic simulation of automata by networks of two-state automata. *Discrete Appl. Math.,* **30** (1991), 77–82.

Z. Ésik [1991b], Results on homomorphic realization of automata by α_0-products. *Theoret. Comput. Sci.,* **87** (1991), 229–249.

Z. Ésik [1992], Varieties of automata and transformation semigroups. *Acta Math. Hungar.,* **59** (1992), 59–74.

Z. Ésik [2000], A proof of the Krohn-Rhodes decomposition theorem. *Theoret. Comput. Sci.,* **234** (2000), 287–300.

Z. Ésik and P. Dömösi [1986], Complete classes of automata for the α_0-product. *Theoret. Comput. Sci.,* **47** (1986), 1–14.

Z. Ésik, P. Dömösi, F. Gécseg, and J. Virágh [1986], Homomorphic realization of automata with compositions. In *Mathematical Foundations of Computer Science*, Lecture Notes in Comput. Sci. **233**, Springer-Verlag, New York, 1986, 299–307.

Z. Ésik and F. Gécseg [1983], General products and equational classes of automata. *Acta Cybernet.,* **6** (1983), 281–284.

Z. Ésik and F. Gécseg [1986], On α_0-products and α_2-products. *Theoret. Comput. Sci.,* **48** (1986), 1–8.

Z. Ésik and F. Gécseg [1989], A decidability result for homomorphic representation of automata by α_0-products. *Acta Math. Hungar.,* **53** (1989), 205–212.

Z. Ésik and Gy. Horváth [1983], The α_2-product is homomorphically general. In *Papers on Automata Theory*, V, Karl Marx University of Economics, Budapest, 1983, 49–62.

Z. Ésik and B. Imreh [1980a], Remarks on finite commutative automata. *Acta Cybernet.,* **5** (1980/81), 143–146.

Z. Ésik and B. Imreh [1980b], Subdirectly irreducible commutative automata. *Acta Cybernet.,* **5** (1980/81), 251–260.

Z. Ésik and J. Virágh [1985], On λ-products of automata. In *Topics in the Theoretical Bases and Applications of Computer Science*, Akadémiai Kiadó, Budapest, 1986, 79–89.

Z. Ésik and J. Virágh [1986], On products of automata with identity. *Acta Cybernet.,* **7** (1986), 299–311.

Z. Ésik and J. Virágh [1987], A note on α $*_0$-products of aperiodic automata. *Acta Cybernet.,* **8** (1987), 41–43.

N. V. Evtušhenko [1979], On the realization of automata by cascade connection of standard automata (in Russian). *Avtomat. i Vyčisl. Tekhn. (Riga),* **13** (1979), 50–53.

F. Fogelman-Soulié [1987], Automata networks and artificial intelligence. In *Automata Networks in Computer Science: Theory and Applications*, F. Fogelman-Soulié, Y. Robert and M. Tchuente, eds., Manchester University Press, Manchester, U.K., Princeton University Press, Princeton, N.J., 1987.

F. Fogelman-Soulié, Y. Robert, and M. Tchuente (eds.) [1987], *Automata Networks in Computer Science*, Manchester University Press, Manchester, U.K., Princeton University Press, Princeton, N.J., 1987.

G. F. Frobenius [1968], *Collected Works*. Vol. I, II, III (in German). Springer-Verlag, Berlin, New York, 1968.

Z. Fülöp and S. Vágvölgyi [1990], A complete rewriting system for a monoid of tree transformation classes. *Inform. and Comput.,* **86** (1990), 195–212.

Z. Fülöp [1991], A complete description for a monoid of deterministic bottom-up tree transformation classes. *Theoret. Comput. Sci.,* **88** (1991), 253–268.

Z. Fülöp and G. Dányi [1998], Compositions with superlinear deterministic top-down tree transformations. *Theoret. Comput. Sci.,* **194** (1998), 57–85.

P. Gács [1985], Reliable computation with cellular automata. *J. Comput. System Sci.,* **32** (1985), 79–90.

E. Galois [1832], On the Condition that an Equation Be Soluble by Radicals (in French). Manuscript, Paris, 1832; in *Journal de Liouville*, Tome XI (1846), 381–444; *Oeuvres Mathématiques,* Paris, 1897, 51–61.

M. Garzon [1995], *Models of Massive Parallelism. Analysis of Cellular Automata and Neural Networks.* Texts Theoret. Comput. Sci. EATCS Ser., Springer-Verlag, Berlin, 1995.

F. Gécseg [1965], On the group of bijective automaton mappings defined by finite automata (in Russian). *Kibernetika,* **1** (1965), 37–39; translated in *Cybernetics,* **1** (1965), 37–39.

F. Gécseg [1965], On the composition of automata without loop (in Russian). *Acta Sci. Math. (Szeged),* **26** (1965), 269–272.

F. Gécseg [1966a], On R-products of automata. I. *Studia Sci. Math. Hungar.,* **1** (1966), 437–441.

F. Gécseg [1966b], On R-products of automata. II. *Studia Sci. Math. Hungar.,* **1** (1966), 443–447.

F. Gécseg [1967], On R-products of automata. III. *Studia Sci. Math. Hungar.,* **2** (1967), 163–166.

F. Gécseg [1968], Metrically complete systems of automata (in Russian). *Kibernetika (Kiev),* **3** (1968), 96–98; translated in *Cybernetics,* **3** (1968), 96–98.

F. Gécseg [1969], On complete systems of automata. *Acta Sci. Math. (Szeged),* **30** (1969), 295–300.

F. Gécseg [1973], Model theoretical methods in the theory of automata. In *Mathematical Foundations of Computer Science: Symposium Proceedings, High Tatras*, 1973, 57–63.

F. Gécseg [1974], Composition of automata. In *Automata, Languages and Programming: 2nd Colloquium, Saarbrücken*, Lecture Notes in Comput. Sci. **14**, Springer-Verlag, New York, 1974, 351–363.

F. Gécseg [1976a], On products of abstract automata. *Acta Sci. Math.,* **38** (1976), 21–43.

F. Gécseg [1976b], Representation of automaton mappings in finite length. *Acta Cybernet.*, **2** (1976), 285–289.

F. Gécseg [1983], On α_i-products of automata–homomorphic representation. In *Algebra, Combinatorics and Logics in Computer Science.* Vol. I, II, *Colloq. Math. Soc. János Bolyai* **42**, North-Holland, Amsterdam, 1986, 403–421.

F. Gécseg [1984], Finite representations and infinite products. In *Automata, Languages and Mathematical Systems: Conference Proceedings,* Karl Marx University of Economics, Budapest, 1984, 55–66.

F. Gécseg [1985a], On ν_1-products of commutative automata. *Acta Cybernet.,* **7** (1985), 51–59.

F. Gécseg [1985b], Metric representations by ν_i-products. *Acta Cybernet.,* **7** (1985), 203–209.

F. Gécseg [1986], *Products of Automata.* EATCS Monogr. Theoret. Comput. Sci. **7**, Springer-Verlag, Berlin, Heidelberg, New York, Tokyo, 1986.

F. Gécseg and B. Imreh [1987a], On metric equivalence of ν_i-products. *Acta Cybernet.,* **8** (1987), 129–134.

F. Gécseg and B. Imreh [1987b], A comparison of α_i-products and ν_i-products. *Found. Control Engrg.,* **12** (1987), 3–9.

F. Gécseg and B. Imreh [1989], On star-products of automata. *Acta Cybernet.,* **9** (1989), 167–172.

F. Gécseg and B. Imreh [1992], Finite isomorphically complete systems. *Discrete Appl. Math.,* **36** (1992), 307–311.

F. Gécseg and B. Imreh [1995a], On completeness of nondeterministic automata. *Acta Math. Hungar.*, **68** (1995), 151–159.

F. Gécseg and B. Imreh [1995b], On the cube-product of nondeterministic automata. *Acta Sci. Math. (Szeged)*, **60** (1995), 321–327.

F. Gécseg and B. Imreh [2001], On isomorphic representations of generalized definite automata. *Acta Cybernet.*, **15** (2001), 33–43.

F. Gécseg, B. Imreh, and A. Pluhár [1998], On the existence of finite isomorphically complete systems of automata. *J. Automata. Lang. Comb.*, **3** (1998), 77–84.

F. Gécseg and H. Jürgensen [1990], Automata represented by products of soliton automata. *Theoret. Comput. Sci.*, **74** (1990), 163–181.

F. Gécseg and H. Jürgensen [1991], On α_0-ν_1-products of automata. *Theoret. Comput. Sci.*, **80** (1991), 35–51.

F. Gécseg and I. Peák [1972], *Algebraic Theory of Automata.* Disquisitiones Mathematicae Hungaricae **2**, Akadémiai Kiadó, Budapest, 1972.

A. Gill [1970], Single-channel and multichannel finite-state machines. *IEEE Trans. Comput.*, **19** (1970), 1073–1078.

S. Ginsburg [1962], *An Introduction to Mathematical Machine Theory.* Addison-Wesley, Reading, Mass., 1962.

S. Ginsburg [1975], *Algebraic and Automata Theoretic Properties of Formal languages.* North-Holland, Amsterdam, 1975.

A. Ginzburg [1966], About some properties of definite, reverse-definite and related automata. *IEEE Trans. Electronic Computers,* **EC-15** (1966), 809–810.

A. Ginzburg [1968], *Algebraic Theory of Automata.* Academic Press, London, New York, 1968.

V. M. Gluškov [1961], Abstract theory of automata (in Russian). *Uspekhi Mat. Nauk,* **16** **(101)** (1961), 3–62; correction, ibid., **17 (104)** (1962), 270.

E. Goles and S. Martinez [1990], *Neural and Automata Networks.* Kluwer Academic Publishers, Dordrecht, Netherlands, 1990.

G. Grätzer [1979], *Universal Algebra*, 2nd ed. Springer-Verlag, New York, Berlin, Heidelberg, Tokyo, 1979.

J. Gruska [1997], *Foundations of Computing.* Thomson International Computer Press, Boston, 1997.

F. Harary [1969], *Graph Theory.* Addison-Wesley, New York, 1969.

J. Hartmanis and R. E. Stearns [1966], *Algebraic Structure Theory of Sequential Machines.* Prentice-Hall, Englewood Cliffs, N.J., 1966.

J. Hartmanis [1966], Loop-free structure of sequential machines. *Inform. and Control,* **5** (1962), 25–43.

K. Henckell, S. Lazurus, and J. L. Rhodes [1988], Prime decomposition theorem for arbitrary semigroups: General holonomy and synthesis theorem. *J. Pure Appl. Algebra,* **55** (1988), 127–172.

T. Hikita and I. G. Rosenberg [1994], Completeness for uniformly delayed circuits. *Acta Appl. Math.*, **52** (1998), 49–61.

W. M. L. Holcombe [1982], *Algebraic Automata Theory.* Cambridge University Press, Cambridge, U.K., 1982.

O. Hölder [1889], Reduction of an arbitrary algebraic integer from a chain of integers (in German). *Math. Ann.*, **34** (1889), 26–56.

J. E. Hopcroft, R. Motwani, and J. D. Ullman [2001], *Introduction to Automata Theory, Languages, and Computation*, 2nd ed. Addison-Wesley, Reading, Mass., 2001.

J. E. Hopcroft and J. D. Ullman [1969], *Formal Languages and Their Relation to Automata.* Addison-Wesley, Reading, Mass., 1969.

J. E. Hopcroft and J. D. Ullman [1979], *Introduction to Automata Theory, Languages, and Computation.* Addison-Wesley, Reading, Mass., 1979.

J. M. Howie [1991], *Automata and Languages.* Oxford University Press, Oxford, U.K., 1991.

J. Hořejš [1978], Mappings defined by finite automata (in Russian). *Problem. Kibernet.,* **9** (1963), 23–26.

B. Imreh [1977], On α_i-products of automata. *Acta Cybernet.,* **3** (1977), 301–307.

B. Imreh [1980a], On isomorphic representations of commutative automata with respect to α_i-products. *Acta Cybernet.,* **5** (1980/81), 21–32.

B. Imreh [1980b], On finite nilpotent automata. *Acta Cybernet.,* **5** (1980/81), 281–293.

B. Imreh [1985], On finite definite automata. *Acta Cybernet.,* **7** (1985), 61–65.

B. Imreh [1988], A note on generalized ν_1-product. *Acta Cybernet.,* **8** (1988), 247–252.

B. Imreh [1994], On complete systems of automata. In *Words, Languages and Combinatorics* II, World Scientific, River Edge, N.J., 1994, 207–215.

B. Imreh [1996], Compositions of nondeterministic automata. In *Semigroups, Formal Languages and Computer Systems. RIMS Sūrikaisenkikenkyūsho Kōkyūroku,* **960** (1996), 44–53.

B. Imreh [2002], On the equivalence of the cube-product and the general product of automata. *Theoret. Comput. Sci.,* **245** (2000), 103–113.

B. Imreh and M. Ito [1997], On α_i-product of nondeterministic automata. *Algebra Colloq.,* **4** (1997), 195–202.

B. Imreh and M. Ito [1999], A note on the star-product. *Acta Cybernet.*, **14** (1999), 99–104.

M. Ito and J. Duske [1983], On cofinal and definite automata. *Acta Cybernet.*, **6** (1983), 181–189.

G. I. Ivanov [1973], Temporal representation of controlled automata. *Izv. AN SSSR, Technicheskaia Kibernetika,* **6** (1973), 106–113.

C. Jordan [1869], Commentary on Galois (in French). *Math. Ann.*, **1** (1869), 141–160.

R. Y. Kain [1972], *Automata Theory: Machines and Languages.* McGraw-Hill, New York, 1972.

L. Kaloujnine (Kalužnin) and M. Krasner [1950], Complete product of permutation groups and group extension, I (in French). *Acta Sci. Math. (Szeged)*, **13** (1950), 208–230.

L. Kaloujnine (Kalužnin) and M. Krasner [1951a], Complete product of permutation groups and group extension, II (in French). *Acta Sci. Math. (Szeged)*, **14** (1951), 36–66

L. Kaloujnine (Kalužnin) and M. Krasner [1951b], Complete product of permutation groups and group extension, III (in French). *Acta Sci. Math. (Szeged)*, **14** (1951), 69–82.

M. Kishinevsky, A. Kondratyev, A. Taubin, and V. Varshavsky [1994a], *Concurrent Hardware: The Theory and Practice of Self-Timed Design.* Wiley Professional Computing, Chichester, U.K., 1994.

M. Kishinevsky, A. Kondratyev, A. Taubin, and V. Varshavsky [1994b], Analysis and identification of speed-independent circuits on an event model. *Form. Methods Syst. Des.*, **4**, (1994), 33–75.

Z. Kohavi [1964], Secondary state assignment for sequential machines. *IEEE Trans. Electronic Computers*, **13** (1964), 193–203.

Z. Kohavi [1970], *Switching and Finite Automata Theory.* McGraw-Hill, New York, 1970.

D. C. Kozen [1997], *Automata and Computability.* Springer-Verlag, New York, 1997.

K. B. Krohn and J. L. Rhodes [1962], Algebraic theory of machines. In *Proc. Symp. Math. Theory of Automata,* J. Fox, ed., Polytechnic Press, Brooklyn, N.Y., 1962, 341–384.

K. B. Krohn and J. L. Rhodes [1965], Algebraic theory of machines, I. Prime decomposition theorem for finite semi-groups and machines. *Trans. Amer. Math. Soc.*, **116** (1965), 450–464.

K. B. Krohn, J. L. Rhodes, and B. R. Tilson [1968], The prime decomposition theorem of the algebraic theory of machines. In *Algebraic Theory of Machines, Languages and Semigroups,* M. Arbib, ed., Academic Press, New York, 1968.

W. Kuich and A. Salomaa [1986], *Semirings, Automata, Languages.* EATCS Monogr. Theoret. Comput. Sci., Springer-Verlag, Berlin, 1986.

K. Kuratowski [1930], On the problem of left curves in topology (in French). *Fund. Math.*, **15** (1930), 271–283.

G. Lallement [1971], On the prime decomposition theorem for finite monoids. *Math. Systems Theory,* **5** (1971), 8–12.

G. Lallement [1979], *Semigroups and Combinatorial Applications.* John Wiley and Sons, New York, 1979.

V. Lashkia [1993], Closures on finite algebras. *C. R. Math. Rep. Acad. Sci. Canada,* **15** (1993), 35–40.

V. Lashkia and A. Nozaki [1998], On completeness of automata. *IEICE Trans. Inf. Syst.*, **11** (1998), 1305–1307.

A. A. Letichevsky [1961], Completeness conditions for finite automata (in Russian). *Ž. Vyčisl. Mat. i Mat. Fiz.*, **1** (1961), 702–710; translated as *Comput. Math. and Math. Phys.*, **1** (1961), 702–710.

H. R. Lewis and C. H. Papadimitriou [1981], *Elements of the Theory of Computation.* Prentice-Hall, Englewood Cliffs, N.J., 1992.

P. M. Lewis, II, D. J. Rosenkranz, and R. E. Stearns [1978], *Compiler Design Theory.* Addison-Wesley, Reading, Mass., 1978.

A. Lindenmayer and G. Rozenberg (eds.) [1976], *Automata, Languages, Development.* North-Holland Publishing, Amsterdam, 1976.

P. Linz [2001], *An Introduction to Formal Languages and Automata*, 2nd ed., D. C. Heath and Company, Lexington, Mass., 1996; Jones and Bartlett, Boston, 2001.

E. Madelaine and D. Vergamini [1989], *AUTO: A Verification Tool for Distributed Systems Using Reduction of Finite Automata Networks.* FORTE, Vancouver, B.C. Canada, 1989, 61–66.

L. Martin, C. Reisher, and I. G. Rosenberg [1978], Completeness problem for switching circuits constructed from delayed gates. In *Proc. Eight International Symposium on Multi-Valued Logic*, IEEE, Long Beach, Calif., 1978, 142–148.

C. Martin-Vide, A. Mateescu, and V. Mitrana [2002], Parallel finite automata systems communicating by states. *Internat. J. Foundations Comput. Sci.*, **13** (2002), 733–749.

C. Martin-Vide and V. Mitrana [2000], Parallel communicating automata systems: A survey. *Korean J. Comput. Appl. Math.* **7** (2000), 237–257.

C. Martin-Vide and V. Mitrana [2001], Some undecidable problems for parallel communicating finite automata systems. *Informat. Proc. Letters,* **77** (2001), 239–245.

C. Martin-Vide and Gh. Păun [1999], Cooperating distributed splicing systems. *J. Automata Lang. Comb.,* **4** (1999), 3–16.

W. S. McCulloch and W. Pitts [1943], A logical calculus of the ideas immanent in nervous activity. *Bull. Math. Biophys.,* **5** (1943), 115–133.

R. McNaughton [1982], *Elementary Computability, Formal Languages, and Automata.* Prentice-Hall, Englewood Cliffs, N.J., 1982.

R. McNaughton and S. Papert [1971], *Counter-Free Automata. MIT Research Monograph* **65**, MIT Press, Cambridge, Mass., 1971.

A. Meduna [2000], *Automata and Languages: Theory and Applications.* Springer-Verlag, London, 2000.

M. L. Minsky [1967], *Computation: Finite and Infinite Machines.* Prentice-Hall, Englewood Cliffs, N.J., 1967.

M. Minsky and S. Papert [1969], *Perceptrons, an Introduction to Computational Geometry.* MIT Press, Cambridge, Mass., 1969.

B. Monien and H. Sudborough [1988], Comparing interconnection network. In *Mathematical Foundations of Computer Science*, Lecture Notes in Comput. Sci. **324,** Springer-Verlag, New York, 1988, 138–153.

K. Nakamura [1974], Asynchronous cellular automata and their computational ability. *Systems, Computers, Controls*, **5** (1974), 58–66.

R. J. Nelson [1968], *Introduction to Automata.* John Wiley and Sons, New York, 1968.

C. L. Nehaniv [1992], *Global Sequential Coordinates on Semigroups, Automata, and Infinite Groups*, Ph.D. thesis in mathematics, University of California, Berkeley, 1992.

C. L. Nehaniv [1995], Cascade decomposition of arbitrary semigroups. In *Semigroups, Formal Languages and Groups*, J. Fountain, ed., Kluwer Academic Publishers, Dordrecht, Netherlands, 1995, 391–425.

C. L. Nehaniv [1996], From relation to emulation: The covering lemma for transformation semigroups. *J. Pure Appl. Algebra,* **107** (1996), 75–87.

C. L. Nehaniv [2002a and in press], Asynchronous automata networks can emulate any synchronous automata network. *Internat. J. Algebra Comput.*, to appear.

C. L. Nehaniv [2002b], Self-reproduction in asynchronous cellular automata. In *Proc.* 2002 *NASA/DoD Conference on Evolvable Hardware*, IEEE Computer Society Press, 2002, 201–209.

C. L. Nehaniv [2002c], Evolution in asynchronous cellular automata. In *Artificial Life* VIII: *Proc.* 8*th International Conference on Artificial Life*, R. K. Standish, M. A. Bedau, and H. A. Hussein, eds., MIT Press, Cambridge, Mass., 2002, 65–73.

H. Neumann [1967], *Varieties of groups*, Ergeb. Math. Grenzgebiete **37**, Springer-Verlag, Berlin, 1967.

J. von Neumann [1966], *Theory of Self Reproducing Automata.* A. W. Burks, ed., University of Illinois Press, Urbana, 1966.

M. Nivat (ed.) [1973], *Automata, Languages and Programming.* Proceedings of a symposium organized by Institut de Recherche d'Informatique et d'Automatique, North-Holland, Amsterdam, 1973.

A. Nozaki [1978], Practical decomposition of automata. *Informat. and Control,* **36** (1978), 275–291.

C. H. Papadimitriou [1994], *Computational Complexity.* Addison-Wesley, Reading, Mass. 1994.

T. W. Parsons [1992], *Introduction to Compiler Construction.* W. H. Freeman, New York, 1992.

T. Petkovič, M. Čirič, and S. Bogdanovič [1998], Decomposition of automata and transformation semigroups. *Acta Cybernet.,* **13** (1998), 385–403.

S. Piccard [1946], *On the Bases of Symmetric Groups and the Pairs of Substitutions Which Regularly Generate a Group* (in French). *Mém, Univ. Neuchâtel* **19**, Librarie Vuibert, Paris, 1946.

G. N. Raney [1958], Sequential functions. *J. Assoc. Comput. Mach.,* **5** (1958), 177–180.

J. L. Rhodes [1986a], Infinite iteration of matrix semigroups I. Structure theorem for torsion semigroups. *J. Algebra,* **98** (1986), 422–451.

J. L. Rhodes [1986b], Infinite iteration of matrix semigroups II. Structure theorem for torsion semigroups. *J. Algebra,* **100** (1986), 25–137.

J. L. Rhodes and B. R. Tilson [1989], The kernel of monoid morphisms. *J. Pure Appl. Algebra,* **62** (1989), 227–268.

E. Roche [1996], Compact factorization of finite-state transducers and finite-state automata. *Nordic J. Comput.,* **4** (1997), 187–216.

D. Rödding [1983], Modular decomposition of automata. In *Foundations of Computation Theory*, M. Karpinsky, ed., Lecture Notes in Comput. Sci. **158**, Springer-Verlag, New York, 1983, 394–412.

I. G. Rosenberg [1970], Complete sets for finite algebras. *Math. Nachr.,* **44** (1970), 253–258.

I. G. Rosenberg [1980], Combinatorial and algebraic aspects of switching circuits. In *Discrete Mathematical Analysis and Combinatorial Computation*, University of New Brunswick, Fredericton, N.B., Canada, 1980, 11–23.

I. G. Rosenberg and T. Hikita [1983], Completeness for uniformly delayed circuits. In *Proc., 13th International Symposium on Multi-Valued Logic*, Kyoto, Japan, 1983, 2–10.

I. G. Rosenberg and T. Hikita [1983], *Supplement to Completeness for Uniformly Delayed Circuits*, manuscript.

T. Saito and H. Nishio [1989], Structural and behavioral equivalence relations in automata networks. *Theoret. Comput. Sci.,* **63** (1989), 223–237.

A. Salomaa [1960a], A theorem concerning the composition of functions of several variables ranging over a finite set. *J. Symbolic Logic,* **25** (1960), 203–208.

A. Salomaa [1960b], *On the Composition of Functions of Several Variables Ranging over a Finite Set.* Ann. Univ. Turkuensis Ser. A I **41**, Turku, Finland, 1960.

A. Salomaa [1962], *Some Completeness Criteria for Sets of Functions over a Finite Domain.* Ann. Univ. Turku Ser. A I **53**, Turku, Finland, 1962.

A. Salomaa [1963a], *On Basic Groups for the Set of Functions over a Finite Domain.* Ann. Acad. Sci. Fenn. Ser. A I **338**, Helsinki, Finland, 1969.

A. Salomaa [1963b], *On Essential Variables of Functions, Especially in the Algebra and Logic.* Ann. Acad. Sci. Fenn. Ser. A I **339**, Helsinki, Finland, 1963.

A. Salomaa [1969], *Theory of Automata.* Pergamon Press, Oxford, U.K., 1969.

A. Salomaa [2003], Composition sequences for functions over a finite domain. *Theoret. Comput. Sci.,* **292** (2003), 263–281.

M. W. Shields [1987], *An Introduction to Automata Theory.* Blackwell Scientific Publications, Oxford, U.K., 1987.

M. Simon [1999], *Automata Theory,* World Scientific, Singapore, 1999.

A. R. Smith, III [1971], Simple computation-universal cellular spaces. *J. Assoc. Comput. Machinery*, **18** (1971) 339–353.

M. Steinby [1969], *On Definite Automata and Related Systems.* Ann. Acad. Sci. Fenn. Ser. A I **444**, Helsinki, Finland, 1969.

J. Stokłosa [1977], On operation preserving functions of shift registers. *Found. Control. Engrg.,* **2** (1977), 211–214.

T. Sudkamp [1988], *Languages and Machines.* Addison-Wesley, Reading, Mass., 1988.

T. Sudkamp [1997], *Languages and Machines: An Introduction to the Theory of Computer Science*, 2nd ed. Addison-Wesley Longman, Inc., Reading, Mass., 1997.

M. Tchuente [1979], Parallel calculation of a linear mapping on a computer network. *Linear Algebra Appl.,* **28** (1979), 223–247.

M. Tchuente [1982], Parallel realization of permutations over trees. *Discrete Math.,* **39** (1982), 211–214.

M. Tchuente [1983], Computation of Boolean functions on networks of binary automata. *J. Comput. System Sci.,* **26** (1983), 269–277.

M. Tchuente [1985], Permutation factorization on star-connected networks of binary automata. *SIAM J. Algebraic Discrete Methods,* **6** (1985), 537–540.

M. Tchuente [1986], Computation on binary tree network. *Discrete Appl. Math.,* **14** (1986), 295–310.

M. Tchuente [1988], Computation on finite networks of automata. In *Automata Networks*, C. Choffrut, ed., Lecture Notes in Comput. Sci. **316,** *Springer-Verlag*, New York, 1988, 53–67.

T. Toffoli [1978], Integration of Phase-Difference Relations in Asynchronous Sequential Networks. In *Automata, Languages, and Programming*, G. Ausiello and C. Bohm, eds., Lecture Notes in Comput. Sci. **62**, Springer-Verlag, New York, 1978, 457–463.

R. Tucci [1988], A note on the decomposition of infinite automata. *Intern. J. Computer Math.,* **24** (1988), 141–149.

R. Tucci [1989], A decomposition theory for a class of infinite transformation semigroups. *Acta Cybernet.,* **9** (1989), 39–54.

T. Toffoli and N. Margolus, *Cellular Automata Machines*, MIT Press, Cambridge, Mass., 1987.

V. I. Varshavsky [1968], Collective behaviour and control problems. *Machine Intell.,* **3** (1968), 217–242.

V. I. Varshavsky [1969], The organization of interaction in collectives of automata. *Machine Intell.,* **4** (1969), 285–311.

V. I. Varshavsky (ed.) [1990], Self-timed control of concurrent processes. In *The Design of Aperiodic Logical Circuits in Computers and Discrete Systems.* Math. Appl. (Soviet Ser.) 52, Kluwer Academic Publishers, Dordrecht, Netherlands, 1990.

V. I. Varshavsky and B. L. Osievich [1965], Networks composed of ternary majority elements. *IEEE Trans. Electronic Computers,* **14** (1965), 730–733.

H. Wielandt [1964], *Permutation Groups.* Academic Press, New York, 1964.

D. Wood [1987], *Theory of Computation.* John Wiley and Sons, New York, 1987.

M. Yoeli [1961], The cascade decomposition of sequential machines. *IRE Trans. Electronic Comput.,* **10** (1961), 587–592.

M. Yoeli [1965], Generalized cascade decompositions of automata. *J. Assoc. Comput. Machinery,* **12** (1965), 411–422.

V. P. Zarovnyĭ [1965], Automaton mappings and wreath products (in Russian). *Kibernet.,* **1** (1965), 29–36; translated in *Cybernet.,* **1** (1965), 29–36.

H. P. Zeiger [1967], Cascade synthesis of finite-state machines. *Inform. and Control,* **10** (1967), 419–433 (plus erratum).

Index